PRENTICE-HALL, INC., *Englewood Cliffs, New Jersey*

DONALD M. SIMMONS

Fairfield-Maxwell, Ltd.

NONLINEAR PROGRAMMING FOR OPERATIONS RESEARCH

PRENTICE-HALL INTERNATIONAL SERIES IN MANAGEMENT

Library of Congress Cataloging in Publication Data

SIMMONS, DONALD M
Nonlinear programming for operations research.

1. Nonlinear programming. I. Title.
T57.8.S55 519.7'6 75-37509
ISBN 0-13-623397-X

Printed in the United States of America

10 9 8 7 6 5 4 3 2 1

Prentice-Hall International, Inc., *London*
Prentice-Hall of Australia Pty. Limited, *Sydney*
Prentice-Hall of Canada, Ltd., *Toronto*
Prentice-Hall of India Private Limited, *New Delhi*
Prentice-Hall of Japan, Inc., *Tokyo*
Prentice-Hall of South-east Asia Private Limited, *Singapore*

For My Parents

Contents

Preface

The field of nonlinear programming suffers from a dichotomy that makes the teaching of it surprisingly awkward. On the one hand, the problem-solving algorithms themselves are not overly sophisticated mathematically, and even those that involve rather intricate computations will yield up their secrets readily enough to patient study. On the other hand, some of the general mathematical properties of the algorithms, such as the conditions required for convergence and the rate at which convergence may be expected to proceed, are substantially more complex and difficult. Establishment of these properties depends critically upon the use of various results from advanced algebra and mathematical analysis and upon the rigorous development of an entire body of algorithmic convergence theory.

A thorough mastery of this convergence theory, though doubtless invigorating, is simply not essential to a sound understanding either of the nonlinear programming algorithms themselves or of the conditions required for optimality and other related theoretical aspects. Indeed, modern convergence theory did not begin being developed until the late 1960's, some fifteen years after the derivation of the Kuhn-Tucker conditions and up to ten years after the development of many of the most important and (still) useful problem-solving methods. One interesting facet of this chronology is that these belated theoretical advances finally allowed researchers to demonstrate whether or not a given algorithm, which may have provided years of faithful and efficient service, would necessarily converge to a locally optimal solution of every possible problem to which it could be applied. When such convergence could not be proven for certain algorithms, the ingenious contrivance of "counter-example" problems which in fact could not be solved by them became a popular sub-discipline in the nonlinear programming field. The practical utility of this work seems open to question.

At any rate, it is the purpose of this textbook to develop the theoretical foundation and major algorithms of modern nonlinear programming, while placing only minor emphasis on algorithmic convergence theory. This greatly reduces the mathematical background required of the student. Although such topics as closed sets, local optima, quadratic forms, limit points, and convergence of sequences will be introduced and expounded, it will in general be assumed that the student has no prior familiarity with them. Thus the course prerequisites for the text are limited to linear algebra, linear programming, and differential calculus. In particular, a thorough grasp of duality theory within the context of differential calculus is essential to a full understanding of nonlinear programming. This treatment of duality theory, which was included in the author's earlier book *Linear Programming for Operations Research,* is reviewed here at substantial length in Chapter 2.

Apart from the development of the theoretical foundations of nonlinear programming—duality, Lagrange multipliers, and Kuhn-Tucker theory—the emphasis in this text will be on careful and detailed presentation of a number of solution algorithms that have shown themselves to be of continuing importance and practical utility. For the most part, these methods are capable of arriving at optimal solutions reasonably quickly (with respect to competing algorithms) and with a very small or arbitrarily reducible computational error. Moreover, in most cases the methods presented are relatively simple to work with and to code for computer use. The author's appreciation of this latter virtue stems from his experience in the practice of operations research. In general, simplicity and economy in human affairs are less valued by those who design than by those who execute. Nonlinear programming is no exception.

Thus, in summary, the material chosen for presentation in this textbook is oriented toward the future practitioner of operations research. The book is intended to accompany a one-semester course in nonlinear programming in engineering, management, economics, or other related curricula, including those multi-disciplinary course programs that emphasize mathematical modelling and problem-solving. The presentation would be appropriate for master's or doctoral level study in those fields. Where interpretations of mathematical concepts are offered, the tendency is to frame them in economic or resource-allocation terms. This is the natural context of most decision problems in any professional discipline; furthermore, experience shows that students find these explanations easiest to understand.

In order to help the student firm up his grasp of the solution algorithms presented, most of them are accompanied by detailed discussions of example problems, which typically require working through two or three iterations.

In addition, some 240 exercises, segmented by sections for easy reference, are furnished at the ends of the various chapters. Apart from the example problems and exercises, a large number of computational topics are discussed here and there throughout the text, including simultaneous equation solution, treatment of degeneracy, restricted basis entry, anti-zigzagging procedures, and many others. Where several algorithms for solving the same problem type are presented, a discussion of their computational advantages and disadvantages in both hand and digital computer solution is also normally included.

I would like to acknowledge a debt of gratitude to Professors Alvin Drake, Gordon Kaufman, John Little, Alan Martin, Felipe Ochoa, and Jeremy Shapiro—all now or formerly at M.I.T.—for the insights, encouragement, and other forms of assistance they provided me during my years there, when my ideas for this book were generated and nurtured. Equally essential, though in a different way, were Connie Spargo and Margaret Whitehouse, who managed to deliver the manuscript from the clutches of my regrettable handwriting.

Donald M. Simmons

1

Introduction to Nonlinear Programming

1.1 REFLECTIONS

Throughout this textbook we shall be concerned with the *mathematical optimization problem* or *mathematical program*, which calls for the maximization or minimization of an algebraic function of one or more variables. The choice of values for the variables will in general be restricted by algebraic equations and/or inequalities called *constraints*, so that the goal is to find not the best possible value of the function but rather the best value permitted by the constraints.

For centuries mathematicians have been formulating and solving optimization problems and studying their theoretical properties. Euclid, for example, knew how to find the point on the straight line $Ax + By = C$ that is closest to the origin; using the compass and straightedge, he was minimizing $(x^2 + y^2)^{1/2}$ subject to $Ax + By = C$. In the seventeenth century Newton and Leibnitz derived the fundamental theorems of differential calculus, with which it became possible to find the maxima and minima of all continuous algebraic functions. By the nineteenth century the more specialized method of Lagrange multipliers had been developed for solving optimization problems subject to equality constraints. The word "solving" is used here in its theoretical sense: A solution was proved to exist, and a well-defined procedure was specified for obtaining it, provided that one had enough time and paper.

However, before the advent of digital computers, an engineer faced with even a moderately large optimization problem, one involving, say, 100 constraints in 200 variables, was simply not able to exploit those theoretical results. It is surely a tribute to classical nineteenth-century mathematics that most of the efficient methods used today for solving simultaneous linear equations were known also to Gauss. Nevertheless, as a practical matter, he could not solve a problem consisting of 1000 such equations—and we can. Or again, if the simultaneous equations were nonlinear, Newton-Raphson iteration would find a solution, provided that there weren't too many. But even 10 or 15 were "too many."

It should therefore be emphasized that most of the improvements in technique discovered by modern applied mathematicians are due less to our keener insight than to our more urgent motivation. Our pursuit has been much more diligent because our computers give us the capability of taking advantage of our findings. This is particularly true of optimization problems subject to *inequality* constraints, which were terribly laborious for classical mathematicians and thus little studied by them.

Another point should be made on behalf of nineteenth-century mathematics: Not only is large-scale constrained optimization electronically feasible today, but in addition it is very much of interest. Modern economics, engineering, and management science have given rise to many large and complicated problems in the past 25 years which simply were not recognized and formulated before. More and more, decision makers are willing to pay to have them solved, as well as to finance research that may lead to still greater efficiency in the future. These are the major reasons *operations research*, that branch of applied mathematics which is concerned with the formulation and solution of optimization problems, has recently become so stylish, so interesting, and—to its more aggressive practitioners—so profitable.

1.2 MATHEMATICAL PROGRAMMING PROBLEMS

Let us turn now to the actual algebraic formulation of the problem we have been discussing. Henceforth we shall use the more precise terms *mathematical programming problem* and *mathematical program*, which are equivalent, to refer to any optimization problem having the following general format: Find those values of the variables $x_1, \ldots, x_n$ that

$$\text{maximize or minimize } f(x_1, \ldots, x_n) \tag{1-1}$$

$$\text{subject to } g_i(x_1, \ldots, x_n) \begin{Bmatrix} \leqq \\ = \\ \geqq \end{Bmatrix} b_i, \qquad i = 1, \ldots, m, \tag{1-2}$$

where the b_i are scalar constants and where the functions f and g_i and the variables x_j all take on real scalar values. In any given problem the various members of (1-2) may have different equality/inequality signs; note that (1-2) includes any nonnegativity restrictions $x_j \geqq 0$ that may be present. For convenience the variables $x_1, \ldots, x_n$ will usually be collected into an n-component vector $\mathbf{x}$. Thus the mathematical program (1-1) and (1-2) can also be represented as follows:

$$\begin{aligned} &\text{Max or Min } f(\mathbf{x}), && \mathbf{x} = (x_1, \ldots, x_n) \\ &\text{subject to } g_i(\mathbf{x})\{\leqq, =, \geqq\}b_i, && i = 1, \ldots, m, \end{aligned}$$

with "Max" and "Min" being abbreviations for "maximize" and "minimize."

The branch of applied mathematics concerned with solving problems of the above form is known, appropriately enough, as *mathematical programming*. It should be mentioned that the latter term is sometimes used interchangeably with *nonlinear programming*. We prefer to make the distinction that "mathematical programming" refers to all problems of the form (1-1) and (1-2), while "nonlinear programming" refers to all such problems in which one or more of the functions f and g_i includes a nonlinear term or terms. Thus the field of mathematical programming can be partitioned into the mutually exclusive subfields of linear and nonlinear programming.

We shall now state some definitions which will be quite familiar to anyone who has studied linear programming. Any particular assignment of numerical values to the variables $x_1, \ldots, x_n$ is called a *solution* to the problem and may be thought of as a point in Euclidean n-space E^n. The equations and/or inequalities (1-2) are the *constraints*, and a solution is said to be *feasible* or *infeasible* according to whether it does or does not satisfy every one of the constraints. The collection of all feasible solutions or points in E^n is known as the *feasible region*. Incidentally, when no constraints at all are present (that is, when $m = 0$) we still call the problem a "mathematical program"; the feasible region then consists of all of E^n, and there are no infeasible solutions.

The function f to be optimized—that is, maximized or minimized—will be referred to in this text as the *objective* or *objective function*. A feasible solution is also an *optimal solution* or an *optimum* if it yields a value of the objective function that is as good (i.e., as large or as small) as or better than the value yielded by every other feasible solution.[1] Thus "optimal" implies "feasible." From the definition it is clear that the optimal solution to a given problem need not be unique, but that the optimal value of the objective function must be.

[1]Actually, an optimal solution as defined here can be described more precisely as a *global optimum*. The distinction between local and global optima will be drawn in Chapter 3.

The goal in solving a mathematical programming problem is, of course, to identify an optimal solution, along with its associated objective value; when there are multiple optima it may or may not be satisfactory to find just one of them. It cannot be assumed, however, that an optimal solution will always exist. It may happen in a given problem that there is no assignment of values to the variables that satisfies all the constraints; in such a case, when no feasible solution exists, the problem itself is said to be *infeasible*. Another possibility is that in a problem having feasible solutions the optimal value of the objective function is positively or negatively infinite; we say that such a problem is *unbounded*, or has an unbounded optimal solution. When this occurs the feasible region itself is usually *unbounded*, here meaning "of infinite extent"; a simple example would be the maximization of $f(x_1) = x_1^3$ subject to $x_1 \geqq 0$. But a problem with a bounded (or *finite*) feasible region can also have an unbounded optimal solution, as would be the case when $f(x_1) = 1/x_1$ is maximized subject to $0 \leqq x_1 \leqq 1$.

1.3 VARIETIES AND CHARACTERISTICS OF MATHEMATICAL PROGRAMS

Various types of mathematical programs are distinguished according to the nature of the functions $f(\mathbf{x})$ and $g_i(\mathbf{x})$. The simplest type, in which all the functions are linear, is called a *linear program* and has presumably been studied by the reader already. All other varieties may be referred to collectively as *nonlinear programs*. It is useful, however, to single out certain problem types within the general nonlinear class for which special solution methods have been devised. The *quadratic program* (or quadratic programming problem), for example, calls for the optimization of a quadratic objective subject to a set of linear constraints. A *convex program* is any problem that can be represented in the form

$$\begin{gathered} \text{Min } f(\mathbf{x}) \\ \text{subject to } g_i(\mathbf{x}) \leqq b_i, \qquad i = 1, \ldots, m, \end{gathered}$$

where the functions f and g_i all have a property known as *convexity*. This property will be discussed in Chapter 3, but in the meantime it is interesting to note that all linear and some quadratic programs are included in this class of problems. Still another type is the *geometric program*,

$$\begin{gathered} \text{Min } g_0(\mathbf{x}) \\ \text{subject to } g_i(\mathbf{x}) \leqq 1, \qquad i = 1, \ldots, m, \\ \text{and } \mathbf{x} \geqq \mathbf{0}, \end{gathered}$$

where $\mathbf{0}$ is the null vector and the $g_i(\mathbf{x})$, $i = 0, 1, \ldots, m$, are *posinomial* functions.[2]

Another useful characterization of mathematical programming problems is according to their *size*, that is, according to their numbers of constraints and variables. A *small-scale* problem can be functionally defined as one that can be solved by hand, or by using a hand calculator, in a "reasonable" amount of time. A nonlinear program satisfying this definition would have no more than about four or five variables and constraints, while a small-scale linear program might have up to twice as many of each. An *intermediate-scale* problem is one that can be attacked and solved directly, regardless (or almost regardless) of its structural details, on a modern digital computer. In this category the upper limits on size might be roughly on the order of 100 constraints and 200 variables for nonlinear programs and 2000 constraints and 10,000 variables for linear programs. Finally, a *large-scale* problem would be one that can be solved with currently available mathematical methods and computer technology only if it possesses certain special structural characteristics that allow it to be approached in some indirect or piecewise manner. The most important large-scale mathematical programming technique is *decomposition*, in which a problem with suitably structured constraints is broken or decomposed into a number of subproblems. The reader may already have encountered this technique in his study of linear programming; its applications to nonlinear programs are more or less analogous. It is important to note that in this textbook we shall be concerned basically with small- and intermediate-scale nonlinear programming, particularly the latter—most problems of practical interest in operations research are intermediate in size, in the sense of the above definitions. Although larger problems do arise in practice (mostly in management contexts), we shall not be giving any coverage to large-scale mathematical programming in the chapters that follow; this material is best treated in a later and more advanced course, using some such text as that of Lasdon [29].

Having described how the different types of mathematical programs can be characterized according to the nature of their objective and constraint functions, or according to size, we now wish to make some remarks that will apply in general to all mathematical programming problems to be considered in this text. First, we shall require that the objective and all constraint func-

[2]A posinomial function is a sum of any number of power terms, each of which is of the form

$$c_i x_1^{a_{i1}} x_2^{a_{i2}} \cdots x_n^{a_{in}} = c_i \prod_{j=1}^{n} x_j^{a_{ij}},$$

where c_i and the a_{ij} are constants, with $c_i > 0$. It is assumed that it can never be optimal for any variable to take on the value zero, which ensures that $x_j^{a_{ij}}$ is defined for negative a_{ij}.

tions be continuous and have continuous first partial derivatives. Although we shall discuss a few solution methods that can be applied to problems *not* satisfying these continuity conditions (see the material on *experimental seeking* in Chapter 5), such problems are not normally classified as mathematical programs.

Next, it should be noted that each constraint of (1-2) must be of one of the three types indicated within the braces; no strict inequalities are permitted. We make this provision as a practical convenience: For any problem it will always be desirable to specify an exact optimal solution, assuming one exists. This would be impossible if, for example, $f(x_1) = x_1^2$ were to be minimized subject to $x_1 > 0$. Moreover, strict inequality constraints simply do not arise naturally in the real world; restrictions imposed by budgets, inventory on hand, available floor space, minimum acceptable performance levels, maximum allowable error rates, and so on can all be expressed within the format of (1-2).

There are other limitations implicit in our definition of a mathematical program. One is that the statement of the problem must be *deterministic*; that is, the exact algebraic form of each function and the value of every constant parameter is assumed to be known with certainty. Thus when the constraint

$$\sum_{i=1}^{n} x_i \leqq 100$$

appears, we are certain—or are adopting the pretense—that the upper bound on the sum is precisely 100 and not 99 or 101. Problems frequently arise in which this is not true: For example, a company may be trying to allocate its capital budget in some optimal manner before it knows exactly what its total available capital will be. Such a *stochastic programming problem* will be amenable to the methods of this text only insofar as it can be reformulated or transformed into a deterministic optimization problem satisfying (1-1) and (1-2), or perhaps into a sequence of them. A simple example will appear as Exercise 1-7 at the end of this chapter.

A much more important limitation is that we shall *not* be considering problems with *integrality* constraints, that is, with one or more variables permitted to take on only integer values. To be sure, such *integer programming problems* do appear in a great variety of real-world contexts. For example, the optimal number of aircraft carriers for the Navy to plan on building when it allocates its budget must obviously be an integer. In other problems involving scheduling, ordering, and sequencing, it is often convenient to make use of a *decision variable* x_{ij}, where either $x_{ij} = 1$ if the ith object or task is to be placed in the jth position or $x_{ij} = 0$ if not. Dozens of additional examples could be cited. The trouble is, however, that integer programming is too rich and too intricate a subject to be profitably treated as a species of mathematical programming; its current solution methods and recent research have little in common with those of continuous-variable problems. Integer programming is, in fact, a fit subject for a textbook of its own.

Finally, *dynamic programming* will also be excluded from this text. The term "dynamic programming" does not refer to a specific problem type at all but is the name given to a general strategy for solving optimization problems by proceeding one stage or one variable at a time. Thus it does not belong in the same semantic grouping with "mathematical programming," "linear programming," and the rest.

1.4 SOME EXAMPLE FORMULATIONS

We remarked in Section 1.1 that the great flowering of operations research over the past quarter-century has been due in large part to the growing awareness that many of the complex problems arising in business, in government, and in scientific research are amenable to quantitative techniques. The problems with which operations research is concerned tend to occur in decision-making contexts: Typically, a decision maker is faced with the question of choosing from among many possible policies or alternatives the one that yields a maximum or minimum value of some numerically measurable criterion of performance (the objective). This necessity of coming to a decision and implementing it in the real world gives focus and meaning to any particular optimization problem and lends a characteristic vitality and flavor to operations research in general.

The actual work of solving a decision problem can be thought of as comprising two stages: formulation and solution. Although the solution phase will occupy our attention for most of this book, it should not be forgotten that the art of formulating or *modeling* real-world situations in terms of mathematical relationships is also of crucial importance. It frequently happens that an operations research analyst working on some project devotes most of his effort to selecting and refining a mathematical model and only a small part to programming his computer to solve it. The process of modeling a problem does not consist solely and simply of determining the exact nature of all the physical and economic relationships involved, as might naïvely be supposed. The analyst must also take into account the relative ease or difficulty with which various types of mathematical programs can be solved. Sometimes an awkward and complicated model that describes a real-world decision problem with great accuracy must be discarded in favor of one that is slightly less realistic but mathematically more tractable. It is therefore very important that in addition to mastering the theoretical aspects of the various solution methods to be presented the reader take note of their computational advantages and disadvantages, which we shall be commenting on from time to time.

In the remainder of this section we shall present several examples illustrating the formulation of mathematical programming problems as they arise in various real-world contexts; others will be found in the exercises at the end of the chapter. The reader is advised to work through them with care, inasmuch as the modeling of optimization problems, like other arts, can be mastered only through experience.

A. Inventory Maintenance

A large apartment building is heated by burning oil at the constant rate of λ gallons per year; the oil is fed into the furnace from a storage tank with a maximum capacity of M gallons. Oil can be replenished only by ordering it from a fuel company, which charges a fixed delivery cost of \$$A$ plus \$$c$ per gallon of oil. Suppose that delivery and refilling of the tank occur instantaneously and that all oil must be paid for in full as soon as it is delivered. If the money tied up in the oil inventory costs \$$r$ per dollar per year, what is the optimal reorder policy?

There is clearly no point in reordering until the supply of oil reaches zero, so a reorder policy is completely described by a single variable y, the size in gallons of each order. The number of orders per year must be λ/y, so that a year's fuel and delivery cost is

$$(A + cy)\frac{\lambda}{y} = \frac{A\lambda}{y} + c\lambda.$$

Since oil on hand decreases smoothly from y to zero and then returns immediately to y, the average number of gallons of oil on hand is $y/2$, and the inventory carrying cost is thus $cr(y/2)$ dollars per year. The overall objective is to minimize annual cost,

$$\text{Min}\ \frac{A\lambda}{y} + cr\frac{y}{2},$$

where the constant term $c\lambda$ has been omitted. The only restrictions on the order quantity are that it be nonnegative and that it be no greater than the tank's capacity:

$$y \leqq M$$

and

$$y \geqq 0.$$

Thus we have a mathematical program in one variable and two constraints—nonnegativity restrictions will usually not be treated differently from other constraints. Note that if $y \geqq 0$ were not included in this formulation, the optimal value of the objective would be achieved by letting y approach negative infinity.

B. Production Scheduling

A metal foundry employs 1000 workers; in one week each of them can either produce two swords from raw materials or convert one sword into a plowshare, but not both. At present the foundry has an ample supply of raw materials on hand but no swords or plowshares. After three weeks have passed all swords and plowshares will be sold for \$20 and \$75 apiece, respectively. How many of each item should be produced during each of the three weeks in order to maximize total sales? Assume for simplicity that a sword can be converted into a plowshare *only* in the week immediately after it is produced.

Let s_i be the number of workers producing swords during the ith week. It is obviously not necessary to allow for the possibility that workers might ever be left idle, since producing an extra couple of swords is always (at least) \$40 better than doing nothing. The number of men converting swords into plowshares in the ith week must therefore be $1000 - s_i$. Since no swords are initially available for conversion during week 1, we must have $s_1 = 1000$, and the *manpower constraints* are simply

$$0 \leqq s_2 \leqq 1000 \tag{1-3}$$

and

$$0 \leqq s_3 \leqq 1000, \tag{1-4}$$

with $s_1 = 1000$ being treated as constant.

We also need to ensure that in the ith week we do not assign more workers to the profitable conversion of swords into plowshares than there were swords produced in week $i - 1$; that is,

$$1000 - s_i \leqq 2s_{i-1} \qquad \text{for } i = 2, 3.$$

Notice furthermore that for $i = 2$ this restriction,

$$1000 - s_2 \leqq 2s_1 = 2000,$$

is automatically satisfied by the fact that s_2 cannot be negative. Thus the only production-linking constraint that must be explicitly stated is

$$1000 - s_3 \leqq 2s_2,$$

which is equivalent to

$$2s_2 + s_3 \geqq 1000. \tag{1-5}$$

The required constraints for this problem are therefore (1-3), (1-4), and (1-5).

As for the objective function, a man working on swords produces 2(\$20) = \$40 worth of goods, while a plowshare worker "destroys" a sword and creates a plowshare, thereby adding a value of \$75 − \$20 = \$55. The objective function is therefore

$$\text{Max } 40(1000 + s_2 + s_3) + 55(0 + 1000 - s_2 + 1000 - s_3)$$

or

$$\text{Max } 150{,}000 - 15s_2 - 15s_3,$$

which is equivalent to

$$\text{Min } s_2 + s_3.$$

This problem is, of course, a linear program.

C. A Geometrical Problem

A chemical company must send 1000 cubic meters of chlorine gas to its research laboratory in another state. Because the gas is extremely dangerous, a special hermetically sealed rectangular railroad car must be built for transporting it. The material from which the top and bottom must be constructed costs \$200 per square meter, while the siding material costs half as much; however, only 50 square meters of siding can be obtained. Moreover, the maximum height of the car permitted by tunnels and other overhead clearances is 3 meters. Regardless of the car's dimensions each round trip to the laboratory and back will cost \$800. Assuming no time limit on the overall procedure, what dimensions minimize the total cost of constructing the car and delivering the gas?

Let d, w, and h be the car's length, width, and height. The objective is to minimize overall cost; that is,

$$\text{Min } 800\left(\frac{1000}{dwh}\right) + 2dw(200) + (2dh + 2wh)(100),$$

where the three terms are contributed by transportation cost, top-and-bottom material, and siding, respectively. The constraints mentioned in the problem are

$$2dh + 2wh \leqq 50$$

and

$$h \leqq 3.$$

Finally, we must eliminate the possibility of negative dimensions:

$$d,\ w,\ h \geqq 0.$$

This happens to be a geometric programming problem; to achieve the precise format given in Section 1.3, we need only divide each constraint by its right-hand-side constant.

D. Optimal Control

An unmanned rocket of mass M is to travel through outer space from one docking station to another that is exactly 1 million miles away. The rocket has two engines, one capable of exerting up to b units of thrust (i.e., force) in the forward direction and the other capable of up to c units in the reverse direction. At discrete points 100,000 miles apart the thrust exerted by these engines may be altered instantaneously by radio control, but between those points no adjustments are permitted. How should the engines of the rocket be controlled in order to bring it to its destination in the shortest possible time? Assume that no frictional or gravitational forces act on the rocket.[3]

Let v_i, $i = 1, \ldots, 11$, be the velocity of the rocket in miles per hour as it passes the ith control point, where

$$v_1 = 0 \quad \text{and} \quad v_{11} = 0$$

are required. From the laws of motion we may write

$$v_{i+1} = v_i + a_i t_i, \qquad i = 1, \ldots, 10, \tag{1-6}$$

where a_i is the rocket's constant acceleration as it moves from point i to $i + 1$, and t_i is the total time it takes to cover that 100,000-mile distance. Because the acceleration is constant, the average velocity between points i and $i + 1$ is

$$\frac{v_i + v_{i+1}}{2} = v_i + \frac{a_i t_i}{2} = \frac{100{,}000}{t_i}.$$

Converting to a quadratic equation and solving for t_i, we find

$$t_i = \frac{-v_i + (v_i^2 + 200{,}000 a_i)^{1/2}}{a_i}.$$

Substitution into (1-6) then yields

$$v_{i+1} = (v_i^2 + 200{,}000 a_i)^{1/2}.$$

It remains only to substitute force for acceleration in accordance with Newton's law:

$$f_i = M a_i,$$

[3]This example is taken from Zangwill [60].

where f_i is the net (algebraic) thrust exerted by the rocket's engines in the forward direction between points i and $i + 1$. The optimal control problem can now be written as a mathematical program:

$$\text{Min} \sum_{i=1}^{10} \left(\frac{-v_i + (v_i^2 + 200{,}000 f_i/M)^{1/2}}{f_i/M} \right)$$

$$\begin{aligned} \text{subject to } v_{i+1} &= (v_i^2 + 200{,}000 f_i/M)^{1/2}, \qquad i = 1, \ldots, 10, \\ v_1 &= 0, \\ v_{11} &= 0, \\ \text{and } -c \leqq f_i &\leqq b, \qquad i = 1, \ldots, 10. \end{aligned}$$

Although control problems of this sort can be solved by the techniques of mathematical programming, they are much better suited to dynamic programming methods.

E. A Minimax Strategy

The peace-loving nation of Pacifica wishes to allocate N antiballistic missiles (ABMs) among its M cities in order to provide optimal deterrence against a possible nuclear attack. The value of the ith city, based on population and industry, is estimated to be V_i. Each ABM can destroy with certainty one attacking missile but can only be used against an attacker aimed at the city where it has been located. Given any deployment of Pacifican ABMs, a potential enemy nation with limited resources could do serious damage merely by buying or building enough offensive missiles to wipe out one city. Since the enemy nation would presumably decide to attack the city at which it could do the most damage per dollar (per offensive missile), how should Pacifica allocate its ABMs in order to minimize this maximum damage per dollar? Treat all variables as continuous.

Let x_i be the number of ABMs deployed at city i; from the problem statement, that city could then be destroyed by $x_i + 1$ offensive missiles, yielding to the attacker a payoff per missile of $V_i/(x_i + 1)$. Pacifica therefore wishes to minimize the maximum of the $V_i/(x_i + 1)$. This can be represented mathematically by letting the variable z equal that (unknown) maximum. The problem then becomes

$$\text{Min } z$$

$$\text{subject to } \frac{V_i}{x_i + 1} \leqq z, \qquad i = 1, \ldots, M,$$

with the additional constraints being

$$\sum_{i=1}^{M} x_i = N$$

and

$$x_i \geqq 0, \qquad i = 1, \ldots, M.$$

In game theory Pacifica's optimal deployment is known as a *minimax* strategy. It should be added that the problem would be rather more difficult if the realistic constraint that all x_i must be integers were included.

1.5 DIFFICULTIES CAUSED BY NONLINEARITY

Before beginning our study of mathematical programming, we shall examine a few simple two-variable examples which should convey a sense of why nonlinear problems are so much more difficult to solve than linear ones. First, consider the following linear program:

$$\begin{aligned} \text{Max } z &= x_1 + 4x_2 \\ \text{subject to } x_1 + x_2 &\leqq 8, \\ -3x_1 + 2x_2 &\leqq 6, \\ \text{and } x_1, x_2 &\geqq 0, \end{aligned}$$

where z is used to represent the (variable) value of the objective function. The feasible region is shown in Fig. 1-1. In solving this problem via the simplex method,[4] we would start at the origin, where $z = 0$, and test for optimality by computing the reduced costs. These would tell us that the value

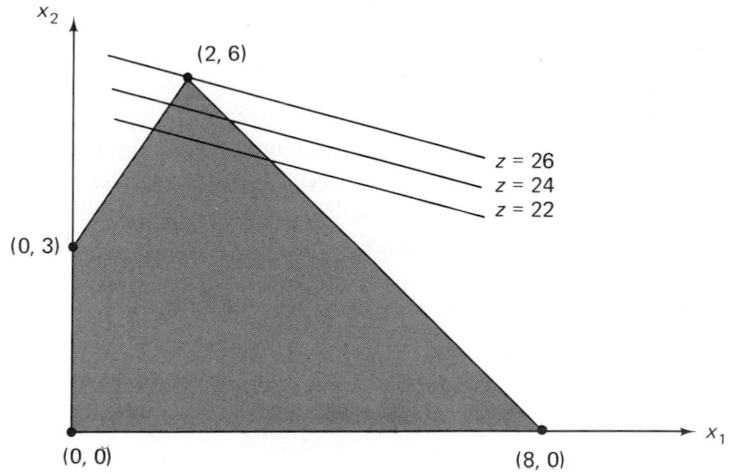

FIGURE 1-1

[4]A review and summary of the simplex method will appear in Chapter 2.

of z could be increased by moving away from the origin in either the x_1- or the x_2-direction (i.e., by pivoting either x_1 or x_2 into the basis). Choosing the x_2-direction, in which the rate of increase of z is greater, we would pivot to the point (0, 3), where again computation of the reduced costs would show that the conditions for optimality are not met. The next pivot step would bring us to the point (2, 6), and this time the reduced-cost computations would indicate that further movement away from (2, 6) in any feasible direction could only decrease the value of the objective function. We would therefore conclude that $x_1 = 2, x_2 = 6, z = 26$ is the optimal solution to the problem. Three indifference surfaces, including the optimal one, are shown in Fig. 1-1 (in linear programming the indifference surfaces are hyperplanes, which become straight lines in the two-variable case).

The ease with which linear programming problems, or LPPs, can be solved is due to three important factors:

1. Given any bounded LPP, at least one "corner" or extreme point of the feasible region must be optimal (and if the LPP is unbounded, an infinite optimal solution can be demonstrated by finding an extreme point at which an infinitely long edge of the feasible region originates). Therefore, to solve any linear program it is only necessary to search over a finite number of feasible solutions, namely, those corresponding to the extreme points.
2. The various extreme-point solutions can be obtained easily and directly, one from another, by means of linear algebraic transformations.
3. When an extreme point has been found such that no movement away from it in any feasible direction can improve the value of the objective funtion, then that extreme point must be the optimal solution to the problem. In other words, any solution that is "locally optimal" must also be optimal over the entire feasible region.

All three of these characteristics of linear programming problems are fully exploited by the well-known and highly efficient simplex algorithm.

Unfortunately, nonlinear programs (NLPs) do not, in general, possess those advantageous properties. As a result they are far more difficult to solve than linear programs. Consider, for example, the following NLP:

$$\text{Min } z = [(x_1 - 8)^2 + (x_2 - 4)^2]^{1/2}$$
$$\text{subject to } x_1 + x_2 \leqq 8,$$
$$-3x_1 + 2x_2 \leqq 6,$$
$$\text{and } x_1, x_2 \geqq 0.$$

The feasible region is the same as in the previous example, and the objective function now asks us to find the feasible point that lies the shortest distance away from the exterior point (8, 4). The problem has been solved graphically

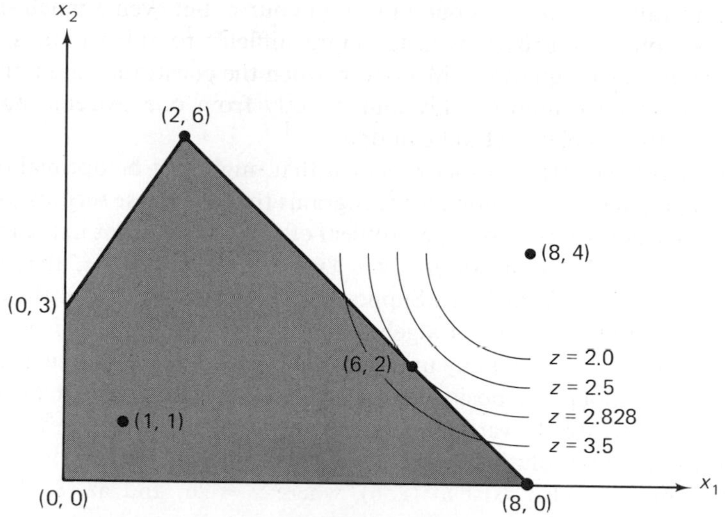

FIGURE 1-2

in Fig. 1-2; as can be seen, the optimal solution is $x_1 = 6$, $x_2 = 2$, where the "indifference circle" $z = 2.828$ is tangent to the boundary of the feasible region. Notice that the optimum does not lie at an extreme point; thus this problem could not have been solved by a simplex-type algorithm capable of examining only the finite set of extreme points.

Furthermore, even if we could develop an algorithm that would search over all *boundary* points of any given feasible region, we could not use it to solve every nonlinear program. For example, such an algorithm would be unable to locate the point in the feasible region of Fig. 1-2 that minimizes $z = (x_1 - 1)^2 + (x_2 - 1)^2$, i.e., that is closest to the point (1, 1). The optimal solution is, of course, $x_1 = 1$, $x_2 = 1$, which is in the interior of the feasible set.

But suppose it were somehow known that the optimal solution of some nonlinear program must lie at a certain extreme point (where two or more constraint surfaces intersect). A great deal of computational labor might still be required before its coordinates could actually be determined. To find the point of intersection of the curves

$$e^{x_1} \sin x_2 - \log^2(x_2 + 1) = 0$$

and

$$x_1 x_2 - x_2^5 + \tan(x_1 + x_2) = 1,$$

for example, one would have to use an iterative numerical procedure which might or might not eventually converge to the desired solution. Most nonlinear constraints are not this outlandish, of course, but even a much simpler system of nonlinear equations is far more difficult to solve than a set of simultaneous linear equations. Moreover, when the constraints are nonlinear there is no way to jump quickly and directly from one extreme point to another, as the simplex method can do.

The possible existence of *local optima* that might not be optimal overall is another characteristic of nonlinear programs that can cause serious trouble. As a simple example, consider the problem of finding the point in the feasible region of Fig. 1-2 that maximizes $z = (x_1 - 1)^2 + (x_2 - 1)^2$, that is, the point that is *farthest* from (1, 1). Suppose we somehow arrived at the origin, where $z = 2$, and by means of algebraic testing discovered that it is a local maximum, in the sense that any movement away from the origin in a feasible direction (i.e., along the positive x_1- or x_2-axis or into the first quadrant) causes z to decrease. Nevertheless, we could not then conclude that (0, 0) must be the optimal solution to the given problem. And, in fact, it turns out that local maxima also exist at (2, 6), where $z = 26$, and at (8, 0), where $z = 50$; the latter is the optimal solution overall.

In general, the methods of nonlinear programming, as we shall see in the chapters to come, are capable of finding local optima only. For certain problems whose objective functions and constraints are such that no local optima other than the overall optimum can exist, this limitation causes no additional difficulties. But to solve any problem that has or might have several local optima, it is necessary to do the extra work of finding them all, or at least of finding all that lie within some portion of the feasible region. By contrast, of course, this issue did not arise in linear programming: An LPP is solved as soon as one local optimum is found.

One further difficulty that can arise in nonlinear programming is worth mentioning: It is quite possible for the feasible region to consist of two or more entirely disconnected sets of points. A simple two-dimensional example is provided by the following:

$$\sin x_1 - x_2 \geqq 0,$$
$$x_1 \geqq 0,$$
$$\text{and} \qquad x_2 \geqq 0.$$

As is evident from Fig. 1-3, the feasible region defined by this set of constraints consists of an infinite number of disconnected parts. This sort of situation can be troublesome whenever we are using the very common type of solution algorithm that begins at a feasible point and moves away from it toward a local optimum in a continuous zigzag path that is never permitted to enter infeasible territory. (Incidentally, the simplex method works in just this way.) Such a path obviously can never cross from one segment of a disconnected feasible region to another—hence we could never arrive at the

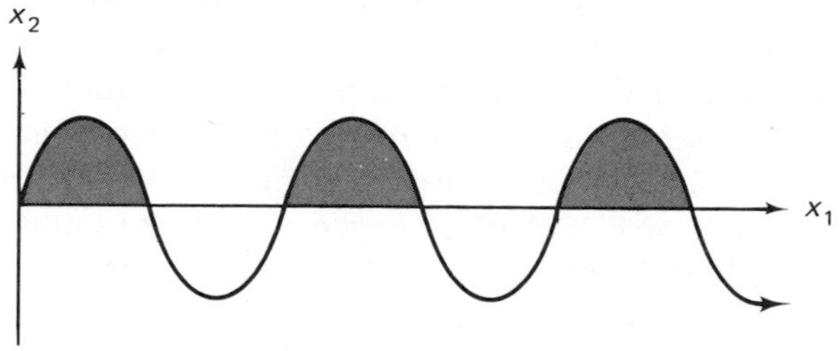

FIGURE 1-3

overall optimal solution unless we happened to choose a starting point in the right segment. It should be added that this difficulty is essentially the same as that which can arise when several different local optima exist within a fully connected feasible region: The particular optimum at which a "path-following" algorithm ends up may depend almost completely on the choice of starting point. But, to repeat a point made earlier, this is another problem that does not arise in linear programming.

1.6 HISTORICAL OUTLINE

Although its antecedents stretch back at least as far as the invention of the calculus in the late seventeenth century, the field of mathematical programming has existed as such only since World War II. The two seminal events were the development of the simplex method by George Dantzig (see his article in [27]) in the late 1940s and the publication of a paper by Kuhn and Tucker [28] in 1951 which established necessary conditions for optimal solutions to nonlinear programming problems. Since that time, the history of mathematical programming can be divided, roughly speaking, into two parallel and occasionally convergent streams of development, one springing from Dantzig's simplex method and the other from Kuhn-Tucker optimality theory.

In the case of the simplex method, the succeeding development was at first rather narrowly confined to the identification and study of various important subclasses of linear programs, a study which culminated in a number of highly efficient algorithms for transportation and network flow problems and in a few contributions to game theory. It was not long, however, before broader applications began to appear. In the mid-1950s a technique called *separable programming* [7] was developed that allowed approximate optimal solutions to be obtained for certain types—later generalized to all types—of nonlinear programs. In this method, piecewise linear approximations are used in place of all nonlinear functions, enabling the problem to be solved by a modified simplex scheme. Then in the late 1950s a confluence of the simplex computational approach and the theory of Kuhn and Tucker produced a number of different algorithms for solving quadratic programming problems (see [4], [30], and [52], among others). Basically, these methods rely on the fact that the first partial derivatives of a quadratic function are linear: This is precisely what makes quadratic programs the most easily handled of all nonlinear problems. Still later, around 1960, some important simplex-related results and algorithms began to appear in integer programming. In fact, much of the recent research in this area, involving algorithms based on Abelian group theory, has a strong "simplicial" flavor.

Although we shall be giving major coverage to both separable programming and quadratic programming, it is the second stream of development, founded on the Kuhn-Tucker theory, that will be of more direct and fundamental concern to us in this text. The Kuhn-Tucker necessary conditions, which are closely bound up with the classical notion of the gradient vector, constitute the basis for identifying local optimal solutions to the mathematical program (1-1) and (1-2). They are therefore an indispensable theoretical underpinning for virtually all the solution methods of nonlinear programming. The most important and useful of these methods were developed primarily in the period 1958–1967. Some can be applied to many or all nonlinear programs (see Zangwill [57], Fiacco and McCormick [13], and Zoutendijk [61]); others are applicable to problems having linear constraints only (Rosen [37], Wolfe [54], and Zangwill [58]); and still others are designed specifically for certain subspecies of the nonlinear program (such as the geometric program [11]). It is interesting to note that several of the algorithms for solving linearly constrained problems work with sequences of basic solutions and employ the pivoting computations of the simplex method.

The years following this very fertile period in the history of nonlinear programming have thus far appeared to be a time of appraisal and consolidation. Generally speaking, the major developments after 1967 have been related to evaluating and improving the efficiency of already-existing solution methods. The rates at which many of the most commonly used algorithms converge to optimal solutions have been subjected to both theoretical and empirical examination. Extra computational devices and tests have been grafted onto the early or the late stages of various iterative procedures (see Fiacco and McCormick [15] and also McCormick [33]), resulting in decreased overall solution times. Another important line of investigation has been a study of the basic theoretical properties of algorithms in general and of the mathematical conditions required for their convergence to optimal solutions (Zangwill [60]). Finally, and most recently, particular attention has been paid to "large-scale mathematical programming," that is, to the difficulties encountered in attempting to solve the very large nonlinear problems that sometimes arise in operations research.

Most of these currently developing lines of research in mathematical programming will not be given full and detailed coverage in this text. In general, they are rather more sophisticated mathematically than the rest of the material we shall be presenting and are thus somewhat beyond the scope of an introductory course in nonlinear programming. The chief concern of this text will be to derive, discuss, and evaluate the many different mathematical programming algorithms that have proved to be useful in solving real-world optimization problems. The overall viewpoint will be pragmatic, with substantial space being devoted to solving example problems and describing the relative advantages and disadvantages of competing algorithms. Students with strongly theoretical interests might therefore wish to do supplementary reading in other texts, such as those of Zangwill [60] and Luenberger [31], both of which contain excellent treatments of algorithmic convergence theory.

1.7 PREVIEW OF THE TEXT

The methods of nonlinear programming to be covered in this text arrange themselves for teaching purposes in an order that happens to correspond, more or less, to the order of their historical appearance. After two chapters devoted to review and mathematical background, we shall present in Chapter 4 a rather detailed development of the theory and computational methods of unconstrained optimization, recapitulating in the process about 300 years of classical mathematics. The major coverage will be given to gradient-following procedures for obtaining local optima, with numerous variations and modifications included. Then, in Chapter 5, we shall discuss various strategies for "experimental seeking," a mode of optimization in which, typically but not always, the goal is to plan an efficient search for the best values of a set of controllable variables (for example, the temperature and vapor pressure of some industrial chemical reaction) by using a limited number of costly or time-consuming experiments.

Following this discussion of solution methods for unconstrained or essentially unconstrained problems, in Chapter 6 we shall develop the theoretical basis of constrained optimization. Although some space will be devoted to establishing the existence of Lagrange multipliers for equality-constrained problems, the pivotal result will be the derivation of the Kuhn-Tucker conditions, a set of algebraic relationships which hold at the optimal solution of any mathematical program. These conditions, along with several related theoretical results, form the foundation for virtually all the problem-solving algorithms presented in the remainder of the text. Chapter 7 is devoted exclusively to the simplest and most important type of nonlinear program, the quadratic program; three major algorithms for solving it are presented there in substantial detail. In Chapter 8 we shall discuss a wide variety of solution methods for nonlinear programs having linear constraints only; these methods are, of course, applicable to quadratic programs (or, for that matter, to linear programs) and are in some cases comparable in efficiency to the methods of Chapter 7. Finally, in Chapter 9 we shall be concerned with solution algorithms that can be applied to the general nonlinear programming problem, regardless of the nature of its objective and constraint functions. We note in passing that, although these algorithms are fully capable of solving nonlinear programs with linear constraints, they are substantially less efficient in doing so than are the methods of Chapter 8, which are designed specifically to exploit that linearity.

To supplement our presentation of nonlinear programming we have also provided two chapters containing mathematical background material. Chapter 2 begins on a formal note with a derivation of the Taylor expansion of a function of n variables about a fixed point in E^n. The rest of the chapter is then devoted to a review of linear programming, especially including duality theory, which will play a crucial role throughout our work (both Lagrange multipliers and the Kuhn-Tucker multipliers are generalized versions of the dual variables of linear programming). Although the simplex method and the linear algebra underlying it will be summarized rather briskly, detailed proofs will be given for several important lemmas and theorems related to duality. No matter how familiar these theorems may appear, the reader is urged to study their proofs with care: A firm grasp of them and of the principles they embody will be essential to a clear understanding of nonlinear programming.

In Chapter 3 we shall then discuss several major background topics that lie at the heart of mathematical programming, including local and global optima, quadratic forms, and convex function theory. Although these topics can be thought of as "background," in the sense that they are fundamental enough to find applications in several different branches of mathematics, we shall assume no familiarity whatever with any of them and will proceed in each case from the most basic definitions. Thus the reader who has mastered all the material covered in Chapter 2 can rest assured that no further demands will be made upon his prior knowledge.

EXERCISES

Section 1.3

1-1. Construct a mathematical programming problem that has exactly five feasible solutions.

1-2. In linear programming we could never completely get rid of inequality constraints. When an inequality constraint

$$a_{i1}x_1 + \cdots + a_{in}x_n \leqq b_i$$

was converted to an equation by means of a slack variable,

$$a_{i1}x_1 + \cdots + a_{in}x_n + s_i = b_i,$$

we were still obliged to include the inequality

$$s_i \geqq 0$$

in the problem formulation. In mathematical programming, however, we can eliminate inequalities altogether by using *slack functions*, which are nonnegative everywhere, rather than slack variables. Show how this can be done.

1-3. (a) Sometimes in the practice of operations research an "either-or" restriction is encountered; in mathematical terms, this requires the values of the variables to be such that either one or the other (or both) of two constraints must be satisfied. Show how the following problem can be reformulated so as to be compatible with the format of (1-1) and (1-2):

$$\text{Max } f(\mathbf{x})$$
$$\text{subject to } \textit{either } g_1(\mathbf{x}) = b_1 \quad \textit{or} \quad g_2(\mathbf{x}) = b_2.$$

Now write a set of constraints that defines the "tic-tac-toe" feasible region shown in Fig. 1-4 (i.e., the feasible region consisting of the four darkened line segments).

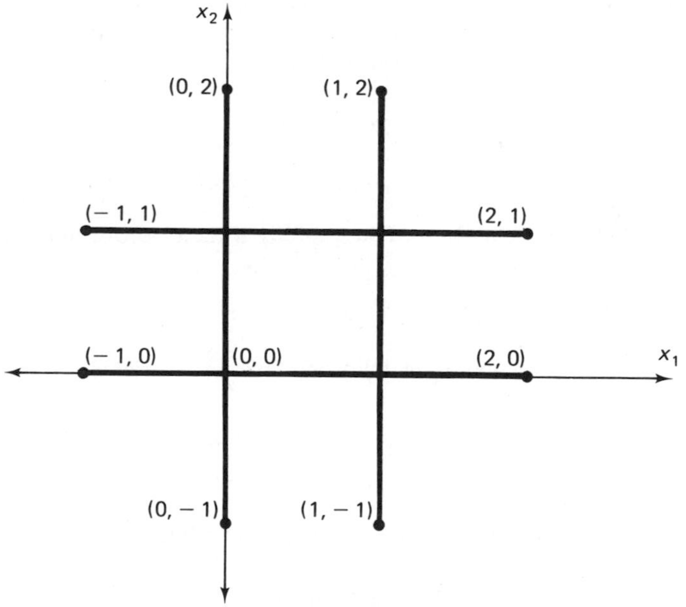

FIGURE 1-4

(b) Reformulate the following as a mathematical program satisfying (1-1) and (1-2):

$$\text{Max } f(\mathbf{x})$$

$$\text{subject to } \textit{either } g_1(\mathbf{x}) \leqq b_1 \quad \textit{or} \quad g_2(\mathbf{x}) \leqq b_2.$$

[*Hint:* Use the result of Exercise 1-2.]

1-4. (a) Given the set of constraints $g_i(\mathbf{x}) \leqq b_i$, $i = 1, \ldots, m$, suppose that you wish to find the maximum feasible value of *either* $f_1(\mathbf{x})$ *or* $f_2(\mathbf{x})$, whichever is greater. That is, you are trying to choose feasible values of the variables $\mathbf{x}$ in such a way as to maximize the function

$$f(\mathbf{x}) \equiv \text{Max}\{f_1(\mathbf{x}), f_2(\mathbf{x})\}.$$

Can you formulate a single mathematical program of format (1-1) and (1-2) to accomplish this? If not, can you manage it with more than one mathematical program?

(b) Repeat part (a) when the function

$$f(\mathbf{x}) \equiv \text{Max}\{f_1(\mathbf{x}), f_2(\mathbf{x})\}$$

is to be *minimized* subject to the same set of constraints.

1-5. In most of our work in mathematical programming we shall require that all algebraic functions have continuous first derivatives; hence we shall not allow $|x_j|$, the *absolute value* of x_j, to appear in either the objective or the constraints. Show how the substitutions $x_j \equiv \alpha_j - \beta_j$, $j = 1, \ldots, n$, can be used to reformulate the following as a legitimate mathematical program:

$$\text{Min } \sum_{j=1}^{n} |x_j|$$

$$\text{subject to } g_i(x_1, \ldots, x_n) \leqq b_i, \qquad i = 1, \ldots, m.$$

Could these transformations be used successfully if the objective function were

$$\text{Max } \sum_{j=1}^{n} |x_j|?$$

1-6. Let a penalty function of the following form be associated with each of the variables x_j, $j = 1, \ldots, n$:

$$p(x_j) = \begin{cases} x_j, & \text{if } x_j \geqq 0, \\ 0, & \text{if } x_j \leqq 0. \end{cases}$$

What transformation of variables would allow us to reformulate the problem

$$\text{Min } \sum_{j=1}^{n} p(x_j)$$

$$\text{subject to } g_i(x_1, \ldots, x_n) \leqq b_i \qquad i = 1, \ldots, m,$$

in such a way that the function $p(x_j)$ does not have to be explicitly used? Again, this function is objectionable because its first derivative is not continuous at $x_j = 0$.

1-7. *A stochastic programming problem:* Suppose a decision maker is trying to choose a nonnegative value of some variable x in order to maximize a certain objective function $f(x)$, where the only other constraint on x is that it cannot exceed a boundary level b. His mathematical program takes a very simple form:

$$\text{Max } f(x)$$
$$\text{subject to } 0 \leqq x \leqq b.$$

Now suppose that the boundary level, which might, for example, be a budget, is not known with certainty. The decision maker might wish to use a very large value of x, but he would then run the risk of ending up with an infeasible solution if b turned out to be unexpectedly small. One approach to this difficulty is to acknowledge that it is permissible to violate the constraint—that is, to choose an x that turns out to exceed b—but in case this happens a penalty cost must be paid. The most obvious and simplest penalty cost would be $k(x - b)$ if x exceeds b and zero if not; if b were a budget, k would represent the per-dollar interest charge for borrowing extra money to make up the shortage. Let $E(x)$ be the expected value of the penalty cost (based on the probability distribution of b) associated with choosing x. Formulate the decision maker's new problem. Then, using a transformation of variables if necessary, formulate an approximated version of this problem in which the random character of b is treated by assuming that it is known with certainty that b will take on its expected value $\bar{b}$. Comment on the relative advantages and disadvantages of the two formulations, and describe circumstances in which the approximation is likely to be desirable or acceptable.

Section 1.4

Exercises 1-8 through 1-14: Formulate each of the following as a mathematical programming problem, indicating which of them are linear programs.

1-8. *The portfolio problem:* The Corny Bernfeld Disaster Relief Fund has decided to invest part or all of its total capital of $\$C$ in the common stocks of N different corporations. All investments are to be liquidated after some fixed time period has elapsed. The fund's research department has calculated that an investment of \$1 in the ith stock over that time period will produce an expected profit of $\$m_i$ with a variance of v_i, for $i = 1, \ldots, N$. Assuming (unrealistically) that all stocks are mutually independent, how much of its capital should the fund invest in each stock if it wishes to minimize its total variance while guaranteeing an overall expected profit of at least $\$M$? (If the random variables $\tilde{x}_1$ and $\tilde{x}_2$ are independent and have means m_1 and m_2 and variances v_1 and v_2, then the random variable $\tilde{y} \equiv \tilde{x}_1 + \tilde{x}_2$ has mean $m_1 + m_2$ and variance $v_1 + v_2$, while the random variable $\tilde{z} \equiv k\tilde{x}_1$, where k is any constant, has mean km_1 and variance $k^2 v_1$.)

1-9. Suppose that in Section 1.4A the apartment building has a second oil storage tank with a capacity of m gallons. It is supposed to be kept full for emergencies, but the apartment may use oil from it by paying an additional cost of $\$d$ per gallon removed (representing the cost of making alternative provisions for emergencies). When fuel is replenished the second tank is automatically filled first. Now what is the optimal reorder policy? How many variables are required to describe it?

1-10. A chemist wishes to dissolve some crystals of potassium chloride (KCl) and sodium iodide (NaI) in a single aqueous solution. Let m_{K}, m_{Cl}, m_{Na}, and m_{I} denote the quantities, in grams, of potassium, chloride, sodium, and iodide that are to be dissolved. When they are introduced into the same solution, there will also be a tendency for potassium iodide and sodium chloride crystals to form. The maximum amount of any salt that can be dissolved in a given volume of water is determined by the *solubility product* of that salt; for example, the product of the concentration (in grams per cubic centimeter of water) of potassium and that of chloride cannot exceed the solubility product S_{KCl} of potassium chloride. What is the minimum volume of water required to dissolve all the chemist's salts so that none of the four possible types of crystals will form?

1-11. A college has 5 men's and 5 women's dormitories in which m_i males and f_i females, $i = 1, \ldots, 5$, are initially housed. All 10 buildings have a maximum capacity of C students, so that

$$m_i \leqq C \quad \text{and} \quad f_i \leqq C, \qquad i = 1, \ldots, 5.$$

At midsemester the administration decides to integrate the sexes by moving men and women into each other's dormitories in such a way that the total population of every dorm will be between 40 and 60% male. Letting d_{ij} be the straight-line distance from male dorm i to female dorm j, how should students be shifted so as to minimize the total moving distance for all students? Note that we have not bothered to give the distances between any two men's dormitories. Why are they irrelevant? That is, why can it never be optimal to shift a man from one male dormitory to another?

1-12. A boy wants to fire a marble from his toy cannon over his backyard fence into the yard of his neighbor (see Fig. 1-5). The cannon has a muzzle velocity, initially imparted to the marble, of v_0 feet per second and can direct the marble at any angle θ from the horizontal. How far from the fence and at what angle should he place the cannon if he wants the marble to land in his neighbor's yard as far from the fence (not the cannon) as possible? The fence is h feet high, the marble is assumed to be fired from ground level, and the relevant laws of motion will be found in your freshman physics notes.

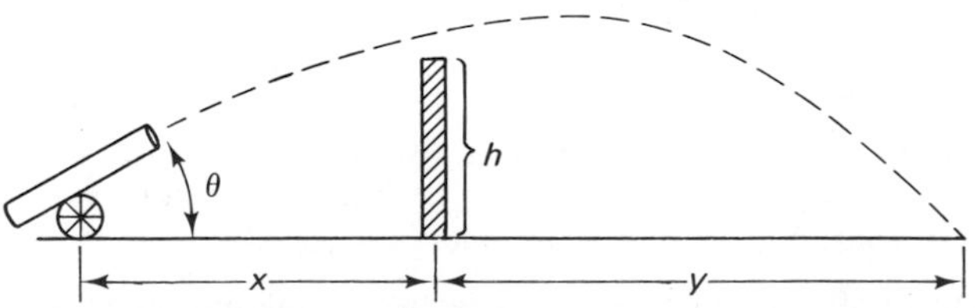

FIGURE 1-5

1-13. *The caterer's problem:* This is a famous problem from the annals of the operations research literature. A professional caterer will require r_j fresh napkins on each of N successive days, $j = 1, \ldots, N$; these napkins may be newly purchased or laundered after a previous use. Suppose the laundry has two types of service, "fast" and "slow." The fast service costs f cents per napkin and requires m days, while the slow service costs s cents and takes n days. Naturally, $m < n$ and $f > s$. If new napkins cost c cents apiece and the caterer starts with none, how can he meet his requirements at minimum total cost? Assume the napkins have no resale value.

1-14. Sir Robinson Smythe-Jones has resolved to set a new land speed record by propelling a skeleton car with a three-stage jet engine across a flat frictionless surface. Car and driver together have mass m. Each of the stages consists of chemical fuel enclosed in a container of negligible mass which can be jettisoned. The fuel to be used, which provides power by undergoing a chemical change that does not alter its mass appreciably, "burns" steadily at the rate of one mass unit per second to produce F units of force per second. How much fuel should Sir Robinson put into each stage in order to maximize his final velocity, given that the total mass of fuel used may not exceed M units? Assume that each stage will be jettisoned as soon as its fuel is spent.

Section 1.5

Exercises 1-15 through 1-19: For each of the following mathematical programming problems, graph or sketch the feasible region and a few indifference curves of the objective function, including the optimal one (if any), and identify the optimal solution or solutions.

1-15. Min $z = 2x_1 - x_2$
subject to $x_1 - x_2 \leqq 2,$
$-2x_1 + x_2 \leqq 3,$
and $x_1, x_2 \geqq 0.$

1-16. Max $z = (x_1^2 + x_2^2)^{-1}$
subject to $x_1 + x_2 \geqq -2,$
$x_1 \leqq 3,$
and $x_2 \leqq 4.$

1-17. Max $z = x_2$
subject to $x_1^2 - x_2 \geqq 0$
and $-2 \leqq x_1 \leqq 2.$

1-18. Max $z = x_1$
subject to $x_1^2 + x_2^2 \leqq 13$
and $x_1^2 x_2^2 \geqq 36.$

1-19. Min $z = x_1 x_2 - x_1 - 2x_2$
subject to $x_1 + x_2 \leqq 3$
and $x_1, x_2 \geqq 0.$

2

Review of Linear Programming

2.1 INTRODUCTION

Before embarking on the summary of linear programming that will constitute the bulk of this chapter, we shall present in Section 2.2 a rather detailed derivation of Taylor's theorem for functions of n variables, with emphasis upon the special case of displacement along a given straight line in E^n. Although this material may already be quite familiar to the reader, its inclusion is justified by the very significant role it plays in mathematical programming. Our discussion of it is placed here, rather than at the beginning of Chapter 3, in order to allow us to use a Taylor expansion in the proof of the duality theorem. Following our review of Taylor we shall turn our attention briefly in Section 2.3 to some results from linear algebra, moving on from there to a summary of the simplex method and, finally, to a thorough discussion of duality theory.

Regarding notation, boldface capital letters will be used to represent matrices and boldface lowercase letters to represent vectors; thus **A** and **b** are, respectively, a matrix and a vector of unspecified dimensions. The phrase "m-by-n" matrix will mean a matrix having m rows and n columns. In addition, we shall adopt the following conventional notation, which is the same as that used in the linear programming texts of Hadley [21] and Simmons [42]:

a_{ij} is the element in row i and column j of the matrix **A**,

$\mathbf{a}_j$ is the jth column of the matrix **A**,

b_i is the ith component of the vector $\mathbf{b}$,

$\mathbf{I}_n$ is the identity matrix of order n,

$\mathbf{e}_j$ is the unit vector having a 1 in the jth position and zeros elsewhere (dimension unspecified), and

$\mathbf{0}$ is either a null vector or a null matrix, depending on context.

Finally, throughout this text, unless stated otherwise, *all vectors will be understood to be of the column variety*; when we need to represent a 1-by-n array of numbers in our theoretical development, we shall frequently use a transposed column vector rather than a row vector.

2.2 TAYLOR'S THEOREM

The proof of Taylor's theorem, as we shall derive it here, depends on a result from elementary calculus known as Rolle's theorem, which in turn depends on the fact that the derivative of any continuous, differentiable function of one variable vanishes at a maximum or minimum. It seems appropriate to begin a textbook on optimization methods by sketching out a proof of this last proposition. [For the moment we shall use the rather imprecise terms "maximum" and "minimum" in their most familiar sense to refer to the peaks and troughs of a continuous curve; that is, the function $f(x)$ takes on a "maximum" at the point X if $f(X) \geqq f(x)$ for all points x within some small positive distance of X. Beginning in Chapter 3, however, we shall distinguish among three different types of maxima, the type under consideration at present being known as an *unconstrained local maximum*.]

Let the function $f(x)$ be continuous at every point in the closed interval $a \leqq x \leqq b$ and let the maximum value of the function over this interval occur at X, where $a < X < b$; that is, $f(X) \geqq f(x)$ for all x such that $a \leqq x \leqq b$. Notice that we are not permitting X to be an end point of the interval. Since X is an internal point, $f(X) \geqq f(x)$ for all x within some suitably small distance δ of X, so that in particular

$$f(X+h) - f(X) \leqq 0 \qquad \text{whenever } 0 < h < \delta.$$

Dividing by $h > 0$, we have

$$\frac{f(X+h) - f(X)}{h} \leqq 0 \qquad \text{whenever } 0 < h < \delta,$$

and taking the limit of the left-hand side as h approaches zero, we can write

$$\frac{df(X)}{dx} \leqq 0 \tag{2-1}$$

provided that this derivative exists at X. By repeating the argument for small negative h, we obtain $df(X)/dx \geqq 0$, and this combines with (2-1) to yield

$$\frac{df(X)}{dx} = 0. \tag{2-2}$$

A similar derivation shows that (2-2) also holds when X is a minimum.

Consider now the following:

Rolle's theorem: If the function f(x) *is continuous at every point in the closed interval* $a \leqq x \leqq b$ *and differentiable at every point in the open interval* $a < x < b$, *and if moreover* $f(a) = f(b) = 0$, *then there exists at least one point* X *such that* $a < X < b$ *and* $df(X)/dx = 0$.

Our proof of Rolle's theorem omits one or two necessary formal steps (see, for example, Thomas [45]). Suppose first that the value of the function is zero throughout the interval; in that case the derivative is identically zero, and the conclusion follows trivially. On the other hand, suppose that $f(x)$ is positive (negative) at some point within the interval, as shown in Fig. 2-1(a). Then, since $f(x)$ is zero at the end points, it must attain its maximum (minimum) over the interval $a \leqq x \leqq b$ at some point X, $a < X < b$. But the derivative is known to exist at X, so it follows that $df(X)/dx = 0$, as desired. Note that the existence of the derivative throughout the interval—or at least at X—is an essential condition: It rules out such functions as the one shown in Fig. 2-1(b), for which there is no point in the interval $a < x < b$ such that $df/dx = 0$.

We are now ready to derive Taylor's formula for functions of a single variable, which is one of the most useful results in all of mathematical analysis. Following standard practice, we shall use f' as an alternative notation for the derivative df/dx; similarly, f'' represents d^2f/dx^2 and $f^{[n]}$ represents d^nf/dx^n.

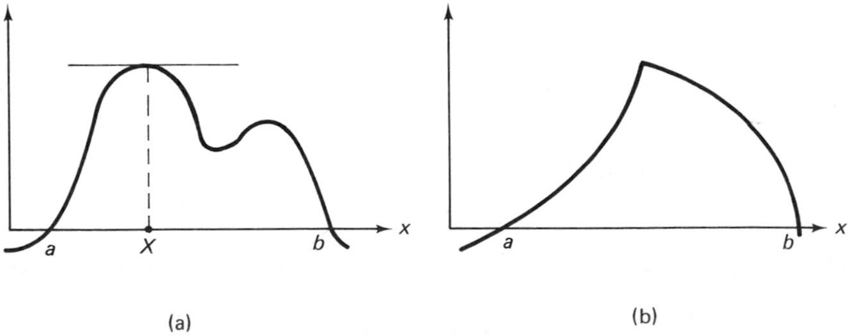

FIGURE 2-1

Theorem 2.1 (Taylor's formula): Let a *and* b *be any two points on the real line, with* $a < b$, *and let* n *be any nonnegative integer. If the function* f(x) *and its first* $n + 1$ *derivatives exist and are continuous in the closed interval* $a \leqq x \leqq b$, *then*

$$f(b) = f(a) + \frac{(b - a)}{1!}f'(a) + \frac{(b - a)^2}{2!}f''(a) + \cdots + \frac{(b - a)^n}{n!}f^{[n]}(a) + R_{n+1},$$

where

$$R_{n+1} = \frac{(b - a)^{n+1}}{(n + 1)!}f^{[n+1]}(X), \qquad a < X < b.$$

In the remainder term R_{n+1}, the $(n + 1)$th derivative is being evaluated at some point X which cannot be specified exactly but which is known to lie between a and b. Incidentally, R_{n+1} is called *Lagrange's form of the remainder*; there are other forms, but the above is by far the most useful for our purposes.

Our derivation of Taylor's formula will be indirect and somewhat artificial, though not particularly difficult. Let a constant K be defined by the following equation, which is linear in K:

$$f(b) = f(a) + \frac{(b - a)}{1!}f'(a) + \frac{(b - a)^2}{2!}f''(a) + \cdots + \frac{(b - a)^n}{n!}f^{[n]}(a) + \frac{K(b - a)^{n+1}}{(n + 1)!}. \tag{2-3}$$

Next, define a new function $F(x)$:

$$F(x) \equiv f(b) - f(x) - \frac{(b - x)}{1!}f'(x) - \frac{(b - x)^2}{2!}f''(x) - \cdots - \frac{(b - x)^n}{n!}f^{[n]}(x) - \frac{K(b - x)^{n+1}}{(n + 1)!}. \tag{2-4}$$

Since $F(x)$ is a sum of continuous functions, it must itself be continuous in the closed interval $a \leqq x \leqq b$. Moreover, we may write

$$F(b) = 0 \quad \text{and} \quad F(a) = 0,$$

with the former holding trivially and the latter following from (2-3). The derivative of $F(x)$ is

$$F'(x) = 0 - f'(x) + f'(x) - (b - x)f''(x) + (b - x)f''(x) - \frac{(b - x)^2}{2!}f'''(x) + \cdots - \frac{(b - x)^n}{n!}f^{[n+1]}(x) + \frac{K(b - x)^n}{n!}.$$

Since all terms except the last two cancel in consecutive pairs, this reduces to

$$F'(x) = \frac{(b - x)^n}{n!}[K - f^{[n+1]}(x)]. \tag{2-5}$$

From our hypothesis we know that this derivative exists everywhere in the interval $a \leqq x \leqq b$. It follows, therefore, that $F(x)$ satisfies all the conditions of Rolle's theorem, and there must accordingly exist a point X in the open interval $a < X < b$ at which the derivative (2-5) vanishes:

$$F'(X) = \frac{(b - X)^n}{n!}[K - f^{[n+1]}(X)] = 0.$$

Since $X \neq b$, we must have

$$K = f^{[n+1]}(X),$$

and substituting this value of K into (2-3) proves the theorem.

Given any point $x = a$ and any integer n, if Theorem 2.1 holds for some other point b, then it must also hold for any point x *between* a and b. Furthermore, provided that the continuity conditions are satisfied, it is easily shown that Taylor's formula also holds when $b < a$, except that the point X in the remainder term must satisfy $b < X < a$ (to prove this, merely trace through the proof of Theorem 2.1, reversing the end points of every interval mentioned). These two observations, in conjunction with Taylor's formula, establish a corollary result which is important enough to be stated as a theorem:

Theorem 2.2 (Taylor's theorem): Let a *be any point on the real line and* n *be any nonnegative integer. Then for any value of* x *such that the function* f *and its first* n + 1 *derivatives exist and are continuous at every point between* a *and* x *inclusive,*

$$\text{(2-6)} \qquad f(x) = f(a) + \frac{(x - a)}{1!}f'(a) + \frac{(x - a)^2}{2!}f''(a) + \cdots + \frac{(x - a)^n}{n!}f^{[n]}(a) + R_{n+1},$$

where

$$R_{n+1} = \frac{(x - a)^{n+1}}{(n + 1)!}f^{[n+1]}(X), \qquad X \text{ between } a \text{ and } x.$$

Taylor's theorem specifies a second algebraic function, a polynomial in $x - a$ with a finite number of terms, which can be used in place of any function $f(x)$ possessing the desired continuity properties.[1] The substitution (2-6), usually with $n = 0$ or $n = 1$, is very useful in the theoretical development of mathematical programming and in other branches of mathematical analysis as well: When one wishes to establish a property or prove a state-

[1] Actually, (2-6) implicitly specifies several polynomials that can be substituted for $f(x)$, since if it holds for some given $n = n_0$, it also holds for any smaller nonnegative value of n.

ment about a general algebraic function, it is often much easier to work with one of its polynomial equivalents.

Suppose now that *all* derivatives of $f(x)$ are continuous in the vicinity of $x = a$, and suppose further that

$$\lim_{n \to \infty} R_{n+1} = 0.$$

It is worth emphasizing that these assumptions are not terribly restrictive: They are satisfied by all the usual algebraic functions (polynomial, trigonometric, exponential, logarithmic, etc.) for "almost all" values of x at which they are defined. It then follows that $f(x)$ can be represented by the infinite series

$$f(x) = \sum_{n=0}^{\infty} \frac{(x-a)^n}{n!} f^{[n]}(a), \tag{2-7}$$

where $0! = 1$ by convention. We say that (2-7) is a *Taylor series* or, more specifically, a *Taylor expansion of* $f(x)$ *about the point* $x = a$. When truncated at a particular n, it allows us to calculate an approximate value for $f(x)$ at *any* point $x = \hat{x}$, provided only that $f, f', \ldots, f^{[n]}$ are continuous in the closed interval $a \leqq x \leqq \hat{x}$ (or $\hat{x} \leqq x \leqq a$). For any given n, in general, the closer $\hat{x}$ is to a, the better the approximation of $f(\hat{x})$ will be. Similarly, for any given $\hat{x}$ and a, the greater the number of terms n, the better the approximation.

As an example, consider the application of Taylor's theorem to the function $f(x) = e^x$, using $a = 0$. Since $f^{[n]}(x) = e^x$ and $f^{[n]}(0) = 1$ for all nonnegative n, and since the function e^x is defined and continuous for all values of x, we may substitute into (2-6) to obtain

$$f(x) = e^x = 1 + x + \frac{x^2}{2!} + \cdots + \frac{x^n}{n!} + R_{n+1}, \tag{2-8}$$

where

$$R_{n+1} = \frac{x^{n+1}}{(n+1)!} e^X, \qquad X \text{ between 0 and } x.$$

The above holds for any particular value of x and for any nonnegative n. But since $X < |x|$, the magnitude of the remainder term cannot exceed $|x|^{n+1} e^{|x|}/(n+1)!$, and it is not difficult to show that

$$\lim_{n \to \infty} R_{n+1} = 0.$$

It follows that the Taylor expansion

$$e^x = \sum_{n=0}^{\infty} \frac{x^n}{n!} = 1 + x + \frac{x^2}{2!} + \cdots + \frac{x^n}{n!} + \cdots$$

holds for all real values of x. Incidentally, any Taylor expansion about the point $a = 0$ is also known as a *Maclaurin series.*

Suppose we wish to use (2-8) to calculate the numerical value of e [that is, the value of the function $f(x) = e^x$ at $x = 1$] to three decimal places, knowing only that $2 < e < 3$. For $x = 1$ the remainder term is

$$R_{n+1} = \frac{e^X}{(n+1)!}, \qquad 0 < X < 1,$$

which implies

$$R_{n+1} < \frac{e^1}{(n+1)!} < \frac{3}{(n+1)!}.$$

Since we want the error caused by ignoring R_{n+1} to be no greater than .0005, we seek the smallest value of n such that $3/(n+1)! < .0005$. This turns out to be $n = 7$; hence

$$e = 1 + 1 + \frac{1}{2!} + \cdots + \frac{1}{7!} \cong 2.718$$

is correct to three decimal places.

Returning to theoretical matters, let us now develop a more general Taylor expansion that will be applicable to functions of any number of variables. As an intermediate step toward this goal, we shall use Theorem 2.2 to obtain an expansion which is valid for all points along a specified straight line in E^n. We begin by assuming that for the function $f(\mathbf{x})$, where $\mathbf{x} \equiv (x_1, \ldots, x_n)$, all the first partial derivatives $\delta f/\delta x_j$, all the second partial derivatives $\delta^2 f/\delta x_i \delta x_j, \ldots$, and all the $(n+1)$th partial derivatives are continuous in the vicinity of a fixed point $\mathbf{x}_0$ in E^n. Imagine now that we are free to move away from $\mathbf{x}_0$ either in some specific direction $\mathbf{s}$ or in the opposite or $(-\mathbf{s})$-direction, thereby generating a straight line whose general point is

$$\mathbf{x} = \mathbf{x}_0 + \theta \mathbf{s}, \tag{2-9}$$

where the parameter θ can take on any real value. The value of the function at any one of these points is $f(\mathbf{x}) = f(\mathbf{x}_0 + \theta \mathbf{s})$, which can be expressed formally as a function of the single parameter θ:

$$F(\theta) \equiv f(\mathbf{x}_0 + \theta \mathbf{s}).$$

Since the various partial derivatives of $f(\mathbf{x})$ are continuous in the vicinity of $\mathbf{x}_0$, so are the first $n+1$ derivatives of $F(\theta)$ in the vicinity of $\theta = 0$, and we may use Theorem 2.2 to write the following Taylor expansion:

$$F(\theta) = F(0) + \frac{\theta}{1!}F'(0) + \frac{\theta^2}{2!}F''(0) + \cdots + \frac{\theta^n}{n!}F^{[n]}(0) + R_{n+1}. \tag{2-10}$$

To express F', F'', etc., in terms of the partial derivatives of f, we recall the *chain rules* of differential calculus, noting that (2-9) implies $dx_j/d\theta = s_j$:

$$\frac{dF}{d\theta} = \sum_{j=1}^{n} \frac{\delta f}{\delta x_j} \cdot \frac{dx_j}{d\theta} = \sum_{j=1}^{n} \frac{\delta f}{\delta x_j} \cdot s_j,$$

$$\frac{d^2F}{d\theta^2} = \frac{d}{d\theta}\left(\sum_{j=1}^{n} \frac{\delta f}{\delta x_j} \cdot s_j\right) = \sum_{i=1}^{n} \frac{\delta}{\delta x_i}\left(\sum_{j=1}^{n} \frac{\delta f}{\delta x_j} \cdot s_j\right)\frac{dx_i}{d\theta}$$

$$= \sum_{i=1}^{n}\sum_{j=1}^{n} \frac{\delta^2 f}{\delta x_i\, \delta x_j} s_j s_i,$$

and so on. Substituting these into (2-10) yields

$$f(\mathbf{x}_0 + \theta\mathbf{s}) = f(\mathbf{x}_0) + \sum_{j=1}^{n} \frac{\delta f(\mathbf{x}_0)}{\delta x_j}\theta s_j + \frac{1}{2!}\sum_{i=1}^{n}\sum_{j=1}^{n} \theta s_i \frac{\delta^2 f(\mathbf{x}_0)}{\delta x_i\, \delta x_j}\theta s_j + \cdots \text{(higher-order terms)}, \tag{2-11}$$

an expansion which will find some very important applications later on.

Equation (2-11) gives a Taylor expansion for the value of $f(\mathbf{x})$ at all points on the particular straight line (2-9). In view of our continuity assumptions, however, there is no reason why we could not have begun by specifying some direction other than $\mathbf{s}$ and therefore some other straight line. Since $\mathbf{s}$ can be freely chosen, we might as well substitute the general point $\mathbf{x}$ in place of $\mathbf{x}_0 + \theta\mathbf{s}$ in (2-11). This implies $\theta s_j = x_j - x_{0j}$, where x_{0j} is the jth component of $\mathbf{x}_0$, and leads to

$$f(\mathbf{x}) = f(\mathbf{x}_0) + \sum_{j=1}^{n} \frac{\delta f(\mathbf{x}_0)}{\delta x_j}(x_j - x_{0j}) + \frac{1}{2!}\sum_{i=1}^{n}\sum_{j=1}^{n} (x_i - x_{0i})\frac{\delta^2 f(\mathbf{x}_0)}{\delta x_i\, \delta x_j}(x_j - x_{0j}) + \cdots \text{(higher-order terms)}. \tag{2-12}$$

In effect, this is a Taylor expansion of the value of $f(\mathbf{x})$ along an imaginary straight line from the fixed point $\mathbf{x}_0$ to *any* other point $\mathbf{x}$.

Inasmuch as (2-12) is rather cumbersome, we shall now introduce some conventional notation that will allow us to write it more compactly. Define an n-component column vector $\mathbf{\nabla f}$ called the *gradient* and an n-by-n matrix called the *Hessian* as follows:

$$\nabla \mathbf{f} \equiv \left(\frac{\delta f}{\delta x_1}, \ldots, \frac{\delta f}{\delta x_n}\right) \tag{2-13}$$

and

$$\mathbf{H} \equiv \begin{bmatrix} \dfrac{\delta^2 f}{\delta x_1^2} & \dfrac{\delta^2 f}{\delta x_1\, \delta x_2} & \cdots & \dfrac{\delta^2 f}{\delta x_1\, \delta x_n} \\ \dfrac{\delta^2 f}{\delta x_2\, \delta x_1} & \dfrac{\delta^2 f}{\delta x_2^2} & \cdots & \dfrac{\delta^2 f}{\delta x_2\, \delta x_n} \\ \vdots & \vdots & & \vdots \\ \dfrac{\delta^2 f}{\delta x_n\, \delta x_1} & \dfrac{\delta^2 f}{\delta x_n\, \delta x_2} & \cdots & \dfrac{\delta^2 f}{\delta x_n^2} \end{bmatrix}. \tag{2-14}$$

We shall use $\nabla \mathbf{f}(\mathbf{x})$ to represent the gradient vector[2] with each of its components evaluated at the point $\mathbf{x}$, and similarly for $\mathbf{H}(\mathbf{x})$. Substituting (2-13) and (2-14) into (2-12) produces

$$f(\mathbf{x}) = f(\mathbf{x}_0) + \nabla \mathbf{f}^T(\mathbf{x}_0) \cdot (\mathbf{x} - \mathbf{x}_0) + \frac{1}{2!}(\mathbf{x} - \mathbf{x}_0)^T \mathbf{H}(\mathbf{x}_0) \cdot (\mathbf{x} - \mathbf{x}_0) + \cdots \text{(higher-order terms)}, \tag{2-15}$$

where the superscript T will be used in this text to denote the transpose.

For the last few paragraphs we have not been writing out the higher-order terms of the expansions, both because they are quite complicated (additional special notation would be required) and because they will not be needed in our work in mathematical programming. We shall, in fact, be using n-variable Taylor expansions of the two lowest orders only, i.e., those for which $n = 0$ and $n = 1$. These two formulas are obtained by truncating (2-15) at the appropriate term, with the gradient or Hessian in the remainder term being evaluated at some unspecified point on the imaginary straight line between $\mathbf{x}_0$ and $\mathbf{x}$:

General first-order Taylor expansion:

$$f(\mathbf{x}) = f(\mathbf{x}_0) + \nabla \mathbf{f}^T[\mathbf{x}_0 + \theta(\mathbf{x} - \mathbf{x}_0)] \cdot (\mathbf{x} - \mathbf{x}_0), \qquad 0 < \theta < 1. \tag{2-16}$$

General second-order Taylor expansion:

$$f(\mathbf{x}) = f(\mathbf{x}_0) + \nabla \mathbf{f}^T(\mathbf{x}_0) \cdot (\mathbf{x} - \mathbf{x}_0) + \tfrac{1}{2}(\mathbf{x} - \mathbf{x}_0)^T \mathbf{H}[\mathbf{x}_0 + \theta(\mathbf{x} - \mathbf{x}_0)] \cdot (\mathbf{x} - \mathbf{x}_0), \qquad 0 < \theta < 1. \tag{2-17}$$

[2]The gradient vector is extremely interesting for various reasons and will play a pivotal role throughout this text. For the moment, however, it is merely a notational convenience.

The "in-between" point $\mathbf{x}_0 + \theta(\mathbf{x} - \mathbf{x}_0)$ is sometimes also represented as $\theta\mathbf{x} + (1 - \theta)\mathbf{x}_0$, where again $0 < \theta < 1$.

2.3 NOTES FROM LINEAR ALGEBRA

The remainder of the chapter will be devoted to a review of linear programming, beginning in this section with a brief compendium of those definitions and results from linear algebra that form the foundation of the simplex method. No proofs will be presented. A more detailed discussion of this material may be found in a reference such as Hadley [20].

Given a set of n-component vectors $\mathbf{u}_1, \mathbf{u}_2, \ldots, \mathbf{u}_m$, the n-component vector

$$\mathbf{u} = \lambda_1\mathbf{u}_1 + \lambda_2\mathbf{u}_2 + \cdots + \lambda_m\mathbf{u}_m = \sum_{i=1}^{m} \lambda_i\mathbf{u}_i,$$

where $\lambda_1, \ldots, \lambda_m$ are any scalar numbers, is called a *linear combination* of $\mathbf{u}_1, \ldots, \mathbf{u}_m$.

A set of n-component vectors $\mathbf{u}_1, \ldots, \mathbf{u}_m$ is said to be *linearly dependent* if and only if there exist scalar numbers $\lambda_1, \ldots, \lambda_m$ *not all zero* such that

$$\sum_{i=1}^{m} \lambda_i\mathbf{u}_i = \mathbf{0}_n,$$

where $\mathbf{0}_n$ is the n-component null vector. But if the only linear combination of $\mathbf{u}_1, \ldots, \mathbf{u}_m$ that adds to the null vector has all the $\lambda_i = 0$, then the set of vectors is said to be *linearly independent*.

A set of n-component vectors is said to *span* E^n, or to be a *spanning set*, if every vector in E^n can be written as a linear combination of them.

Any set of linearly independent vectors that spans E^n constitutes a *basis* for E^n. For example, the set of n-component unit vectors $\mathbf{e}_1, \mathbf{e}_2, \ldots, \mathbf{e}_n$ forms a basis for E^n.

The representation of any vector as a linear combination of a given set of basis vectors is unique.

Any basis for E^n contains exactly n vectors, and any n linearly independent vectors in E^n form a basis. No more than n vectors in E^n can be linearly independent, and no fewer than n can span E^n.

Let $\mathbf{a}_1, \ldots, \mathbf{a}_m$ constitute a basis for E^m and let $\mathbf{b} = \sum_{i=1}^{m} \lambda_i\mathbf{a}_i$ be the unique representation of the m-component vector $\mathbf{b}$ in terms of those basic vectors. If any vector $\mathbf{a}_k$ for which λ_k is nonzero is removed from the basic set and $\mathbf{b}$ is substituted for it, then the resulting set is also a basis for E^m.

Given an n-by-n matrix $\mathbf{A}$, if there exists an n-by-n matrix $\mathbf{B}$ such that

$$\mathbf{AB} = \mathbf{BA} = \mathbf{I}_n,$$

where $\mathbf{I}_n$ is the identity matrix of order n, then $\mathbf{B}$ is called the *inverse* of $\mathbf{A}$. We shall use the symbol $\mathbf{A}^{-1}$ to denote the inverse of $\mathbf{A}$. The inverse of any n-by-n matrix, if it exists, is unique (note that only a square matrix can have an inverse). The matrix $\mathbf{A}$ is said to be *nonsingular* if $\mathbf{A}^{-1}$ exists and *singular* if it does not.

If either the columns or the rows of the square matrix $\mathbf{A}$ form a linearly independent set of vectors, then $\mathbf{A}$ has an inverse. If $\mathbf{A}$ has an inverse, then both the rows of $\mathbf{A}$ and the columns of $\mathbf{A}$ are linearly independent.

If $\mathbf{A}$ is an m-by-n matrix with $m \leqq n$ and the rows of $\mathbf{A}$ are linearly independent, then there exists at least one set of m linearly independent columns of $\mathbf{A}$.

Consider a set or *system* of m simultaneous linear equations in the unknowns $x_1, x_2, \ldots, x_n$:

$$\text{(2-18)} \qquad \begin{cases} a_{11}x_1 + a_{12}x_2 + \cdots + a_{1n}x_n = b_1 \\ a_{21}x_1 + a_{22}x_2 + \cdots + a_{2n}x_n = b_2 \\ \quad \vdots \qquad\qquad\qquad\qquad\qquad \vdots \\ a_{m1}x_1 + a_{m2}x_2 + \cdots + a_{mn}x_n = b_m, \end{cases}$$

where the a_{ij} and b_i are scalar constants. Any assignment of values to the variables $x_1, \ldots, x_n$ that satisfies all the equations is called a *solution* to the system. If no solution exists, the equations are said to be *inconsistent*, while if more than one exists, the system is *indeterminate*.

Define a row vector

$$\mathbf{s}_i \equiv [a_{i1}, a_{i2}, \ldots, a_{in}, b_i]$$

for each equation in the system (2-18), $i = 1, \ldots, m$. The kth of these equations is said to be *redundant* if $\mathbf{s}_k$ is exactly equal to a linear combination of all the other $\mathbf{s}_i$, $i = 1, \ldots, k-1, k+1, \ldots, m$. We also say that the system (2-18) is itself redundant if any of its member equations are redundant.

Let the simultaneous equation system (2-18) be represented by the matrix/vector equation

$$\text{(2-19)} \qquad \mathbf{A}\mathbf{x} \equiv x_1\mathbf{a}_1 + x_2\mathbf{a}_2 + \cdots + x_n\mathbf{a}_n = \mathbf{b},$$

where $\mathbf{A}$ is an m-by-n matrix, $\mathbf{a}_i$ is the ith column of $\mathbf{A}$, and $\mathbf{b}$ and $\mathbf{x}$ are column vectors having m and n components, respectively. Assume that $m \leqq n$ and that the rows of $\mathbf{A}$ are linearly independent, and suppose that m linearly independent columns $\mathbf{a}_\alpha$, $\mathbf{a}_\beta$, ... have been selected from $\mathbf{A}$. If we set the $n - m$ variables *not* associated with these columns equal to zero, then the unique solution to the resulting system of m equations in the m unknowns

$x_\alpha, x_\beta, \ldots$ is called a *basic solution.* The columns $\mathbf{a}_\alpha, \mathbf{a}_\beta, \ldots$ are the *basic columns,* while $x_\alpha, x_\beta, \ldots$ are the *basic variables.* If any one (or more) of the basic variables has the value zero, then the basic solution is said to be *degenerate.*

2.4 THE SIMPLEX METHOD

From linear algebra let us now move on to linear programming and a brisk review of the simplex method. The material in this section is excerpted from Simmons [42]; another suitable reference is Hadley [21]. We begin by recalling that a *linear programming problem* is, by definition, a mathematical program having a linear objective function and linear constraints:

$$\text{(2-20)} \qquad \text{Max or Min } z = c_1x_1 + c_2x_2 + \ldots + c_px_p$$

$$\text{(2-21)} \qquad \text{subject to } a_{i1}x_1 + a_{i2}x_2 + \cdots + a_{ip}x_p \begin{Bmatrix} \leqq \\ = \\ \geqq \end{Bmatrix} b_i, \qquad i = 1, \ldots, m,$$

where the a_{ij}, the b_i, and the c_j are scalar constants. Notice that nonnegativity restrictions $x_j \geqq 0$, which are simply a type of linear constraint, are included within the general format (2-21).

Before the Simplex method can be applied to a linear program, we must first convert it into what is known as *standard form*[3]:

$$\begin{aligned} \text{Max } z &= c_1x_1 + c_2x_2 + \cdots + c_nx_n \\ \text{subject to } a_{i1}x_1 + a_{i2}x_2 + \cdots + a_{in}x_n &= b_i, \qquad i = 1, \ldots, m, \\ \text{and } x_j &\geqq 0, \qquad j = 1, \ldots, n, \end{aligned}$$

where the number of variables n may or may not be the same as before. The process of conversion may require several steps:

1. If the linear program as originally formulated calls for the minimization of $z = c_1x_1 + \cdots + c_px_p$, we can instead substitute the equivalent objective

$$\text{Max } z' = d_1x_1 + \cdots + d_px_p,$$

where $d_j \equiv -c_j$, all j.

2. If any variable x_j is *not* restricted to nonnegative values, it can be eliminated via the transformation

$$x_j \equiv x'_j - x''_j,$$

[3]Our choice of a maximization objective for the standard format is at variance with certain other writers, who prefer minimization.

where $x'_j, x''_j \geqq 0$. Note that every real value of x_j can be expressed by nonnegative values of x'_j and x''_j.

3. Finally, any inequality constraints in the original formulation can be converted to equations by the addition of nonnegative *slack* or *surplus variables*; thus the constraints

$$a_{11}x_1 + a_{12}x_2 + \cdots + a_{1p}x_p \leqq b_1$$

and

$$a_{21}x_1 + a_{22}x_2 + \cdots + a_{2p}x_p \geqq b_2$$

would become

$$a_{11}x_1 + a_{12}x_2 + \cdots + a_{1p}x_p + x_{p+1} \qquad = b_1$$

and

$$a_{21}x_1 + a_{22}x_2 + \cdots + a_{2p}x_p \qquad - x_{p+2} = b_2,$$

$$\text{with } x_{p+1}, x_{p+2} \geqq 0,$$

where x_{p+1} is a slack and x_{p+2} a surplus variable. The new variables would, of course, be assigned cost coefficients of zero.

In matrix/vector notation the standard form of the linear programming problem is written as follows:

$$\text{(2-22)} \qquad \begin{cases} \text{Max } z = \mathbf{c}^T\mathbf{x} \\ \text{subject to } \mathbf{Ax} = \mathbf{b} \\ \text{and } \mathbf{x} \geqq \mathbf{0}, \end{cases}$$

where $\mathbf{A}$ is m-by-n, $\mathbf{b}$ is m-by-1, and $\mathbf{c}$ and $\mathbf{x}$ are n-by-1. We remind the reader that the superscript T denotes the transpose of a vector or matrix.

We are now ready to introduce the *simplex method*, which, in conjunction with certain auxiliary procedures, is capable of solving any linear program in standard form, and therefore ultimately any linear program. Leaving aside the auxiliary procedures for the moment, however, the term "simplex method" refers specifically to an iterative computational algorithm that can obtain the optimal solution of any standard-form problem for which the following two conditions hold:

1. None of the constraints are redundant, and
2. A *basic feasible solution*—that is, a nonnegative basic solution to the constraint set $\mathbf{Ax} = \mathbf{b}$—has already been found.

It can be shown that any LPP that has a feasible solution (and no redundant constraints) also has a basic feasible solution (BFS). This means that the simplex method is potentially capable of solving any feasible linear program, provided that computational procedures are available for eliminating redundant constraints and identifying an initial BFS. Such procedures do, in fact, exist and will be reviewed later in this section.

A third condition might also have been added to conditions 1 and 2 above, namely, that the number of variables n in the standard-form LPP must exceed the number of constraints m in order to make the problem "interesting." If n were less than m, then a basic feasible solution—as required by condition 2—would simply not exist, by definition. And if $n = m$ the absence of redundant constraints, as dictated by condition 1, would guarantee that only one feasible solution could exist: $\mathbf{x} = \mathbf{A}^{-1}\mathbf{b}$. In such a case there would be no need to choose an optimum, and hence no need to apply the simplex method.

Each of the basic feasible solutions to a linear programming problem corresponds to some *extreme point* or corner[4] of the feasible region; conversely, each extreme point corresponds to *at least* one BFS and may correspond to more if they are degenerate. By using a geometric argument, one can easily show that an optimal solution to any feasible and bounded linear program lies at one of the extreme points. It therefore follows that an exhaustive search over all basic feasible solutions must yield an optimal solution to any (bounded) LPP. This is precisely how the simplex method operates: Beginning at some extreme point/BFS, it identifies and *pivots* to another extreme point adjacent to the first but having a better objective value. The cycle is repeated either until the optimum is reached or until it is discovered that the problem is unbounded.

Before getting into the computational details of the simplex method, we must establish a certain amount of notation. Suppose that we are solving the standard-form linear programming problem (2-22) and that some current basis or basic feasible solution—whether it be the initial BFS or one subsequently generated—includes the ordered set of m variables $x_\alpha, x_\beta, \ldots$. Let these *basic variables* be collected into an m-component column vector $\mathbf{x}_B$. The ordering of the variables in $\mathbf{x}_B$ is not determined by their original order in $\mathbf{x}$; that is, while the first two components of $\mathbf{x}_B$ are $x_{B1} = x_\alpha$ and $x_{B2} = x_\beta$, it is not necessarily true that $\alpha < \beta$. In an analogous manner, let the components of $\mathbf{c}$ and the columns $\mathbf{a}_j$ of $\mathbf{A}$ that are associated with the m basic variables be assembled, respectively, into the m-by-1 vector $\mathbf{c}_B$ and the m-by-m *basis matrix* $\mathbf{B}$. The ordering of the components of $\mathbf{c}_B$, denoted by c_{Bi}, and of the *basic columns* of $\mathbf{B}$, denoted by $\mathbf{b}_i$, must exactly correspond to the ordering of the basic variables in $\mathbf{x}_B$, so that $c_{B1} = c_\alpha$, $\mathbf{b}_1 = \mathbf{a}_\alpha$, and so on.

[4]Recall that a point $\mathbf{x}$ is an *extreme point* of a set S if it lies in S but does *not* lie between any two other points in S. More precisely, $\mathbf{x}$ in S is an extreme point of S if there do not exist two different points $\mathbf{x}_1$ and $\mathbf{x}_2$ in S such that $\mathbf{x} = \lambda\mathbf{x}_1 + (1 - \lambda)\mathbf{x}_2$, where λ is any number satisfying $0 < \lambda < 1$.

The values of the basic variables in the current BFS are not difficult to find. We rewrite the constraint set of (2-22) as follows:

$$\text{(2-23)} \qquad \mathbf{Ax} = x_1\mathbf{a}_1 + x_2\mathbf{a}_2 + \cdots + x_n\mathbf{a}_n = \mathbf{b}.$$

After crossing out the terms in (2-23) corresponding to the nonbasic variables, whose value is zero, we can reorder the remaining terms to produce

$$x_\alpha\mathbf{a}_\alpha + x_\beta\mathbf{a}_\beta + \cdots = x_{B1}\mathbf{b}_1 + x_{B2}\mathbf{b}_2 + \cdots \equiv \mathbf{Bx}_B = \mathbf{b}.$$

Since the basic columns are linearly independent, it follows that $\mathbf{B}^{-1}$ exists and that

$$\text{(2-24)} \qquad \mathbf{x}_B = \mathbf{B}^{-1}\mathbf{b}.$$

Thus the components of the basic feasible solution can be evaluated directly, provided that the inverse of the associated basis matrix is known. Similarly, the current value of the objective function

$$z = c_1x_1 + \cdots + c_nx_n$$

is given by

$$z = c_\alpha x_\alpha + c_\beta x_\beta + \cdots = \mathbf{c}_B^T\mathbf{x}_B = \mathbf{c}_B^T\mathbf{B}^{-1}\mathbf{b}.$$

At this point, to determine whether or not the current BFS is optimal, we compute for every variable x_j (or column $\mathbf{a}_j$) the quantities

$$\text{(2-25)} \qquad \mathbf{y}_j = \mathbf{B}^{-1}\mathbf{a}_j$$

and

$$\text{(2-26)} \qquad z_j - c_j = \mathbf{c}_B^T\mathbf{y}_j - c_j,$$

noting that whenever x_j is a basic variable $\mathbf{y}_j$ is a unit vector and $z_j - c_j = 0$. The $z_j - c_j$ values, called the *reduced costs*, are now examined. If it is found that $z_j - c_j \geqq 0$ for every nonbasic variable x_j, then the current solution $\mathbf{x}_B$ is optimal. If $z_k - c_k < 0$ for some k and the vector $\mathbf{y}_k$ is nonpositive in every component, then the problem has an unbounded optimal solution. Finally, if $z_j - c_j < 0$ for one or more j, and if for every such j at least one component of $\mathbf{y}_j$ is positive, then it is necessary to *pivot* to a new basic feasible solution.

A pivot is a computational procedure in which one variable is removed from the basic set and another takes its place, occupying the same position in the ordering. The new basic variable is chosen according to the *simplex entry criterion*: The nonbasic variable x_k enters the basis if and only if $z_k - c_k < 0$ and

$$z_k - c_k = \min_j (z_j - c_j). \tag{2-27}$$

The variable that will leave the basis is then determined by the *simplex exit criterion*: Given that x_k is to enter the basis, the variable x_{Br} must leave, where

$$\frac{x_{Br}}{y_{rk}} = \min_i \left(\frac{x_{Bi}}{y_{ik}}, \ y_{ik} > 0 \right). \tag{2-28}$$

In applying these two criteria, any ties that occur may be broken arbitrarily.

When the new set of basic variables has been determined, the new values of the $\mathbf{y}_j$ and $z_j - c_j$ are computed and the cycle is repeated. Fortunately, it is not necessary to calculate these new values via (2-25) and (2-26). It can be shown that when x_k replaces x_{Br} in the basis the following transformation formula holds for all j, $j = 1, \ldots, n$:

$$\begin{bmatrix} \hat{\mathbf{y}}_j \\ \hat{z}_j - c_j \end{bmatrix} = \begin{bmatrix} \mathbf{y}_j \\ z_j - c_j \end{bmatrix} + y_{rj} \mathbf{\Phi}, \tag{2-29}$$

where $\hat{\mathbf{y}}_j$ and $\hat{z}_j - c_j$ are the new values; $\mathbf{y}_j$, $z_j - c_j$, and y_{rj} are old values; and $\mathbf{\Phi}$ is the $(m + 1)$-component column vector

$$\mathbf{\Phi} = \left(-\frac{y_{1k}}{y_{rk}}, \ldots, -\frac{y_{r-1,k}}{y_{rk}}, \frac{1}{y_{rk}} - 1, -\frac{y_{r+1,k}}{y_{rk}}, \ldots, -\frac{y_{mk}}{y_{rk}}, -\frac{z_k - c_k}{y_{rk}} \right). \tag{2-30}$$

The transformation formula for the values of the basic variables and the objective function is similar:

$$\begin{bmatrix} \hat{\mathbf{x}}_B \\ \hat{z} \end{bmatrix} = \begin{bmatrix} \mathbf{x}_B \\ z \end{bmatrix} + x_{Br} \mathbf{\Phi}, \tag{2-31}$$

where carets again denote the new values.

It turns out that there is a rather straightforward way of handling all these calculations, whether they are being done by hand or by computer. The current values of the relevant variables and quantities are stored in a skeleton diagram called a *simplex tableau*, as shown in Fig. 2-2. All variables x_j, basic or not, are listed across the top of the tableau, while the values of the basic variables and the objective function (which is here being maximized) are entered in the first column. Each of the remaining columns "belongs to" or is associated with some variable x_j and contains its $\mathbf{y}_j$ vector and its $z_j - c_j$.

	x_1	x_2	$\cdots$	x_n
x_{B1}	y_{11}	y_{12}	$\cdots$	y_{1n}
x_{B2}	y_{21}	y_{22}	$\cdots$	y_{2n}
$\vdots$	$\vdots$	$\vdots$		$\vdots$
x_{Bm}	y_{m1}	y_{m2}	$\cdots$	y_{mn}
Max z	$z_1 - c_1$	$z_2 - c_2$	$\cdots$	$z_n - c_n$

FIGURE 2-2. THE SIMPLEX TABLEAU.

At any pivot the values in the current tableau are scanned and the entering and exiting variables selected; the newly calculated values are then entered in a new simplex tableau.

Incidentally, although we have restricted the discussion thus far to maximization problems, the simplex method can, of course, be used to solve *minimization* problems as well. One obvious way of minimizing $z = \mathbf{c}^T\mathbf{x}$ is to maximize $z' = -z = (-\mathbf{c})^T\mathbf{x}$ instead, following the steps exactly as outlined above. If, however, it is desired to adapt the simplex method for solving minimization problems directly, two modifications to those steps are necessary, both stemming from the fact that it is now *positive* rather than negative values of $z_j - c_j$ that are favorable for basis entry:

1. If at some stage all $z_j - c_j$ are nonpositive, then the current basic feasible solution is optimal.
2. The entry criterion is exactly the opposite of (2-27): x_k enters the basis if $z_k - c_k > 0$ and

$$z_k - c_k = \max_j (z_j - c_j).$$

It should be noted that the exit criterion (2-28) is not affected by the switch to minimization.

The pivoting computations discussed above will now be illustrated by an example. Suppose that the following tableau represents some stage in the application of the simplex method to a maximization linear programming problem having six variables and three constraints:

	x_1	x_2	x_3	x_4	x_5	x_6
$x_{B1} = x_2 = 0$	1.2	1	0	−1	0	4.8
$x_{B2} = x_5 = 2$	−2.4	0	0	3	1	1.6
$x_{B3} = x_3 = 3$	2	0	1	5	0	−1
Max $z = 6$	2	0	0	−3	0	−1

The current basic feasible solution is $x_{B1} = x_2 = 0$, $x_{B2} = x_5 = 2$, $x_{B3} = x_3 = 3$, and it is degenerate. Since some of the $z_j - c_j$ are negative, optimality has not yet been achieved, and we choose x_4 to enter the basis. Applying the exit criterion,

$$\min_i\left(\frac{x_{Bi}}{y_{ik}}, y_{ik} > 0\right) = \min\left(\frac{x_{B2}}{y_{24}}, \frac{x_{B3}}{y_{34}}\right) = \min\left(\frac{2}{3}, \frac{3}{5}\right) = \frac{3}{5},$$

so x_3 leaves the basis. Using $k = 4$ and $r = 3$ in (2-30) yields

$$\mathbf{\Phi} = (+.2, -.6, -.8, +.6),$$

and we now use the transformation formulas (2-29) and (2-31) to fill in the new tableau:

	x_1	x_2	x_3	x_4	x_5	x_6
$x_{B1} = x_2 = .6$	1.6	1	.2	0	0	4.6
$x_{B2} = x_5 = .2$	−3.6	0	−.6	0	1	2.2
$x_{B3} = x_4 = .6$	.4	0	.2	1	0	− .2
Max $z = 7.8$	3.2	0	.6	0	0	−1.6

The objective value has improved from 6 to 7.8, but optimality has still not been attained.

As mentioned earlier, the simplex pivoting procedure can be applied to a linear program in standard form only after all redundant constraints have been eliminated and an initial basic feasible solution found. Problems that do not immediately satisfy these conditions—as most do not—can nevertheless be solved via the so-called "two-phase" algorithm, in which the simplex method is applied twice, rather than once. Given any standard-form linear program (2-22), let every constraint whose right-hand side b_i is negative be multiplied through by -1, to produce the following:

$$\text{(2-32)} \qquad \begin{cases} \text{Max } z = \mathbf{c}^T\mathbf{x} \\ \text{subject to } \mathbf{Ax} = \mathbf{b} \\ \text{and } \quad \mathbf{x} \geqq \mathbf{0}, \\ \text{with } \quad \mathbf{b} \geqq \mathbf{0}. \end{cases}$$

We next form the auxiliary problem

$$\text{(2-33)} \qquad \begin{cases} \text{Max } z' = -\mathbf{1} \cdot \mathbf{w} = -\sum_{i=1}^{m} w_i \\ \text{subject to } \mathbf{Ax} + \mathbf{I}_m\mathbf{w} = \mathbf{b} \\ \text{and } \qquad \mathbf{x}, \mathbf{w} \geqq \mathbf{0}, \end{cases}$$

where $\mathbf{w}$ is an m-component column vector of *artificial variables* and $\mathbf{1}$ is an m-component row vector of 1s. Observe that the constraints of (2-33) cannot be redundant, since a different artificial variable was added to each; moreover, an initial basic feasible solution can be identified immediately:

$$\text{(2-34)} \qquad \mathbf{x}_B = \mathbf{w} = \mathbf{b} \geqq \mathbf{0}, \qquad \text{with } \mathbf{x} = \mathbf{0}.$$

We can therefore solve problem (2-33) via the simplex pivoting method; this, in fact, constitutes phase 1 of the two-phase algorithm.

The solving of (2-33) must lead to one of three mutually exclusive outcomes:

1. *The optimal value of z' is negative*, in which case the original problem (2-32) has no feasible solution.

2. *The optimal value of z' is zero and the optimal basic solution $\mathbf{x}_{B^*}$ includes no artificial variables.* This implies that $\mathbf{x}_{B^*}$ is a legitimate BFS (possibly degenerate) for the original problem (2-32) and that none of the original constraints were redundant. We are now ready to enter phase II, in which (2-32) is solved via the simplex method, using $\mathbf{x}_{B^*}$ as an initial BFS. Certain modifications of the optimal phase I tableau are required in order to prepare it for phase II. Inasmuch as the artificial variables w_i do not appear in problem (2-32), the tableau columns associated with them should simply be deleted. Then, since the objective function of phase II is different from that of phase I, it is necessary to recompute the values of z and of the various $z_j - c_j$; the columns $\mathbf{y}_j$ and $\mathbf{x}_B$, however, do not have to be changed at all. Simplex pivoting can now proceed in the usual manner until an optimal solution to (2-32) is obtained.

3. *The optimal value of z' is zero and $\mathbf{x}_{B^*}$ includes one or more artificial variables,* whose values must all be zero. In this case we have found a feasible solution to the original problem (2-32), although not yet a basic feasible solution. Without shifting to phase II, we now attempt to remove all the zero-

valued artificial variables from the basis by pivoting in real variables x_j to replace them (at this stage the $z_j - c_j$ are being ignored). Note that the values of the basic variables, and therefore the solution itself, are not changed by these pivots, since $x_{B'r}$ is always zero in the transformation formula (2-31). If we eventually succeed in removing all the artificial variables from the basis, then the problem belongs in the domain of case 2, as discussed above. If not, the constraints of problem (2-32) are redundant (it can be shown, in fact, that if the artificial variable w_s cannot be expelled from the basis, except by pivoting in another artificial variable to replace it, then the sth constraint of $\mathbf{Ax} = \mathbf{b}$ is redundant). Nevertheless, (2-32) can still be solved: It is only necessary to carry the artificial basic variables on into phase II, during which they will never be removed from the basis nor take on positive values. Note that the values of z and of the $z_j - c_j$ must be recomputed, as in case 2, before phase II pivoting can begin.

2.5 THE DUAL PROBLEM

For the rest of the chapter we shall be concerned with the duality theory of linear programming. The notion of duality, in generalized form, plays a central and unifying role throughout all of optimization theory, not only in mathematical programming, but in the classical method of Lagrange multipliers as well. It is chiefly through duality theory that the linear program can be viewed in its proper perspective, i.e., as one of the many different subspecies of the general class of problems known as mathematical programs. The fact that the linear program can be solved via the highly efficient simplex algorithm makes it uniquely important from the practical and computational points of view, but it has no particular theoretical importance apart from that which it shares with all other mathematical programs, namely, its *coexistence with a dual problem* whose optimal solution is closely related to its own in a number of interesting ways.

We are aware that the reader will already have studied the duality theory of linear programming, perhaps quite thoroughly. Nevertheless, because this material receives only partial coverage in many linear programming courses, and because it will be crucial in establishing and illuminating the Kuhn-Tucker conditions, we shall develop the results we need formally and in some detail. When clarification or fuller explanation is desired for any part of the presentation below, the reader should consult Chapter 3 of Simmons [42].

We begin by defining a new algebraic format for the representation of linear programming problems. A linear program is said to be in *canonical form* if it is written as follows:

$$\text{(2-35)} \qquad \begin{cases} \text{Max } z = \mathbf{c}^T\mathbf{x} \\ \text{subject to } \mathbf{Ax} \leqq \mathbf{b} \\ \text{and } \quad \mathbf{x} \geqq \mathbf{0}, \end{cases}$$

where, as before, $\mathbf{A}$ is an m-by-n matrix. Having shown in the previous section that any LPP can be represented in standard form, we can prove that any LPP can also be written in canonical form by demonstrating the transformation from standard to canonical. But this merely requires observing that the standard-form constraint

$$a_{i1}x_1 + a_{i2}x_2 + \cdots + a_{in}x_n = b_i$$

is equivalent to the pair of inequalities

$$a_{i1}x_1 + a_{i2}x_2 + \cdots + a_{in}x_n \leqq b_i$$

and

$$a_{i1}x_1 + a_{i2}x_2 + \cdots + a_{in}x_n \geqq b_i.$$

When the second is multiplied through by -1, canonical form is achieved. More generally, the standard-form linear program

$$\text{Max } z = \mathbf{c}^T\mathbf{x} \text{ subject to } \mathbf{Ax} = \mathbf{b} \text{ and } \mathbf{x} \geqq \mathbf{0}$$

can be rewritten in canonical form as follows:

$$\begin{aligned} \text{Max } z &= \mathbf{c}^T\mathbf{x} \\ \text{subject to } \mathbf{Ax} &\leqq \mathbf{b}, \\ -\mathbf{Ax} &\leqq -\mathbf{b}, \\ \text{and } \mathbf{x} &\geqq \mathbf{0}. \end{aligned}$$

Thus we have shown that every LPP has an equivalent canonical representation: Any theorem proved about canonical problems can therefore be extended to linear programs in general.

We now introduce a very important definition: *Given a linear programming problem P in canonical form,*

$$\text{(2-35)} \qquad \begin{cases} \text{Max } z = \mathbf{c}^T\mathbf{x} \\ \text{subject to } \mathbf{Ax} \leqq \mathbf{b} \\ \text{and } \quad \mathbf{x} \geqq \mathbf{0}, \end{cases}$$

where $\mathbf{A}$ is m-by-n, *the following linear program is called its dual or its dual problem D*:

$$\text{(2-36)} \qquad \begin{cases} \text{Min } z' = \mathbf{b}^T\mathbf{u} \\ \text{subject to } \mathbf{A}^T\mathbf{u} \geqq \mathbf{c} \\ \text{and } \quad \mathbf{u} \geqq \mathbf{0}, \end{cases}$$

where $\mathbf{u}$ is an m-component column vector of unknowns and $\mathbf{A}$, $\mathbf{b}$, and $\mathbf{c}$ are taken from the given problem P. When speaking of P in this context of having a dual, we shall refer to it as the *primal problem* or, more simply, the primal. The components of $\mathbf{u}$ are the *dual variables*, while the inequalities $\mathbf{A}^T\mathbf{u} \geqq \mathbf{c}$ are the *dual constraints*; their primal counterparts are similarly named.

Observe that there are exactly as many dual constraints as there are primal variables and as many dual variables as primal constraints. Since each primal variable x_j is associated with the column $\mathbf{a}_j$ of $\mathbf{A}$, it is also associated with the jth row of $\mathbf{A}^T$ and therefore with the jth constraint of the dual problem. In an analogous fashion the ith primal constraint is associated with or corresponds to the ith dual variable and vice versa.

Inasmuch as any linear programming problem can be converted into canonical form, it must be true that *every LPP has a dual problem*, namely, the dual of its canonical equivalent. The route by which the dual of a non-canonical linear program can be found is illustrated in the proof of the following result:

Theorem 2.3: The dual of the dual is the primal.

The dual problem (2-36) can be represented in canonical form as follows:

$$\begin{aligned} \text{Max } z'' &= (-\mathbf{b})^T\mathbf{u} \\ \text{subject to } -\mathbf{A}^T\mathbf{u} &\leqq -\mathbf{c} \\ \text{and} \qquad \mathbf{u} &\geqq \mathbf{0}. \end{aligned}$$

Applying the definition directly, we find that the dual of the original dual is

$$\text{(2-37)} \qquad \begin{cases} \text{Min } z''' = (-\mathbf{c})^T\mathbf{x} \\ \text{subject to } -\mathbf{A}\mathbf{x} \geqq -\mathbf{b} \\ \text{and} \qquad \mathbf{x} \geqq \mathbf{0}, \end{cases}$$

where we have used the fact that the transpose of $\mathbf{A}^T$ is $\mathbf{A}$. Finally, we need only observe that (2-37) is equivalent to (2-35). Q.E.D.

In a similar manner we can shown that the standard-form linear program

$$\text{(2-22)} \qquad \text{Max } z = \mathbf{c}^T\mathbf{x} \text{ subject to } \mathbf{A}\mathbf{x} = \mathbf{b} \text{ and } \mathbf{x} \geqq \mathbf{0}$$

has the following dual problem:

$$\text{(2-38)} \qquad \text{Min } z' = \mathbf{b}^T\mathbf{u} \text{ subject to } \mathbf{A}^T\mathbf{u} \geqq \mathbf{c},$$

where the variables $\mathbf{u}$ may take on any real values. Dual linear programs can also be written for primal problems having unrestricted variables or having both inequality and equality constraints; the rules for forming them can be stated in a few sentences, as follows (these rules can all be established via proofs of the type given for Theorem 2.3). The dual of a *maximization* linear program may be written directly provided that none of its constraints is of the form[5]

$$a_{i1}x_1 + a_{i2}x_2 + \cdots + a_{in}x_n \geqq b_i. \tag{2-39}$$

The dual will have a nonnegative variable associated with each primal inequality ($\leqq$) constraint and an unrestricted variable for each primal equality constraint; it will have an inequality ($\geqq$) constraint associated with each nonnegative primal variable and an equality constraint for each unrestricted primal variable. Similarly, the dual of a *minimization* problem may be written directly provided that none of its constraints is of the form

$$a_{i1}x_1 + a_{i2}x_2 + \cdots + a_{in}x_n \leqq b_i. \tag{2-40}$$

Again, unrestricted variables are associated with equality constraints and nonnegative variables with inequalities. In either of the two cases, if a "forbidden" constraint appears in the primal, it may simply be multiplied on both sides by -1, allowing the dual to be written directly.

The rules for forming the dual problem can be summarized in a mnemonic table as follows:

Primal	*Dual*
Maximize	Minimize
Constraint $\leqq$	Variable $u_i \geqq 0$
Constraint $=$	Variable unrestricted
Variable $x_j \geqq 0$	Constraint $\geqq$
Variable unrestricted	Constraint $=$

The headings Primal and Dual are, of course, arbitrary and could have been reversed. The table is quite easy to use; for example, if the primal problem is

[5]This is not quite an accurate statement. It is perfectly possible to write the dual directly from a primal maximization problem that includes constraints of the type (2-39); such constraints give rise to *nonpositive* dual variables. However, nonpositive variables seem physically unrealistic and are, in any case, of no particular theoretical interest. They can be avoided if all constraints of the form (2-39) are multiplied through by -1 before the dual is formed.

$$\begin{aligned} \text{Max } z &= x_1 + 3x_2 \\ \text{subject to } x_1 + 2x_2 + 3x_3 &\leqq 9, \\ x_1 \qquad - x_3 &= -5, \\ \text{and} \qquad x_2, x_3 &\geqq 0, \end{aligned}$$

the dual can be written from it immediately, as follows:

$$\begin{aligned} \text{Min } z' &= 9u_1 - 5u_2 \\ \text{subject to} \quad u_1 + u_2 &= 1, \\ 2u_1 \qquad &\geqq 3, \\ 3u_1 - u_2 &\geqq 0, \\ \text{and} \qquad u_1 &\geqq 0. \end{aligned}$$

2.6 THE DUALITY THEOREM

In this section and the next we shall establish several important properties of primal and dual linear programs. Proofs will be stated for the problems P and D that appeared in our original definition of the dual [see (2-35) and (2-36)]. The results, however, will apply to *all* pairs of primal and dual problems, since they can be shown by the methods of the previous section to be equivalent to P and D.

We first present two lemmas, which are attributed to Gale, Kuhn, and Tucker (see Chapter 19 of [27]).

Duality Lemma I: *If* $\mathbf{x}_0$ *is a feasible solution to the primal* P *and* $\mathbf{u}_0$ *is a feasible solution to the dual* D, *then* $\mathbf{c}^T\mathbf{x}_0 \leqq \mathbf{b}^T\mathbf{u}_0$.

The proof is quite simple. Since $\mathbf{x}_0$ is a feasible solution to P, we have $\mathbf{A}\mathbf{x}_0 \leqq \mathbf{b}$; the nonnegativity of $\mathbf{u}_0$ then allows us to write

$$\mathbf{u}_0^T\mathbf{A}\mathbf{x}_0 \leqq \mathbf{u}_0^T\mathbf{b} = \mathbf{b}^T\mathbf{u}_0, \tag{2-41}$$

where we have transposed the scalar $\mathbf{u}_0^T\mathbf{b}$. Similarly, we are given $\mathbf{A}^T\mathbf{u}_0 \geqq \mathbf{c}$, which transposes to $\mathbf{u}_0^T\mathbf{A} \geqq \mathbf{c}^T$; this implies

$$\mathbf{u}_0^T\mathbf{A}\mathbf{x}_0 \geqq \mathbf{c}^T\mathbf{x}_0. \tag{2-42}$$

Taken together, (2-41) and (2-42) establish the lemma.

Duality Lemma II: *If* $\mathbf{x}_0$ *and* $\mathbf{u}_0$ *are feasible solutions to* P *and* D *and* $\mathbf{c}^T\mathbf{x}_0 = \mathbf{b}^T\mathbf{u}_0$, *then both* $\mathbf{x}_0$ *and* $\mathbf{u}_0$ *are optimal solutions.*

If $\hat{\mathbf{x}}$ is any other feasible solution to P, we know from the previous lemma that

$$\mathbf{c}^T\hat{\mathbf{x}} \leqq \mathbf{b}^T\mathbf{u}_0 = \mathbf{c}^T\mathbf{x}_0.$$

Since P is a maximization problem, $\mathbf{x}_0$ must be an optimal solution. A parallel argument shows that $\mathbf{u}_0$ is an optimal solution to D.

The foundation has now been laid for the proof of the duality theorem. This is the single most important theoretical result in the field of linear programming and probably in all of mathematical programming as well. It may be stated as follows:

Theorem 2.4 (the duality theorem): A feasible solution $\mathbf{x}_0$ *to the primal problem* P *is optimal if and only if there exists a feasible solution* $\mathbf{u}_0$ *to the dual problem* D *such that* $\mathbf{c}^T\mathbf{x}_0 = \mathbf{b}^T\mathbf{u}_0$.

This theorem was originally stated by John von Neumann, and a proof first appeared in 1956 which made use of an old (1902) result from linear algebra known as *Farkas' lemma* or the Minkowski-Farkas theorem. By taking a corollary of that theorem and applying more linear algebra, it was possible to construct a rather unedifying proof of the duality theorem. In 1962 an entirely different derivation by Dreyfus and Freimer appeared as an appendix to a textbook in dynamic programming [10]. This is the proof we now present.

Assume $\mathbf{u}_0$ is a feasible solution to D and $\mathbf{c}^T\mathbf{x}_0 = \mathbf{b}^T\mathbf{u}_0$. By hypothesis $\mathbf{x}_0$ is a feasible solution to P; therefore, from Duality Lemma II, $\mathbf{x}_0$ is optimal. This, of course, is trivial; it is the other direction, the "only if," that is of interest to us.

Recalling that our primal problem P is

$$\text{(2-35)} \qquad \begin{cases} \text{Max } z = \mathbf{c}^T\mathbf{x} \\ \text{subject to } \mathbf{A}\mathbf{x} \leqq \mathbf{b} \\ \text{and} \quad \mathbf{x} \geqq \mathbf{0}, \end{cases}$$

let us imagine that $\mathbf{A}$ and $\mathbf{c}$ are held fixed and define the scalar function $f(\boldsymbol{\beta})$ to be the maximum possible value of the objective when the right-hand side vector of the constraint equation takes on the value $\boldsymbol{\beta}$. Since $\mathbf{x}_0$ is by hypothesis the optimal solution to P, we have $f(\mathbf{b}) = \mathbf{c}^T\mathbf{x}_0$.

We now perturb this optimal solution by increasing its jth component x_{0j} by a small *positive* amount θ; this yields the new solution

$$\text{(2-43)} \qquad \hat{\mathbf{x}} = \mathbf{x}_0 + \theta\mathbf{e}_j,$$

where $\mathbf{e}_j$ is a unit vector with a 1 in the jth position. Since $\mathbf{x}_0$ was a feasible solution to P, it is easily verified that $\hat{\mathbf{x}}$ is a feasible solution to the linear program

$$\text{(2-44)} \qquad \begin{cases} \text{Max } z = \mathbf{c}^T\mathbf{x} \\ \text{subject to } \mathbf{Ax} \leqq \mathbf{b} + \theta\mathbf{a}_j \\ \text{and } \quad \mathbf{x} \geqq \mathbf{0}, \end{cases}$$

where $\mathbf{a}_j$ is the jth column of $\mathbf{A}$. Note that $\hat{\mathbf{x}}$ has an objective value of

$$\mathbf{c}^T(\mathbf{x}_0 + \theta\mathbf{e}_j) = f(\mathbf{b}) + \theta c_j.$$

It is not necessarily true, however, that $\hat{\mathbf{x}}$ is the optimal solution to the new problem (2-44); all we can say is that

$$\text{(2-45)} \qquad f(\mathbf{b} + \theta\mathbf{a}_j) \geqq f(\mathbf{b}) + \theta c_j.$$

At this point let us perform a Taylor expansion on the left-hand side of (2-45); using (2-11), we may write

$$f(\mathbf{b}) + \sum_{i=1}^{m} \frac{\delta f(\mathbf{b})}{\delta \beta_i}\theta a_{ij} + (\text{terms in } \theta^2 \text{ and higher}) \geqq f(\mathbf{b}) + \theta c_j.$$

If we adopt the more convenient notation

$$\text{(2-46)} \qquad \frac{\delta f}{\delta b_i} \equiv \frac{\delta f(\mathbf{b})}{\delta \beta_i},$$

the expansion becomes

$$\text{(2-47)} \qquad f(\mathbf{b}) + \theta \sum_{i=1}^{m} \frac{\delta f}{\delta b_i} a_{ij} + (\text{terms in } \theta^2 \text{ and higher}) \geqq f(\mathbf{b}) + \theta c_j.$$

For sufficiently small perturbations θ, division of (2-47) by θ yields

$$\text{(2-48)} \qquad \sum_{i=1}^{m} \frac{\delta f}{\delta b_i} a_{ij} \geqq c_j \qquad \textit{whenever } \theta \textit{ is positive.}$$

Now suppose that the value x_{0j} of the jth variable in the optimal solution to P is greater than zero. We can then permit θ to be a small *negative* number, and (2-43) will still be a feasible solution to (2-44). Repeating the above derivation, we arrive again at (2-47). But this time division by θ reverses the inequality, and we find that

$$\text{(2-49)} \qquad \sum_{i=1}^{m} \frac{\delta f}{\delta b_i} a_{ij} \leqq c_j \qquad \textit{whenever } \theta \textit{ is negative.}$$

Since perturbation in both directions is possible when x_{0j} is greater than zero, we conclude from (2-48) and (2-49) that

$$\sum_{i=1}^{m} \frac{\delta f}{\delta b_i} a_{ij} = c_j \qquad \text{if } x_{0j} > 0, \tag{2-50}$$

whereas

$$\sum_{i=1}^{m} \frac{\delta f}{\delta b_i} a_{ij} \geqq c_j \qquad \text{if } x_{0j} = 0. \tag{2-51}$$

At this point we observe that the maximum value of the objective function of problem P cannot be decreased by an algebraic increase in $\mathbf{b}$ or by an isolated increase in any one of its components. Such increases exclude no solutions from the feasible set and may admit new ones. This implies that

$$\frac{\delta f}{\delta b_i} \geqq 0, \qquad i = 1, \ldots, m, \tag{2-52}$$

and it follows immediately from (2-50) and (2-51) that the vector $\mathbf{u}_0$ whose ith component is

$$u_{0i} \equiv \frac{\delta f}{\delta b_i} \tag{2-53}$$

is a feasible solution to the dual problem D.

We must still show that the values of the primal and dual objectives are equal. The dual constraints $\mathbf{A}^T\mathbf{u} \geqq \mathbf{c}$ must hold for the feasible solution $\mathbf{u}_0$; transposing and multiplying by the nonnegative vector $\mathbf{x}_0$ gives

$$\mathbf{u}_0^T\mathbf{A}\mathbf{x}_0 \geqq \mathbf{c}^T\mathbf{x}_0.$$

Component-wise, this is the same as

$$\sum_{j=1}^{n} (\mathbf{u}_0^T\mathbf{A})_j x_{0j} \geqq \sum_{j=1}^{n} c_j x_{0j}, \tag{2-54}$$

where $(\mathbf{u}_0^T\mathbf{A})_j$ is the jth component of the row vector $\mathbf{u}_0^T\mathbf{A}$:

$$(\mathbf{u}_0^T\mathbf{A})_j \equiv \mathbf{u}_0^T\mathbf{a}_j \equiv \sum_{i=1}^{m} u_{0i}a_{ij}. \tag{2-55}$$

But, from (2-50) and (2-51), either $x_{0j} = 0$ or $(\mathbf{u}_0^T\mathbf{A})_j = c_j$ for each j; therefore equality obtains term by term in (2-54), and

$$\mathbf{u}_0^T\mathbf{A}\mathbf{x}_0 = \mathbf{c}^T\mathbf{x}_0. \tag{2-56}$$

One final step remains. Since we have assumed that $\mathbf{x}_0$ is a feasible and optimal solution to P, the primal constraints $\mathbf{Ax} \leqq \mathbf{b}$ and the nonnegativity of $\mathbf{u}_0$ allow us to write

$$\mathbf{u}_0^T\mathbf{Ax}_0 \leqq \mathbf{u}_0^T\mathbf{b}.$$

Representing this as a sum of terms,

$$\sum_{i=1}^{m} u_{0i}(\mathbf{Ax}_0)_i \leqq \sum_{i=1}^{m} u_{0i}b_i. \tag{2-57}$$

Now suppose that $\mathbf{x}_0$ satisfies the ith primal constraint as an *inequality*:

$$a_{i1}x_{01} + a_{i2}x_{02} + \cdots + a_{in}x_{0n} < b_i. \tag{2-58}$$

In this case the ith primal constraint is *not binding*: It is not preventing the objective function from attaining a greater value. Whenever this is true—that is, whenever $(\mathbf{Ax}_0)_i < b_i$—we argue from economics and common sense that a slight increase in b_i will not increase the objective value obtainable. Moreover, a small decrease in b_i will obviously not decrease the optimal objective value, since the constraint (2-58) will remain satisfied. This means that for small perturbations

$$\frac{\delta f}{\delta b_i} \equiv u_{0i} = 0 \qquad \text{when } (\mathbf{Ax}_0)_i < b_i. \tag{2-59}$$

We have shown that for the ith primal constraint, $i = 1, \ldots, m$, either $(\mathbf{Ax}_0)_i = b_i$ or (2-59) holds; it therefore follows that (2-57) must be an equality, and

$$\mathbf{u}_0^T\mathbf{Ax}_0 = \mathbf{u}_0^T\mathbf{b}. \tag{2-60}$$

Finally, from (2-56) and (2-60) we may write

$$\mathbf{u}_0^T\mathbf{b} = \mathbf{c}^T\mathbf{x}_0,$$

as was to be shown.

Having begun with the hypothesis that $\mathbf{x}_0$ is a feasible and optimal solution to the primal problem P, we have constructed $\mathbf{u}_0$, a feasible solution to D, and have shown that their objective values in their respective problems are equal. This proves the duality theorem. Incidentally, we also know from Duality Lemma II that $\mathbf{u}_0$ is an optimal solution to D.

2.7 THE EXISTENCE THEOREM AND COMPLEMENTARY SLACKNESS

Strictly speaking, the duality theorem just proved is an existence theorem: It guarantees the existence of a feasible solution to the dual problem whenever the primal has a finite optimum. However, the result to be derived below, which covers all types of linear programs—including infeasible and unbounded problems—is more or less generally known as *the* existence theorem of linear programming. It is, in its own right, an extremely important and far-reaching result; however, because it derives so easily from the duality theorem, it is more accurately a corollary rather than a theorem of the first magnitude. Nevertheless,

Theorem 2.5 (the existence theorem):

> *2.5A: A linear program has a finite optimal solution if and only if both it and its dual have feasible solutions.*
>
> *2.5B: If the primal problem has an unbounded maximum, then the dual problem has no feasible solution.*
>
> *2.5C: If the dual problem has no feasible solution but the primal problem has, then the primal problem has an unbounded maximum.*

Proof of 2.5A

If the primal has an optimal feasible solution, then the duality theorem guarantees that the dual has a feasible solution.

Going the other way, assume that the primal and dual have feasible solutions $\hat{\mathbf{x}}$ and $\hat{\mathbf{u}}$; therefore $\mathbf{c}^T\hat{\mathbf{x}}$ and $\mathbf{b}^T\hat{\mathbf{u}}$ are finite. From Duality Lemma I, $\mathbf{b}^T\hat{\mathbf{u}}$ serves as an upper bound on the primal objective $\mathbf{c}^T\mathbf{x}$, although not necessarily a least upper bound. At any rate, the primal must have a finite optimum. Q.E.D.

Proof of 2.5B

Let the primal be unbounded, in which case it has a feasible solution. If the dual also had a feasible solution, then according to Theorem 2.5A the primal would have to have a finite optimum, which is a contradiction.

Proof of 2.5C

Suppose $\hat{\mathbf{x}}$ is a feasible solution to the primal. It cannot also be optimal because, if it were, the duality theorem would dictate that a feasible solution to the dual problem must exist, contrary to our hypothesis. It follows that no feasible solution to the primal can be optimal, so that the primal is unbounded.

These results are summarized in the following table:

	Primal problem is feasible	Primal problem is not feasible
Dual problem is feasible	Both optima exist	Dual problem unbounded
Dual problem is not feasible	Primal problem unbounded	May occur

As an example of how primal unboundedness is associated with dual infeasibility, consider the following pair of linear programs:

$$P: \quad \text{Max } z = x_1 + x_2 \quad \text{subject to} \quad x_1 - x_2 = -1, \quad -2x_1 + x_2 \leqq -2, \quad \text{and} \quad x_1, x_2 \geqq 0$$

$$D: \quad \text{Min } z' = -u_1 - 2u_2 \quad \text{subject to} \quad u_1 - 2u_2 \geqq 1, \quad -u_1 + u_2 \geqq 1, \quad \text{and} \quad u_2 \geqq 0$$

It is not difficult to show graphically that P has an unbounded maximum while D is infeasible.

The remainder of this section will be devoted to an important set of algebraic relationships, known collectively as *complementary slackness*, which prevail among the individual components of any pair of optimal primal and dual solutions. We begin by recalling from Section 2.4 that an inequality constraint can be transformed into an equality by the introduction of a nonnegative slack or surplus variable. Let us return to the familiar canonical primal problem P and apply that transformation to each of its constraints:

$$(2\text{-}61) \qquad \begin{cases} \text{Max } z = \mathbf{c}^T\mathbf{x} \\ \text{subject to } \mathbf{A}\mathbf{x} + \mathbf{x}_s = \mathbf{b} \\ \text{and} \quad \mathbf{x}, \mathbf{x}_s \geqq \mathbf{0}, \end{cases}$$

where $\mathbf{x}_s$ is an m-component column vector of slack variables. The dual problem D may also be converted to standard form, by means of surplus variables:

$$(2\text{-}62) \qquad \begin{cases} \text{Min } z' = \mathbf{b}^T\mathbf{u}, \\ \text{subject to } \mathbf{A}^T\mathbf{u} - \mathbf{u}_s = \mathbf{c} \\ \text{and} \quad \mathbf{u}, \mathbf{u}_s \geqq \mathbf{0}, \end{cases}$$

with the n nonnegative surplus variables being collected into $\mathbf{u}_s$.

At this point we observe that the constraints of (2-61) imply

$$\mathbf{u}^T\mathbf{A}\mathbf{x} + \mathbf{u}^T\mathbf{x}_s = \mathbf{u}^T\mathbf{b}$$

for any feasible primal solution $(\mathbf{x}, \mathbf{x}_s)$ and any m-by-1 vector of values $\mathbf{u}$; this is equivalent to

$$\mathbf{x}^T\mathbf{A}^T\mathbf{u} + \mathbf{u}^T\mathbf{x}_s = \mathbf{b}^T\mathbf{u}, \tag{2-63}$$

where two scalar numbers have been transposed. Similarly, using the constraints of (2-62), we can write

$$\mathbf{x}^T\mathbf{A}^T\mathbf{u} - \mathbf{x}^T\mathbf{u}_s = \mathbf{c}^T\mathbf{x} \tag{2-64}$$

for any feasible dual solution $(\mathbf{u}, \mathbf{u}_s)$ and any n-component vector $\mathbf{x}$.

Suppose now that optimal solutions $(\mathbf{x}_0, \mathbf{x}_{0s})$ and $(\mathbf{u}_0, \mathbf{u}_{0s})$ exist for the primal and dual problems, where $\mathbf{x}_{0s}$ and $\mathbf{u}_{0s}$ are the optimal values of the primal slack vector $\mathbf{x}_s$ and dual surplus vector $\mathbf{u}_s$. Theorem 2.4 and Duality Lemma I together ensure that their objective values are equal:

$$\mathbf{c}^T\mathbf{x}_0 = \mathbf{b}^T\mathbf{u}_0 \tag{2-65}$$

(cost coefficients of all slack and surplus variables are zero). Since optimal solutions are by definition feasible, (2-63) and (2-64) must hold for them:

$$\mathbf{x}_0^T\mathbf{A}^T\mathbf{u}_0 + \mathbf{u}_0^T\mathbf{x}_{0s} = \mathbf{b}^T\mathbf{u}_0 \tag{2-66}$$

and

$$\mathbf{x}_0^T\mathbf{A}^T\mathbf{u}_0 - \mathbf{x}_0^T\mathbf{u}_{0s} = \mathbf{c}^T\mathbf{x}_0. \tag{2-67}$$

Then because of (2-65) we can equate (2-66) and (2-67), arriving at

$$\mathbf{u}_0^T\mathbf{x}_{0s} + \mathbf{x}_0^T\mathbf{u}_{0s} = 0. \tag{2-68}$$

But since all components of all four vectors are nonnegative, we can permit none of the $m + n$ terms in the sum (2-68) to be nonzero. The implications are stated in the theorem of *complementary slackness*:

Theorem 2.6: Given any pair of optimal solutions to a linear programming problem and its dual, the following hold:

> *2.6A: For each* i, i = 1, . . ., m, *the product of the* ith *primal slack variable and* ith *dual variable is zero; that is, if either is positive, the other must be zero.*
>
> *2.6B: For each* j, j = 1, . . . , n, *the product of the* jth *primal variable and the* jth *dual surplus variable is zero.*

(In this theorem and below, every allusion to a variable's being positive or zero is understood to mean that the variable's *optimal value* is positive or zero.) If 2.6A and 2.6B both hold, we say that complementary slackness prevails. It is important to observe that if, say, a primal slack variable is zero, it does *not* follow that the corresponding dual variable must be positive. Both may be zero, since the requirement is simply that their product be zero.

One final result remains to be derived:

Theorem 2.7: If $(\mathbf{x}_0, \mathbf{x}_{0s})$ *and* $(\mathbf{u}_0, \mathbf{u}_{0s})$ *are feasible solutions to the primal and dual problems and complementary slackness prevails between them, then each is an optimal solution to its respective problem.*

The proof reverses the derivation of the complementary slackness conditions 2.6A and 2.6B. We are given

$$\mathbf{u}_0^T\mathbf{x}_{0s} + \mathbf{x}_0^T\mathbf{u}_{0s} = 0,$$

which yields

$$\mathbf{u}_0^T\mathbf{x}_{0s} = -\mathbf{u}_{0s}^T\mathbf{x}_0. \tag{2-69}$$

Adding $\mathbf{u}_0^T\mathbf{A}\mathbf{x}_0$ to both sides of (2-69) produces

$$\mathbf{u}_0^T(\mathbf{A}\mathbf{x}_0 + \mathbf{x}_{0s}) = (\mathbf{u}_0^T\mathbf{A} - \mathbf{u}_{0s}^T)\mathbf{x}_0. \tag{2-70}$$

Now, since $(\mathbf{x}_0, \mathbf{x}_{0s})$ and $(\mathbf{u}_0, \mathbf{u}_{0s})$ are feasible solutions to (2-61) and (2-62), respectively, we may substitute the right-hand sides of the primal and dual constraints into (2-70); thus

$$\mathbf{u}_0^T\mathbf{b} = \mathbf{c}^T\mathbf{x}_0.$$

By Duality Lemma II both solutions are optimal.

Although this result has no bearing on the solution of linear programming problems via the simplex method, it does form the theoretical foundation of certain other solution algorithms, including the so-called "primal-dual" and "out-of-kilter" algorithms. These methods proceed by setting up a linear program and its dual and then operating on both of them simultaneously, perturbing first one variable, then another, and aiming for a pair of feasible solutions that satisfy complementary slackness. Theorem 2.7 then guarantees that such a pair of solutions optimizes both the primal and the dual problems.

2.8 INTERPRETATIONS OF DUALITY

In the past three sections we have taken the general linear programming problem in canonical form, used parameters from it to construct a second or "dual" LPP, and established several interesting and useful theoretical results involving these two problems and their optimal solutions. However, before we move ahead into nonlinear programming, where these results will be generalized and exploited in various computational algorithms, we shall devote several pages to a "common-sense," economic explanation of various aspects of duality theory. Our purpose in doing so is to enrich the student's understanding of these abstract mathematical concepts and to show how they actually represent real-world situations and relationships that are encountered over and over again in the practice of operations research.

We begin by outlining an ordinary resource allocation problem of the type that is frequently faced by plant managers and corporate planners. A certain factory can produce n different products or *outputs*, using a total of m different raw materials as *inputs*. Each unit of the jth product, $j = 1, \ldots, n$, can be sold for $\$c_j$ and requires for its manufacture the amounts a_{ij} of the various inputs, $i = 1, \ldots, m$. The total amount of the ith input on hand is b_i. Using only the available raw materials, how much of each output should the plant manager plan to produce in order to maximize his overall gross income from sales?

The formulation is not difficult. Letting x_j be the number of units of the jth output to be produced, $j = 1, \ldots, n$, the objective function is simply

$$\text{Max } z = c_1x_1 + c_2x_2 + \cdots + c_nx_n.$$

The limited availability of the raw materials imposes a resource constraint

$$a_{i1}x_1 + a_{i2}x_2 + \cdots + a_{in}x_n \leqq b_i$$

for each input, $i = 1, \ldots, m$. Finally, negative production levels are prohibited:

$$x_j \geqq 0, \qquad j = 1, \ldots, n.$$

Thus we have a linear program; it may be represented in matrix/vector form as follows:

$$\begin{aligned} \text{Max } z &= \mathbf{c}^T\mathbf{x} \\ \text{subject to } \mathbf{Ax} &\leqq \mathbf{b} \\ \text{and } \quad \mathbf{x} &\geqq 0, \end{aligned}$$

where $\mathbf{A}$ is m-by-n, $\mathbf{x}$ and $\mathbf{c}$ are n-by-1, and $\mathbf{b}$ is m-by-1.

Suppose we now consider a different problem faced by the plant manager, or perhaps by the vice-president in charge of finance. To protect the dollar income which will ultimately be realized when the finished products are sold, the raw materials on hand should be insured against fire, theft, and so on. Insurance policies cost money, and it will be desirable to place small per-unit valuations on the various raw materials; on the other hand, the valuations should be large enough to provide full compensation—that is, to replace the income that would otherwise accrue from sales—in the event of any accident or disaster. Subject to these considerations, what is the optimal insurance scheme?

Let u_i be the valuation placed on each unit of the ith input, $i = 1, \ldots, m$, so that the overall amount of the ith input currently on hand is given a total valuation of $b_i u_i$. No raw material can have a negative value, since at worst it could simply be discarded; hence

$$u_i \geqq 0, \qquad i = 1, \ldots, m.$$

If we take the cost of insurance to be a simple multiple of the total valuation of all items insured, then the objective of minimizing insurance cost is

$$\text{Min } z' = b_1 u_1 + b_2 u_2 + \cdots + b_m u_m.$$

But how can the vice-president decide how large his valuations need to be in order to guarantee full compensation for any loss? Not having solved the linear program we formulated above, he has no way of knowing how the various raw materials will be used. That is, if one unit of the sixth input were destroyed by fire, he doesn't know exactly how much less dollar income would then be realized—and that income differential is precisely the insurance valuation u_6 that he wishes to place on the sixth input.

However, the vice-president should reflect that the value of a unit of any raw material exists only insofar as it can be converted into a salable product. Whatever the individual values of the various inputs may be, it is clear that the combined values of all inputs required to produce exactly one unit of the first product (that is, a_{11} units of the first input, a_{21} units of the second, and so on) must be at least c_1:

$$a_{11} u_1 + a_{21} u_2 + \cdots + a_{m1} u_m \geqq c_1.$$

The inequality is appropriate because, while *at the very least* that set of inputs could be assembled into one unit of product 1 and sold for c_1, it is also possible that they could be used cleverly in other ways (perhaps requiring that production levels of certain other outputs be modified) in order to *further* increase total sales income. Inasmuch as raw materials can be made to yield income in only n different ways, i.e., via the production of n different products, the vice-president can ensure adequate valuations for insurance purposes simply by requiring that

$$a_{1j} u_1 + a_{2j} u_2 + \cdots + a_{mj} u_m \geqq c_j, \qquad j = 1, \ldots, n.$$

We have thus arrived at a second linear program. Placing its matrix formulation side by side with that of the first, we can see that the two problems are a primal/dual pair.

"Production Problem"	*"Insurance Problem"*
Max $z = \mathbf{c}^T\mathbf{x}$	Min $z' = \mathbf{b}^T\mathbf{u}$
subject to $\mathbf{A}\mathbf{x} \leqq \mathbf{b}$	subject to $\mathbf{A}^T\mathbf{u} \geqq \mathbf{c}$
and $\mathbf{x} \geqq 0$	and $\mathbf{u} \geqq 0$

Whereas the primal problem is one of determining how a fixed supply of resources should be allocated to the production of a number of possible outputs of known value, the dual problem is concerned with finding correct per-unit valuations or "prices" for those resources, with the price of each reflecting implicitly the value of the contribution it makes to the outputs. In accordance with this rather classical model, dual variables are generally known to economists as *shadow prices*. Note that the optimal primal and dual objective values, which represent, respectively, the total gross sales income from all products and the total valuation placed on all raw-material inputs, should, within the context of this example, be exactly the same. This coincides with the theoretical results obtained in the past three sections.

Before going on, we should mention that while the dual problem arose quite naturally in this resource-allocation context and made good economic sense, a great many linear programs are encountered in operations research and elsewhere whose duals do not lend themselves quite so readily to clear and logical interpretation. Usually the dual variables can be viewed in some way as implicit values or prices, as in the example just discussed, but sometimes it is not at all obvious what the dual objective represents or how the dual constraints can be understood within the framework of the primal problem. The student will be given several opportunities at the end of the chapter to exercise his imagination along these lines.

A more general mathematical interpretation of the dual variables of *any* linear programming problem is suggested by our proof of the duality theorem and by

$$u_{0i} \equiv \frac{\delta f}{\delta b_i} \tag{2-53}$$

in particular. *The optimal values of the dual variables are in fact the rates of change of the optimum attainable primal objective value with respect to changes in the right-hand-side elements of the primal constraints.* Suppose the value of the ith dual variable in the optimal solution to the dual problem were u_{0i}. This would tell us that if the right-hand side b_i of the ith primal constraint were increased by a sufficiently small amount ϵ, then the optimal value of the primal objective would be increased by ϵu_{0i} (except for certain values of $\mathbf{b}$ at which the partial derivative $\delta f/\delta b_i$ is discontinuous—to be discussed below). In terms of our resource-allocation example, if one extra small unit ($\epsilon = 1$) of the ith input were available, then an extra $\$u_{0i}$ of sales income could be realized by the factory. Therefore u_{0i} is the maximum price that the plant manager would just be willing to pay for an additional unit of the ith input; this maximum price is also known as the *marginal value* or, as noted above, the shadow price of that input. Equivalently, u_{0i} can also be thought of as the loss incurred when one unit of the available ith input is lost or destroyed; hence it gives the correct insurance valuation for the *last* unit of input i (after one unit is lost the primal problem would be different, and so might the optimal value of the ith dual variable).

Our remarks can be illustrated with a diagram of a two-variable problem. Suppose that the (primal) linear program of Fig. 2-3 has three constraints, labeled numerically, as well as the usual nonnegativity restrictions. In particular, suppose constraint 3 is

$$x_1 + x_2 \leqq b_3 \tag{2-71}$$

and let the objective be to maximize $z = x_2$. When b_3 is quite small, so that this constraint is in the position $3a$ as shown in Fig. 2-3, the optimal solution occurs at ($x_1 = 0, x_2 = b_3$), where $3a$ intersects the x_2-axis. Here the constraint is satisfied as an equality and is therefore said to be *binding*[6]: It binds

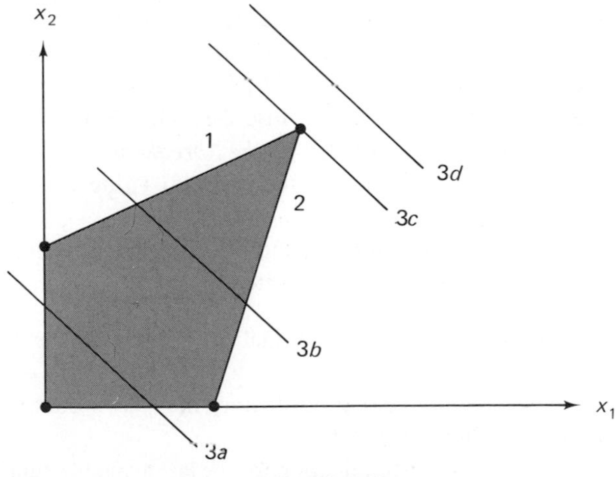

FIGURE 2-3

[6]By definition, a constraint is binding if and only if the optimal value of its slack or surplus variable is zero. An equality constraint is considered to be automatically binding.

down the objective function and prevents it from assuming a greater value, which it would be free to do in the absence of 3*a*. Similarly, we say that (when b_3 is small) constraints 1 and 2 are *not binding*.

Observing Fig. 2-3, it is clear that if (2-71) were *relaxed*—that is, if b_3 were increased—the feasible region would be enlarged, and the objective function could take on a greater value than before. For a while, as b_3 increases, the optimal point will simply climb up the x_2-axis; since the optimal objective value will continue to be $z = x_2 = b_3$, it will evidently increase one unit for each unit of increase in b_3. Therefore, over this range of b_3-values, the marginal value u_{03} will equal 1. For somewhat larger values of b_3, when constraint (2-71) has reached, say, position 3*b*, the optimum will lie at the intersection of constraints 1 and 3, both of which will then be binding. As b_3 increases further, the growth of the optimal objective value will continue in linear proportion, at a rate given by u_{03} (which will now be less than 1). However, when the constraint reaches position 3*c*, further relaxation can no longer raise the optimal value of z; constraints 1 and 2 have met and completely bounded the feasible region, so that additional increases in b_3 cannot enlarge it. Thus when constraint (2-71) is, say, in position 3*d*, the optimal value of the third dual variable u_{03} (that is, the marginal value of the third input) must be zero; in this position the constraint is *superfluous*, in the sense of not forming any part of the boundary of the feasible region.[7] Similarly, when (2-71) was in its original position 3*a*, the optimal values of the first and second dual variables were both zero.

If we imagine that this entire linear programming problem (i.e., maximize $z = x_2$ subject to nonnegativity and to constraints 1, 2, and 3) is held fixed with the single exception that b_3 is permitted to vary, we can plot $z_{\text{opt}} \equiv f(\mathbf{b})$, the optimal value of z, as a function of b_3. This is done in Fig. 2-4; the labeled points are the values of b_3 for which the constraint (2-71) assumes the various positions shown in Fig. 2-3. The slope of this piecewise linear function at any point b_3, *except at the "break points" where line segments meet*, is just dz_{opt}/db_3, which is the optimal value of the third dual variable.

With the aid of Fig. 2-4 the student should be able to see that each of the following statements is almost but not quite correct:

1. The optimal value u_{03} of the third dual variable equals the change in z_{opt} per "unit" change in b_3.

2. u_{03} is the maximum price that our factory manager would just be willing to pay for an extra "unit" of input 3.

3. u_{03} is the correct insurance valuation for the last available "unit" of input 3.

[7]It would be careless usage to call such a constraint *redundant*, because in Section 2.3 we defined as redundant any constraint that could be expressed as a linear combination of all the others. This is not true of constraint 3 in the example we are considering.

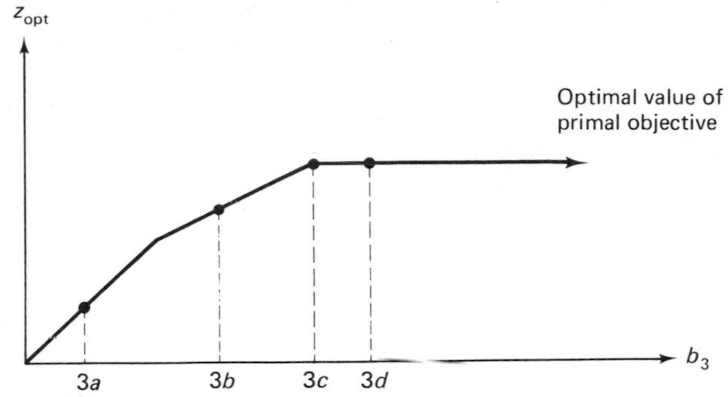

FIGURE 2-4

In all three cases the inaccuracy lies in the fact that, for any unit of finite size, a one-unit change in b_3 might result in a shift from one segment of the piecewise linear function to another; if this happened the change in z_{opt} would not be equal to u_{03}. The optimal value of the third dual variable,

$$u_{03} \equiv \frac{dz_{opt}}{db_3} \equiv \frac{\delta f(\mathbf{b})}{\delta b_3},$$

should instead be regarded as an instantaneous rate of change, with the further understanding that u_{03} is not uniquely defined at the break points between linear segments (one such point is $3c$). This is obvious from Fig. 2-4: The first derivative of the function is discontinuous at those points. It turns out that when the constraint (2-71) of our linear program is in the break-point position $3c$, the dual problem has more than one optimal solution; in particular, u_{03} can then be equal to the slope of *either* of the two line segments that meet at position $3c$ in Fig. 2-4. This phenomenon of alternative dual optima will be illustrated in Exercise 2-21.

Figure 2-4 makes good sense from an economic point of view. Returning to our factory manager's problem, suppose we held everything fixed except the available amount b_3 of the third input, and suppose we then graphed the maximum obtainable sales income as a function of b_3. We would in fact obtain a piecewise linear curve similar to that of Fig. 2-4. If input 3 were essential to all products and none of it were on hand, then no output whatsoever could be produced by the factory and no income would accrue (in general, of course, some products might not require any of input 3, so we can say only that $z_{opt} \geqq 0$ for $b_3 = 0$). If a small amount of input 3 were available in inventory, it would be used along with certain other inputs in the most profitable combination possible, say, to manufacture product Q. The marginal value u_{03} of input 3 would then be relatively high, since each additional unit of it would yield a relatively large increment of sales income. As b_3 increased, however, it would eventually become impossible to continue deriving product Q from input 3, because one of the other necessary inputs would be totally depleted. At this break point further increments of input 3 would have to be devoted to the production of some less profitable output, and z_{opt} would begin increasing at a slower rate. Thus the marginal value of one of these latter units of input 3 would be less than the marginal value of the first unit. By this argument we can see that for each input i, $i = 1, \ldots, m$, the piecewise linear function we have been discussing must continue to flatten out as the abscissa b_i increases: The slope can never increase and must eventually fall to zero.[8]

In view of the foregoing remarks, the economic implications of complementary slackness should be fairly obvious. Consider Theorem 2.6A, for example: If the ith primal slack variable is positive, then the optimal production scheme leaves the ith input unexhausted. In these circumstances a little bit more of it would clearly be useless: Its marginal value and the value of its associated dual variable must be zero. Conversely, if the ith dual variable is positive, then the ith input must have some marginal value. If any of it were found to be available, it would be immediately used, in order to extract that value. Hence in the optimal production scheme it must have been completely exhausted, so the ith primal slack is zero. Notice that it is possible for the ith dual variable to be zero while the ith primal constraint is binding. This could occur if there were several other binding constraints; in that case a slight increment of input i might not increase profits because the other binding constraints would continue to "freeze" the optimal solution.

EXERCISES

Section 2.2

2-1. Write a Taylor expansion of each of the following functions about the point $x = 0$:

(a) $f(x) = \cos x$
(b) $f(x) = \log(1 + x)$
(c) $f(x) = (1 + x)^m$

2-2. Use Taylor's theorem to compute the sine of 1 radian (that is, $\sin x$ for $x = 1$) to three decimal places.

2-3. Consider the general Taylor expansion of the function $f(x)$ about the point $x = a$:

$$f(x) = f(a) + \frac{(x-a)}{1!}f'(a) + \frac{(x-a)^2}{2!}f''(a) + \frac{(x-a)^3}{3!}f'''(a) + \cdots.$$

[8]This statement is not true for inputs which are themselves salable outputs. If some particular input can be sold directly, in the absence of all other inputs, at a price of θ per unit, then the slope of its piecewise linear function will eventually fall, not to zero, but to θ.

Using this formula, show that $f'(a) = 0$ is a necessary but not a sufficient condition for $f(x)$ to take on a minimum at $x = a$. State a set of sufficient conditions involving $f'(a)$ and $f''(a)$. Suppose now that $f'(a) = f''(a) = \cdots = f^{[n]}(a) = 0$ but $f^{[n+1]}(a) \neq 0$; under what further conditions would $x = a$ be a minimum? Based on your answer, does $f(x) = x^{13}$ take on a minimum at $x = 0$? Does $f(x) = x^{16}$?

2-4. Use Eq. (2-12) to verify that the following formula gives a valid Taylor expansion of the function of two variables $f(x, y)$ in the neighborhood of the point (x_0, y_0):

$$f(x_0 + h, y_0 + k) = f(x_0, y_0) + h\frac{\delta f}{\delta x} + k\frac{\delta f}{\delta y}$$
$$+ \frac{1}{2}\left(h^2\frac{\delta^2 f}{\delta x^2} + 2hk\frac{\delta^2 f}{\delta x\,\delta y} + k^2\frac{\delta^2 f}{\delta y^2}\right) + \cdots,$$

where all partial derivatives are understood to be evaluated at (x_0, y_0). Show that if

$$\frac{\delta f}{\delta x} = 0, \quad \frac{\delta f}{\delta y} = 0, \quad \frac{\delta^2 f}{\delta x^2} > 0, \quad \text{and} \quad \frac{\delta^2 f}{\delta x^2}\cdot\frac{\delta^2 f}{\delta y^2} > \left(\frac{\delta^2 f}{\delta x\,\delta y}\right)^2$$

all hold at the point (x_0, y_0), then (x_0, y_0) is a minimum of $f(x, y)$.

2-5. For students who are familiar with quadratic forms: State a set of sufficient conditions based on the formula (2-15) for the point $\mathbf{x}_0$ in E^n to be a minimum of the function $f(\mathbf{x})$.

Section 2.3

2-6. The following four vectors form a basis in E^4: $\mathbf{a}_1 = (1, 0, 0, 0)$, $\mathbf{a}_2 = (0, 3, 4, 2)$, $\mathbf{a}_3 = (-1, -2, 0, 0)$, and $\mathbf{a}_4 = (0, 0, 0, 1)$. For which of them could each of the following vectors be substituted so as to preserve the basis?

(a) $\mathbf{b}_1 = (0, 1, 4, -3)$
(b) $\mathbf{b}_2 = (0, 3, 4, 2)$
(c) $\mathbf{b}_3 = (-1, 4, 8, 5)$
(d) $\mathbf{b}_4 = (0, 0, 0, 0)$
(e) $\mathbf{b}_5 = (-2, -1, 4, 2)$

2-7. Tell whether each of the following sets of simultaneous linear equations is inconsistent, indeterminate, and/or redundant. How many different solutions and how many different basic solutions exist for each?

(a)
$$\begin{aligned} x_1 + x_2 \qquad\quad &= 10 \\ 2x_1 \qquad - x_3 &= 1 \\ x_1 + x_2 + x_3 &= 19 \end{aligned}$$

(b)
$$\begin{aligned} x_1 - x_2 + 2x_3 &= 2 \\ -x_1 + 3x_2 - 2x_3 &= -2 \\ 3x_1 - x_2 + 6x_3 &= 16 \end{aligned}$$

(c)
$$\begin{aligned} x_1 + x_2 \qquad\quad &= 0 \\ 2x_1 - x_2 + 3x_3 &= 0 \\ -x_1 \qquad + x_3 &= 0 \end{aligned}$$

(d)
$$\begin{aligned} x_1 + x_2 + x_3 &= 0 \\ 2x_1 - x_2 \qquad &= 0 \\ x_1 - 2x_2 - x_3 &= 0 \end{aligned}$$

(e)
$$\begin{aligned} x_1 + 2x_2 \qquad\quad &= 1 \\ x_1 \qquad + x_3 &= -1 \\ 2x_2 + x_3 &= 10 \\ x_1 - 2x_2 - x_3 &= -4 \end{aligned}$$

(f)
$$\begin{aligned} x_1 - 2x_2 - x_3 &= 4 \\ 2x_1 + x_2 - 5x_3 &= 7 \\ 4x_1 - 3x_2 - 7x_3 &= 15 \\ x_1 + 3x_2 - 4x_3 &= 3 \end{aligned}$$

Section 2.4

Exercises 2-8 through 2-11: Solve each of the following linear programs via the two-phase simplex algorithm.

2-8. Max $z = 3x_1 + 2x_2 + 2x_3 + x_4$

$$\begin{aligned} \text{subject to}\quad x_1 - 4x_2 - x_3 + 5x_4 &= 7,\\ 2x_1 + x_2 + 3x_3 + x_4 &= 16,\\ 3x_1 - 2x_2 - x_3 + 2x_4 &= 0,\\ \text{and}\quad x_i &\geqq 0, \quad \text{all } i. \end{aligned}$$

2-9. Max $z = 2x_1 + x_2 - x_3 + 2x_4$

$$\begin{aligned} \text{subject to}\quad 2x_1 - x_2 - x_3 + x_4 &= 3,\\ -x_1 + 3x_2 + x_3 - x_4 &= 2,\\ x_1 + 3x_2 + 2x_3 + x_4 &= 5,\\ \text{and}\quad x_i &\geqq 0, \quad \text{all } i. \end{aligned}$$

2-10. Max $z = x_1 + 2x_2 + x_3 - 3x_4$

$$\begin{aligned} \text{subject to}\quad x_1 - x_2 + x_3 - x_4 &= 4,\\ 2x_1 - 4x_2 - 2x_3 + 2x_4 &= -6,\\ x_1 + 2x_2 - 3x_3 - x_4 &= -5,\\ \text{and}\quad x_i &\geqq 0, \quad \text{all } i. \end{aligned}$$

2-11. Max $z = 3x_1 + 2x_2 + 2x_3 + x_4$

$$\begin{aligned} \text{subject to}\quad x_1 + 3x_2 + x_3 + 2x_4 &= 12,\\ 2x_1 - x_2 + 3x_3 - x_4 &= 7,\\ 3x_1 - 5x_2 + 5x_3 - 4x_4 &= 2,\\ \text{and}\quad x_i &\geqq 0, \quad \text{all } i. \end{aligned}$$

2-12. (a) Suppose we have a feasible solution $\mathbf{x}_B$, with associated basis matrix $\mathbf{B}$, to some linear program P:

$$\begin{aligned} &\text{Max} && z = \mathbf{c}^T\mathbf{x}\\ &\text{subject to} && \mathbf{Ax} = \mathbf{b}\\ &\text{and} && \mathbf{x} \geqq \mathbf{0}. \end{aligned}$$

If $\hat{\mathbf{x}}$ is any other feasible solution (not necessarily basic), prove that

$$x_{Bi} = \sum_{j=1}^{n} \hat{x}_j y_{ij}, \qquad i = 1, \ldots, m,$$

where y_{ij} is the ith component of the vector $\mathbf{y}_j = \mathbf{B}^{-1}\mathbf{a}_j$.
(b) Use part (a) to prove that a basic feasible solution to problem P is optimal if $z_j - c_j \geqq 0$ for every nonbasic variable x_j.

2-13. Show how it is possible, using artificial variables and a system of "penalties," to solve the standard-form linear program (2-22) in one phase rather than two. How will infeasibility and unboundedness be discovered? Use this one-phase approach to solve the problem of Exercise 2-11.

Section 2.5

2-14. Write the duals of the following linear programs:

(a) Max $z = -2x_1 - x_2$
subject to $x_1 + x_2 \leqq 10$,
$x_1 - 2x_2 = -8$,
$x_1 + 3x_2 \geqq 9$,
and $x_1 \geqq 0$,

(b) Min $z = 2x_1 - x_2 + x_3$
subject to $2x_1 + x_2 - x_3 \leqq 8$,
$-x_1 + x_3 \geqq 1$,
$x_1 + 2x_2 + 3x_3 = 9$,
and $x_1, x_2 \geqq 0$.

(c) Max $z = x_1$
subject to $x_1 \leqq 9$,
and $x_1 \geqq 4$.

(d) Min $z = 3x_1 - 2x_2$
subject to $-4x_1 + 3x_2 \leqq 10$,
and $x_1 \geqq 0$.

(e) Min $z = x_1$
subject to $x_1 + x_2 + x_3 = 13$,
$x_2 \geqq x_3$,
$5 - x_3 \leqq x_1$,
$x_1 \leqq 0$,
and $x_3 \leqq 10$.

2-15. Using the definition of the dual problem, prove that the dual of the standard-form linear program (2-22) is given by (2-38).

2-16. Show how the dual of the linear program

$$\text{Max } z = \mathbf{c}^T\mathbf{x} \text{ subject to } \mathbf{Ax} \geqq \mathbf{b} \text{ and } \mathbf{x} \geqq \mathbf{0}$$

can be written in terms of *nonpositive* variables.

Section 2.6

2-17. Use the duality theorem to prove the following: If at any iteration of the simplex method $z_j - c_j \geqq 0$ for all nonbasic variables x_j (in a maximization problem), then the current basic feasible solution is optimal.

2-18. Use the simplex optimality theorem, as stated in Exercise 2-17, to prove the duality theorem.

Section 2.7

2-19. Suppose that $x_1 = x_3 = 0$, $x_2 = 10.4$, $x_4 = 0.4$ is given to be the optimal solution to the linear program

$$\begin{aligned}
\text{Max } z &= 2x_1 + 4x_2 + 3x_3 + x_4 \\
\text{subject to } 3x_1 + x_2 + x_3 + 4x_4 &\leqq 12, \\
x_1 - 3x_2 + 2x_3 + 3x_4 &\leqq 7, \\
2x_1 + x_2 + 3x_3 - x_4 &\leqq 10, \\
\text{and} \quad x_1, x_2, x_3, x_4 &\geqq 0,
\end{aligned}$$

Use this information to find the optimal solution to the dual problem.

2-20. For any m-by-n matrix $\mathbf{A}$ and any n-component vector $\mathbf{c}$, prove that there exists some vector of values $\mathbf{x}$ satisfying $\{\mathbf{Ax} \leqq \mathbf{0}, \mathbf{c}^T\mathbf{x} = 1\}$ if and only if there exists *no* vector $\mathbf{u}$ satisfying $\{\mathbf{A}^T\mathbf{u} = \mathbf{c}, \mathbf{u} \geqq \mathbf{0}\}$.

Section 2.8

2-21. Diagram and solve graphically the following primal linear program:

$$\begin{aligned} &\text{Max } z = x_2 \\ \text{subject to} \quad & x_1 + x_2 \leqq 3, \\ & x_2 \leqq 2, \\ & -x_1 + x_2 \leqq 1, \\ \text{and} \quad & x_1, x_2 \geqq 0. \end{aligned}$$

Calculate the rate of change of the optimal value of z with respect to (a) increases and (b) decreases in the right-hand-side element $b_1 = 3$ of the first constraint. Now write the dual problem and obtain *all* optimal solutions. Discuss the relationship between dz_{opt}/db_1 and the optimal value(s) of the first dual variable.

2-22. (a) Suppose that k units of some commodity, where $k > 0$, must be shipped from node A to node D along the one-way arcs shown in the accompanying network diagram. The label on each arc gives the per-unit cost of shipping commodity

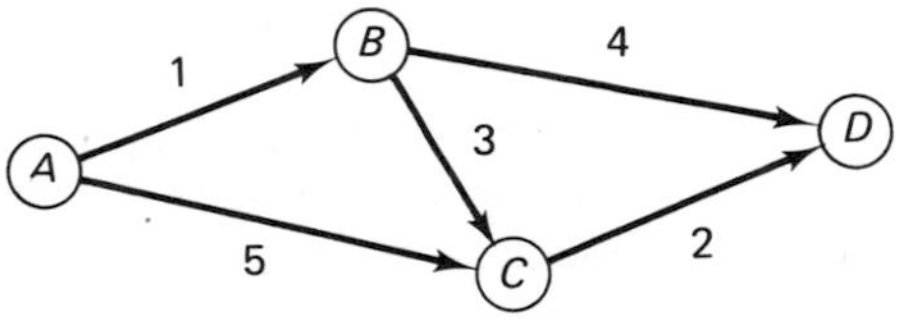

along it; no charge is assessed for shipping through the intermediate nodes B and C. If the number of units to be shipped from A to B is represented by x_{AB}, and so on, verify that the shipping plan that moves the k units to D at minimum cost is obtained by solving the following linear program:

$$\begin{aligned} &\text{Min } z = x_{AB} + 5x_{AC} + 3x_{BC} + 4x_{BD} + 2x_{CD} \\ \text{subject to} \quad & -x_{AB} - x_{AC} = -k, \\ & x_{AB} - x_{BC} - x_{BD} = 0, \\ & x_{AC} + x_{BC} - x_{CD} = 0, \\ & x_{BD} + x_{CD} = k, \\ \text{and all} \quad & x_{IJ} \geqq 0. \end{aligned}$$

Will all k units necessarily be shipped over the same route from A to D?

(b) Now write the dual of this LPP and suggest an interpretation of the dual variables, which are called *node numbers*. What are the implications of the dual constraints and objective function and of complementary slackness? How does your interpretation account for the equality of the optimal values of the primal and dual objectives?
(c) Show that the primal constraints as stated above are redundant, and discuss the impact of this fact upon the dual problem and its optimal solution.

2-23. Suppose the network of Exercise 2-22 is now *capacitated*: that is, the number of units shipped over any arc $I - J$ is required to be no greater than some upper bound c_{IJ}, which is called the *capacity*. The constraints

$$x_{IJ} \leqq c_{IJ} \qquad \text{for all arcs } I - J$$

must now be added to the linear program. Write the dual of the new LPP and discuss the dual variables, complementary slackness, etc.

2-24. Consider a simple zero-sum, two-person *game* in which each player has two possible *strategies* or "moves." A single play of the game consists of each player choosing simultaneously one of his two strategies. If Row chooses his ith strategy and Column chooses his jth, Row collects from Column a payment a_{ij} as given in the following *payoff matrix*:

		Column's strategies 1	Column's strategies 2
Row's strategies	1	$a_{11} = 1$	$a_{12} = 4$
	2	$a_{21} = 3$	$a_{22} = 2$

The game is to be repeated many times. If Row always chooses 2, Column will soon deduce this and choose 2, holding Row to winnings of two units per play. If Row always chooses 1, he will do even worse. John von Neumann showed that it is optimal for Row to play a *mixed strategy*: At every play he uses a random device that chooses strategy 1 or 2 with probabilities x_1 and x_2, respectively. Define V, the *value of the game*, to be the maximum expected winnings per play that Row can achieve, given that Column is playing his own best strategy also. Show that Row can obtain his optimal probabilities, or optimal mixed strategy, by solving the following linear program:

$$\text{Max } z = V$$

$$\begin{aligned} \text{subject to } \quad V - x_1 - 3x_2 &\leqq 0, \\ V - 4x_1 - 2x_2 &\leqq 0, \\ x_1 + x_2 &= 1, \\ \text{and} \quad x_1, x_2 &\geqq 0. \end{aligned}$$

Now derive and interpret the dual problem.

2-25. Repeat Exercise 2-24 when the payoff matrix is

	1	2
1	1	2
2	4	3

Discuss as completely as possible. In this example Row's strategy 2 is called a *dominant* strategy. How would the game be played if both players had a dominant strategy?

3

Further Mathematical Background

3.1 INTRODUCTION

In this chapter we shall discuss several rather basic topics from classical mathematics, including local and global optima, quadratic forms, and convex function theory, each of which will be of major importance in our development of nonlinear programming. Although the reader may already have encountered some or all of this material in earlier course work, we shall assume that he has forgotten everything and proceed rigorously in each case from the fundamental definitions. As in Chapter 2, we urge every reader to study this background material with care, regardless of how familiar or innocuous it may appear to be.

3.2 MAXIMA AND MINIMA

Before we begin considering solution methods for mathematical optimization problems, we must be careful to specify exactly what an optimum is. There are, in fact, three kinds of optima to be defined below, and the distinctions among them are quite important. Throughout the present section we shall speak only of maxima; it will be understood that analogous definitions and properties exist for minima. Note also that the term "optimum," however it may be modified, always means "maximum" or "minimum."

We first present several definitions and simple results from set theory; no previous knowledge will be assumed. This material will help to provide a clear spatial understanding of local and global maxima, as well as a more or less rigorous foundation upon which their definitions can be based. The reader who is meeting these set-theoretic concepts for the first time may rest assured that, in general, they exactly coincide with what his intuition will tell him. Further details can be found in Rudin [39].

1. The *distance* between two points $\mathbf{x} = (x_1, \ldots, x_n)$ and $\mathbf{y} = (y_1, \ldots, y_n)$ in Euclidean n-space E^n, usually denoted $|\mathbf{x} - \mathbf{y}|$, is defined to be

$$\left[\sum_{j=1}^{n} (x_j - y_j)^2\right]^{1/2},$$

which, of course, corresponds to the ordinary notion of distance.

2. A point $\mathbf{x}$ is an *interior point* of a set S in E^n if $\mathbf{x}$ is in S and there exists some positive number ϵ such that all points in E^n within a distance ϵ of $\mathbf{x}$ are also in S. A point $\mathbf{y}$ is a *boundary point* of a set S in E^n if for any $\epsilon > 0$ there exist within a distance ϵ of $\mathbf{y}$ at least one point in S and at least one point not is S. Note that a boundary point of S need not be in S and that any point in S is either an interior point or a boundary point of S. For example, consider the following set of points in E^1 (the real line):

$$S = \{x \mid 0 \leqq x < 1\}.$$

Both $x = 0$ and $x = 1$ are boundary points, although the latter is not in S, and any point x such that $0 < x < 1$ is an interior point of S.

3. A *closed* set is a set that contains all of its own boundary points. Thus the set of real numbers $\{x \mid 0 \leqq x \leqq 1\}$ is a closed set, while the set $\{x \mid 0 \leqq x < 1\}$ is not.

4. If $g_i(\mathbf{x})$ is a continuous function, then the collection of points $\mathbf{x}$ in E^n that satisfy $g_i(\mathbf{x}) \leqq b_i$ constitutes a closed set. Closed sets are also defined by $g_i(\mathbf{x}) \geqq b_i$ and by $g_i(\mathbf{x}) = b_i$.

5. The intersection of a finite number of closed sets is a closed set (by definition, a point $\mathbf{x}$ lies in the *intersection* of the sets $X_1, X_2, \ldots$ if and only if it lies in every one of those sets).

6. The universe of all points $\mathbf{x}$ in E^n is a closed set.

Consider now the constraint set of the general mathematical programming problem (maximization version), as defined in Chapter 1:

(3-1) maximize $f(\mathbf{x})$, $\mathbf{x} = (x_1, \ldots, x_n)$,

(3-2) subject to $g_i(\mathbf{x}) \{\leqq, =, \geqq\} b_i$, $i = 1, \ldots, m$.

By restricting the constraints to the three types indicated above, we ensure that the feasible region must be a closed set; this follows from propositions 4 and 5 above. Inasmuch as proposition 6 guarantees that even the unconstrained optimization problem has a closed set of feasible solutions, we have established that *the feasible region of any mathematical program* (3-1) *and* (3-2) *is a closed set.*

Now, the goal in solving a maximization problem is to identify the *global* or *absolute maximum* from among the feasible points.

A global maximum of the function f(**x**) *over a closed set* S *occurs at the point* **x*** *in* S *if and only if* f(**x**) $\leqq$ f(**x***) *for all points* **x** *in* S. *We then say that* **x*** *maximizes* f(**x**) *in* S, *or that* f(**x**) *takes on a global maximum over* S *at the point* **x***.

It is clear from the definition that global maxima may occur at more than one point in *S*, although the maximum feasible value of $f(\mathbf{x})$ must be unique.

If the closed set *S* is bounded, it seems obvious—but is not at all easy to prove—that some point in *S* must in fact be a global maximum of $f(\mathbf{x})$. This is true even when the maximum value of $f(\mathbf{x})$ in *S* is infinite. Taking the closed set *S* to be the feasible region (3-2), we shall state without proof an existence theorem: *For any mathematical programming problem having a bounded, nonempty feasible region, at least one feasible point must be a global maximum.* If the feasible region is unbounded, the maximum may either occur at some specifiable finite point or may instead lie "infinitely far away," being approached as one or more of the components of **x** become positively or negatively infinite. (The reader should recall at this juncture the distinction drawn in Section 1.2 between an unbounded feasible region and an unbounded optimal solution, which may occur at a finite point.)

There exist, in addition to the global maximum, two other types, both known as *local maxima*; to distinguish them we shall refer to the "constrained" and "unconstrained" varieties. The *unconstrained local maximum* is, intuitively, the highest point anywhere in the surrounding neighborhood, for a neighborhood that may be very small:

Let f(**x**) *be defined for all* **x** *in some set* T. *Then an unconstrained local* (*or relative*) *maximum of* f(**x**) *occurs at* **x*** *in* T *provided that there exists some* $\delta > 0$ *such that if* **x** *is within a distance* δ *of* **x***, *then* **x** *is in* T *and* f(**x**) $\leqq$ f(**x***).

An unconstrained local or relative maximum can be thought of as the top of any hill in a region containing one, two, or several hills. By definition it can occur only in the interior of *T*, never on the boundary. For example, in Fig. 3-1 we take *T* to be the closed interval $0 \leqq x \leqq 1$ on the real line; $f(x)$ is graphed above it. The points $x = a$, $x = b$, and $x = c$ are unconstrained local maxima of $f(x)$, but $x = 0$ and $x = 1$ are not.

Note that the global maximum in Fig. 3-1 occurs at $x = 0$, which is *not* an unconstrained local maximum. It *is*, however, a local maximum of the other variety:

Let f(**x**) *be defined for all* **x** *in some set* U. *Then a constrained local* (*or relative*) *maximum of* f(**x**) *with respect to the set* U *occurs at* **x*** *in* U *provided that there exists some* $\delta > 0$ *such that, if* **x** *is in* U *and lies within a distance* δ *of* **x***, *then* f(**x**) $\leqq$ f(**x***).

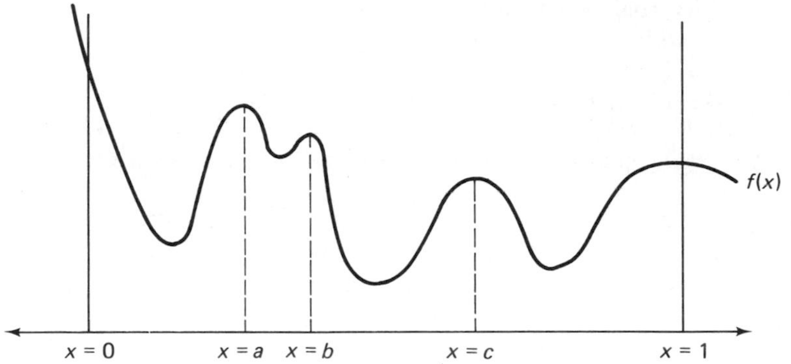

FIGURE 3-1

Notice the difference: This definition, unlike the one preceding, allows $\mathbf{x}^*$ to be a boundary point. It is not difficult to see that *any unconstrained local maximum is also a constrained local maximum*, although the reverse is not true.

The unconstrained local maximum is a familiar concept from classical mathematics; it is what the freshman calculus student has in mind when he seeks to determine the relative maxima and minima of a function by setting its first partial derivatives equal to zero. However, when inequality constraints are introduced, as in a mathematical programming problem, the objective function is then being optimized over a bounded or partially bounded feasible region rather than over all of E^n. Under these circumstances the global maximum of the objective function may well occur at some boundary point at which our definition of a constrained local maximum is satisfied but at which the partial derivatives do not necessarily all vanish.

We have remarked that any unconstrained local maximum is also a constrained local maximum. It is also true that *any (finite) point that constitutes a global maximum of $f(\mathbf{x})$ over a set S must be a constrained local maximum as well.* This follows trivially from the definitions—all points $\mathbf{x}$ in S within *any* distance of a global maximum $\mathbf{x}^*$ must satisfy $f(\mathbf{x}) \leqq f(\mathbf{x}^*)$. The proposition just stated has important implications in mathematical programming: Although the problem-solving algorithms that have been devised thus far are in general capable only of finding constrained local optima, we can at least identify the global optimal solution as being the best of the local optima (provided, of course, that all of the latter have been found).

As a final note let us emphasize that a constrained local optimum $\mathbf{x}^*$ lying on the boundary of a set U depends in a very direct way on the precise location of that boundary. If the boundary is perturbed slightly, so is the location of $\mathbf{x}^*$. In general, computation of the coordinates of $\mathbf{x}^*$ requires some algebraic manipulation of the constraint functions that define the relevant portion of the boundary. It is clear that unconstrained local optima do not have this same type of dependence on the location of boundaries: They can be found simply via manipulation of the partial derivatives of the objective function.

In the rest of this chapter we shall be concerned principally with the theory of convex and concave functions. These are two classes of ordinary mathematical functions characterized by certain defining properties that will be discussed in detail below. Convex function theory plays a crucial role in mathematical programming; it will be used repeatedly later on to establish optimality conditions and solution algorithms for various types of optimization problems.

Before getting into convex function theory, however, it is convenient for us to present a brief discussion of *quadratic forms.* The characterization of any given quadratic function as convex, concave, or neither depends on an understanding of quadratic forms and their properties. In addition, alternative definitions of convexity and concavity, valid for all algebraic functions, can be stated in terms of quadratic forms. We shall therefore discuss this topic first and then proceed with an uninterrupted development of convex function theory.

3.3 QUADRATIC FORMS

Although the great majority of problems arising in engineering, economics, management, and statistics are represented (often improperly) by linear models and solved by linear mathematical methods, it is fair to say that quadratic models and techniques are firmly established in second place. Some obvious and important examples include the familiar notion of statistical variance, nonlinear estimation and curve-fitting via parabolic approximation, the quadratic programming problem (maximize a quadratic function subject to linear constraints—to be discussed in Chapter 7), and the quadratic assignment problem of combinatorial mathematics (e.g., assign 10 facilities to 10 different fixed locations so as to minimize the total distance walked back and forth by all users of the facilities).

We begin the development of this section with the following basic definition:

A quadratic form is any scalar-valued function, defined for all $\mathbf{x}$ *in* E^n, *that takes the following form:*

$$q(\mathbf{x}) = \sum_{i=1}^{n} \sum_{j=1}^{n} a_{ij} x_i x_j, \tag{3-3}$$

where each a_{ij} *is a real number (possibly zero).*

Notice that this representation includes two terms in $x_h x_k$ for all pairs of subscripts h and k such that $h \neq k$. A quadratic form differs from a quadratic *function* in that it does not include any linear or scalar terms; thus

$$\begin{aligned} q(x_1, x_2) &= 7x_1^2 + x_1x_2 - 3x_2x_1 - 4x_2^2 \\ &= 7x_1^2 - 2x_1x_2 - 4x_2^2 \end{aligned}$$

is a quadratic form in two variables, while

$$f(x_1, x_2) = 7x_1^2 - 2x_1x_2 - 4x_2^2 - 6x_1 + x_2 + 7$$

is a quadratic function.

Quadratic forms are introduced in linear algebra textbooks because the theorems and techniques of linear algebra can be applied to gain some useful results about them. Notice that any given quadratic form $q(\mathbf{x})$ can be expressed in matrix notation as

$$q(\mathbf{x}) \equiv \mathbf{x}^T\mathbf{A}\mathbf{x}, \tag{3-4}$$

where the elements a_{ij} of the n-by-n matrix $\mathbf{A}$ are simply the scalar coefficients in (3-3). It is clear that, inasmuch as

$$a_{ij}x_ix_j + a_{ji}x_jx_i = (a_{ij} + a_{ji})x_ix_j \qquad \text{for all } i \neq j,$$

a given quadratic form $q(\mathbf{x})$ can be equivalently represented by many different matrices $\mathbf{A}$ or sets of coefficients a_{ij}. However, for any given $q(\mathbf{x})$ there exists only one *symmetric* matrix[1] $\mathbf{D}$ that satisfies $q(\mathbf{x}) \equiv \mathbf{x}^T\mathbf{D}\mathbf{x}$, namely, that matrix $\mathbf{D}$ whose elements are defined by

$$d_{ij} = d_{ji} = \left(\frac{a_{ij} + a_{ji}}{2}\right) \tag{3-5}$$

for all subscript pairs i and j (if $i = j$ the relation reduces to $d_{ii} = 2a_{ii}/2 = a_{ii}$). The student may write out the component terms to check that $q(\mathbf{x}) \equiv \mathbf{x}^T\mathbf{D}\mathbf{x}$ for all $\mathbf{x}$ provided that (3-5) is satisfied; it will then be obvious that the symmetric matrix $\mathbf{D}$ is unique. Henceforth, when we use the expression $\mathbf{x}^T\mathbf{D}\mathbf{x}$ for a quadratic form, the matrix $\mathbf{D}$ will be understood to be symmetric.

Turning to some examples, the most general possible quadratic form in two variables is

$$q(x_1, x_2) = ax_1^2 + bx_1x_2 + cx_2^2,$$

[1] Recall that the square matrix $\mathbf{D}$ is *symmetric* if it equals its own transpose, $\mathbf{D} = \mathbf{D}^T$.

which can be written in matrix form as

$$q(x_1, x_2) \equiv \mathbf{x}^T\mathbf{D}\mathbf{x} \equiv [x_1 \quad x_2]\begin{bmatrix} a & b/2 \\ b/2 & c \end{bmatrix}\begin{bmatrix} x_1 \\ x_2 \end{bmatrix}.$$

No other symmetric matrix can be used to represent $q(x_1, x_2)$. The quadratic form associated with the nonsymmetric matrix

$$\mathbf{A} = \begin{bmatrix} 2 & -2 \\ 8 & -1 \end{bmatrix}$$

is

$$q(x_1, x_2) = 2x_1^2 - 2x_1x_2 + 8x_2x_1 - x_2^2 = 2x_1^2 + 6x_1x_2 - x_2^2,$$

but it is also true that

$$q(x_1, x_2) \equiv 2x_1^2 + 3x_1x_2 + 3x_2x_1 - x_2^2,$$

which has a symmetric set of coefficients. As a final example,

$$q(x_1, x_2, \ldots, x_n) = ax_1^2$$

is a legitimate quadratic form for any positive integer n; thus, for $n = 2$ we would have $q(x_1, x_2) = ax_1^2 + 0 \cdot x_1x_2 + 0 \cdot x_2^2$. The unique symmetric matrix associated with $q(x_1, x_2, \ldots, x_n)$ has the element a in its upper left corner and 0s everywhere else.

We now wish to define some properties that certain quadratic forms may possess. Consider the quadratic form $q(x_1, x_2) = 2x_1^2 - 8x_1x_2 + 8x_2^2$, which may be expressed equivalently as $q(x_1, x_2) = 2(x_1 - 2x_2)^2$. No matter what real values we choose for x_1 and x_2, the value of $q(x_1, x_2)$ obviously cannot be negative, since the function can be represented as a positive multiple of a squared term. The value can, however, be zero, not only when $x_1 = x_2 = 0$, but also whenever $x_1 = 2x_2$—thus, for example, $q(6, 3) = 0$. Such quadratic forms are said to be *positive semidefinite*; by contrast, those that vanish only when $\mathbf{x} = \mathbf{0}$ (and are positive for all other real $\mathbf{x}$) are called *positive definite* forms. Note that $\mathbf{x} = \mathbf{0}$ must always be an exceptional case, inasmuch as any quadratic form $q(\mathbf{x})$ equals zero when evaluated at $\mathbf{x} = \mathbf{0}$.

Let us state these definitions precisely:

The quadratic form $q(\mathbf{x}) \equiv \mathbf{x}^T\mathbf{D}\mathbf{x}$ *is positive definite if* $q(\mathbf{x}) > 0$ *for all* $\mathbf{x} \neq \mathbf{0}$ *in* E^n.

The quadratic form $q(\mathbf{x}) \equiv \mathbf{x}^T\mathbf{D}\mathbf{x}$ *is positive semidefinite if* $q(\mathbf{x}) \geqq 0$ *for all* $\mathbf{x}$ *in* E^n *but* $q(\mathbf{x})$ *is not positive definite.*

Negative definite and semidefinite forms are analogously defined. The matrix representation $\mathbf{x}^T\mathbf{D}\mathbf{x}$ is included in these definitions because the unique symmetric matrix $\mathbf{D}$ associated with a positive semidefinite (or whatever) quadratic form is also said to be positive semidefinite. Finally, a quadratic form satisfying none of the above four definitions is an *indefinite* form; that is, $q(\mathbf{x})$ is indefinite if $q(\mathbf{x}_1) > 0$ for some $\mathbf{x}_1$ in E^n and $q(\mathbf{x}_2) < 0$ for some $\mathbf{x}_2$.

A few examples should serve to clarify the definitions. The quadratic form $q(x_1, x_2) = x_1^2 + 10x_2^2$ is positive definite, as is its associated matrix

$$\mathbf{D} = \begin{bmatrix} 1 & 0 \\ 0 & 10 \end{bmatrix}$$

Similarly, $q(x_1, x_2) = -x_1^2 - 10x_2^2$ is negative definite. However, the quadratic form $q(x_1, x_2, x_3) = x_1^2 + 10x_2^2$ is positive *semi*definite, because $q(\mathbf{x}) \geqq 0$ for all $\mathbf{x}$ in E^3 but $q(0, 0, 1) = 0$. Choosing some slightly more complicated examples, the quadratic form

$$q(x_1, x_2) = 2x_1^2 - 8x_1x_2 + 9x_2^2 = 2(x_1 - 2x_2)^2 + x_2^2$$

is positive definite—the proof is trivial—but

$$q(x_1, x_2, x_3) = 2(x_1 - 2x_2)^2 + x_3^2$$

is positive semidefinite because $q(2, 1, 0) = 0$.

A very simple property of quadratic forms may be noted at this point: *If* $q(\mathbf{x})$ *is positive definite* (*or semidefinite*), *then* $-q(\mathbf{x})$ *is negative definite* (*or semidefinite*), and vice versa. This follows immediately from the definitions. In addition, several other elementary but important properties of positive definite quadratic forms are collected in the following lemma, which will be most useful to us in our discussion of quadratic programming in Chapter 7:

Lemma 3.1: Let the symmetric n-*by*-n *matrix* $\mathbf{D}$ *be positive* (*negative*) *definite. Then*

(a) *The inverse* $\mathbf{D}^{-1}$ *exists,*
(b) $\mathbf{D}^{-1}$ *is positive* (*negative*) *definite, and*
(c) $\mathbf{A}\mathbf{D}\mathbf{A}^{\mathrm{T}}$ *is positive* (*negative*) *semidefinite for any* m-*by*-n *matrix* $\mathbf{A}$.

To prove part (a), let $\mathbf{d}_1, \ldots, \mathbf{d}_n$ be the columns of $\mathbf{D}$ and assume they are linearly dependent; then there exist scalars $\lambda_1, \ldots, \lambda_n$ not all zero such that

$$\sum_{i=1}^{n} \lambda_i \mathbf{d}_i \equiv \mathbf{D}\boldsymbol{\lambda} = \mathbf{0},$$

where $\boldsymbol{\lambda}$ is the column vector $(\lambda_1, \ldots, \lambda_n)$. This implies $\boldsymbol{\lambda}^T\mathbf{D}\boldsymbol{\lambda} = 0$ for some $\boldsymbol{\lambda} \neq \mathbf{0}$, which contradicts our hypothesis that $\mathbf{D}$ is positive definite. Hence the assumption was false, the columns of $\mathbf{D}$ are linearly independent, and $\mathbf{D}^{-1}$ exists.

The proofs for parts (b) and (c) are scarcely more difficult. For any $\mathbf{x}$ in E^n we may write

$$\mathbf{x}^T\mathbf{D}^{-1}\mathbf{x} = \mathbf{x}^T\mathbf{D}^{-1}(\mathbf{D}\mathbf{D}^{-1})\mathbf{x} = \mathbf{x}^T(\mathbf{D}^T)^{-1}\mathbf{D}\mathbf{D}^{-1}\mathbf{x} = (\mathbf{D}^{-1}\mathbf{x})^T\mathbf{D}(\mathbf{D}^{-1}\mathbf{x}),$$

where the second equality follows from the fact that $\mathbf{D}$ is symmetric. If $\mathbf{D}$ is positive definite, then $\mathbf{x}^T\mathbf{D}^{-1}\mathbf{x} \equiv (\mathbf{D}^{-1}\mathbf{x})^T\mathbf{D}(\mathbf{D}^{-1}\mathbf{x})$ is nonnegative for all $\mathbf{x}$ in E^n and is zero only for those $\mathbf{x}$ that satisfy $\mathbf{D}^{-1}\mathbf{x} = \mathbf{0}$. But $\mathbf{x} = \mathbf{0}$ is the only such $\mathbf{x}$; hence $\mathbf{D}^{-1}$ is positive definite. Q.E.D.

Similarly, if $\mathbf{D}$ is positive definite, then for any m-by-n matrix $\mathbf{A}$

$$\mathbf{x}^T(\mathbf{A}\mathbf{D}\mathbf{A}^T)\mathbf{x} = (\mathbf{A}^T\mathbf{x})^T\mathbf{D}(\mathbf{A}^T\mathbf{x}) \geqq 0 \qquad \text{for all } \mathbf{x} \text{ in } E^m.$$

However, in general $\mathbf{A}^T\mathbf{x}$ may equal $\mathbf{0}$ even when $\mathbf{x}$ does not, so we can only conclude that $\mathbf{A}\mathbf{D}\mathbf{A}^T$ must be positive semidefinite. The proofs of parts (b) and (c) for negative definite $\mathbf{D}$ would, of course, be analogous to those above.

In the examples presented above the reader doubtless noticed that a certain amount of algebraic manipulation—completing the squares and so on—was necessary in order to classify a quadratic form of just two or three variables as positive definite, positive semidefinite, or whatever. As the number of variables increases, the labor required by this approach rapidly becomes prohibitive. We therefore state without proof a family of results from linear algebra which provide a well-defined and reasonably efficient procedure for classifying quadratic forms.

A quadratic form $q(\mathbf{x}) \equiv \mathbf{x}^T\mathbf{D}\mathbf{x}$, *where* $\mathbf{D}$ *is the associated symmetric matrix, is positive semidefinite if and only if the variables can be ordered so that* d_{11} *is positive and the following determinants are all nonnegative:*

$$\begin{vmatrix} d_{11} & d_{12} \\ d_{21} & d_{22} \end{vmatrix} \geqq 0; \quad \begin{vmatrix} d_{11} & d_{12} & d_{13} \\ d_{21} & d_{22} & d_{23} \\ d_{31} & d_{32} & d_{33} \end{vmatrix} \geqq 0; \quad \ldots; \quad |\mathbf{D}| \geqq 0,$$

where the d_{ij} *are the elements of* $\mathbf{D}$.

If the various determinants are all strictly positive, then the quadratic form is positive definite. The set of necessary and sufficient conditions for $\mathbf{x}^T\mathbf{D}\mathbf{x}$

to be *negative* semidefinite is quite similar:

The quadratic form $q(\mathbf{x}) \equiv \mathbf{x}^T\mathbf{D}\mathbf{x}$, $\mathbf{x}$ *in* E^n, *is negative semidefinite if and only if the variables can be ordered so that* d_{11} *is negative and*

$$\begin{vmatrix} d_{11} & d_{12} \\ d_{21} & d_{22} \end{vmatrix} \geqq 0; \quad \begin{vmatrix} d_{11} & d_{12} & d_{13} \\ d_{21} & d_{22} & d_{23} \\ d_{31} & d_{32} & d_{33} \end{vmatrix} \leqq 0; \quad \ldots; \quad (-1)^n|\mathbf{D}| \geqq 0.$$

These conditions can be derived directly from those given for positive semidefinite forms by using the fact that the determinant of $-\mathbf{D}$, where $\mathbf{D}$ is an r-by-r matrix, equals $(-1)^r$ times the determinant of $\mathbf{D}$.

3.4 CONVEX AND CONCAVE FUNCTIONS

We are now ready to take up the main theme of this chapter, convex function theory, which is concerned with two special classes of mathematical functions and their properties. We shall be particularly interested in the maximization and minimization of these functions over what are called *convex sets* (the material in the first few paragraphs of this section is covered in most linear programming courses and therefore may be quite familiar to the reader).

A set X in E^n is *convex* if, for any two points $\mathbf{x}_1$ and $\mathbf{x}_2$ in X and any scalar value of λ satisfying $0 \leqq \lambda \leqq 1$, the point

$$\mathbf{x} = \lambda\mathbf{x}_1 + (1 - \lambda)\mathbf{x}_2 \tag{3-6}$$

also lies in X.

The expression $\lambda\mathbf{x}_1 + (1 - \lambda)\mathbf{x}_2$, where $0 \leqq \lambda \leqq 1$, is known as a *convex combination* of the points $\mathbf{x}_1$ and $\mathbf{x}_2$; thus a set is convex if and only if every convex combination of any two of its points is also in the set. More generally, given a set of points $\mathbf{x}_1, \ldots, \mathbf{x}_m$ in E^n, any summation of the form

$$\sum_{i=1}^{m} \lambda_i \mathbf{x}_i, \qquad \text{where } \sum_{i=1}^{m} \lambda_i = 1 \text{ and all } \lambda_i \geqq 0, \tag{3-7}$$

is called a *convex combination* of those points. A convex combination is a special kind of linear combination.

A more important geometrical characterization of convex sets follows from the fact that $\lambda\mathbf{x}_1 + (1 - \lambda)\mathbf{x}_2$, $0 \leqq \lambda \leqq 1$, represents the straight-line segment between the points $\mathbf{x}_1$ and $\mathbf{x}_2$. The definition (3-6) therefore implies that a straight line drawn between any two points in a convex set lies entirely within the set. For example, a sphere, a triangle, the entire space E^n, a straight line, and a single point are convex sets, while a doughnut, the letter L, and the surface of a sphere (which is a set of points in E^3) are not. Incidentally, there is no such thing as a "concave set."

One type of convex set that is of great interest in linear programming is the *hyperplane*. A hyperplane in E^n is the set of all points $\mathbf{x} = (x_1, \ldots, x_n)$ satisfying

$$c_1x_1 + c_2x_2 + \cdots + c_nx_n = z,$$

where z and $c_1, \ldots, c_n$ are a set of scalar constants (with at least one $c_j \neq 0$). In vector notation, a hyperplane is given by

$$\{\mathbf{x} \mid \mathbf{c}^T\mathbf{x} = z\} \tag{3-8}$$

for some specific column vector $\mathbf{c}$ and scalar z. In E^2 the set of points (3-8) is just a straight line, $c_1x_1 + c_2x_2 = z$; in E^3 it corresponds to our ordinary geometrical notion of a plane.

By simply applying the definitions of this section, we can easily prove that *any hyperplane is a convex set*. Suppose $\mathbf{x}_1$ and $\mathbf{x}_2$ lie on the hyperplane $\mathbf{c}^T\mathbf{x} = z$ in E^n. For any point $\hat{\mathbf{x}} = \lambda\mathbf{x}_1 + (1 - \lambda)\mathbf{x}_2$, $0 \leqq \lambda \leqq 1$, on the straight line between them, we have

$$\mathbf{c}^T\hat{\mathbf{x}} = \lambda\mathbf{c}^T\mathbf{x}_1 + (1 - \lambda)\mathbf{c}^T\mathbf{x}_2 = \lambda z + (1 - \lambda)z = z.$$

Therefore $\hat{\mathbf{x}}$ also lies on the hyperplane, which accordingly satisfies the definition of a convex set. By substituting inequalities as required, it can also be shown that the sets of points in E^n

$$\{\mathbf{x} \mid \mathbf{c}^T\mathbf{x} \leqq z\} \quad \text{and} \quad \{\mathbf{x} \mid \mathbf{c}^T\mathbf{x} \geqq z\},$$

which are called *closed half-spaces*, are convex sets as well.

The role of convex sets in mathematical programming depends almost completely on the existence and properties of the following two types of functions:

The scalar function $f(\mathbf{x})$ is a *convex function* defined over a convex set X in E^n if for any two points $\mathbf{x}_1$ and $\mathbf{x}_2$ in X,

$$f(\lambda\mathbf{x}_1 + (1 - \lambda)\mathbf{x}_2) \leqq \lambda f(\mathbf{x}_1) + (1 - \lambda)f(\mathbf{x}_2) \tag{3-9}$$

for all λ such that $0 \leqq \lambda \leqq 1$. Similarly, $f(\mathbf{x})$ is a *concave function* over X if for any $\mathbf{x}_1$ and $\mathbf{x}_2$ in X and any λ satisfying $0 \leqq \lambda \leqq 1$,

$$f(\lambda\mathbf{x}_1 + (1 - \lambda)\mathbf{x}_2) \geqq \lambda f(\mathbf{x}_1) + (1 - \lambda)f(\mathbf{x}_2). \tag{3-10}$$

Note that (3-9) and (3-10) must hold as equalities whenever $\lambda = 0$ or $\lambda = 1$. We also say that $f(\mathbf{x})$ is *strictly* convex or concave over X if (3-9) or (3-10) is satisfied as a strict inequality for every $\mathbf{x}_1$ and $\mathbf{x}_2$ in X and for all λ in $0 < \lambda < 1$; by definition, any strictly convex function is also a convex function. Finally, when we say that a function is convex without specifying a set

X, we shall mean that it is convex throughout all of E^n (which is itself a convex set).

In the above definitions we have followed the common practice of including the provision that convexity or concavity must prevail over a convex set. The technical reason for this is that if X were not a convex set, the point $\lambda\mathbf{x}_1 + (1 - \lambda)\mathbf{x}_2$ would not necessarily lie in X, and we could not be certain that the function could even be evaluated there. In any case, convex and concave functions defined over *non*convex sets are of no particular interest in mathematical programming.

Convex and concave functions have a simple and informative geometrical characterization. Any function $f(\mathbf{x})$ of points in E^n may be thought of as a *hypersurface* $x_{n+1} = f(x_1, \ldots, x_n)$ suspended in $(n + 1)$-dimensional space over (and/or under) the hyperplane $x_{n+1} = 0$; the height of the surface over the hyperplane at any point $\mathbf{x}$ in E^n gives the value of the function at that point. Recall from linear algebra that $\lambda\mathbf{x}_1 + (1 - \lambda)\mathbf{x}_2$, $0 \leqq \lambda \leqq 1$, defines a straight-line segment between the points $\mathbf{x}_1$ and $\mathbf{x}_2$. The definition (3-9) of a *convex* function can then be given the following interpretation. For any two points $\mathbf{x}_1$ and $\mathbf{x}_2$ in E^n, let a straight-line segment be drawn in E^{n+1} between the two points $(\mathbf{x}_1, f(\mathbf{x}_1))$ and $(\mathbf{x}_2, f(\mathbf{x}_2))$ on the hypersurface $f(\mathbf{x})$. Then the right-hand side of (3-9) gives the height of that line segment at any point $\mathbf{x} = \lambda\mathbf{x}_1 + (1 - \lambda)\mathbf{x}_2$, and the definition evidently requires that the entire line segment lie on or above the hypersurface $f(\mathbf{x})$. In the case of a *concave* function, the line must lie on or below the hypersurface.

If all this is difficult to visualize, a study of the two-dimensional case should clarify matters. Consider the convex function $f(x)$ graphed in Fig. 3-2(a); it is defined for points x on the real line E^1. Between the two points $(a, f(a))$ and $(b, f(b))$ on the curve $f(x)$ a straight-line segment has been drawn, and it lies entirely above (or, rather, not below) the curve. The figure shows that at the general point $x = \lambda a + (1 - \lambda)b$ the definition (3-9) is satisfied, since the segment $\overline{BC}$ is not greater in length than $\overline{AC}$.

Several additional examples of convex and concave functions of one variable are displayed in Fig. 3-2. An interesting general property can be observed in these diagrams: As x increases, the slope of each convex (concave) function appears to be uniformly nondecreasing (nonincreasing). This property can actually be proved directly (see Exercise 3-13), and a generalized n-dimensional version of it, to be stated formally as Theorem 3.10, will be derived at the end of the chapter.

Based on the definitions (3-9) and (3-10), a number of theoretical results can be obtained that allow us to identify various types of mathematical functions as convex or concave.

Theorem 3.1: Any linear function $f(\mathbf{x}) = \mathbf{c}^T\mathbf{x} \equiv \sum_{j=1}^{n} c_j x_j$ *is both convex and concave over all of* E^n.

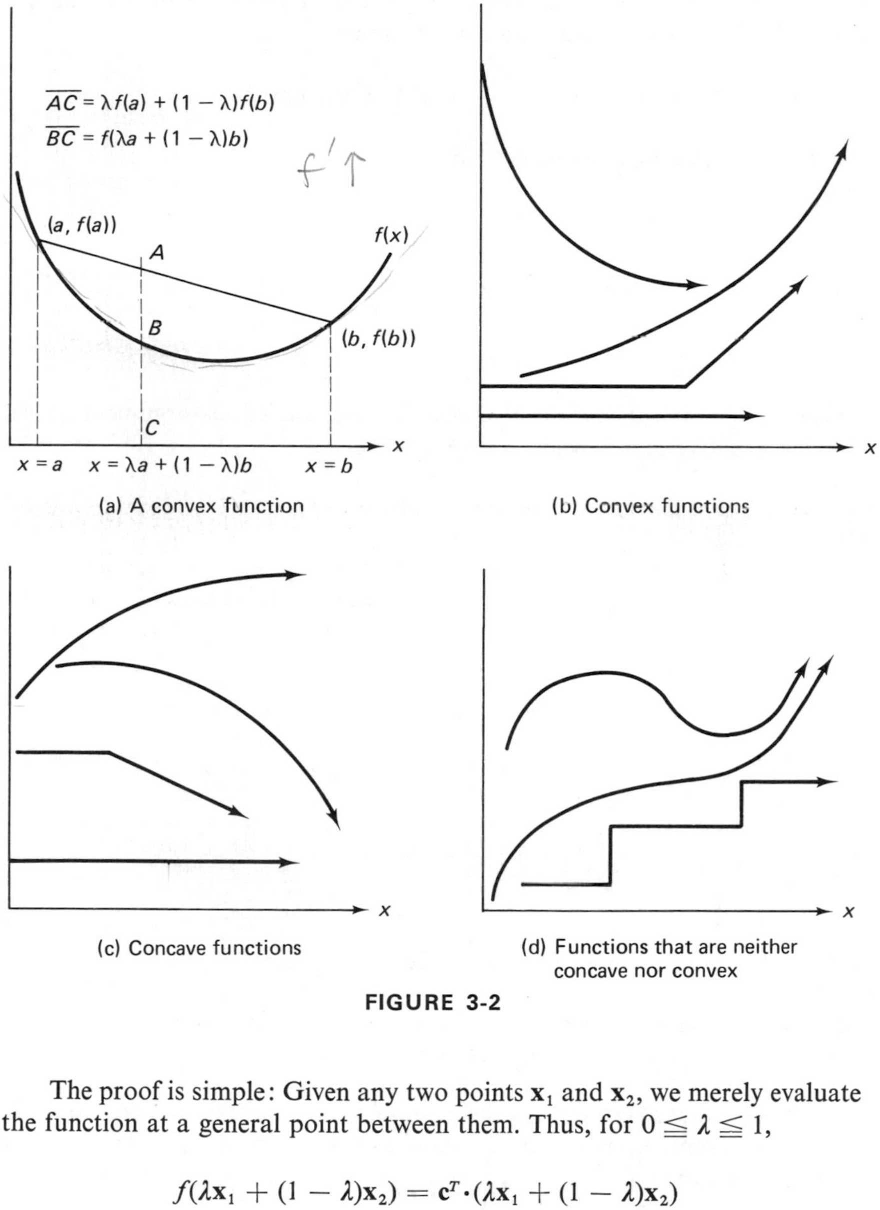

FIGURE 3-2

The proof is simple: Given any two points $\mathbf{x}_1$ and $\mathbf{x}_2$, we merely evaluate the function at a general point between them. Thus, for $0 \leqq \lambda \leqq 1$,

$$\begin{aligned} f(\lambda\mathbf{x}_1 + (1-\lambda)\mathbf{x}_2) &= \mathbf{c}^T \cdot (\lambda\mathbf{x}_1 + (1-\lambda)\mathbf{x}_2) \\ &= \lambda\mathbf{c}^T\mathbf{x}_1 + (1-\lambda)\mathbf{c}^T\mathbf{x}_2 \\ &= \lambda f(\mathbf{x}_1) + (1-\lambda)f(\mathbf{x}_2). \end{aligned}$$

Since equality prevails, both (3-9) and (3-10) are satisfied, as was to be shown. Exercise 3-8 asks the student to prove the converse, namely, that any function that is both convex and concave must be linear.

Theorem 3.2: If f($\mathbf{x}$) *is convex, then* $-$f($\mathbf{x}$) *is concave, and vice versa.*

If $f(\mathbf{x})$ is convex, then by definition

$$f(\lambda \mathbf{x}_1 + (1 - \lambda)\mathbf{x}_2) \leqq \lambda f(\mathbf{x}_1) + (1 - \lambda)f(\mathbf{x}_2)$$

for $\mathbf{x}_1$ and $\mathbf{x}_2$ in E^n and for $0 \leqq \lambda \leqq 1$. Multiplying through by -1, we get

$$-f(\lambda \mathbf{x}_1 + (1 - \lambda)\mathbf{x}_2) \geqq \lambda(-f(\mathbf{x}_1)) + (1 - \lambda)(-f(\mathbf{x}_2)),$$

which satisfies the definition of concavity. An analogous argument shows that the negative of a concave function is convex.

Theorem 3.3: The sum of two or more convex (concave) functions is convex (concave).

This theorem is proved in the same way as the last two, the idea being to manipulate the given information in order to produce either (3-9) or (3-10). Let $f(\mathbf{x}) \equiv \sum_{i=1}^{k} g_i(\mathbf{x})$, where all the $g_i(\mathbf{x})$ are convex functions. Then

$$\begin{aligned} f(\lambda \mathbf{x}_1 + (1 - \lambda)\mathbf{x}_2) &\equiv \sum_{i=1}^{k} g_i(\lambda \mathbf{x}_1 + (1 - \lambda)\mathbf{x}_2) \\ &\leqq \sum_{i=1}^{k} \{\lambda g_i(\mathbf{x}_1) + (1 - \lambda)g_i(\mathbf{x}_2)\} \\ &= \lambda \sum_{i=1}^{k} g_i(\mathbf{x}_1) + (1 - \lambda) \sum_{i=1}^{k} g_i(\mathbf{x}_2) \\ &\equiv \lambda f(\mathbf{x}_1) + (1 - \lambda) f(\mathbf{x}_2), \end{aligned}$$

so that the sum of convex functions is likewise convex.

Theorem 3.4: Any positive semidefinite quadratic form q($\mathbf{x}$) $\equiv \mathbf{x}^T\mathbf{D}\mathbf{x}$, *where* $\mathbf{D}$ *is symmetric, is a convex function over all of* E^n.

To show that $\mathbf{x}^T\mathbf{D}\mathbf{x}$ satisfies the definition of a convex function, we again choose two points $\mathbf{x}_1$ and $\mathbf{x}_2$ in E^n and consider any point $\mathbf{x}_0$ on the line segment between them:

$$\mathbf{x}_0 = \lambda \mathbf{x}_1 + (1 - \lambda)\mathbf{x}_2,$$

where $0 \leqq \lambda \leqq 1$. Then, since this yields

$$\mathbf{x}_0 = \mathbf{x}_2 + \lambda(\mathbf{x}_1 - \mathbf{x}_2),$$

we may write

$$\begin{aligned}\mathbf{x}_0^T\mathbf{D}\mathbf{x}_0 &= \mathbf{x}_2^T\mathbf{D}\mathbf{x}_2 + \lambda\mathbf{x}_2^T\mathbf{D}(\mathbf{x}_1 - \mathbf{x}_2) + \lambda(\mathbf{x}_1 - \mathbf{x}_2)^T\mathbf{D}\mathbf{x}_2 \\ &\quad + \lambda^2(\mathbf{x}_1 - \mathbf{x}_2)^T\mathbf{D}(\mathbf{x}_1 - \mathbf{x}_2) \\ &= \mathbf{x}_2^T\mathbf{D}\mathbf{x}_2 + 2\lambda(\mathbf{x}_1 - \mathbf{x}_2)^T\mathbf{D}\mathbf{x}_2 + \lambda^2(\mathbf{x}_1 - \mathbf{x}_2)^T\mathbf{D}(\mathbf{x}_1 - \mathbf{x}_2),\end{aligned}$$

using the fact that $\mathbf{D}$ is symmetric. Now, the third term is nonnegative because $\mathbf{D}$ is positive semidefinite; moreover, the limits placed on λ guarantee that $k\lambda^2$ is less than or equal to $k\lambda$ for any nonnegative number k. We therefore have

$$\begin{aligned}\mathbf{x}_0^T\mathbf{D}\mathbf{x}_0 &\leqq \mathbf{x}_2^T\mathbf{D}\mathbf{x}_2 + 2\lambda(\mathbf{x}_1 - \mathbf{x}_2)^T\mathbf{D}\mathbf{x}_2 + \lambda(\mathbf{x}_1 - \mathbf{x}_2)^T\mathbf{D}(\mathbf{x}_1 - \mathbf{x}_2) \\ &= \mathbf{x}_2^T\mathbf{D}\mathbf{x}_2 + \lambda(\mathbf{x}_1 - \mathbf{x}_2)^T\mathbf{D}\mathbf{x}_2 + \lambda(\mathbf{x}_1 - \mathbf{x}_2)^T\mathbf{D}\mathbf{x}_1 \\ &= \mathbf{x}_2^T\mathbf{D}\mathbf{x}_2 - \lambda\mathbf{x}_2^T\mathbf{D}\mathbf{x}_2 + \lambda\mathbf{x}_1^T\mathbf{D}\mathbf{x}_1\end{aligned}$$

(to obtain the last equality we canceled $\lambda\mathbf{x}_1^T\mathbf{D}\mathbf{x}_2 - \lambda\mathbf{x}_2^T\mathbf{D}\mathbf{x}_1$, which equals zero when $\mathbf{D}$ is symmetric). It has thus been shown that

$$q(\mathbf{x}_0) \equiv q(\lambda\mathbf{x}_1 + (1 - \lambda)\mathbf{x}_2) \leqq \lambda q(\mathbf{x}_1) + (1 - \lambda)q(\mathbf{x}_2),$$

which establishes our result.

We can immediately state as a corollary that *any negative semidefinite quadratic form is a concave function.* This is easily shown: If the quadratic form $q(\mathbf{x})$ is negative semidefinite, then $p(\mathbf{x}) \equiv -q(\mathbf{x})$ is positive semidefinite and therefore a convex function, and it follows from Theorem 3.2 that $q(\mathbf{x})$ must be concave. Another corollary may also be stated: A *positive definite* quadratic form is a *strictly convex* function.

Any example of a positive or negative semidefinite form, such as those given in Section 3.3, will serve as an example of a convex or concave function. Thus

$$q(x_1, x_2, \ldots, x_n) = -3x_1^2$$

is a concave function in any Euclidean space E^n, but only in E^1 is it strictly concave as well (because in E^1 it is negative definite rather than merely negative semidefinite). Again, the quadratic form

$$q(\mathbf{x}) = \sum_{j=1}^{n} x_j^2$$

is a convex function—in fact, it is strictly convex—in any space E^n.

Because of Theorems 3.3 and 3.4 we are now in a position to decide whether any given quadratic *function* is a concave or a convex function. A quadratic function $f(\mathbf{x})$ may be represented, after collecting terms, as

$$f(\mathbf{x}) \equiv q(\mathbf{x}) + \mathbf{c}^T\mathbf{x} + c_0,$$

where $q(\mathbf{x})$ is a quadratic form, $\mathbf{c}^T\mathbf{x}$ is a sum of linear terms, and c_0 is a scalar constant. We know from Theorem 3.1 that any function $\mathbf{c}^T\mathbf{x} + c_0$ is both concave and convex (the addition of a scalar constant is irrelevant). Therefore all we need to do is to examine $q(\mathbf{x})$: If it is positive semidefinite (or definite), then $f(\mathbf{x})$ must be a convex function because it is the sum of two convex functions. For example,

$$f(x_1, x_2) = 10x_1 + 10x_2 + 100 - x_1^2 - 2x_1x_2 - x_2^2$$

is a concave function because its quadratic terms constitute a negative semidefinite form. But the quadratic function

$$f(x_1, x_2) = x_1^2 + 2x_1x_2 + 100x_1 + 100x_2$$

is neither convex nor concave because its quadratic part $q(x_1, x_2) = x_1^2 + 2x_1x_2$ is neither positive nor negative (semi)definite: Note that $q(1, 5)$ is positive and that $q(1, -5)$ is negative. Other examples will be found in the exercises.

3.5 CONVEX FEASIBLE REGIONS AND OPTIMAL SOLUTIONS

We have seen that all hyperplanes and closed half-spaces are convex sets. These results, which depend on the fact that linear functions are both concave and convex, are generalized in the following:

Theorem 3.5: If $f(\mathbf{x})$ *is a convex function over* E^n, *then the set of points*

$$S \equiv \{\mathbf{x} \mid f(\mathbf{x}) \leqq b\},$$

where b *is any real number, is a convex set. Similarly, if* $f(\mathbf{x})$ *is concave, then*

$$T \equiv \{\mathbf{x} \mid f(\mathbf{x}) \geqq b\}$$

is a convex set.

To prove this theorem we must show that if $\mathbf{x}_1$ and $\mathbf{x}_2$ are in S, then so is any convex combination of them. Using the definition (3-9) of a convex function, we have

$$f(\lambda\mathbf{x}_1 + (1 - \lambda)\mathbf{x}_2) \leqq \lambda f(\mathbf{x}_1) + (1 - \lambda)f(\mathbf{x}_2) \leqq \lambda b + (1 - \lambda)b = b,$$

as was to be shown. The proof that T is a convex set is analogous.

For example, since $f(x_1, x_2) = x_1^2 + x_2^2$ is a convex function, the unit circle in E^2, which contains the points $\{\mathbf{x} \mid x_1^2 + x_2^2 \leqq 1\}$, must be a convex set. As a simpler example, the set of points $\{x \mid x^2 \leqq 1\}$ on the real line is also a convex set in E^1, consisting simply of the line segment from $x = -1$ to $x = +1$. By contrast, the set

$$T = \{x \mid x^2 \geqq 1\},$$

which consists of two disconnected rays, is *not* convex; this is what would be expected, because $f(x) = x^2$ is not a concave function, as required by Theorem 3.5. Similarly, $T = \{\mathbf{x} \mid x_1^2 + x_2^2 \geqq 1\}$, which consists of the x_1-x_2 plane with a circular "hole" at the origin, is also not a convex set.

We now observe that *the intersection of any two convex sets must itself be convex.* To prove this, consider any two points $\mathbf{x}_1$ and $\mathbf{x}_2$ within the intersection of the convex sets X_1 and X_2 in E^n. Because $\mathbf{x}_1$ and $\mathbf{x}_2$ lie in X_1, so does the straight-line segment between them, but that line segment also lies in X_2, so it must lie in the intersection as well. Hence the intersection is convex. By extension, the intersection of any finite number of convex sets is convex.

At this point it should be clear that the mathematical programming problem

$$\text{(3-11)} \qquad \begin{cases} \text{Max } f(\mathbf{x}), & \mathbf{x} = (x_1, \ldots, x_n), \\ \text{subject to } g_i(\mathbf{x})\{\leqq, =, \geqq\}b_i, & i = 1, \ldots, m, \end{cases}$$

can easily have a convex feasible region. All that is required is that each of the individual constraints generate a convex set, in which case their intersection will be convex. Insofar as the inequality constraints are concerned, each function $g_i(\mathbf{x})$ need only be concave or convex according to the conditions of Theorem 3.5. What of the equality constraints? It is sufficient to stipulate that for every constraint $g_i(\mathbf{x}) = b_i$ in (3-11) the function $g_i(\mathbf{x})$ must be linear; the set of points $\mathbf{x}$ satisfying the constraint must then be a hyperplane, which is a convex set. From another point of view, the requirement $g_i(\mathbf{x}) = b_i$ is equivalent to the pair of constraints

$$g_i(\mathbf{x}) \leqq b_i \quad \text{and} \quad g_i(\mathbf{x}) \geqq b_i,$$

in the sense that these two inequalities jointly define the same set of feasible points as does the single equation $g_i(\mathbf{x}) = b_i$. In light of Theorem 3.5, then, we should stipulate that $g_i(\mathbf{x})$ be both convex and concave, that is, linear. We have proved the following:

Theorem 3.6: The feasible region of the mathematical program (3-11) *is a convex set if the following three sufficient conditions are met:*

(a) *For all constraints* $g_i(\mathbf{x}) \leqq b_i$, *the function* g_i *is convex;*
(b) *For all constraints* $g_i(\mathbf{x}) \geqq b_i$, *the function* g_i *is concave; and*
(c) *For all constraints* $g_i(\mathbf{x}) = b_i$, *the function* g_i *is linear.*

Inasmuch as linear constraints are both concave and convex, one immediate corollary of Theorem 3.6 is that *the feasible region of any linear program is a convex set.* We could, of course, have obtained this last result without the aid of convex function theory, merely by observing that the feasible region of an LPP is the intersection of hyperplanes and closed half-spaces, which, as we have seen, are convex sets.

Although the conditions stated in Theorem 3.6 are sufficient to establish the convexity of the feasible region, they are *not* necessary; that is, a feasible region may be convex even though one or more of them is violated. For example, the points satisfying the following constraints form a convex set (see Fig. 3-3):

$$x_1 \leqq \pi,$$
$$\sin x_1 - x_2 \geqq 0,$$
$$x_1^2 - x_2 \geqq -2,$$
and
$$x_1, x_2 \geqq 0.$$

Yet the second of them, involving $\sin x_1$, is neither convex nor concave over all of E^2, while the third is of the form $g_i(\mathbf{x}) \geqq b_i$ with $g_i(\mathbf{x})$ convex (being the sum of two convex functions), which violates condition (b). The conditions of Theorem 3.6 would have been both necessary *and* sufficient if we had specified that each constraint function need be convex or concave in accordance with (a), (b), and (c) *only* over that portion of its length that forms part of the boundary of the feasible region (thus no restriction at all would be placed on an inoperative constraint such as $x_1^2 - x_2 \geqq -2$ in Fig. 3-3). For problems of any size at all, however, such necessary conditions would simply not be very useful; too much computation would be required to establish the feasible region, determine which constraint hypersurfaces bound it, identify their various points and surfaces of intersection, and so on. It is much more convenient simply to remark that the feasible region of a mathematical programming problem whose constraints do not all satisfy Theorem 3.6 may in fact be convex anyway. The student should be alive to this possibility: For small problems he may be able to demonstrate convexity without a prohibitive amount of computation. This is particularly true when there are only two variables and the feasible region can be plotted.

We come now to a pivotal theorem, one that gathers together the various notions of convex and concave functions, convex sets, and local and global optima. In Chapter 6 we shall derive a fundamental set of algebraic relations,

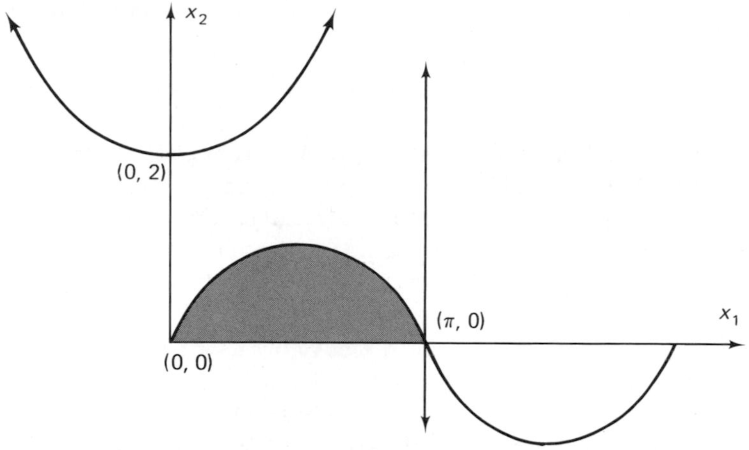

FIGURE 3-3

called the Kuhn-Tucker conditions, that must hold at each constrained local optimum of any mathematical programming problem. We shall then be able, in theory, to solve any given problem by writing out those conditions and obtaining all the points $\mathbf{x}_0$ that satisfy them (they may also be satisfied by certain exceptional points that are not local optima). One of the $\mathbf{x}_0$ must then be the desired global optimum, assuming that the latter occurs at a finite point—this was established in Section 3.2.

Under certain circumstances, however, only one constrained local optimum will exist, and it will of necessity be the global optimum. These favorable circumstances are described by the following:

Theorem 3.7: If the function f($\mathbf{x}$) *is defined and convex over the closed convex set* X *in* E^n, *then any constrained local minimum of* f($\mathbf{x}$) *in* X *is a global minimum over* X. *If* f($\mathbf{x}$) *is concave over the closed convex set* X, *then any constrained local maximum in* X *is a global maximum over* X.

To prove this theorem, suppose that $\mathbf{x}_0$ is a constrained local minimum but not a global minimum, so that there exists some point $\mathbf{x}^*$ in X such that $f(\mathbf{x}^*) < f(\mathbf{x}_0)$. Then for any λ satisfying $0 < \lambda < 1$ the convexity of $f(\mathbf{x})$ implies

$$\begin{aligned} f(\lambda\mathbf{x}^* + (1-\lambda)\mathbf{x}_0) &\leqq \lambda f(\mathbf{x}^*) + (1-\lambda)f(\mathbf{x}_0) \\ &< \lambda f(\mathbf{x}_0) + (1-\lambda)f(\mathbf{x}_0) = f(\mathbf{x}_0) \end{aligned} \tag{3-12}$$

(the strict inequality can be used since $\lambda = 0$ is prohibited). Now consider the straight-line segment from $\mathbf{x}_0$ to $\mathbf{x}^*$, which must lie entirely in X. For any small positive δ there exists some $\lambda > 0$ defining a point on that line segment

$$\hat{\mathbf{x}} = \lambda \mathbf{x}^* + (1 - \lambda)\mathbf{x}_0$$

that lies in X at a distance δ away from $\mathbf{x}_0$. But from (3-12) we have $f(\hat{\mathbf{x}}) < f(\mathbf{x}_0)$. Inasmuch as this holds for any $\delta > 0$, the point $\mathbf{x}_0$ cannot satisfy the definition of a constrained local minimum (recall Section 3.2). This contradiction verifies the first part of the theorem; the second part can be proved in an analogous manner.

Theorem 3.7 has an important direct implication: In solving a mathematical program whose objective is to minimize a convex function or to maximize a concave function and whose constraints satisfy Theorem 3.6, the first constrained local optimum discovered must be the desired solution. Problems satisfying this description constitute an important class known as *convex programs*; they will be explored more fully in later chapters.

We may as well present at this point a related but secondary result. Having developed a theorem about the minima of convex functions, what can we say about the maxima of such functions?

Theorem 3.8: If the function f(**x**) *is convex (concave) over the bounded set* X *in* E^n, *then at least one global maximum (minimum) of* f(**x**) *over* X *occurs at a boundary point of* X.

This is not a terribly important result, and its proof will be left to the ingenuity of the student (see Exercise 3-17).

We close this section with an example illustrating Theorem 3.7. Consider the following mathematical programming problem:

$$\begin{aligned} \text{Min } f(x_1, x_2) &= x_1^2 - 4x_1x_2 + 5x_2^2 + 2x_1 - 6x_2 \\ \text{subject to } x_1^2 + x_2^2 &\leqq 9, \\ x_1 + x_2 &\leqq 3, \\ \text{and } x_2 &\geqq 0. \end{aligned}$$

The first constraint requires that a convex function be less than or equal to a constant; from Theorem 3.5 the set of points satisfying it constitutes a convex set. Since the other two constraints, being linear, also define convex sets, the feasible region must itself be convex. The objective function can be expressed as

$$f(x_1, x_2) = (x_1 - 2x_2)^2 + x_2^2 + 2x_1 - 6x_2,$$

and since the quadratic portion is positive definite, the problem is evidently one of minimizing a convex function over a convex set.

Although we have not yet studied the Kuhn-Tucker conditions, we can at least see if there are any *unconstrained* local minima of $f(x_1, x_2)$ in the feasible region. The partial derivatives are

$$\frac{\delta f}{\delta x_1} = 2x_1 - 4x_2 + 2$$

and

$$\frac{\delta f}{\delta x_2} = -4x_1 + 10x_2 - 6,$$

both of which vanish at the point (1, 1). This point satisfies all the constraints, and a check of the second derivatives shows that it is an unconstrained local minimum,[2] and therefore a constrained local minimum. Then by Theorem 3.7 it must be the (global) optimal solution to the problem—there is no need to search further.

3.6 ALTERNATIVE CHARACTERIZATIONS OF CONVEX AND CONCAVE FUNCTIONS

In solving mathematical programming problems, and in other forms of analysis as well, it is frequently necessary or desirable to determine whether a given function is convex or concave. In Section 3.4 we derived rules for classifying linear and quadratic functions directly, but for anything more complicated we have no choice at present but to write out $f(\lambda\mathbf{x}_1 + (1 - \lambda)\mathbf{x}_2)$ and $\lambda f(\mathbf{x}_1) + (1 - \lambda)f(\mathbf{x}_2)$ and manipulate algebraically, hoping to establish either (3-9) or (3-10). The reader can get some idea of how difficult this procedure generally is by trying to show that the rather simple function $f(x) = e^x$ satisfies the definition (3-9) of convexity.

We shall therefore develop a very useful theorem that will allow us to classify any function as concave or convex (or neither) by performing a test on its second derivatives. Although the test will not be practical for extremely messy functions or for large numbers of variables, it will often be quite easy to apply. It is particularly simple in the one-variable case: All we need to do is determine whether the second derivative is negative or positive over the interval of interest. For example, since the second derivative of $f(x) = e^x$ is nonnegative for all x, that function must be convex everywhere. It is also true that if $f(x)$ is convex, then $d^2f/dx^2 \geqq 0$ for all x; in other words, the

[2] We have gotten slightly ahead of ourselves in this example. Sufficient conditions for a point $\mathbf{x}^*$ to be an unconstrained local minimum (which include the vanishing of the first partial derivatives) will be discussed in Section 4.2. However, they have already been derived for functions of two variables by those students who solved Exercise 2-4.

second-derivative test will essentially provide an alternative definition or characterization of convex and concave functions.

To prove these results, we must first establish a certain lemma, which in fact amounts to still another definition of concavity and convexity. Because it will be useful in several places, we state it as a theorem:

Theorem 3.9: Let the function f(**x**) *have continuous first partial derivatives. Then* f(**x**) *is concave over some convex region* R *in* E^n *if and only if*

$$f(\mathbf{x}) \leqq f(\mathbf{x}^*) + \nabla \mathbf{f}^T(\mathbf{x}^*)\cdot(\mathbf{x} - \mathbf{x}^*) \tag{3-13}$$

for any two points **x** *and* **x*** *in* R, *where* $\nabla \mathbf{f}(\mathbf{x}^*)$ *is the column vector*

$$\left(\frac{\delta f}{\delta x_1}, \ldots, \frac{\delta f}{\delta x_n}\right) \tag{3-14}$$

with each component evaluated at **x***. *Similarly,* g(**x**) *is convex if and only if*

$$g(\mathbf{x}) \geqq g(\mathbf{x}^*) + \nabla \mathbf{g}^T(\mathbf{x}^*)\cdot(\mathbf{x} - \mathbf{x}^*). \tag{3-15}$$

The vector (3-14), known as the *gradient* of *f*, was introduced in Section 2.2, and will reappear frequently throughout the text. To prove (3-13), assume first that $f(\mathbf{x})$ is concave; then by definition

$$f(\lambda \mathbf{x} + (1 - \lambda)\mathbf{x}^*) \geqq \lambda f(\mathbf{x}) + (1 - \lambda)f(\mathbf{x}^*)$$

for all λ, $0 \leqq \lambda \leqq 1$, or

$$f(\mathbf{x}^* + \lambda(\mathbf{x} - \mathbf{x}^*)) \geqq f(\mathbf{x}^*) + \lambda[f(\mathbf{x}) - f(\mathbf{x}^*)].$$

We now perform a first-order Taylor expansion on the left side of this inequality; using $\mathbf{x}^* + \lambda(\mathbf{x} - \mathbf{x}^*)$ and $\mathbf{x}^*$ in place of $\mathbf{x}$ and $\mathbf{x}_0$ in (2-16), we have

$$f(\mathbf{x}^*) + \nabla \mathbf{f}^T[\mathbf{x}^* + \theta\lambda(\mathbf{x} - \mathbf{x}^*)]\cdot\lambda(\mathbf{x} - \mathbf{x}^*) \geqq f(\mathbf{x}^*) + \lambda[f(\mathbf{x}) - f(\mathbf{x}^*)]$$

for some θ, $0 < \theta < 1$. For any positive λ this yields

$$\nabla \mathbf{f}^T[\mathbf{x}^* + \theta\lambda(\mathbf{x} - \mathbf{x}^*)]\cdot(\mathbf{x} - \mathbf{x}^*) \geqq f(\mathbf{x}) - f(\mathbf{x}^*),$$

and letting λ approach zero verifies (3-13).

Going the other way, let us assume that (3-13) holds everywhere in *R*; the proof follows Zangwill [60]. Choose any two points $\mathbf{x}_1$ and $\mathbf{x}_2$ in *R* and any value of λ satisfying $0 \leqq \lambda \leqq 1$, and define

$$\mathbf{x}^* = \lambda \mathbf{x}_1 + (1 - \lambda)\mathbf{x}_2. \tag{3-16}$$

Because *R* is a convex region, $\mathbf{x}^*$ is in *R*. Using $\mathbf{x}_1$ in place of $\mathbf{x}$ in (3-13) and substituting (3-16) selectively for $\mathbf{x}^*$ yields

$$f(\mathbf{x}_1) \leqq f(\mathbf{x}^*) + \nabla \mathbf{f}^T(\mathbf{x}^*)\cdot(\mathbf{x}_1 - \lambda \mathbf{x}_1 - (1 - \lambda)\mathbf{x}_2)$$

or

$$f(\mathbf{x}_1) \leqq f(\mathbf{x}^*) + (1 - \lambda)\, \nabla \mathbf{f}^T(\mathbf{x}^*) \cdot (\mathbf{x}_1 - \mathbf{x}_2). \tag{3-17}$$

If instead we use $\mathbf{x}_2$ in place of $\mathbf{x}$ in (3-13), we can similarly derive

$$f(\mathbf{x}_2) \leqq f(\mathbf{x}^*) - \lambda\, \nabla \mathbf{f}^T(\mathbf{x}^*) \cdot (\mathbf{x}_1 - \mathbf{x}_2). \tag{3-18}$$

Now multiply (3-17) by λ and (3-18) by $1 - \lambda$ and add; since λ and $1 - \lambda$ are nonnegative, the inequalities are preserved and we obtain

$$\lambda f(\mathbf{x}_1) + (1 - \lambda) f(\mathbf{x}_2) \leqq \lambda f(\mathbf{x}^*) + (1 - \lambda) f(\mathbf{x}^*) = f(\mathbf{x}^*).$$

This is precisely the definition of concavity. Q.E.D. The proof of (3-15) is, of course, analogous.

Notice the difference between the original definition of a concave function and this new representation. Equation (3-10) states that a straight-line segment between any two points on the concave hypersurface $f(\mathbf{x})$ must lie entirely *below* the surface (or on it). On the other hand, (3-13) requires that a hyperplane that is tangent to a concave hypersurface at any point $\mathbf{x}^*$ must lie entirely *above* it. [The last statement is easily verified. Let $h(\mathbf{x}) = \mathbf{c}^T\mathbf{x}$ be the hyperplane that is tangent to $f(\mathbf{x})$ at the point $\mathbf{x}^*$, so that $f(\mathbf{x}^*) = h(\mathbf{x}^*) = \mathbf{c}^T\mathbf{x}^*$ and

$$\frac{\delta f(\mathbf{x}^*)}{\delta x_j} = \frac{\delta h(\mathbf{x}^*)}{\delta x_j} = c_j, \qquad j = 1, \ldots, n.$$

Then, substituting into (3-13), we find that for any other point $\mathbf{x}$

$$f(\mathbf{x}) \leqq \mathbf{c}^T\mathbf{x}^* + \mathbf{c}^T \cdot (\mathbf{x} - \mathbf{x}^*) = \mathbf{c}^T\mathbf{x} = h(\mathbf{x}),$$

as was to be shown.] The two different linear approximations are illustrated in Fig. 3-4 for a concave function of one variable.

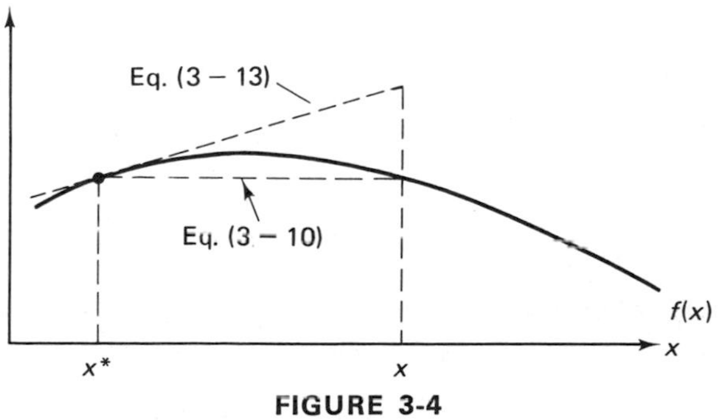

FIGURE 3-4

We can now use Theorem 3.9 to develop the convexity/concavity test alluded to at the beginning of this section. The test will be based on the Hessian matrix of second partial derivatives: Recall from Section 2.2 that the Hessian associated with any function $f(\mathbf{x})$, $\mathbf{x}$ in E^n, is

$$\mathbf{H} \equiv \begin{bmatrix} \frac{\delta^2 f}{\delta x_1^2} & \frac{\delta^2 f}{\delta x_1\, \delta x_2} & \cdots & \frac{\delta^2 f}{\delta x_1\, \delta x_n} \\ \frac{\delta^2 f}{\delta x_2\, \delta x_1} & \frac{\delta^2 f}{\delta x_2^2} & \cdots & \frac{\delta^2 f}{\delta x_2\, \delta x_n} \\ \cdot & \cdot & & \cdot \\ \cdot & \cdot & & \cdot \\ \cdot & \cdot & & \cdot \\ \frac{\delta^2 f}{\delta x_n\, \delta x_1} & \frac{\delta^2 f}{\delta x_n\, \delta x_2} & \cdots & \frac{\delta^2 f}{\delta x_n^2} \end{bmatrix}$$

We shall use $\mathbf{H}(\mathbf{x})$ to indicate the Hessian matrix evaluated at the point $\mathbf{x}$.

Theorem 3.10: Let the function $f(\mathbf{x})$ *have continuous second partial derivatives. Then* $f(\mathbf{x})$ *is concave (convex) over some convex region* R *in* E^n *if and only if its Hessian matrix is negative (positive) definite or semidefinite for all* $\mathbf{x}$ *in* R.

Note that $\mathbf{H}(\mathbf{x})$ is symmetric, so that the definitions of definite and semidefinite quadratic forms are directly applicable. Before proving anything we perform a second-order Taylor expansion of $f(\mathbf{x})$ about a general point $\mathbf{x}^*$ in R; for any other $\mathbf{x}$ in R we have, using $\mathbf{x}^*$ in place of $\mathbf{x}_0$ in (2-17),

$$\text{(3-19)} \qquad f(\mathbf{x}) = f(\mathbf{x}^*) + \nabla \mathbf{f}^T(\mathbf{x}^*)\cdot(\mathbf{x} - \mathbf{x}^*) + \tfrac{1}{2}(\mathbf{x} - \mathbf{x}^*)^T \mathbf{H}(\mathbf{x}^* + \theta(\mathbf{x} - \mathbf{x}^*))\cdot(\mathbf{x} - \mathbf{x}^*)$$

for some θ, $0 < \theta < 1$. Because R is a convex region, the point

$$\mathbf{x}^* + \theta(\mathbf{x} - \mathbf{x}^*) = \theta\mathbf{x} + (1 - \theta)\mathbf{x}^*$$

must lie within it.

We shall now prove the theorem for concave functions. Suppose first that the Hessian is negative (semi)definite throughout R; then the last term on the right-hand side of (3-19) is nonpositive, and we have

$$\text{(3-13)} \qquad f(\mathbf{x}) \leqq f(\mathbf{x}^*) + \nabla \mathbf{f}^T(\mathbf{x}^*)\cdot(\mathbf{x} - \mathbf{x}^*)$$

for any two points $\mathbf{x}$ and $\mathbf{x}^*$ in R. It follows immediately from Theorem 3.9 that the function $f(\mathbf{x})$ must be concave.

Going in the other direction, let us assume that $f(\mathbf{x})$ is concave throughout R but that the Hessian matrix is *not* negative definite or semidefinite at some point $\mathbf{x}^*$ in R. There must then exist a vector of values $\mathbf{y}$ such that

$$\mathbf{y}^T\mathbf{H}(\mathbf{x}^*)\cdot\mathbf{y} > 0. \tag{3-20}$$

Note that our assumption implies that (3-20) must hold for some *interior point* $\mathbf{x}^*$ of the region R; if the original $\mathbf{x}^*$ is a boundary point, we can choose another $\mathbf{x}^*$ very close to it but inside R, and because the Hessian is continuous, (3-20) will still hold. If we now define $\mathbf{x}_0 = \mathbf{x}^* + \mathbf{y}$, the inequality (3-20) can be written as

$$(\mathbf{x}_0 - \mathbf{x}^*)^T\mathbf{H}(\mathbf{x}^*)\cdot(\mathbf{x}_0 - \mathbf{x}^*) > 0, \tag{3-21}$$

where $\mathbf{x}_0$ is not necessarily in R. But consider another point

$$\mathbf{x} = \mathbf{x}^* + p(\mathbf{x}_0 - \mathbf{x}^*),$$

where p is any real number other than 0. It follows from (3-21) that

$$(\mathbf{x} - \mathbf{x}^*)^T\mathbf{H}(\mathbf{x}^*)\cdot(\mathbf{x} - \mathbf{x}^*) > 0 \tag{3-22}$$

for any p, because the left-hand side of (3-22) is just a positive multiple p^2 of the left-hand side of (3-21). Inasmuch as the Hessian is continuous, we may choose an $\mathbf{x}$ so close to $\mathbf{x}^*$ (i.e., choose a value of p so small) that $\mathbf{x}$ is in R and

$$(\mathbf{x} - \mathbf{x}^*)^T\mathbf{H}(\mathbf{x}^* + \theta(\mathbf{x} - \mathbf{x}^*))\cdot(\mathbf{x} - \mathbf{x}^*) > 0 \tag{3-23}$$

for *all* θ, $0 \leqq \theta \leqq 1$.

Now, by hypothesis $f(\mathbf{x})$ is concave over R, so that (3-13) holds for the points $\mathbf{x}$ and $\mathbf{x}^*$ currently under consideration. But so does the Taylor expansion (3-19), and subtracting (3-13) from (3-19) produces

$$0 \geqq \tfrac{1}{2}(\mathbf{x} - \mathbf{x}^*)^T\mathbf{H}(\mathbf{x}^* + \theta(\mathbf{x} - \mathbf{x}^*))\cdot(\mathbf{x} - \mathbf{x}^*)$$

for some value of θ between 0 and 1. This contradicts (3-23), so our assumption was false and the theorem is established. The proof for convex functions would be entirely analogous.

Having proved Theorem 3.10, we should now mention that it was not quite properly stated. A function may be convex or concave in R while its Hessian *vanishes* at certain points in R: for example, this happens in the case of linear functions at every point in E^n. A matrix whose elements are all zero is not ordinarily said to be positive or negative semidefinite and docs not satisfy the alternative definitions stated in terms of determinants at the end of Section 3.3. Thus Theorem 3.10 should be amended to read that a function is concave over R if and only if its Hessian is negative definite or semidefinite *or vanishes* at every point in R. The changes required in the proof are trivial.

Let us now write out explicitly the one- and two-variable versions of Theorem 3.10, with the amendment just discussed.

One Variable: $f(x)$ *is concave (convex) over an interval* I *of the real line if and only if*

$$\frac{d^2f}{dx^2} \leqq 0 \qquad (\geqq 0)$$

at all points x *in* I.

Two Variables: $f(x_1, x_2)$ *is concave (convex) over a region* R *in* E^2 *if and only if*

$$\frac{\delta^2 f}{\delta x_1^2} \leqq 0 \qquad (\geqq 0)$$

and

$$\frac{\delta^2 f}{\delta x_1^2} \cdot \frac{\delta^2 f}{\delta x_2^2} - \left[\frac{\delta^2 f}{\delta x_1\, \delta x_2}\right]^2 \geqq 0$$

for all (x_1, x_2) *in* R.

For example, is the function

$$f(x_1, x_2) = x_1 e^{x_1 + x_2}$$

convex over all nonnegative values of x_1 and x_2? The value of

$$\frac{\delta^2 f}{\delta x_1^2} = (x_1 + 2)e^{x_1 + x_2}$$

is nonnegative, as required; however,

$$\frac{\delta^2 f}{\delta x_1^2} \cdot \frac{\delta^2 f}{\delta x_2^2} - \left[\frac{\delta^2 f}{\delta x_1\, \delta x_2}\right]^2 = e^{2(x_1+x_2)}(x_1^2 + 2x_1 - x_1^2 - 2x_1 - 1) = -e^{2(x_1+x_2)}$$

which is negative everywhere. Thus the Hessian is neither positive nor negative semidefinite, and the given function is neither convex nor concave over the given region (nor over any other region in E^2).

EXERCISES

Section 3.2

3-1. Let S be the set of all points in E^2 satisfying

$$-1 \leqq x_1 \leqq 3 \quad \text{and} \quad -1 \leqq x_2 \leqq 3.$$

Find the global maximum and minimum and all constrained and unconstrained local maxima and minima of the function $f(x_1, x_2) = x_1^2 + x_2^2$ in the region S.

3-2. Is the point $x_1 = 1$, $x_2 = 0$ a constrained local maximum of $f(x_1, x_2) = x_1^2$ with respect to the feasible region

$$\{(x_1, x_2) \mid x_1^2 + x_2^2 = 1\}?$$

Is it a global maximum over that feasible region? An unconstrained local maximum?

3-3. In each of the following cases tell whether the point P is a global maximum or minimum or a constrained or unconstrained local maximum or minimum (more than one of these may apply):
(a) The feasible region consists of a single point P in E^n.
(b) The feasible region consists of 10 isolated points in E^n, one of which is P, but P has neither the greatest nor the smallest objective value.
(c) All points in E^n, including P, are feasible and have the same objective value.

Section 3.3

3-4. Tell whether each of the following quadratic forms is positive definite, positive semidefinite, or neither:
(a) $q(x_1, x_2) = 2x_1^2 + x_2^2$
(b) $q(x_1, x_2) = 2x_1^2 - x_2^2$
(c) $q(x_1, x_2, x_3) = 2x_1^2 + x_2^2$
(d) $q(x_1, x_2) = x_1x_2$
(e) $q(x_1, x_2) = x_1^2 + x_1x_2 + x_2^2$
(f) $q(x_1, x_2) = x_1^2 + 2x_1x_2 + x_2^2$
(g) $q(x_1, x_2) = x_1^2 + 3x_1x_2 + x_2^2$
(h) $q(x_1, x_2) = x_1^2 + 7x_1x_2 + 13x_2^2$
(i) $q(x_1, x_2, x_3) = 4x_1^2 + x_1x_2 + 2x_2^2 + 4x_1x_3 + 2x_2x_3 + x_3^2$

3-5. Tell whether each of the following quadratic forms is positive or negative definite, positive or negative semidefinite, or none of these:
(a) $q(x_1, x_2, x_3) = -x_1^2 + 2x_1x_2 - 2x_2^2 - x_3^2$
(b) $q(x_1, x_2, x_3) = x_1^2 + 2x_1x_2 + x_2^2 + 2x_2x_3 + x_3^2$
(c) $q(x_1, x_2, x_3) = -(x_1 - 2x_2)^2 - (x_1 - 2x_3)^2 - (x_2 - 2x_3)^2$
(d) $q(x_1, x_2, x_3) = 4x_1^2 + 4x_1x_2 + 2x_2^2 + 4x_1x_3 + 2x_2x_3 + x_3^2$
(e) $q(x_1, x_2, x_3) = -x_1^2 + 2x_1x_2 - 4x_2^2 + 3x_1x_3 + 6x_2x_3 - 9x_3^2$

3-6. In specifying those necessary and sufficient conditions for positive definiteness at the end of Section 3.3, we stipulated that the matrix **D** had to be in its unique symmetric form. Cite an example to show that if some quadratic form is represented as

$$q(\mathbf{x}) \equiv \mathbf{x}^T\mathbf{A}\mathbf{x}$$

with **A** not necessarily symmetric, then **A** can satisfy the conditions for positive definiteness,

$$a_{11} > 0, \quad \begin{vmatrix} a_{11} & a_{12} \\ a_{21} & a_{22} \end{vmatrix} > 0, \ldots, |\mathbf{A}| > 0,$$

without $q(\mathbf{x})$ being positive definite.

Section 3.4

3-7. Tell whether each of the following functions is convex, concave, or neither (over all $\mathbf{x}$ in E^n):
(a) $f(x) = 6x - x^2 - 3$
(b) $f(x_1, x_2) = x_1 - x_2$
(c) $f(x_1, x_2) = x_1 x_2 + x_1 + x_2$
(d) $f(x_1, x_2) = 8$
(e) $f(x_1, \ldots, x_n) = -\sum_{i=1}^{n} x_i^2$
(f) $f(x_1, \ldots, x_n) = -x_1^2$
(g) $f(x_1, x_2) = kx_1^2 + kx_1x_2 + kx_2^2$, where $k > 0$
(h) $f(x_1, x_2) = x_1^2 + 2x_1x_2 + 2x_2^2 - 10x_1 - 10x_2$
(i) $f(x_1, x_2) = x_1^2 + 3x_1x_2 + 2x_2^2 - 10x_1 - 10x_2$
(j) $f(x_1, x_2, x_3) = -x_1^2 + 2x_1x_2 - 4x_2^2 + 3x_1x_3 + 6x_2x_3 - 9x_3^2$

3-8. (a) Prove that if $f(\mathbf{x})$ is both concave and convex over all of E^n, then $f(\mathbf{x})$ must be linear.
(b) Prove that a positive multiple of a convex function is convex.

3-9. (a) Prove that if $g(x_1)$ is not a convex function, then $f(x_1, x_2) \equiv g(x_1) + h(x_2)$ cannot be convex.
(b) If $g(x_1)$ is not convex, can $f(x_1, x_2) \equiv g(x_1) + h(x_1, x_2)$ be convex? Either give an example or prove it to be impossible.
(c) If $g(x_1)$ and $h(x_2)$ are convex functions, is $f(x_1, x_2) \equiv g(x_1) + h(x_2)$ necessarily convex? Prove or give a counterexample.

3-10. Prove that if $f(\mathbf{x}) > 0$ and $f(\mathbf{x})$ is concave for all $\mathbf{x}$ in R, then the function $g(\mathbf{x}) \equiv 1/f(\mathbf{x})$ is convex over R.

3-11. Let $g(\mathbf{x}) \equiv \text{Max}\,\{f_1(\mathbf{x}), f_2(\mathbf{x}), \ldots, f_k(\mathbf{x})\}$ be defined for all $\mathbf{x}$ in E^n. Prove or disprove each of the following two statements:
(a) If every $f_i(\mathbf{x})$, $i = 1, \ldots, k$, is a convex function, then so is $g(\mathbf{x})$.
(b) If every $f_i(\mathbf{x})$, $i = 1, \ldots, k$, is a concave function, then so is $g(\mathbf{x})$.

3-12. Let $f(x)$ be a concave function defined for all real numbers x. Show that $g(y) \equiv f(2 - y)$ is a concave function for all y.

3-13. Suppose $f(x)$ is defined for all x in some closed interval I of the real line. Prove that $f(x)$ is a convex function if and only if $d^2f/dx^2 \geqq 0$ for all x in I. [*Hint:* The "if half" can be proved using the mean-value theorem from elementary calculus; to prove the other half, choose any x_1 and x_2, define an intermediate point $y = \lambda x_1 + (1 - \lambda)x_2$, and manipulate the definition (3-9).]

3-14. Consider the linear programming problem

$$\begin{aligned} &\text{Max } z = \mathbf{c}^T\mathbf{x} \\ &\text{subject to } \mathbf{A}\mathbf{x} \leqq \mathbf{b} \\ &\text{and} \qquad\quad \mathbf{x} \geqq \mathbf{0}. \end{aligned}$$

(a) Let $G(\mathbf{c})$ be the optimal value of z considered as a function of the cost vector $\mathbf{c}$,

where all components of $\mathbf{A}$ and $\mathbf{b}$ are assumed to be held fixed. Determine whether G is a concave or a convex function.
(b) Let $F(\mathbf{b})$ be the optimal value of z as a function of $\mathbf{b}$, with $\mathbf{A}$ and $\mathbf{c}$ held fixed. Show that F is concave.

3-15. Given a set of points $\mathbf{x}_1, \ldots, \mathbf{x}_m$ in the convex set X in E^n, show that any convex combination of them is also in X.

Section 3.5

3-16. Which of the following sets of points are convex?
(a) $\{(x_1, x_2) \mid x_1^2 + x_2^2 \leqq 10, x_1 \leqq 2, x_2 \geqq 0\}$
(b) $\{(x_1, x_2) \mid x_1^2 + x_2^2 \leqq 10, x_1 = 2, x_2 \geqq 0\}$
(c) $\{(x_1, x_2) \mid x_1^2 + x_2^2 = 10, x_1 \leqq 2, x_2 \geqq 0\}$
(d) $\{(x_1, x_2) \mid x_1 - (2 - x_2)^2 \leqq 2, x_1 + x_2 \leqq 2, x_1 \geqq 0, x_2 \geqq 0\}$
(e) $\{(x_1, x_2) \mid -x_1^2 + 4x_1x_2 - 5x_2^2 \geqq -10, x_1^2 + x_1x_2 + x_2^2 - 100x_1 \leqq 1\}$
(f) $\{(x_1, x_2) \mid (x_1 - 1)^2 + x_2^2 \leqq 1, x_1^2 + x_2^2 \geqq 4\}$
(g) $\{(x_1, x_2) \mid x_1^2 + x_2^2 \leqq 1$ *or* $(x_1 - 1)^2 + x_2^2 \leqq 1\}$
(h) $\{\mathbf{x} \mid \mathbf{x}^T\mathbf{D}\mathbf{x} = 100, \mathbf{D}$ positive definite$\}$
(i) $\{\mathbf{x} \mid \mathbf{x}^T\mathbf{D}\mathbf{x} = 0, \mathbf{D}$ negative definite$\}$
(j) $\{\mathbf{x} \mid x_1^2 \geqq x_2^2 \geqq \cdots \geqq x_n^2\}$
(k) $\{x_1 \mid x_1$ an integer$\}$
(l) The empty set (i.e., the set that contains *no* points in E^n)

3-17. Prove Theorem 3.8.

3-18. (a) Let $f(\mathbf{x})$ be defined over a convex set S in E^n. We say that $f(\mathbf{x})$ is *quasi-concave* over S if for any two points $\mathbf{x}_1$ and $\mathbf{x}_2$ in S,

$$f(\lambda\mathbf{x}_1 + (1 - \lambda)\mathbf{x}_2) \geqq \min\{f(\mathbf{x}_1), f(\mathbf{x}_2)\}$$

for all λ, $0 \leqq \lambda \leqq 1$. Show that any concave function is quasi-concave.
(b) Is a quasi-concave curve always unimodal or "one-humped?" Give an example of a nonlinear function of one variable x that is both convex and quasi-concave for all x.
(c) Prove that a function $f(\mathbf{x})$ is quasi-concave if and only if the set of points

$$S = \{\mathbf{x} \mid f(\mathbf{x}) \geqq b\},$$

where b is any real number, is a convex set.
(d) Must the sum of two quasi-concave functions necessarily be quasi-concave? Prove or give a counterexample.

3-19. (a) Let S be a set of points $\mathbf{x}$ in E^n. Suppose that for any two points $\mathbf{x}_1$ and $\mathbf{x}_2$ in S there exists some λ, where $0 < \lambda < 1$, such that the point

$$\mathbf{x}_3 = \lambda\mathbf{x}_1 + (1 - \lambda)\mathbf{x}_2$$

is also in S. Show that S is not necessarily a convex set.

(b) Let S be a set of points $\mathbf{x}$ in E^n. Suppose that there exists some particular λ_0, where $0 < \lambda_0 < 1$, such that for any two points $\mathbf{x}_1$ and $\mathbf{x}_2$ in S the point

$$\mathbf{x}_3 = \lambda_0\mathbf{x}_1 + (1 - \lambda_0)\mathbf{x}_2$$

is also in S. Show that S is not necessarily a convex set.

Section 3.6

3-20. Tell whether each of the following functions is convex, concave, or neither over the indicated region:
(a) $f(x) = x^{-2}$ for $x > 0$
(b) $f(x) = +x^{1/2}$ for $x > 0$
(c) $f(x_1, x_2) = +(x_1 + x_2)^{1/2}$ for $x_1, x_2 > 0$
(d) $f(x_1, x_2) = x_1x_2$ for all (x_1, x_2) in E^2
(e) $f(x_1, x_2) = x_1^2x_2^2$ for all (x_1, x_2) in E^2
(f) $f(x_1, x_2) = (x_1 + x_2)e^{x_1+x_2}$ for $x_1, x_2 > 0$
(g) $f(x_1, x_2) = x_1^2(x_2^2 + 4)$ for all (x_1, x_2) such that $x_1^2 + x_2^2 \leqq 4$

3-21. Over what interval on the real line is the "unit normal" probability density function $f(x) = (2\pi)^{-1/2}e^{-x^2/2}$ concave?

3-22. Provide an example in E^1 of each of the following (the interval over which convexity or concavity holds need not include all of E^1):
(a) A concave function $f(x)$ whose reciprocal $g(x) \equiv 1/f(x)$ is concave.
(b) A concave function whose reciprocal is convex.
(c) A convex function whose reciprocal is convex.

3-23. Suppose $f(x)$ is a convex function defined over all of E^1, and let $g(x) \equiv f(f(x))$. Under what circumstances is $g(x)$ convex? Test your answer for $f(x) = x^2$, $f(x) = e^{-x}$, and $f(x) = 1 - x$.

4

Classical Unconstrained Optimization

4.1 VARIETIES OF MATHEMATICAL PROGRAMS

For the rest of this book we shall be concerned with the general mathematical programming problem:

$$\text{(4-1)} \qquad \text{Max or Min } f(\mathbf{x}), \qquad \mathbf{x} = (x_1, \ldots, x_n),$$

$$\text{(4-2)} \qquad \text{subject to } g_i(\mathbf{x})\{\leqq, =, \geqq\}b_i, \qquad i = 1, \ldots, m,$$

where f and the g_i are scalar-valued functions and the b_i are real numbers. If any nonnegativity restrictions $x_j \geqq 0$ are present, they are included in (4-2); depending on the problem-solving algorithm being used, they may or may not receive the same treatment as the other constraints. Optimization problems having no constraints—that is, $m = 0$ in the definition above—are also of interest and will be discussed in some detail in this chapter, but it should be noted that many writers reserve the term "mathematical program" for constrained optimization problems only.

The reader is reminded that, except in the discussion of experimental seeking in Chapter 5, we shall be excluding from consideration all problems that involve integer-valued or discrete-valued variables. In addition, *all functions $f(\mathbf{x})$ and $g_i(\mathbf{x})$ will be assumed to be continuous and to have continuous first partial derivatives.* Fortunately, the functional relationships that arise in real-world engineering and management problems almost always satisfy these continuity conditions anyway, and those that do not can sometimes be judiciously transformed. For example, the constraint

$$g_i(x_1, x_2) = \min(x_1, 5) - x_2 \geqq 0,$$

whose partial derivative with respect to x_1 is discontinuous at $x_1 = 5$, is equivalent to the pair

$$g_i'(x_1, x_2) = x_1 - x_2 \geqq 0$$

and

$$g_i''(x_1, x_2) = 5 - x_2 \geqq 0.$$

As a second example, the constraint

$$\max(x_1, 5) - x_2 \geqq 0,$$

which requires that

$$\textit{either } x_1 - x_2 \geqq 0 \tag{4-3}$$

$$\textit{or } 5 - x_2 \geqq 0 \tag{4-4}$$

(or both) can be dealt with by solving the optimization problem twice, once using (4-3) and once using (4-4), and choosing the better of the two optimal solutions.

Depending on the nature of their constraints, mathematical programs may be divided into three broad classes:

1. Unconstrained optimization problems,
2. Problems subject only to equality constraints, and
3. Problems subject to inequality constraints (with or without equality constraints as well).

This classification would not have been particularly useful in the special case of linear programming. An unconstrained linear program would, of course, be trivially unbounded, while all problems in the second and third categories are treated in exactly the same way; i.e., they are converted to standard form (2-22) and then solved by the simplex method.

For optimization problems in general, the three classes are arranged in order of increasing difficulty. The first type should be quite familiar to the reader, having been introduced in freshman calculus; as we shall see, however, it is not necessarily trivial. The second type can, in theory, be reduced to the first. This is done by solving a constraint equation for one variable in terms of the others, using the resulting expression to eliminate the chosen variable from the objective and all other constraints, and repeating this process until no constraints remain. However, these successive substitutions usually lead to extremely complicated or unmanageable equations and are in any case quite tedious. Lagrange multipliers, to be discussed in the early sections of Chapter 6, provide an easier and more efficient means of attacking problems with equality constraints only.

The third type of optimization problem, involving inequality constraints, is the most difficult to deal with, both theoretically and computationally. The solution methods that have been developed during the post-World War II period have almost all hinged in some way upon a generalization of the classical concept of Lagrange multipliers—or, viewed in another way, upon a generalization of the modern concept of dual variables. Inequality-constrained optimization problems are by far the most frequently occurring type and are thus the principal concern of mathematical programming; they will occupy our attention almost exclusively throughout the last several chapters of this text.

4.2 UNCONSTRAINED OPTIMIZATION

For the remainder of this chapter we shall be considering the general unconstrained optimization problem

$$\text{Max } f(\mathbf{x}) \text{ over all } \mathbf{x} \text{ in } E^n. \tag{4-5}$$

Although most problems arising in operations research have at least a few constraints—with nonnegativity restrictions being the type most often encountered—there are certain contexts in which the problem (4-5) does occur. Probably the most important direct application for unconstrained optimization is the fitting of statistical data to *linear* and *nonlinear regression* models. Suppose we wish to examine the effects of one or more independent variables $x_1, \ldots, x_n$ on a single dependent variable y; for example, y might be the total annual sales of some product and the x_j the dollar amounts spent annually on advertising in each of several different media. Assume that we have available as data m different joint observations of the set of variables $[y, x_1, \ldots, x_n]$, with the ith set of observed values being denoted $[Y_i, X_{i1}, \ldots, X_{in}]$, and assume further that we are willing (provisionally, at least) to postulate a specific functional relationship among the variables:

$$y = f(x_1, \ldots, x_n, \beta_1, \ldots, \beta_p) + u,$$

where the β_j are constant parameters whose values are to be estimated, and u is a random disturbance term with an expected value of zero. The above relationship is known to statisticians and economists as a regression model; it is frequently though not always linear in form:

$$y = \beta_0 + \beta_1 x_1 + \cdots + \beta_n x_n + u.$$

Given some particular regression model with values $\beta_1, \dots, \beta_p$ specified for the parameters, the expected value of y, based on the model, that should have accompanied the ith set of observations $X_{i1}, \dots, X_{in}$ would be $\hat{Y}_i = f(X_{i1}, \dots, X_{in}, \beta_1, \dots, \beta_p)$; in the case of the linear model this would be

$$\hat{Y}_i = \beta_0 + \beta_1 X_{i1} + \cdots + \beta_n X_{in}.$$

However, the value actually observed was Y_i, which implies an error or *residual* of

$$e_i = Y_i - \hat{Y}_i.$$

Roughly speaking, the "better" the choice of values for the parameters β_j, the smaller in magnitude these residuals will tend to be. It can be argued that for any given set of data the best parameter values are those that cause the sum of the squares of the residuals to be as small as possible. These *least-squares* parameter values are thus obtained by minimizing, in the case of the linear model,

$$\sum_{i=1}^{m} e_i^2 \equiv \sum_{i=1}^{m} (Y_i - \beta_0 - \beta_1 X_{i1} - \cdots - \beta_n X_{in})^2,$$

which is a function of the β_j only. If no restrictions are placed on the choice of parameter values, this is an unconstrained optimization problem; one computational algorithm for solving it is described in [32].

Having presented one example of a naturally occurring type of unconstrained optimization problem, we hasten to add that such problems do not in fact arise very frequently in the practice of operations research and its related disciplines. The major importance of the theory and techniques of unconstrained optimization derives from the fundamental role they play in a great many solution algorithms for *constrained* mathematical programs. A substantial number of these algorithms, for example, are built around a simple iterative strategy in which a crucial unconstrained optimization technique—maximization along a straight line—is employed at each stage. Other important solution methods attack constrained mathematical programs by converting them into sequences of unconstrained problems. The point we wish to make at this juncture is that the theoretical and computational aspects of unconstrained optimization are important not only for themselves but also for their applications throughout all of mathematical programming. This is the reason for the fairly extensive coverage to be given to them in the following sections.

Let us now return to the unconstrained problem in its general form,[1]

(4-5) $$\text{Max } f(\mathbf{x}) \text{ over all } \mathbf{x} \text{ in } E^n.$$

[1]As usual, we do not lose generality by restricting the discussion to maximization, inasmuch as we can minimize any $f(\mathbf{x})$ by maximizing its negative.

The distinguishing characteristic of this problem is that any $\mathbf{x}$ in E^n is a feasible point; there are no boundary surfaces whose presence might create constrained local maxima. It then follows that *constrained and unconstrained local maxima are equivalent*, so that the optimal solution to the problem (4-5), provided that it occurs at a finite point, must occur at an unconstrained local maximum.

We stated in Section 4.1 that of the three classes of mathematical programs those having no constraints are the least difficult to solve. In general this is true because all first partial derivatives of the objective function $f(\mathbf{x})$ are known to vanish at any local maximum, and therefore at the (finite) global maximum. The proof of this statement is precisely analogous to our proof at the beginning of Section 2.2 that the derivative of a function of one variable vanishes at an optimum. We have justified the following result:

Theorem 4.1: If $\mathbf{x}_0$ *is a (finite) optimal solution to the unconstrained maximization problem* (4-5), *then*

$$\frac{\delta f(\mathbf{x}_0)}{\delta x_j} = 0, \qquad j = 1, \ldots, n. \tag{4-6}$$

Moreover, (4-6) *holds if* $\mathbf{x}_0$ *is any local optimum of* $f(\mathbf{x})$.

The various points $\mathbf{x}_0$ that satisfy (4-6) are called the *critical points* of the function $f(\mathbf{x})$; they include local optima of both kinds as well as *saddle points*, which are maxima with respect to movement in some of the x_j-directions and minima with respect to movement in others.

Equation (4-6) gives a set of *necessary* conditions for $\mathbf{x}_0$ to be an unconstrained local maximum (or minimum) of $f(\mathbf{x})$. *Sufficient* conditions, which are somewhat more complicated, are stated in the following:

Theorem 4.2: Let the function $f(\mathbf{x})$, $\mathbf{x}$ *in* E^n, *have continuous second partial derivatives. Then if all first partial derivatives of* $f(\mathbf{x})$ *vanish at* $\mathbf{x}_0$ *and if the Hessian matrix evaluated at* $\mathbf{x}_0$ *is negative (positive) definite, then* $\mathbf{x}_0$ *is an unconstrained local maximum (minimum) of* $f(\mathbf{x})$.

These sufficient conditions for a local maximum are deduced rather easily from the Taylor expansion of $f(\mathbf{x})$ in the vicinity of $\mathbf{x}_0$:

$$\begin{aligned} f(\mathbf{x}) = {} & f(\mathbf{x}_0) + \nabla \mathbf{f}^T(\mathbf{x}_0) \cdot (\mathbf{x} - \mathbf{x}_0) \\ & + \tfrac{1}{2}(\mathbf{x} - \mathbf{x}_0)^T \mathbf{H}(\mathbf{x}_0) \cdot (\mathbf{x} - \mathbf{x}_0) + \cdots \text{ (higher-order terms)}, \end{aligned} \tag{2-15}$$

where the gradient $\nabla \mathbf{f}$ and Hessian $\mathbf{H}$ are given by (2-13) and (2-14). By hypothesis

$$\nabla \mathbf{f}(\mathbf{x}_0) \equiv \left(\frac{\delta f(\mathbf{x}_0)}{\delta x_1}, \ldots, \frac{\delta f(\mathbf{x}_0)}{\delta x_n}\right) = \mathbf{0}$$

and (2-15) yields

(4-7) $$f(\mathbf{x}) - f(\mathbf{x}_0) = \tfrac{1}{2}(\mathbf{x} - \mathbf{x}_0)^T\mathbf{H}(\mathbf{x}_0) \cdot (\mathbf{x} - \mathbf{x}_0) + \cdots \text{ (higher-order terms)}.$$

If we consider points $\mathbf{x}$ within a sufficiently small distance δ of $\mathbf{x}_0$, the "higher-order terms"—in each of which the factor $\mathbf{x} - \mathbf{x}_0$ appears three or more times—are dominated by the second-order term and can be ignored. Then, because $\mathbf{H}(\mathbf{x}_0)$ is negative definite, the right-hand side of (4-7) is negative for any $\mathbf{x}$ close to $\mathbf{x}_0$; thus $f(\mathbf{x}) < f(\mathbf{x}_0)$ and $\mathbf{x}_0$ satisfies the definition of an unconstrained local maximum. The proof for a local minimum would be analogous.

Let us write out the conditions of Theorem 4.2 for the cases of one and two variables.

One Variable: The point x_0 *is an unconstrained local maximum (minimum) of the function* $f(x)$ *if*

$$\frac{df}{dx} = 0$$

and

$$\frac{d^2f}{dx^2} < 0 \qquad (> 0)$$

hold at x_0.

Two Variables: The point $\mathbf{x}_0$ *is an unconstrained local maximum (minimum) of the function* $f(x_1, x_2)$ *if*

$$\frac{\delta f}{\delta x_1} = \frac{\delta f}{\delta x_2} = 0$$

$$\frac{\delta^2 f}{\delta x_1^2} < 0 \qquad (> 0)$$

and

$$\frac{\delta^2 f}{\delta x_1^2} \cdot \frac{\delta^2 f}{\delta x_2^2} - \left[\frac{\delta^2 f}{\delta x_1\, \delta x_2}\right]^2 > 0$$

all hold at $\mathbf{x}_0$.

These conditions are usually presented in introductory calculus courses; they are quite useful in solving small optimization problems by hand.

We now return to the general n-variable case for further consideration of the conditions established in Theorem 4.2. If $\mathbf{x}_0$ is to be a local maximum of $f(\mathbf{x})$, then from Theorem 4.1 the gradient $\nabla\mathbf{f}(\mathbf{x}_0)$ must vanish, and it is clear from what remains of the Taylor expansion—see (4-7)—that

(4-8) $$(\mathbf{x} - \mathbf{x}_0)^T\mathbf{H}(\mathbf{x}_0) \cdot (\mathbf{x} - \mathbf{x}_0) \leqq 0$$

for all $\mathbf{x}$ within a small distance of $\mathbf{x}_0$, and therefore for all values of $\mathbf{x} - \mathbf{x}_0$. This simple corollary result is important enough to be stated as a theorem:

Theorem 4.3: If $\mathbf{x}_0$ *is an unconstrained local maximum (minimum) of* $f(\mathbf{x})$, $\mathbf{x}$ *in* E^n, *then the Hessian matrix evaluated at* $\mathbf{x}_0$ *must be negative (positive) definite or semidefinite.*

As an example of the semidefinite case, the function $f(x_1, x_2) = (x_1 - x_2)^2$ has a local minimum at the origin (and at any other point satisfying $x_1 = x_2$); however, its Hessian matrix at every point in E^2 is

$$\mathbf{H}(\mathbf{x}) = \begin{bmatrix} 2 & -2 \\ -2 & 2 \end{bmatrix},$$

which is positive semidefinite.

Based on Theorems 4.2 and 4.3, we can construct the following logical table, which we present in terms of unconstrained local maxima only:

	$\mathbf{x}_0$ *is a local maximum*	$\mathbf{x}_0$ *is not a local maximum*
$\mathbf{H}(\mathbf{x}_0)$ *negative definite and* $\nabla \mathbf{f}(\mathbf{x}_0) = \mathbf{0}$	Can occur	Impossible
$\mathbf{H}(\mathbf{x}_0)$ *negative semidefinite and* $\nabla \mathbf{f}(\mathbf{x}_0) = \mathbf{0}$	Can occur	Can occur
$\mathbf{H}(\mathbf{x}_0)$ *indefinite and* $\nabla \mathbf{f}(\mathbf{x}_0) = \mathbf{0}$	Impossible	Can occur

The table would be equally valid if "positive" and "minimum" were substituted for "negative" and "maximum" wherever those words appear.

It should be noted that the Hessian can be negative semidefinite at critical points that are *not* local maxima. To see why this is true, suppose we assume that for some point $\hat{\mathbf{x}}$ the gradient vanishes and the Hessian is negative semidefinite. Following the proof of Theorem 4.2, we write out the Taylor expansion (2-15) and reduce it to

$$f(\mathbf{x}) - f(\hat{\mathbf{x}}) = \tfrac{1}{2}(\mathbf{x} - \hat{\mathbf{x}})^T\mathbf{H}(\hat{\mathbf{x}}) \cdot (\mathbf{x} - \hat{\mathbf{x}}) + \cdots \text{(higher-order terms)}.$$

But by hypothesis there exists some $\mathbf{y} \neq \mathbf{0}$ such that $\mathbf{y}^T\mathbf{H}(\hat{\mathbf{x}})\mathbf{y} = 0$. Therefore for any $\mathbf{x} = \hat{\mathbf{x}} + \epsilon\mathbf{y}$, where $\epsilon \neq 0$, the second-order term vanishes and does *not* dominate the higher-order terms, and the proof of Theorem 4.2 breaks down. In such a case it is the third-order term which dominates (unless it, too, vanishes) and which must be tested in order to determine whether or not $\mathbf{x}_0$ is a local optimum.

For example, the function $f(x_1, x_2) = -x_1^2 + x_2^3$ has

$$\nabla \mathbf{f}(\mathbf{x}) = (-2x_1, 3x_2^2) \quad \text{and} \quad \mathbf{H}(\mathbf{x}) = \begin{bmatrix} -2 & 0 \\ 0 & 6x_2 \end{bmatrix}.$$

At $x_1 = 0$, $x_2 = 0$, the gradient vanishes and the Hessian is negative semidefinite, yet the origin is *not* a local maximum—any movement away from it in the positive x_2-direction increases f. A simpler example is the general power function

$$f(x) = x^k, \qquad k > 2, k \text{ an integer.}$$

Under what circumstances is the critical point $x = 0$ a local minimum? The Taylor expansion of $f(x)$ about $x = 0$ (i.e., the Taylor expansion in powers of x) is simply the function itself, and the "dominant" term, by default, is x^k. This term is nonnegative for all x in the vicinity of $x = 0$ if and only if the integer k is *even.* Thus $x = 0$ is an unconstrained local minimum for $f(x) = x^{14}$, but not for $f(x) = x^{15}$.

In light of Theorem 3.10, which establishes that a function is concave wherever its Hessian matrix is negative (semi)definite, Theorem 4.3 suggests another important result:

Theorem 4.4: Any function $f(\mathbf{x})$, $\mathbf{x}$ *in* E^n, *having continuous first and second partial derivatives is concave (convex) in a sufficiently small region around*[2] *any unconstrained local maximum (minimum)* $\mathbf{x}_0$.

From Theorem 4.3, the Hessian $\mathbf{H}(\mathbf{x}_0)$ must be negative definite or semidefinite. In the former case, which is by far the more common, the proof is straightforward. We have $\mathbf{y}^T\mathbf{H}(\mathbf{x}_0)\mathbf{y} < 0$ for every $\mathbf{y} \neq \mathbf{0}$ in E^n, and because the Hessian is continuous, $\mathbf{y}^T\mathbf{H}(\mathbf{x})\mathbf{y}$ must also be negative for all $\mathbf{x}$ within a sufficiently small distance of $\mathbf{x}_0$. Therefore $\mathbf{H}(\mathbf{x})$ is negative definite over a region including $\mathbf{x}_0$, and the concavity of $f(\mathbf{x})$ in that region follows from Theorem 3.10. If $\mathbf{H}(\mathbf{x}_0)$ is negative *semi*definite, the proof is more difficult and will not be given here.

Returning now to the unconstrained optimization problem (4-5), we can outline a rather elementary "algorithm" for solving it that should be quite familiar to the reader:

1. Test to see whether $f(\mathbf{x})$ can be made to approach infinity by displacing $\mathbf{x}$ infinitely far in any fixed direction. If it can, the problem is solved; if not, go on to step 2.

2. Since $f(\mathbf{x})$ is assumed to be continuous, its maximum value must be finite. Obtain all the first partial derivatives of $f(\mathbf{x})$ and solve the following system of n nonlinear equations in the n unknowns x_j:

$$\frac{\delta f(\mathbf{x})}{\delta x_j} = 0, \qquad j = 1, \ldots, n. \tag{4-9}$$

[2]When we speak of a function being concave or having some other property in a region "around" a point $\mathbf{x}_0$, we mean that $\mathbf{x}_0$ must be an *interior point* of that region. Thus the function $f(x_1, x_2) = x_1^3 + x_2^3$ cannot be said to be convex in any region around the origin, because convexity holds only in the first quadrant.

The solution to (4-9) that yields the greatest value of $f(\mathbf{x})$ is the desired global maximum.

Recall that the various solutions of (4-9) are known as the *critical points* of $f(\mathbf{x})$; Theorem 4.1 guarantees that the global maximum will be included among them. This general approach, in which the problem of finding the optimal value of an objective function is transformed into one of solving a set of simultaneous equations, is classical in origin, being primarily associated with the name of Euler (1707–1783). Note that the sufficiency conditions of Theorem 4.2 need not be applied at any stage, unless for some reason it is desired to identify all local maxima. In fact, as we shall see, sufficiency tests are not required by any of the widely used algorithms for solving the unconstrained optimization problem (4-5).

We shall not dwell on the question of how to determine in step 1 that there is no direction in which infinite displacement of $\mathbf{x}$ can lead to an unbounded optimum. For most problems arising in the practice of operations research, optimal solutions are finite and the question is academic. Problems in which the global optimum has a finite objective value but lies at an infinitely distant point are perhaps slightly less unusual. However, in such cases the partial derivatives will all approach zero as the optimum is approached, so that the solution will be obtained in step 2. An example would be the maximization of $f(x) = 1 - e^{-x}$; the global maximum of $f(x)$ is 1, attained as x approaches infinity, and this would be discovered by setting $df/dx = e^{-x} = 0$.

As a more straightforward example of the general methodology, let us find the values of x_1 and x_2 that minimize

$$f(x_1, x_2) = e^{x_1 - x_2} + x_1^2 + x_2^2.$$

Since $f(x_1, x_2)$ can never be less than zero, its minimum value is finite and the optimal solution must lie at a point that satisfies the necessary conditions (4-9):

$$\frac{\delta f}{\delta x_1} = e^{x_1 - x_2} + 2x_1 = 0$$

and

$$\frac{\delta f}{\delta x_2} = -e^{x_1 - x_2} + 2x_2 = 0.$$

These equations lead to

$$e^{2x_1} = -2x_1, \tag{4-10}$$

and, using a log table, we discover by trial and error that

$$x_1 = -.2836, \qquad x_2 = .2836, \qquad f(x_1, x_2) = .7281.$$

Since this is the only critical point, it must be the global minimum.

4.3 SIMULTANEOUS NONLINEAR EQUATIONS

Although the procedure just described for solving the unconstrained optimization problem (4-5) is perfectly satisfactory in theory, it leaves an important computational question unanswered: Exactly how do we obtain the solutions to a system of simultaneous nonlinear equations? Even *one* equation in *one* unknown cannot always be solved analytically; this was true, for example, of Eq. (4-10). In such cases the value of the variable must be obtained by numerical iteration.

Several classical procedures for solving systems of nonlinear equations are presented in [23] and [36]. Typical of them is the *Newton-Raphson method*: Given a set of equations

$$f_i(\mathbf{x}) = 0, \qquad i = 1, \ldots, n, \tag{4-11}$$

where $\mathbf{x} = (x_1, \ldots, x_n)$, choose a starting point $\mathbf{x}_0$ and perform a Taylor expansion of each function about that point, truncating after the first-order terms:

$$\begin{cases} f_1(\mathbf{x}_0) + \dfrac{\delta f_1(\mathbf{x}_0)}{\delta x_1}\Delta x_1 + \dfrac{\delta f_1(\mathbf{x}_0)}{\delta x_2}\Delta x_2 + \cdots + \dfrac{\delta f_1(\mathbf{x}_0)}{\delta x_n}\Delta x_n = 0, \\ f_2(\mathbf{x}_0) + \dfrac{\delta f_2(\mathbf{x}_0)}{\delta x_1}\Delta x_1 + \dfrac{\delta f_2(\mathbf{x}_0)}{\delta x_2}\Delta x_2 + \cdots + \dfrac{\delta f_2(\mathbf{x}_0)}{\delta x_n}\Delta x_n = 0, \\ \cdots\cdots\cdots\cdots \\ f_n(\mathbf{x}_0) + \dfrac{\delta f_n(\mathbf{x}_0)}{\delta x_1}\Delta x_1 + \dfrac{\delta f_n(\mathbf{x}_0)}{\delta x_2}\Delta x_2 + \cdots + \dfrac{\delta f_n(\mathbf{x}_0)}{\delta x_n}\Delta x_n = 0, \end{cases} \tag{4-12}$$

where Δx_j has been used in place of $x_j - x_{0j}$ in (2-12). Since the values of all functions and all partial derivatives at $\mathbf{x}_0$ can be computed, this is a system of n simultaneous *linear* equations in the n unknowns Δx_j. These first-order differences are the adjustments to be made on the starting point $\mathbf{x}_0$ in order to produce its successor $\mathbf{x}_1$:

$$x_{1j} = x_{0j} + \Delta x_j, \qquad j = 1, \ldots, n.$$

In effect, we are treating each first-order Taylor expansion as if it were an exact representation of its function—i.e., as if the value of its remainder term were zero—at all points $\mathbf{x}$. If that were true, the point $\mathbf{x}_1$ would satisfy (4-11) exactly; in reality, of course, this will not happen, but hopefully $\mathbf{x}_1$

will be closer to the desired solution than was its predecessor $\mathbf{x}_0$. The process is then repeated, with $\mathbf{x}_1$ being used in place of $\mathbf{x}_0$ in (4-12), and repeated again, until the corrections Δx_j are found to be sufficiently close to zero or, equivalently, until a point $\mathbf{x}_k$ is obtained which approximately satisfies (4-11). If we know or suspect that there may be other solutions to (4-11), we choose another starting point and apply the Newton-Raphson method all over again, continuing to do so until we are confident that all solutions—or at least all solutions that could possibly be of interest—have been found.

As an example, consider the following unconstrained optimization problem:

$$\text{(4-13)} \qquad \text{Min } f(x_1, x_2) = x_1^2 + x_1^2x_2^2 + 3x_2^4.$$

Since $f(x_1, x_2)$ can never be less than zero, we know that its minimum value is finite and must occur at a point where the first partial derivatives vanish:

$$f_1(x_1, x_2) \equiv \frac{\delta f(x_1, x_2)}{\delta x_1} = 2x_1 + 2x_1x_2^2 = 0$$

and

$$f_2(x_1, x_2) \equiv \frac{\delta f(x_1, x_2)}{\delta x_2} = 2x_1^2x_2 + 12x_2^3 = 0.$$

The optimal solution, of course, lies at the origin, but let us solve this pair of nonlinear equations via Newton-Raphson iteration. The first partial derivatives of f_1 and f_2 are

$$\frac{\delta f_1}{\delta x_1} = 2 + 2x_2^2 \qquad \frac{\delta f_1}{\delta x_2} = 4x_1x_2$$

$$\frac{\delta f_2}{\delta x_1} = 4x_1x_2 \qquad \frac{\delta f_2}{\delta x_2} = 2x_1^2 + 36x_2^2$$

If we take $\mathbf{x}_0 = (1, 1)$ as our starting point, the Taylor-expansion equations (4-12) become

$$4 + 4\Delta x_1 + 4\Delta x_2 = 0$$

and

$$14 + 4\Delta x_1 + 38\Delta x_2 = 0.$$

Solving, we find

$$\Delta x_1 = -12/17, \qquad \Delta x_2 = -5/17,$$

and the new point is therefore

$$\mathbf{x}_1 = (.294, \ \ .706), \qquad \text{with } f(\mathbf{x}_1) = .874.$$

The next few iterations produce

$$\mathbf{x}_2 = (.064, .477), \qquad f(\mathbf{x}_2) = .160;$$
$$\mathbf{x}_3 = (.008, .318), \qquad f(\mathbf{x}_3) = .031;$$

and

$$\mathbf{x}_4 = (.000, .212), \qquad f(\mathbf{x}_4) = .006.$$

After 19 iterations a point $\mathbf{x}_{19} = (.000, .000)$ is obtained that is correct to three decimal places; six further iterations are required to achieve four-place accuracy.

Unfortunately, the method given above for solving simultaneous nonlinear equations has a rather severe computational disadvantage: Unless the starting point $\mathbf{x}_0$ is chosen quite close to one of the solutions, the entire process may fail to converge.[3] As an example, consider the problem of finding the solution to the system of one "simultaneous" nonlinear equation

$$f_1(x) = xe^{-x} = 0$$

by Newton-Raphson iteration. The function is graphed in Fig. 4-1 and has only a single finite root at the origin. The derivative is

$$\frac{df_1}{dx} = (1 - x)e^{-x},$$

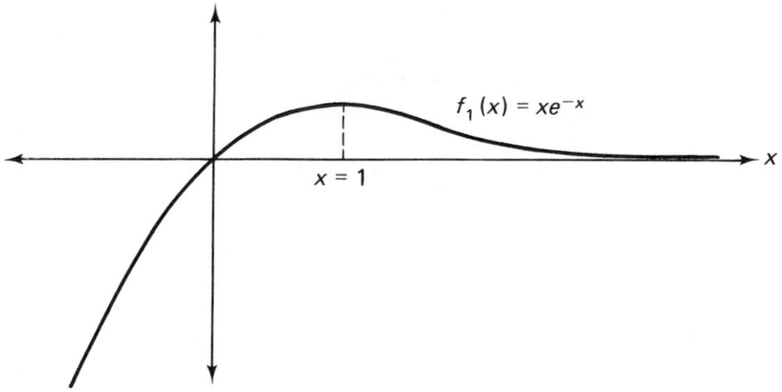

FIGURE 4-1

[3]Given an optimization problem or a set of simultaneous equations having an exact but unknown solution $\mathbf{x}^*$, any iterative computational process that generates a sequence of points $\mathbf{x}_1, \mathbf{x}_2, \ldots$ is said to *converge* to the solution $\mathbf{x}^*$ if, for any positive number δ, there exists an integer p such that $\mathbf{x}_i$ is within a distance δ of $\mathbf{x}^*$ for all $i \geqq p$.

so the first-order Taylor expansion (4-12) becomes

$$x_0e^{-x_0} + (1 - x_0)e^{-x_0}\,\Delta x = 0.$$

Given any current point x_0, the correction in x will be

$$\Delta x = \frac{x_0}{x_0 - 1}. \tag{4-14}$$

Evidently, if the starting point x_0 is greater than 1, the first correction will be positive, so that the new value x_1 will exceed the old; the next correction will again be positive, and so on, with the kth value x_k diverging toward infinity as k increases. On the other hand, if the starting point is less than 1 (including all negative values), the process will eventually converge to the root $x = 0$. The reader can verify this by repeated application of (4-14).

Even when convergence to a local optimum occurs in an orderly and well-behaved manner, however, Newton-Raphson iteration is a very laborious procedure for obtaining points at which the partial derivatives of a function vanish. This is principally because each iteration requires the inversion of an n-by-n matrix. For unconstrained optimization problems of even moderate size, say, 10 variables or more, it is better to forget about solving simultaneous equations and to use instead one of the family of computational approaches known as *direct climbing methods*. These methods, also classical in origin, provide more direct and more efficient means for searching out solutions. They depend crucially on an understanding of the gradient vector, to which we now turn our attention.

4.4 THE GRADIENT VECTOR

The gradient is a concept of cardinal importance in optimization theory and will play a prominent role throughout the rest of this book. Suppose that $f(x_1, x_2, x_3)$ is a scalar-valued function defined at every point in three-dimensional Euclidean space E^3, and consider the vector

$$\nabla \mathbf{f} \equiv \left(\frac{\delta f}{\delta x_1}, \frac{\delta f}{\delta x_2}, \frac{\delta f}{\delta x_3}\right) = \frac{\delta f}{\delta x_1}\mathbf{e}_1 + \frac{\delta f}{\delta x_2}\mathbf{e}_2 + \frac{\delta f}{\delta x_3}\mathbf{e}_3, \tag{4-15}$$

where $\mathbf{e}_i$ is the unit vector pointing in the positive x_i-direction. At any point P in E^3 the component of $\nabla\mathbf{f}$ in the x_i-direction is thus the rate of change of f with respect to x_i, evaluated at P. Consider now a differential displacement away from P in some direction $\mathbf{r}$, defined by

$$\mathbf{dr} \equiv dx_1\mathbf{e}_1 + dx_2\mathbf{e}_2 + dx_3\mathbf{e}_3, \tag{4-16}$$

with dx_i being the component of **dr** in the x_i-direction. Let the length of **dr** be ρ. From analytic geometry the component of $\nabla\mathbf{f}$ in the **r**-direction—that is, the length of the projection of $\nabla\mathbf{f}$ upon **dr**—is their dot product divided by the length of **dr**,

$$\frac{1}{\rho}(\nabla\mathbf{f} \cdot \mathbf{dr}) = \frac{1}{\rho}\left(\frac{\delta f}{\delta x_1}dx_1 + \frac{\delta f}{\delta x_2}dx_2 + \frac{\delta f}{\delta x_3}dx_3\right). \tag{4-17}$$

It is not difficult to show that (4-17) is also equal to the rate of change of f with respect to displacement in the **r**-direction. Suppose we start at P and move a length dx_1 in the direction of $\mathbf{e}_1$, then a length dx_2 in the $\mathbf{e}_2$-direction, and finally a length dx_3 in the $\mathbf{e}_3$-direction. The total displacement equals the vector sum of the three moves, which is

$$dx_1\mathbf{e}_1 + dx_2\mathbf{e}_2 + dx_3\mathbf{e}_3 = \mathbf{dr},$$

and this displacement is known to have length ρ. We next obtain the total albegraic change in f by adding up the individual changes caused by those three moves. Assuming they were so small that the directional rates of change $\delta f/\delta x_i$ remained approximately what they were at P, the total change in f is given by

$$\frac{\delta f}{\delta x_1}dx_1 + \frac{\delta f}{\delta x_2}dx_2 + \frac{\delta f}{\delta x_3}dx_3. \tag{4-18}$$

Division of (4-18) by the length ρ of the overall displacement yields (4-17), which must therefore represent the change in f per unit of displacement in the **r**-direction.

The vector $\nabla\mathbf{f}$ in (4-15) is known as the *gradient* of the function $f(x_1, x_2, x_3)$; it is sometimes denoted by *grad f*. More generally, if $f(\mathbf{x})$ is a scalar-valued function of the variables $x_1, \ldots, x_n$, then its gradient vector is

$$\nabla\mathbf{f} = \left(\frac{df}{\delta x_1}, \ldots, \frac{\delta f}{\delta x_n}\right) = \sum_{i=1}^{n} \frac{\delta f}{\delta x_i}\mathbf{e}_i, \tag{4-19}$$

where the $\mathbf{e}_i$ are unit vectors in E^n. Unless otherwise indicated, the gradient will be understood in this text to be a column vector; also, we shall use $\nabla\mathbf{f}(\hat{\mathbf{x}})$ to denote the gradient of f evaluated at some specific point $\hat{\mathbf{x}}$.

Generalizing from our discussion of the three-variable case, the gradient $\nabla\mathbf{f}(\hat{\mathbf{x}})$ must have the property that *its component in any given direction is equal to the rate of change of $f(\mathbf{x})$ with respect to displacement from $\hat{\mathbf{x}}$ in that direction.* Now, the component of a vector of length ρ in its own direction is ρ, while its component in any other direction is less than ρ. Therefore at any given point **x** *the gradient $\nabla\mathbf{f}$ must point in the direction in which the rate of change of f is greatest and must have length equal to that maximum rate of* *change*. Furthermore, the component of any vector in a perpendicular direction is zero. It follows that $\nabla \mathbf{f}$ is at right angles to any direction in which the (instantaneous) rate of change of f is zero; that is, $\nabla \mathbf{f}$ is perpendicular to all surfaces of the form

$$f(\mathbf{x}) = f(x_1, \ldots, x_n) = k,$$

where k is a constant.

As an example, if f measures distance from the origin in E^3, that is, if

$$f(x_1, x_2, x_3) = x_1^2 + x_2^2 + x_3^2,$$

then the gradient is simply

$$\nabla \mathbf{f} = (2x_1, 2x_2, 2x_3).$$

At any point (x_1, x_2, x_3) it is clear that this vector points directly away from the origin. This makes it perpendicular to all the spheres

$$f(x_1, x_2, x_3) = x_1^2 + x_2^2 + x_3^2 = k,$$

which agrees with our theoretical results.

Another extremely important property of the gradient follows from the fact that the partial derivatives of $f(\mathbf{x})$ are all zero at any unconstrained local optimum. *If $f(\mathbf{x})$ takes on an unconstrained local maximum or minimum at $\mathbf{x}_0$, then its gradient vanishes at $\mathbf{x}_0$; that is, $\nabla \mathbf{f}(\mathbf{x}_0) = \mathbf{0}$.* Of course, by definition the gradient $\nabla \mathbf{f}$ vanishes at all critical points of $f(\mathbf{x})$, saddle points included.

4.5 DIRECT CLIMBING ALGORITHMS

In Section 4.2 we described a general approach for solving the unconstrained optimization problem

$$\text{(4-5)} \qquad \text{Max } f(\mathbf{x}) \text{ over all } \mathbf{x} \text{ in } E^n$$

that was quite elegant and simple, at least in theory: Set the partial derivatives $\delta f/\delta x_j$, $j = 1, \ldots, n$, equal to zero and solve the resulting system of n simultaneous nonlinear equations in the unknowns $x_1, \ldots, x_n$. One of the solutions must then be the global optimum. This type of approach is said to be *indirect*, in the sense that it backs away from the actual problem (4-5) and instead generates an entirely different (although related) system of equations. The optimization problem is then solved by working with these equations rather than with $f(\mathbf{x})$—the general feeling is one of entering through the back door.

By contrast, a *direct* method grapples with the objective function itself in a very straightforward manner. In general, it proceeds by selecting, one by one, a sequence of points in E^n that terminates at or converges to an unconstrained local optimum of $f(\mathbf{x})$. The first point in the sequence is usually chosen by the analyst in accordance with whatever bits of information he may have about the function or the probable location of its optimum [e.g., the general shape of $f(\mathbf{x})$, locations of previously discovered local extrema, and so on]. In the absence of such information the analyst may play a hunch or may simply direct his computer to make a random selection. Thereafter, each point is chosen on the basis of information provided by its predecessor (or predecessors) and in light of some routinized *strategy* or rule for using that information. The overall process thus takes the form of a search which, in the case of maximization, can be thought of as being directed toward the top of a hill. Typically, the "hill climbing" begins at a preselected starting point $\mathbf{x}_0$, proceeds in some straight-line direction to a higher and better point $\mathbf{x}_1$, then in a different direction from $\mathbf{x}_1$ to $\mathbf{x}_2$, and so on, always gaining altitude, until the top of the hill is reached. Inasmuch as there may be several hills, the procedure will in general lead to an unconstrained local maximum that may or may not be the global maximum.

This dichotomy of direct versus indirect approaches will also appear when we study solution methods for *constrained* optimization problems. By introducing an extra set of variables called Lagrange multipliers, it is possible to write a system of simultaneous nonlinear equations that must hold at the optimal solution of any equality-constrained mathematical program. When inequality constraints are present, the optimum must still satisfy a somewhat different set of algebraic relations known as the Kuhn-Tucker conditions. Thus in either case it is possible in theory to obtain the optimal solution indirectly, by working with the necessary conditions rather than with the objective function itself.

In practice, however, indirect approaches are relatively unattractive. The iterative numerical procedures generally available for solving simultaneous nonlinear equations, such as the Newton-Raphson method, tend to be computationally inefficient and in many cases—depending on the character of the functions involved and the choice of starting point—may never converge to a solution at all. Most of the widely used solution methods in mathematical programming take the direct or hill-climbing approach. The ascent to a local maximum is not so simple in an inequality-constrained problem as it is when no constraints are present, because in the former case the movement from one point to the next is usually restricted by the necessity of staying within the feasible region. Thus special computational procedures must be devised for choosing a feasible direction and then for ensuring that displacement in that direction is not carried too far. We shall begin to study these methods in Chapter 7; in the meantime, note that the simplex method takes precisely this hill-climbing approach in solving a linear program.

Let us translate our description of hill climbing into more concrete mathematical language; we restrict our attention to the unconstrained optimization problem (4-5), although our entire discussion will be applicable, with obvious adjustments, to minimization as well as maximization. Given a starting point $\mathbf{x}_0$, a *direct climbing* algorithm seeks an unconstrained local maximum of the function $f(\mathbf{x})$, $\mathbf{x}$ in E^n, by generating a sequence of points $\{\mathbf{x}_i\} \equiv \mathbf{x}_0, \mathbf{x}_1, \mathbf{x}_2, \ldots$ according to some recursion rule

$$\mathbf{x}_{i+1} = \mathbf{x}_i + \theta_i \mathbf{s}_i, \tag{4-20}$$

where $\mathbf{s}_i$ is an n-component vector defining the direction of movement from $\mathbf{x}_i$ to $\mathbf{x}_{i+1}$, and θ_i is a nonnegative scalar that determines the distance moved in the $\mathbf{s}_i$-direction (the actual distance from $\mathbf{x}_i$ to $\mathbf{x}_{i+1}$ is, of course, $\theta_i\,|\mathbf{s}_i|$, where $|\mathbf{s}_i|$ is the magnitude of $\mathbf{s}_i$). At each stage, after the point $\mathbf{x}_i$ has been determined, values are computed for the direction vector $\mathbf{s}_i$ and then for the distance parameter θ_i according to some prespecified general procedure. Usually $\mathbf{s}_i$ is an algebraic function of the first and/or second partial derivatives of $f(\mathbf{x})$ at $\mathbf{x}_i$, whereas θ_i may be determined either via substitution into a formula or via an iterative computational scheme. The choices of $\mathbf{s}_i$ and θ_i are designed to guarantee that the objective value improves at each stage; that is, in the case of maximization, $f(\mathbf{x}_{i+1}) > f(\mathbf{x}_i)$ for all i. Incidentally, we adopt the convention that $\mathbf{x}_i$ is the point determined on the ith iteration.

In any actual application of a direct climbing algorithm, termination occurs at a point $\mathbf{x}_p$, which will hopefully prove to be a local maximum, when the magnitude of the next direction vector $\mathbf{s}_p$ is found to be sufficiently close to zero, i.e., less than some small positive number δ previously chosen by the analyst. In most of the algorithms commonly used for solving unconstrained optimization problems, the direction vector $\mathbf{s}_i$ is either the gradient $\nabla\mathbf{f}(\mathbf{x}_i)$ or a matrix multiple of it. Thus the condition for termination is tantamount to the approximate vanishing of the gradient—recall from Theorem 4.1 that $\nabla\mathbf{f}$ must vanish at every unconstrained local optimum.[4] The criterion for termination is referred to as the *stopping rule*, and when termination occurs both the algorithm and the sequence of points $\{\mathbf{x}_i\}$ are said to have *converged* (this term will be more carefully and rigorously defined in the next section).

A climbing procedure of the type just described will in general, though not always, lead to an unconstrained local maximum of the function $f(\mathbf{x})$.

[4]Sometimes an additional stopping rule of the following form may also be imposed: Terminate at the point $\mathbf{x}_p$ when the distance from $\mathbf{x}_{p-1}$ to $\mathbf{x}_p$ is less than some small positive number ϵ. Such a rule prevents the computer from generating an infinite number of iterations in cases where the sequence $\mathbf{x}_0, \mathbf{x}_1, \mathbf{x}_2, \ldots$ approaches a limiting point $\mathbf{x}^*$ at which the gradient does *not* vanish. This type of pathology is of little practical interest in unconstrained optimization. It does, however, pose serious difficulties in the solving of *constrained* mathematical programs; in the latter connection it is associated with the phenomenon of "zigzagging" or "jamming," which will be discussed in later chapters.

To solve the optimization problem (4-5), however, it is necessary to obtain *all* local maxima of $f(\mathbf{x})$—or, at least, to obtain all those that cannot be logically or mathematically eliminated. Therefore, unless the function is concave, in which case the first local maximum discovered is necessarily the global maximum, the climbing procedure will have to be employed repeatedly, using different starting points each time. If the problem is being solved by hand or by man-machine interaction—a mode of investigation in which the analyst chooses a starting point, presses the console button to allow his computer to climb to a local maximum, then chooses another starting point and repeats—the search can be conducted intelligently, exploiting at each stage the information already gained about the objective function and its behavior. In general, the best tactics are to examine closely those regions where the largest objective values have been observed and to try to eliminate other regions where the global maximum cannot be; this type of approach thus affords great scope for the analyst's deductive powers and ingenuity. If, on the other hand, the problem is being solved entirely by computer, the starting points must be selected automatically. Some computer codes are designed to explore more intensively those regions in which large values of the objective function are discovered; other codes merely proceed from a preselected pattern of starting points, and some even choose starting points at random.

4.6 CONVERGENCE OF DIRECT CLIMBING ALGORITHMS

We have not yet touched upon a very important computational aspect of these direct solution methods. In general, it cannot automatically be assumed that, for any function $f(\mathbf{x})$, a given climbing algorithm will necessarily arrive within an arbitrarily small distance of a local optimum after a finite number of iterations. It is not at all uncommon, for example, for the sequence of points $\{\mathbf{x}_i\} \equiv \mathbf{x}_0, \mathbf{x}_1, \mathbf{x}_2, \ldots, \mathbf{x}_i, \mathbf{x}_{i+1}, \ldots$ generated by a computational algorithm to diverge toward infinity—we saw in Section 4.3 how this could occur in the case of Newton-Raphson iteration. Another possibility is that, although a finite solution is obtained, it may prove to be some type of saddle point rather than a local optimum. In fact, we cannot even assume that any finite solution obtained will at least be a critical point—rigorous theoretical proofs must be furnished.

To discuss all these matters properly, we must begin with two important definitions from mathematical analysis:

Let $\{\mathbf{x}_i\} \equiv \mathbf{x}_0, \mathbf{x}_1, \mathbf{x}_2, \ldots$ be any infinite sequence of points in E^n. Then the point $\mathbf{x}^*$ in E^n is a *limit point* of $\{\mathbf{x}_i\}$ if, for any $\delta > 0$, there exists a positive integer N such that $|\mathbf{x}^* - \mathbf{x}_i| < \delta$ for all $i \geqq N$, where $|\mathbf{x}^* - \mathbf{x}_i|$ is the Euclidean distance between $\mathbf{x}^*$ and $\mathbf{x}_i$.

The infinite sequence $\{\mathbf{x}_i\}$ is said to *converge*, or to converge to the point $\mathbf{x}^*$, if it possesses a limit point $\mathbf{x}^*$.

Intuitively, as $i \longrightarrow \infty$ the points of the sequence $\{\mathbf{x}_i\}$ are getting closer and closer to $\mathbf{x}^*$; moreover, there are an infinite number of them within any given positive distance of $\mathbf{x}^*$. The reader should have little difficulty in proving that no sequence can have more than one limit point.[5]

A sequence that does not converge is said to *diverge*. This may mean either that its terms become infinite in magnitude (i.e., that $|\mathbf{x}_i| \longrightarrow \infty$ as $i \longrightarrow \infty$) or that after a finite number of terms the sequence begins to *oscillate*, or cycle, through two or more endlessly repeated points. For example, the sequence

$$\mathbf{x}_0 = (3, 3), \mathbf{x}_1 = (1, 2), \mathbf{x}_2 = (0, 1), \mathbf{x}_3 = (1, 0), \mathbf{x}_4 = (0, 1), \mathbf{x}_5 = (1, 0), \ldots$$

begins to oscillate with the term $\mathbf{x}_4$; it has no limit point—no positive N exists for, say, $\delta = .01$—and is therefore nonconvergent. In theory, oscillating sequences cannot be generated by direct climbing algorithms, whose rules are intended to ensure that the sequence of objective values $f(\mathbf{x}_0)$, $f(\mathbf{x}_1)$, $f(\mathbf{x}_2), \ldots$ increases or decreases monotonically. In practice, however, the computational approximations that are usually employed may be just inaccurate enough to cause a worsening of the objective value at some stage, thereby raising the possibility of oscillating or cycling.

Having defined what we mean by the convergence of a sequence of points, we now turn our attention to sequences generated by direct climbing algorithms. In general, as was noted earlier, whenever a direct climbing algorithm is used to solve an actual problem, an artificial stopping rule is superimposed on it that causes termination when the magnitude of the direction vector $|\mathbf{s}_i|$ becomes sufficiently small. To consider these algorithms from a theoretical point of view, however, it is essential to examine the sequences of points $\{\mathbf{x}_i\}$ that they generate in the absence of "extra" stopping conditions. That is, we wish to consider direct climbing algorithms as they would operate with the following "ideal" stopping rule: Terminate at the point $\mathbf{x}_k$ if and only if the direction vector $\mathbf{s}_k$ vanishes—exactly, not approximately—at $\mathbf{x}_k$. Suppose, then, that an objective function $f(\mathbf{x})$ and a starting point $\mathbf{x}_0$ have been chosen, and let a direct climbing algorithm be applied, generating a sequence of points $\{\mathbf{x}_i\} \equiv \mathbf{x}_0, \mathbf{x}_1, \mathbf{x}_2, \ldots$ in accordance with the recursion rule

$$\mathbf{x}_{i+1} = \mathbf{x}_i + \theta_i \mathbf{s}_i. \tag{4-20}$$

[5]It should be pointed out that many writers prefer to call $\mathbf{x}^*$ a limit point of $\{\mathbf{x}_i\}$ if for any $\delta > 0$ there are an infinite number of terms within a distance δ of $\mathbf{x}^*$. This definition allows oscillating sequences and certain other types to have multiple limit points. However, insofar as the sequences generated by mathematical programming algorithms are concerned, the two definitions are equivalent.

Use of the ideal stopping rule described above would then lead to one of the following four outcomes:

1. The sequence of points $\{\mathbf{x}_i\}$ generated by the algorithm might terminate after a finite number of iterations at a point $\mathbf{x}_k$ where the direction vector $\mathbf{s}_k$ vanishes exactly,
2. The sequence $\{\mathbf{x}_i\}$ might terminate after a finite number of iterations at a point $\mathbf{x}_k$ where $\mathbf{s}_k \neq \mathbf{0}$ but where an infinite value is computed for θ_k in the recursion relation (4-20),
3. The sequence $\{\mathbf{x}_i\}$ might have an infinite number of members or *terms* and converge to a limit point $\mathbf{x}^*$, or
4. The sequence $\{\mathbf{x}_i\}$ might have an infinite number of terms and fail to converge.

In case 1 the algorithm terminates at a solution point $\mathbf{x}_k$ where $\mathbf{s}_k = \mathbf{0}$. In general, as was mentioned earlier, this immediately implies $\nabla \mathbf{f}(\mathbf{x}_k) = \mathbf{0}$; the point $\mathbf{x}_k$ is then known either as a *critical point* or, in the context of direct climbing, as a *stationary point* of $f(\mathbf{x})$. Note that we do not specify that $\mathbf{x}_k$ must be a local optimum: It is quite possible for any of the climbing algorithms in common use to terminate at stationary points (i.e., critical points) other than local optima. For example, one obvious way of doing so would be to choose a starting point $\mathbf{x}_0$ that happened to be a saddle point; if the gradient were being used as the direction vector, termination would then occur immediately.

Termination in case 2 generally implies the existence of an unbounded optimal solution of the form

$$\lim_{\theta \to \infty} \mathbf{x}_k + \theta \mathbf{s}_k.$$

Of course, the limiting value of $f(\mathbf{x}_k + \theta \mathbf{s}_k)$ as θ increases without bound is not necessarily infinite.

Although finite termination is very convenient and easily dealt with, it is much more common for the sequence of points $\{\mathbf{x}_i\}$ generated by a direct climbing algorithm (with the ideal stopping rule) to have infinitely many terms. When the sequence converges to a limit point $\mathbf{x}^*$—case 3—it is generally true that $\mathbf{x}^*$ must be a stationary point, although this cannot be asserted without formal proof. Moreover, even when $\mathbf{x}^*$ can be proved to be a stationary point in case 3, it need not be a local optimum. Case 4, nonconvergence of the sequence, is another manifestation of unboundedness; in this case the sequence of objective values $\{f(\mathbf{x}_i)\}$ increases monotonically, but both $\mathbf{x}_i$ and $f(\mathbf{x}_i)$ remain finite for every finite i.

Based on the foregoing discussion, the termination characteristics that would be the most desirable and convenient for a direct climbing algorithm to possess are summarized in the following definition of *algorithmic convergence*:

A direct climbing algorithm for unconstrained optimization problems is said to *converge*, or to be *theoretically convergent*, provided that, when the ideal stopping rule is used, the following three conditions are satisfied:

1. If the algorithm terminates at a point $\mathbf{x}_k$ after a finite number of iterations, then either $\mathbf{x}_k$ is a stationary point or evidence of an unbounded solution is provided;
2. If the sequence of points $\{\mathbf{x}_i\}$ generated by the algorithm converges to a limit point $\mathbf{x}^*$ after an infinite number of iterations, then $\mathbf{x}^*$ is a stationary point; and
3. If the sequence $\{\mathbf{x}_i\}$ fails to converge, an unbounded optimal solution exists.

Note that the theoretical convergence of an algorithm does *not* require the convergence of all possible sequences $\{\mathbf{x}_i\}$; divergence is permitted provided that it implies the existence of an unbounded optimum. It is regrettable that the term "convergence" is used both for a property of sequences and for a property of mathematical programming algorithms. The situation is further beclouded by the informal but very common usage of the term in yet a third sense: An engineer or operations research analyst might report that a certain iterative computational procedure successfully "converged" to a solution of some problem he was working on, meaning simply that a solution satisfying his stopping rule was obtained after a finite number of iterations. We, too, shall frequently employ the term without modifiers in this computational sense, relying on context for clarification.

A theoretically convergent algorithm, with any reasonable stopping rule adjoined, has the highly desirable property of guaranteeing that either a stationary point will be obtained or evidence of unboundedness provided within a finite number of iterations (in case 2, of course, a stationary point will only approximately be obtained, because the stopping rule will terminate the sequence $\{\mathbf{x}_i\}$ before the limit point is reached). This means, in particular, that if unboundedness is ruled out—as it can be in most well-posed problems in operations research and engineering—a convergent algorithm can be relied upon to produce a stationary-point solution within a finite number of iterations. For a *non*convergent algorithm, however, the possibility of computational disaster can arise when the sequence $\{\mathbf{x}_i\}$ fails to converge, or *seems* to be failing to converge, even though a bounded optimal solution is known to exist. The analyst is then faced with the prospect not only of ending up without a solution but also of having wasted a considerable amount of expensive computer time in the process. It is therefore not surprising that, whenever possible, analysts and engineers prefer to use mathematical programming algorithms—for constrained as well as unconstrained problems—for which convergence is guaranteed, either rigorously, via a comprehensive theoretical proof, or empirically, in the sense that, while some type of infinite cycling or oscillation may be possible in theory, it has never been encountered in practice (recall that this is true of the simplex method as it is usually coded). Other solution algorithms in common use guarantee convergence if and only if the starting point $\mathbf{x}_0$ is "reasonably close," in some specified sense, to a local optimum.

As might be imagined, the proof of convergence is the most difficult aspect of virtually every iterative computational algorithm, both for the researcher to discover and for the student to understand. For this reason, and because they do not shed much light on how the algorithms actually operate, formal convergence proofs will not usually be presented in this text, although conditions required for convergence and other matters of computational importance will be discussed as appropriate. For a thorough treatment of mathematical convergence theory at a somewhat advanced level, the student should consult the text of Zangwill [60] or that of Luenberger [31]; the former in particular contains major original contributions to the field.

4.7 THE OPTIMAL GRADIENT METHOD

Having addressed ourselves thus far to the direct climbing approach in general, we shall focus our attention during the next few sections on a number of different solution methods that employ it. The various climbing algorithms differ in the answers they give to the following two questions, which must be confronted at every iteration: (1) In what direction do we move next? (2) How far? We can state these questions more precisely: Given a sequence of points whose latest member is $\mathbf{x}_i$, how should $\mathbf{s}_i$ and θ_i be chosen in the iterative step

$$\mathbf{x}_{i+1} = \mathbf{x}_i + \theta_i \mathbf{s}_i \tag{4-20}$$

in order to generate the next point $\mathbf{x}_{i+1}$?

The obvious and most natural choice for $\mathbf{s}_i$—although, as we shall see, not necessarily the best—is the direction of the gradient vector evaluated at $\mathbf{x}_i$,

$$\mathbf{s}_i = \nabla \mathbf{f}(\mathbf{x}_i). \tag{4-21}$$

Assuming that we are trying to find a local maximum of $f(\mathbf{x})$ as quickly as possible, the direction (4-21) is intuitively appealing because it is the direction in which the immediate rate of increase of $f(\mathbf{x})$ is greatest. Similar considerations would suggest the choice of

$$\mathbf{s}_i = -\nabla \mathbf{f}(\mathbf{x}_i) \tag{4-22}$$

when $f(\mathbf{x})$ is being minimized.[6] Climbing algorithms that use (4-21) or (4-22) are known as *gradient methods* or *methods of steepest ascent* (or *descent*). The general approach was first suggested by the mathematician Cauchy in 1847, but despite its nineteenth-century pedigree it is still in wide use today.

Once a gradient method has been chosen, there remains the question of how far to move in the gradient direction, that is, what value of θ to use in

$$\hat{\mathbf{x}} = \mathbf{x} + \theta\, \nabla \mathbf{f}(\mathbf{x}). \tag{4-23}$$

(Note that we are dropping the subscript i in order to simplify notation; the current point $\mathbf{x}_i$ and the new point $\mathbf{x}_{i+1}$ will now be represented by $\mathbf{x}$ and $\hat{\mathbf{x}}$. The reader should bear in mind throughout the ensuing discussion that the symbol $\mathbf{x}$ may therefore represent, depending on the context, either the vector of variables or the latest point generated by the climbing algorithm.) Assuming that a local maximum is being sought, one rather obvious tactic for choosing the *step length* θ—but again not the best—is the tactic employed by the *optimal gradient method*: Move in the gradient direction until the value of $f(\mathbf{x})$ stops increasing and begins to decrease. This is accomplished by maximizing the function

$$g(\theta) \equiv f(\mathbf{x} + \theta\, \nabla \mathbf{f}(\mathbf{x})) \tag{4-24}$$

with respect to θ; the smallest positive θ satisfying

$$\frac{dg(\theta)}{d\theta} = 0 \tag{4-25}$$

is the desired step length.[7] Denoting that maximizing value of θ by θ_m, the new point is then given by

$$\hat{\mathbf{x}} = \mathbf{x} + \theta_m\, \nabla \mathbf{f}(\mathbf{x}).$$

One interesting property of the optimal gradient method is that the gradient vectors at successive iterations are perpendicular to each other; this follows from the fact that the rate of change of f with respect to movement in the

[6]This assumption that the directions of steepest ascent and steepest descent are collinear is legitimate because we have restricted our discussion to functions whose first derivatives are continuous. To observe the effect of removing this restriction, the reader should compare the directions of steepest ascent and descent from the origin for the function $f(x_1, x_2) = x_1 + \max(0, x_2)$.

[7]It is somewhat deceptive to call θ the step length, inasmuch as the actual length of the "step" is θ times the magnitude of the gradient vector. Thus a step length of 1 (i.e., $\theta = 1$) in the gradient direction (10, 10) covers a much greater distance than a step length of 2 in the direction (1, 1). The term "step length" would become more appropriate if the gradient vector were automatically normalized to a magnitude of 1, but in practice this conversion would be a waste of time and is not performed.

direction $\nabla \mathbf{f}(\mathbf{x})$ has a value of zero at $\hat{\mathbf{x}}$, so that the component of $\nabla \mathbf{f}(\hat{\mathbf{x}})$ in the direction $\nabla \mathbf{f}(\mathbf{x})$ must be zero.

It should be noted that we are *not* choosing the value of θ that maximizes $f(\mathbf{x} + \theta \nabla \mathbf{f}(\mathbf{x}))$ over all nonnegative θ. That is, instead of trying to determine the global maximum, which might be a rather difficult task if the function had (or *could* have) several local maxima, we are simply taking the first local maximum we come to. Of course, if the function $g(\theta)$ is concave, only a single local maximum will exist, so that the first one encountered, at $\theta = \theta_m$, will necessarily be the global maximum. This will be the case, for example, when we use a quadratic approximation of (4-24) in calculating θ_m.

Before turning to the question of how the maximizing value θ_m is actually computed, however, we shall discuss the theoretical convergence properties of the optimal gradient method, *assuming that the exact value of* θ_m *is obtained and used at every iteration.* Algorithmic convergence of the optimal gradient method was first proved by Curry [8] in 1944. Condition 1 of the definition, which deals with finite termination of the algorithm, holds trivially: Termination can occur at $\mathbf{x}_k$ only if $\nabla \mathbf{f}(\mathbf{x}_k) = \mathbf{0}$, implying that $\mathbf{x}_k$ is a stationary point, or if $\theta_m = \infty$ at the kth stage, implying an unbounded solution. Condition 3, which states that nonconvergence of the sequence $\{\mathbf{x}_i\}$ must imply the existence of an unbounded optimal solution, seems intuitively quite appealing [in view of the fact that $f(\mathbf{x}_{i+1}) > f(\mathbf{x}_i)$ is guaranteed for all i by the rules of the optimal gradient method], but a rigorous proof is rather subtle and will not be presented here. We are left with condition 2, which we now proceed to establish by contradiction, following the argument of Curry.

Suppose the optimal gradient method is used to seek a local maximum of the function $f(\mathbf{x})$, $\mathbf{x}$ in E^n, and suppose further that the resulting sequence $\{\mathbf{x}_i\}$ has an infinite number of terms and converges to a limit point $\mathbf{x}^*$ at which the gradient $\nabla \mathbf{f}(\mathbf{x}^*)$ does *not* vanish. Let the scalar variable $d(\mathbf{x})$ denote the magnitude or length of the gradient $\nabla \mathbf{f}$ at $\mathbf{x}$, and define a normalized n-component vector

$$\mathbf{z}(\mathbf{x}) \equiv \frac{1}{d(\mathbf{x})} \nabla \mathbf{f}(\mathbf{x})$$

which is of unit length and points in the direction of the gradient. Our assumption for this proof is that $d(\mathbf{x}^*) > 0$. Let any small positive number ϵ be chosen. Then, because the first partial derivatives of f are continuous, there must exist a small number $\delta > 0$ such that, for any point $\mathbf{x}$ within a distance δ of $\mathbf{x}^*$,

$$|\nabla \mathbf{f}(\mathbf{x}) - \nabla \mathbf{f}(\mathbf{x}^*)| < \epsilon\, d(\mathbf{x}^*). \tag{4-26}$$

Recall now the *triangle inequality*: For any two vectors $\mathbf{a}$ and $\mathbf{b}$,

$$|\mathbf{a} + \mathbf{b}| \leqq |\mathbf{a}| + |\mathbf{b}|.$$

Substituting $\mathbf{a} = \nabla \mathbf{f}(\mathbf{x}) - \nabla \mathbf{f}(\mathbf{x}^*)$ and $\mathbf{b} = \nabla \mathbf{f}(\mathbf{x}^*)$, we find

$$d(\mathbf{x}) - d(\mathbf{x}^*) \leqq |\nabla \mathbf{f}(\mathbf{x}) - \nabla \mathbf{f}(\mathbf{x}^*)| < \epsilon\, d(\mathbf{x}^*),$$

which implies

$$d(\mathbf{x}) < (1 + \epsilon)\, d(\mathbf{x}^*).$$

Using slightly different substitutions, we can derive a second inequality that combines with the above to yield

$$(1 - \epsilon)\, d(\mathbf{x}^*) < d(\mathbf{x}) < (1 + \epsilon)\, d(\mathbf{x}^*)$$

or

$$\left|1 - \frac{d(\mathbf{x})}{d(\mathbf{x}^*)}\right| < \epsilon$$

for any $\mathbf{x}$ within a distance δ of $\mathbf{x}^*$.

Now, from Eq. (4-26) and the definition of $\mathbf{z}(\mathbf{x})$, we have

$$|d(\mathbf{x}) \cdot \mathbf{z}(\mathbf{x}) - d(\mathbf{x}^*) \cdot \mathbf{z}(\mathbf{x}^*)| < \epsilon\, d(\mathbf{x}^*),$$

which leads to

$$\left|\frac{d(\mathbf{x})}{d(\mathbf{x}^*)}\mathbf{z}(\mathbf{x}) - \mathbf{z}(\mathbf{x}^*)\right| < \epsilon.$$

But we may also write

$$\left|\mathbf{z}(\mathbf{x}) - \frac{d(\mathbf{x})}{d(\mathbf{x}^*)}\mathbf{z}(\mathbf{x})\right| \equiv \left|1 - \frac{d(\mathbf{x})}{d(\mathbf{x}^*)}\right| \cdot |\mathbf{z}(\mathbf{x})| < \epsilon,$$

where we have used the fact that the magnitude of $\mathbf{z}(\mathbf{x})$ is 1 by definition. Application of the triangle inequality then yields

$$|\mathbf{z}(\mathbf{x}) - \mathbf{z}(\mathbf{x}^*)| \leqq \left|\frac{d(\mathbf{x})}{d(\mathbf{x}^*)}\mathbf{z}(\mathbf{x}) - \mathbf{z}(\mathbf{x}^*)\right| + \left|\mathbf{z}(\mathbf{x}) - \frac{d(\mathbf{x})}{d(\mathbf{x}^*)}\mathbf{z}(\mathbf{x})\right| < 2\epsilon$$

for any $\mathbf{x}$ within a distance δ of $\mathbf{x}^*$.

Suppose now that some point $\mathbf{x}_k$ in the sequence $\{\mathbf{x}_i\}$ is within a distance δ of $\mathbf{x}^*$, so that

$$|\mathbf{z}(\mathbf{x}_k) - \mathbf{z}(\mathbf{x}^*)| < 2\epsilon.$$

Because the optimal gradient method provides that $\mathbf{z}(\mathbf{x}_k)$ and $\mathbf{z}(\mathbf{x}_{k+1})$ must be at right angles,

$$|\mathbf{z}(\mathbf{x}_k) - \mathbf{z}(\mathbf{x}_{k+1})| = \sqrt{2}.$$

Hence for sufficiently small ϵ it is impossible that

$$|\mathbf{z}(\mathbf{x}_{k+1}) - \mathbf{z}(\mathbf{x}^*)| < 2\epsilon$$

(otherwise a violation of the triangle inequality could easily be demonstrated) and therefore impossible that $\mathbf{x}_{k+1}$ is within a distance δ of $\mathbf{x}^*$. But then no integer N can exist such that all points $\mathbf{x}_i$, $i \geqq N$, are within a distance δ of $\mathbf{x}^*$, and it follows that $\mathbf{x}^*$ cannot be a limit point of the sequence $\{\mathbf{x}_i\}$. A contradiction has finally been obtained, and we conclude that if $\{\mathbf{x}_i\}$ converges, the limit point must be a stationary point. Q.E.D.

Although the foregoing rather elementary convergence proof relied on the fact that the gradient vectors computed at any two successive iterations must be perpendicular, other proofs of convergence have been devised for the optimal gradient method that do not make use of this property; see, for example, Zangwill [60]. In fact, it can be shown that any climbing algorithm that generates points via

$$\mathbf{x}_{i+1} = \mathbf{x}_i + \theta_i \, \nabla\mathbf{f}(\mathbf{x}_i)$$

must eventually converge to a stationary point provided only that $f(\mathbf{x}_{i+1}) > f(\mathbf{x}_i)$ at every iteration. That is, no restriction need be placed on the choice of θ_i except that it must yield an improvement in $f(\mathbf{x})$ at every stage. As stated earlier, however, we shall not present actual convergence proofs for most of the algorithms to be covered in this text, preferring to concentrate instead on the conditions required to ensure convergence and on other matters of computational importance.

One such matter that may as well be discussed before we proceed further is the possibility that the optimal gradient method might terminate at a stationary point that is not a local maximum. Although this possibility has been kept in view, at least formally, throughout the past two sections, it is not really a very important practical consideration. In fact, it can occur only in rather unlikely or contrived situations, generally stemming from the choice of a starting point $\mathbf{x}_0$ at which one or more of the partial derivatives $\delta f/\delta x_j$ vanishes. In the most trivial instance, $\mathbf{x}_0$ might itself be a saddle point (or, for that matter, a local *minimum*), in which case it would be found that $\nabla\mathbf{f}(\mathbf{x}_0) = \mathbf{0}$, and termination would occur immediately. As a somewhat less trivial example, consider the problem of maximizing the function $f(x_1, x_2) = x_1^2 - x_2^2$, starting at $\mathbf{x}_0 = (0, 1)$. The gradient is $\nabla\mathbf{f}(\mathbf{x}) = (2x_1, -2x_2)$, so $\nabla\mathbf{f}(\mathbf{x}_0) = (0, -2)$, and (4-24) becomes

$$g(\theta) \equiv f(\mathbf{x}_0 + \theta \, \nabla\mathbf{f}(\mathbf{x}_0)) = f(0, 1 - 2\theta) = -(1 - 2\theta)^2.$$

This is evidently maximized at $\theta_m = .5$, and we have

$$\mathbf{x}_1 = \mathbf{x}_0 + \theta_m \, \nabla\mathbf{f}(\mathbf{x}_0) = \begin{bmatrix} 0 \\ 1 \end{bmatrix} + .5\begin{bmatrix} 0 \\ -2 \end{bmatrix} = \begin{bmatrix} 0 \\ 0 \end{bmatrix}.$$

But $\nabla\mathbf{f}(\mathbf{x}_1) = (0, 0)$, so the algorithm terminates at the origin, which is a stationary point of $f(x_1, x_2)$ but *not* a local maximum.

Notice how in both of these examples the termination at a saddle point was due to the selection of a starting point at which one or more of the partial derivatives $\delta f/\delta x_j$ already had a value of zero. By merely avoiding such starting points, we can virtually eliminate the possibility of saddle-point termination. Alternatively, or perhaps in addition, we can attach a test at the end of the climbing process that will determine whether the point $\mathbf{x}^*$ at which the algorithm has terminated is in fact a local maximum. This can be done most simply by evaluating the gradient $\nabla\mathbf{f}$ at various nearby points, each obtained by perturbing a single component of $\mathbf{x}^*$, and verifying that in each case the gradient vector points back toward $\mathbf{x}^*$. If any one of these perturbation tests fails, $\mathbf{x}^*$ cannot be a local maximum. This type of testing would obviously be very well suited to any computer-coded direct climbing algorithm.

4.8 COMPUTING THE STEP LENGTH IN THE OPTIMAL GRADIENT METHOD

We come now to the question of how the value θ_m that maximizes

$$g(\theta) \equiv f(\mathbf{x} + \theta \, \nabla\mathbf{f}(\mathbf{x})) \tag{4-24}$$

is to be computed (again, note that we are omitting the subscript of the current point $\mathbf{x}$). There are three basic approaches, each of which will be discussed in this section:

1. Differentiation of (4-24) with respect to θ, followed by direct solution—either analytic or iterative—of the resulting equation (4-25),
2. Trial and error, or "sequential search," and
3. Use of a quadratic approximation of (4-24).

The most straightforward of these methods, at least from a theoretical point of view, is the first: Write out explicitly the derivative $dg(\theta)/d\theta$, or encode it in a computer subroutine, and then solve the equation

$$\frac{dg(\theta)}{d\theta} = 0. \tag{4-25}$$

In general, (4-25) will be an extremely complicated equation and will have to be solved by Newton-Raphson iteration or some other numerical technique (using $\theta = 0$ as a starting point), although on rare occasions analytic solution may be possible.

Let us apply the optimal gradient method, using this computational approach, to the following very simple problem:

$$\text{Min } f(x_1, x_2) = (x_1 - 3x_2)^2 + (x_2 - 1)^2.$$

By inspection the optimal solution is $x_1 = 3$, $x_2 = 1$, with $f(3, 1) = 0$, but suppose we begin at the point $\mathbf{x}_0 = (2, 2)$, which has an objective value of 17. The gradient vector is given by

$$\nabla \mathbf{f}(\mathbf{x}) = \begin{pmatrix} 2x_1 - 6x_2 \\ -6x_1 + 20x_2 - 2 \end{pmatrix},$$

and its value at the starting point is

$$\nabla \mathbf{f}(\mathbf{x}_0) = \begin{pmatrix} -8 \\ 26 \end{pmatrix}.$$

Because we are minimizing, we want to move in the direction opposite to the gradient, so

$$\mathbf{x}_1 = \mathbf{x}_0 - \theta_0 \nabla \mathbf{f}(\mathbf{x}_0) = \begin{pmatrix} 2 \\ 2 \end{pmatrix} - \theta_0 \begin{pmatrix} -8 \\ 26 \end{pmatrix} = \begin{pmatrix} 2 + 8\theta_0 \\ 2 - 26\theta_0 \end{pmatrix},$$

where a positive value of θ_0 is being sought. Given any θ_0, the value of the objective is

$$g(\theta_0) \equiv f(\mathbf{x}_0 - \theta_0 \nabla \mathbf{f}(\mathbf{x}_0)) = f(2 + 8\theta_0, 2 - 26\theta_0) = 17 - 740\theta_0 + 8072\theta_0^2.$$

To find the value of θ_0 at which $g(\theta_0)$ stops decreasing, we set its derivative equal to zero:

$$\frac{dg(\theta_0)}{d\theta_0} = -740 + 16{,}144\theta_0 = 0.$$

This yields a step length of $\theta_0 = .0458$, so that

$$\mathbf{x}_1 = \begin{pmatrix} 2.3667 \\ .8082 \end{pmatrix}.$$

The value of the objective has decreased from $f(\mathbf{x}_0) = 17$ to $f(\mathbf{x}_1) = .0402$.

The gradient at $\mathbf{x}_1$ is

$$\nabla \mathbf{f}(\mathbf{x}_1) = \begin{pmatrix} -.1160 \\ -.0357 \end{pmatrix},$$

which, as can be readily verified, is perpendicular to $\nabla \mathbf{f}(\mathbf{x}_0)$. Because the gradient still differs significantly from **0**, the current point $\mathbf{x}_1$ is not a local minimum and we must proceed further. The next point in the sequence is of the form

$$\mathbf{x}_2 = \mathbf{x}_1 - \theta_1 \nabla \mathbf{f}(\mathbf{x}_1) = \begin{pmatrix} 2.3667 + .1160\theta_1 \\ .8082 + .0357\theta_1 \end{pmatrix},$$

which has an objective value given by

$$g(\theta_1) = .0402 - .0147\theta_1 + .00135\theta_1^2.$$

This function is minimized at $\theta_1 = 5.4413$, yielding

$$\mathbf{x}_2 = \begin{pmatrix} 2.9976 \\ 1.0024 \end{pmatrix},$$

at which point the objective value is $f(\mathbf{x}_2) = .000094$ and the gradient is

$$\nabla \mathbf{f}(\mathbf{x}_2) = \begin{pmatrix} -.0188 \\ .0612 \end{pmatrix}.$$

A further iteration then produces

$$\mathbf{x}_3 = \begin{pmatrix} 2.9985 \\ .9995 \end{pmatrix}, \quad f(\mathbf{x}_3) = 2.2 \times 10^{-7}, \quad \text{and} \quad \nabla \mathbf{f}(\mathbf{x}_3) = \begin{pmatrix} -.0003 \\ -.0001 \end{pmatrix}.$$

This point would be likely to satisfy whatever criterion is being used for terminating the algorithm, either because $\mathbf{x}_3$ is so close to $\mathbf{x}_2$ or because the magnitude of the vector $\nabla \mathbf{f}(\mathbf{x}_3)$ is so small.

The progress of the optimal gradient method in converging to the local maximum $\mathbf{x}_3$ is summarized in the following table, where $|\mathbf{x}_i - \mathbf{x}_{i-1}|$ is the Euclidean distance from $\mathbf{x}_{i-1}$ to $\mathbf{x}_i$ (recall that this distance equals the product of the step length θ_{i-1} and the magnitude of the gradient vector $\nabla \mathbf{f}(\mathbf{x}_{i-1})$):

$\mathbf{x}_i$	$\lvert \mathbf{x}_i - \mathbf{x}_{i-1} \rvert$	$f(\mathbf{x}_i)$	$\nabla \mathbf{f}(\mathbf{x}_i)$
$\mathbf{x}_0$ = (2.0000, 2.0000)		17.0000	(−8.0000, 26.0000)
$\mathbf{x}_1$ = (2.3667, .8082)	1.246	0.0402	(−.1160, −.0357)
$\mathbf{x}_2$ = (2.9976, 1.0024)	.660	$\sim 10^{-4}$	(−.0188, .0612)
$\mathbf{x}_3$ = (2.9985, .9995)	.003	$\sim 10^{-7}$	(−.0003, −.0001)

There are several points to be noted in connection with this example. First, the computational development was fairly typical of what happens during any gradient ascent or descent to a local optimum. The earlier steps tend to cover greater distances and bring greater improvements in the objective function than the later ones; moreover, the gradient vector can be expected to decrease in magnitude as the iterations proceed. (Of course, these statements merely describe general tendencies and should by no means be interpreted as hard and fast rules.) Second, all direct climbing methods have the convenient property of being immune to errors in arithmetic; a blunder in computing some $\mathbf{x}_k$, although it is likely to prolong the search, will automatically be rectified when the new direction vector, in this case the gradient, is calculated—in effect, the erroneous $\mathbf{x}_k$ can be thought of as a new starting point. Finally, it should be emphasized that the example solved above was much simpler than any real-world problem that is likely to require the services of an operations research analyst. In particular, it is most unusual for the equation

$$\frac{dg(\theta)}{d\theta} = 0 \tag{4-25}$$

to be capable of exact analytic solution: For the most part this is possible only when the objective function $f(\mathbf{x})$ is quadratic, as in the example presented above (and in such cases the stationary point can be found without recourse to direct climbing techniques anyway, simply by solving the linear system of equations $\delta f/\delta x_j = 0, j = 1, \ldots, n$). In virtually all other cases (4-25) must be solved by means of Newton-Raphson iteration or some other iterative numerical procedure, which at best becomes quite laborious and at worst might introduce convergence problems of its own.

Thus, although the value $\theta = \theta_m$ that maximizes

$$g(\theta) \equiv f(\mathbf{x} + \theta\, \nabla \mathbf{f}(\mathbf{x})) \tag{4-24}$$

can in theory always be found by differentiation followed by solution of (4-25), in practice a more efficient approach is needed. One such approach aims at obtaining θ_m—or a value reasonably close to it—by simple trial and error, working directly with Eq. (4-24). If the problem is being solved by hand, it is possible to proceed in a highly informal manner, simply computing $f(\mathbf{x} + \theta\, \nabla \mathbf{f}(\mathbf{x}))$ for several values of θ, one after another, jumping back and forth in an ad hoc fashion, and eventually "zeroing in" somewhere close to θ_m. In the early stages it does not pay to spend too much time in pursuit of minute improvements, since the maximum being sought is probably still far away and in an ill-defined direction. Later on, of course, more pains must be taken.

By contrast, to program a *search* procedure of this sort for a digital computer, a generalized strategy must be decided upon in advance. One very simpleminded but perfectly adequate approach would be the following.

Given some current point **x**, the point $\mathbf{x} + \epsilon\,\nabla\mathbf{f}(\mathbf{x})$ is examined, where ϵ is a positive number input by the analyst using the computer routine. If $f(\mathbf{x} + \epsilon\,\nabla\mathbf{f}(\mathbf{x})) > f(\mathbf{x})$, another ϵ-step is taken in the direction of the gradient and $f(\mathbf{x} + 2\epsilon\,\nabla\mathbf{f}(\mathbf{x}))$ is compared to $f(\mathbf{x} + \epsilon\,\nabla\mathbf{f}(\mathbf{x}))$. This continues until one of these steps causes a decrease in the function f. From the point of the last successful step the process may now be repeated, using steps of length $\mu\epsilon$, where μ is also an input and $0 < \mu < 1$; in any further repetitions the step length would continue to decrease by powers of μ. In this way a value of θ can be found that is as close to the maximizing value θ_m as desired. Note that any number of modifications can be devised that would improve the procedure to some extent; for example, during the early iterations only the long ϵ-steps might be used. It should be mentioned, however, that there are other search strategies that are somewhat more efficient than the rather unsophisticated step-by-step testing just described; some of these strategies, which depend on such classical concepts as the Fibonacci numbers and the "golden section," will be discussed in Section 5.4.

Both of the approaches presented thus far for determining the "optimal gradient" value θ_m—ordinary differentiation of $g(\theta)$ and sequential trial and error—involve iterative computation [except when Eq. (4-25) produced by differentiation can be solved analytically], and the iterative steps are relatively time-consuming. In general, it would be more efficient to have available an explicit formula or function that would permit the direct calculation of some approximate value of θ_m. Such a formula can in fact be written—although for any given problem it might prove so grossly inaccurate as to be unusable—if in place of the function

(4-24) $$g(\theta) \equiv f(\mathbf{x} + \theta\,\nabla\mathbf{f}(\mathbf{x}))$$

we use a quadratic approximation

(4-27) $$g(\theta) \approx A + B\theta + C\theta^2,$$

where A, B, and C are scalar constants that can be calculated in different ways, depending on what data are available. Note that in any case A must equal the value of the objective function at the current point **x**, because, from (4-24), $g(0) = f(\mathbf{x})$. As θ is increased from zero, causing a displacement in the gradient direction, the value of $g(\theta)$ begins to increase. But, provided that $f(\mathbf{x})$ is bounded *and* that (4-27) is a reasonable approximation to it, at least locally, $g(\theta)$ cannot increase forever and must take on a local maximum at some positive value of θ. The derivative of (4-27) is

$$\frac{dg(\theta)}{d\theta} \approx B + 2C\theta,$$

and the value that (approximately) maximizes $g(\theta)$ is therefore

$$\theta_m = -\frac{B}{2C}.$$

When the second partial derivatives of f are continuous and not too difficult to work with, the most convenient quadratic approximation is given by the second-order Taylor expansion "along a line." Using the current point $\mathbf{x}$ in place of $\mathbf{x}_0$ in (2-11), the expansion becomes

$$f(\mathbf{x} + \theta\mathbf{s}) = f(\mathbf{x}) + \sum_{j=1}^{n} \frac{\delta f(\mathbf{x})}{\delta x_j}\theta s_j + \frac{1}{2!}\sum_{i=1}^{n}\sum_{j=1}^{n} \theta s_i \frac{\delta^2 f(\mathbf{x})}{\delta x_i\,\delta x_j}\theta s_j, \tag{4-28}$$

ignoring the third-order and higher terms. In the optimal gradient method the direction vector $\mathbf{s}$ is $\nabla\mathbf{f}(\mathbf{x})$, with components $s_j = \delta f(\mathbf{x})/\delta x_j$; thus

$$f(\mathbf{x} + \theta\mathbf{s}) = f(\mathbf{x} + \theta\,\nabla\mathbf{f}(\mathbf{x})) \equiv g(\theta),$$

and, matching (4-28) against (4-27), we can see that $\theta_m = -B/2C$ implies

$$\theta_m = \frac{-\sum_{j=1}^{n} (\delta f(\mathbf{x})/\delta x_j)^2}{\sum_{i=1}^{n}\sum_{j=1}^{n} [\delta^2 f(\mathbf{x})/(\delta x_i\,\delta x_j)]\cdot(\delta f(\mathbf{x})/\delta x_i)\cdot(\delta f(\mathbf{x})/\delta x_j)}. \tag{4-29}$$

This value can be computed after all the first and second partial derivatives of f have been evaluated at the current point $\mathbf{x}$. Note that if the function f were being *minimized*, the step length θ_m would be computed in exactly the same way but would have a negative value.

If the second derivatives of f are not available, we can still use other data to construct the parabolic fit

$$g(\theta) \approx A + B\theta + C\theta^2. \tag{4-27}$$

We have observed that $A = f(\mathbf{x})$, a value that is readily obtainable. From the chain rule the derivative of (4-24) evaluated at $\theta = 0$ is

$$\frac{dg(0)}{d\theta} = \frac{d}{d\theta}[f(\mathbf{x} + \theta\,\nabla\mathbf{f}(\mathbf{x}))]_{\theta=0} = \sum_{j=1}^{n}\left[\frac{\delta f(\mathbf{x} + \theta\,\nabla\mathbf{f}(\mathbf{x}))}{\delta x_j}\cdot\frac{dx_j}{d\theta}\right]_{\theta=0}$$

$$= \sum_{j=1}^{n}\left(\frac{\delta f(\mathbf{x})}{\delta x_j}\right)^2,$$

and this must be the value of B, because from (4-27)

$$\frac{dg(0)}{d\theta} \approx (B + 2C\theta)_{\theta=0} = B.$$

The only remaining unknown parameter is C, and it can be determined if we compute $g(\theta) \equiv f(\mathbf{x} + \theta\, \nabla \mathbf{f}(\mathbf{x}))$ at one other value of θ and substitute into (4-27). This completely specifies the parabola (4-27), and the step length $\theta_m = -B/2C$ can again be used.

We shall now solve the following example problem via the optimal gradient method, using the second-order Taylor approximation of $g(\theta)$:

$$\text{(4-30)} \qquad \text{Max } f(x_1, x_2) = (2x_1 - x_2)^2 - e^{x_1^2} - e^{x_2^2}.$$

The first and second partial derivatives are

$$\frac{\delta f}{\delta x_1} = 8x_1 - 4x_2 - 2x_1 e^{x_1^2},$$

$$\frac{\delta f}{\delta x_2} = 2x_2 - 4x_1 - 2x_2 e^{x_2^2},$$

$$\frac{\delta^2 f}{dx_1^2} = 8 - (4x_1^2 + 2)e^{x_1^2},$$

$$\frac{\delta^2 f}{\delta x_2^2} = 2 - (4x_2^2 + 2)e^{x_2^2},$$

and

$$\frac{\delta^2 f}{\delta x_1\, \delta x_2} = -4,$$

and at any stage the optimizing value of θ is given by (4-29). If we begin with a starting point of $\mathbf{x}_0 = (2, -2)$, the following table of values is generated, where $|\mathbf{x}_i - \mathbf{x}_{i-1}|$ is the distance from $\mathbf{x}_{i-1}$ to $\mathbf{x}_i$:

$\mathbf{x}_i$	$\lvert\mathbf{x}_i - \mathbf{x}_{i-1}\rvert$	$f(\mathbf{x}_i)$	$\nabla \mathbf{f}(\mathbf{x}_i)$
$\mathbf{x}_0$ = (2.0000, −2.0000)		−73.1963	(−194.392, 206.392)
$\mathbf{x}_1$ = (1.8004, −1.7881)	.2912	−20.9963	(− 70.522, 76.715)
$\mathbf{x}_2$ = (1.6059, −1.5765)	.2873	− 2.2627	(− 23.194, 28.282)
$\mathbf{x}_3$ = (1.4434, −1.3783)	.2563	+ 3.4750	(− 6.125, 9.897)
$\mathbf{x}_4$ = (1.3448, −1.2191)	.1873	+ 4.7565	(− .776, 2.959)
$\mathbf{x}_5$ = (1.3207, −1.1270)	.0952	+ 4.9179	(− .038, .490)
$\mathbf{x}_6$ = (1.3190, −1.1054)	.0216	+ 4.9234	(− .052, .016)
$\mathbf{x}_7$ = (1.3177, −1.1050)	.0004	+ 4.9234	(+ .004, .013)
$\mathbf{x}_8$ = (1.3178, −1.1045)	.0005	+ 4.9234	(− .005, .001)
$\mathbf{x}_9$ = (1.3177, −1.1045)	.00004	+ 4.9234	(+ .000, .001)

A gradient vector of magnitude less than .001 was required for termination; using this stopping rule, convergence occurred in 9 iterations. We remark that if the starting point $\mathbf{x}_0$ is taken to be (3, −3), the number of iterations required for convergence, assuming the same stopping rule, is 16, while for $\mathbf{x}_0 = (5, -5)$ the number is 33.

As this example illustrates, the optimal gradient method, using a quadratic approximation of $g(\theta)$ to compute estimates of θ_m, can be a very efficient procedure for obtaining unconstrained local optima—when it works. Unfortunately, because the approximation of $g(\theta)$ may be grossly inaccurate, the estimate of θ_m may be so far from the true maximizing value that at some stage the value of the objective function may actually be worsened rather than improved. Because of this possibility, the theoretical convergence proofs for the optimal gradient method do not, in general, apply. Moreover, in any given practical problem it is usually rather difficult to predict whether or not a solution will be obtained: This depends both on the degree to which the parabola (4-27) fits the function (4-24) and on the choice of starting point.

For example, if we try to solve the above problem (4-30) with a starting point of $\mathbf{x}_0 = (.5, .5)$, we find that convergence does *not* occur. The initial gradient vector is $\nabla \mathbf{f}(\mathbf{x}_0) = (.72, -2.28)$, but the maximizing value of θ, as given by (4-29), turns out to be negative: $\theta_m = -1.03$. This is due to the denominator of (4-29) being positive at $\mathbf{x}_0$, which implies that C is positive in the approximating function

$$g(\theta) \approx A + B\theta + C\theta^2. \tag{4-27}$$

Thus $d^2g/d\theta^2 \approx 2C > 0$, so that the local optimum $\theta = \theta_m$ must be a minimum rather than a maximum. In fact, we find that the value of the objective decreases from $f(\mathbf{x}_0) = -2.32$ to $f(\mathbf{x}_1) = f(-0.24, 2.86) \cong -3500$—not an auspicious beginning for a maximization procedure. Several iterations later a point is reached at which $x_1 = 8.64$; the first component of the gradient vector then takes on a value of about -10^{34}, and when an "optimal" step length of $\theta_m \cong 1.0$ is calculated the value of x_1 plunges to -10^{34}. Thereafter the numbers being generated exceed the tolerance of the computer.

In this example convergence failed because a point $\mathbf{x} = (8.64, -1.59)$ was generated that was simply too far away from the optimum. In general, it is true of all solution methods relying on approximated functions that the closer the starting point or current point is to the optimum, the greater is the likelihood of convergence. There are no hard and fast rules, of course; in our example problem (4-30), convergence to the local maximum $(1.32, -1.10)$ was achieved from the starting point $(5, -5)$ but not from $(.5, .5)$, which is substantially closer. It should also be noted that a decrease in the objective value at some stage of a maximizing process, as was encountered above in the first iteration from the starting point $\mathbf{x}_0 = (.5, .5)$, does not of itself guarantee that convergence will never occur. This is borne out by still another test on problem (4-30): When the starting point $\mathbf{x}_0 = (1, -1)$ is used, the objective value drops from 3.56 to -11.55 at the first step, but the process then goes on to converge to the same local maximum after only six more iterations. Behavior of this sort does not make it any easier to estimate the likelihood of convergence from some given starting point, nor to devise rules for deciding, after a number of iterations, when to abandon hope and select another starting point. In this area experience is often the best guide.

We have seen that convergence of the optimal gradient method cannot be guaranteed when a quadratic approximation to $g(\theta) \equiv f(\mathbf{x} + \theta\, \nabla \mathbf{f}(\mathbf{x}))$ is used to compute the step length θ_m. There is also a danger of nonconvergence in trying to obtain the exact value of θ_m at each stage by solving the equation

$$\frac{dg(\theta)}{d\theta} = 0 \tag{4-25}$$

via Newton-Raphson iteration, which may itself fail to converge. Perhaps not surprisingly, the safest procedure for calculating the step length in the optimal gradient method is the humble and unsophisticated trial-and-error technique described earlier in this section. Although it may not be the fastest approach nor the most elegant, it has the crucial advantage of providing guaranteed convergence, so that a computer program using some simple automated version of it can be counted upon to climb from any starting point to a stationary point without human intervention. This frees the analyst from the sorts of headanches that inevitably result whenever an iterative computational procedure fails to converge: Should other starting points in the same region be tried or should the search be carried on elsewhere? Might the nonconvergent sequence of points generated thus far have "jumped over" a local optimum at some stage? When should the analyst abandon his efforts and switch to a slower but surer algorithm? These questions are seldom easy to answer and can consume substantial amounts of the analyst's time.

The foregoing paragraph may seem to have dealt rather generously with the trial-and-error approach for obtaining θ_m. It would be appropriate at this point for the author, as a practitioner of operations research, to acknowledge his bias in favor of simple and reliable computational methods, and to remark that in his experience the value of the man-hours expended in pursuit of modest increases in computational efficiency far exceeds the value of the computer hours ultimately saved thereby.

4.9 MODIFICATIONS TO THE OPTIMAL GRADIENT METHOD

Although the optimal gradient method (OGM) is the best known and most thoroughly studied of all direct climbing algorithms, there are others—in fact, other gradient methods—that are superior to it in various circumstances. In this section we shall discuss three or four procedures that are basically variations on the OGM, while in the next section we shall present a different approach that does not use the gradient direction of movement at all. Summaries of the various climbing algorithms that are in general use, along with extensive lists of references, can be found in two excellent review articles by Spang [43] and Wolfe [54].

Given that we have chosen a gradient method, that is, a method that generates points according to the rule

$$\mathbf{x}_{i+1} = \mathbf{x}_i + \theta\, \nabla\mathbf{f}(\mathbf{x}_i),$$

it may seem reasonable to conjecture that the best choice for θ should be the maximizing distance θ_m used by the optimal gradient method. In general, however, this is not true, as the following example illustrates:

$$\text{Min } f(x_1, x_2) = \frac{x_1^2}{a^2} + \frac{x_2^2}{b^2},$$

where a and b are real constants, with $a^2 > b^2$. The optimum lies at the origin, and several of the indifference curves or contours, which are ellipses, are shown in Fig. 4-2. At any point $\mathbf{x}_i$ the gradient vector is perpendicular

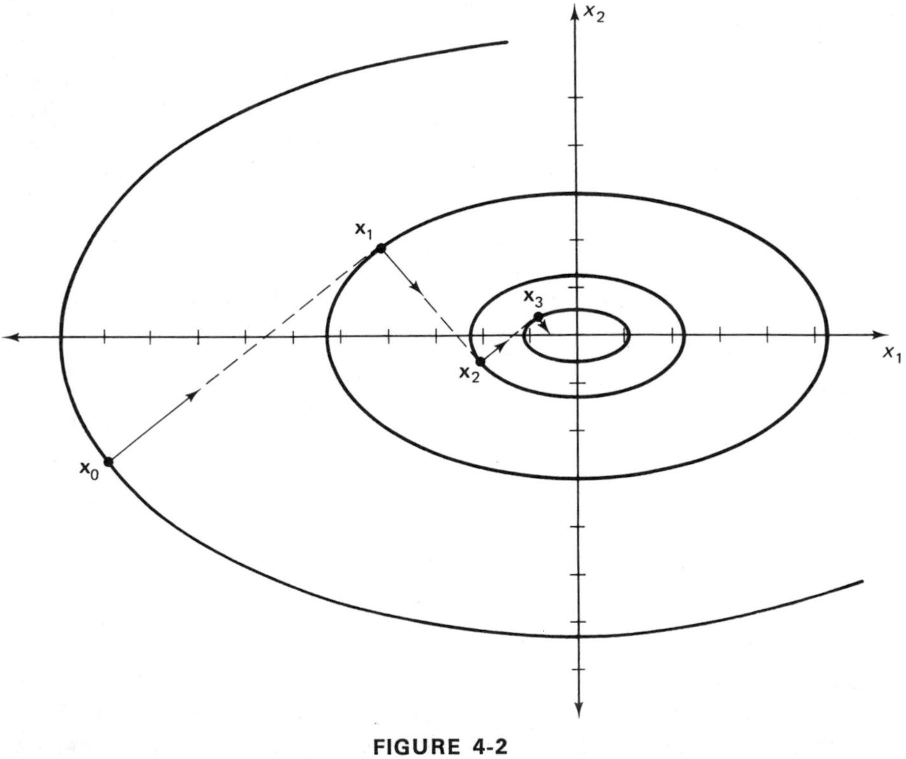

FIGURE 4-2

to the contour that passes through $\mathbf{x}_i$. The strategy of the OGM at each stage is to move along the gradient $\nabla \mathbf{f}(\mathbf{x}_i)$—but in the negative gradient direction, since we are minimizing—until a point $\mathbf{x}_{i+1}$ is reached such that $\nabla \mathbf{f}(\mathbf{x}_i)$ is tangent to the indifference contour through $\mathbf{x}_{i+1}$; the new direction of movement is then perpendicular to the old. The path generated by the OGM for this problem, starting at the point $\mathbf{x}_0$, is shown in Fig. 4-2. Notice how the path zigzags back and forth across the x_1-axis. If we had been able at any stage to stop exactly on the axis, the next gradient vector would have been collinear with the axis and the ensuing step would have carried us straight to the optimum. This observation suggests that, by using at each stage a step length of $\theta = \rho\theta_m$, where $0 < \rho < 1$, instead of the maximizing distance $\theta = \theta_m$, we might have obtained faster convergence.

It turns out that the behavior illustrated in Fig. 4-2 is not an isolated phenomenon but is, in fact, characteristic of almost all types of nonlinear objective functions; in general, the optimal gradient method "overshoots." [The major class of exceptions are the spherical functions

$$f(\mathbf{x}) = \sum_{j=1}^{n} x_j^2,$$

whose indifference contours are circles, spheres, or hyperspheres; the reader can easily verify that such functions are minimized in one optimal gradient step, regardless of starting point.] The tactic of using the shortened step length $\rho\theta_m$ to counteract the tendency of the OGM to overshoot is called *relaxation* or *subrelaxation.* Although the best value of ρ differs according to the form of $f(\mathbf{x})$, the value $\rho = .9$ appears to yield generally good results, with overall rates of convergence often being increased by 100% or more over those of the OGM.

Other modifications of the optimal gradient method that vary the direction of movement at certain iterations are even more successful in speeding convergence. One such procedure is identical to the OGM except that from an occasional point $\mathbf{x}_k$ it uses the "diagonal" or "triangular" direction

$$\mathbf{s}_k = \mathbf{x}_k - \mathbf{x}_{k-2} \tag{4-31}$$

instead of $\mathbf{s}_k = \nabla \mathbf{f}(\mathbf{x}_k)$. The step length θ_k in the diagonal direction is chosen "optimally," that is, to maximize $f(\mathbf{x}_k + \theta \mathbf{s}_k)$. It should be evident that the direction (4-31) is the vector sum of the last two optimal gradient steps,

$$\mathbf{s}_k = \theta_{k-2} \nabla \mathbf{f}(\mathbf{x}_{k-2}) + \theta_{k-1} \nabla \mathbf{f}(\mathbf{x}_{k-1}),$$

and that it is collinear with the hypotenuse of the right triangle whose vertices are $\mathbf{x}_{k-2}$, $\mathbf{x}_{k-1}$, and $\mathbf{x}_k$, as shown in Fig. 4-3(a). The best results seem to be obtained by taking one diagonal step for every six or seven optimal gradient

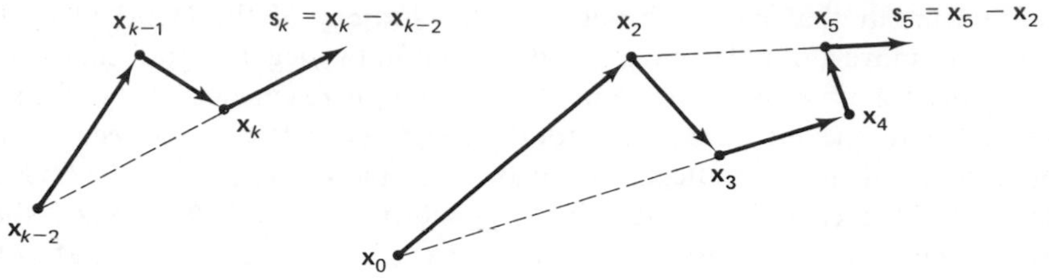

(a) A "diagonal" step (b) The first few "parallel tangents" steps

FIGURE 4-3

steps; convergence can be speeded by as much as a factor of 20. Note that a diagonal step at any stage would have been very helpful in solving the ellipsoidal problem of Fig. 4-2—in fact, such a step would have led directly to the optimum, inasmuch as the line through $\mathbf{x}_0, \mathbf{x}_2, \mathbf{x}_4, \ldots$ passes through the origin, as does the line through $\mathbf{x}_1, \mathbf{x}_3, \ldots$.

Another variation on the OGM is the so-called *method of parallel tangents* [41], in which the gradient direction is alternated with the direction

$$\mathbf{s}_k = \mathbf{x}_k - \mathbf{x}_{k-3}. \tag{4-32}$$

Again, the maximizing step length is chosen at each iteration. The procedure begins with either two or three optimal gradient steps—in the former case the first three points are labeled $\mathbf{x}_0$, $\mathbf{x}_2$, and $\mathbf{x}_3$, with no $\mathbf{x}_1$ being generated—and then uses the direction (4-32) at every other iteration. The first few iterations in a typical problem are mapped in Fig. 4-3(b).

Detailed accounts of computational experience with the OGM modifications discussed above—subrelaxation, diagonal steps, and parallel tangent steps—are included among the references listed in Spang [43] and Wolfe [54].

4.10 NEWTON'S METHOD

In this final section we turn our attention from gradient and gradient-based methods to another type of climbing algorithm that uses a different direction of movement. Although from any given point $\mathbf{x}_k$ the best direction in which to start climbing is $\nabla\mathbf{f}(\mathbf{x}_k)$, it ceases to be best as soon as displacement has begun. Moreover, the farther out along $\nabla\mathbf{f}(\mathbf{x}_k)$ we move, the worse that direction becomes, until (in the optimal gradient method) it differs by a full 90 degrees from the locally best direction. The computational penalty that must be paid for pursuing this ever-worsening direction varies greatly from problem to problem, with the result that the OGM, with or without modifications, yields highly unpredictable rates of convergence. For example, it can be shown that a simple change of scale in a single variable (e.g., measuring some commodity in pounds rather than cubic feet) can significantly alter the rate of convergence.[8] A further disadvantage of the gradient direction is that it becomes particularly inefficient in the vicinity of the optimum: As the components $\delta f/\delta x_j$ approach zero, they tend to oscillate between positive and negative values, causing the gradient direction to vary erratically from one iteration to the next.

A better direction of movement suggests itself when we reflect that the choice of the gradient direction was based on a linear or first-order Taylor approximation to $f(\mathbf{x})$ in the vicinity of the current point $\mathbf{x}_k$:

$$f_L(\mathbf{x}) = f(\mathbf{x}_k) + \nabla \mathbf{f}^T(\mathbf{x}_k) \cdot (\mathbf{x} - \mathbf{x}_k). \tag{4-33}$$

For any fixed length of displacement $|\mathbf{x} - \mathbf{x}_k|$, the approximation (4-33) is maximized by choosing the new point $\mathbf{x}$ in such a way that $\mathbf{x} - \mathbf{x}_k$ is a multiple of the gradient vector $\nabla \mathbf{f}(\mathbf{x}_k)$. It is reasonable to expect that we can improve on this first-order approach by choosing the direction of displacement on the basis of a *second*-order Taylor expansion:

$$f_Q(\mathbf{x}) = f(\mathbf{x}_k) + \nabla \mathbf{f}^T(\mathbf{x}_k) \cdot (\mathbf{x} - \mathbf{x}_k) + \tfrac{1}{2}(\mathbf{x} - \mathbf{x}_k)^T \mathbf{H}(\mathbf{x}_k) \cdot (\mathbf{x} - \mathbf{x}_k). \tag{4-34}$$

Suppose that the Hessian $\mathbf{H}(\mathbf{x}_k)$ is negative definite, perhaps because the current point $\mathbf{x}_k$ is sufficiently close to an unconstrained local maximum (recall the theorems of Section 4.2). Then the value of $\mathbf{x}$ that maximizes the quadratic approximation (4-34) is its single critical point, which is obtained in the usual manner by setting all the partial derivatives $\delta f_Q/\delta x_j$ equal to zero[9]:

$$\nabla \mathbf{f}_Q(\mathbf{x}) = \nabla \mathbf{f}(\mathbf{x}_k) + \mathbf{H}(\mathbf{x}_k) \cdot (\mathbf{x} - \mathbf{x}_k) = \mathbf{0}. \tag{4-35}$$

[8]Consider the problem of minimizing the function $f(x, y) = x^2 + y^2$, starting from the point (2, 2). The initial gradient vector is $\nabla \mathbf{f}(\mathbf{x}) = (2x, 2y) = (4, 4)$, so the direction of movement is $(-4, -4)$, and the optimum, which lies at the origin, is reached in a single step. Now let $z = y/2$, so that we seek to minimize $g(x, z) = x^2 + 4z^2$. The starting point is $(x, z) = (2, 1)$, but the gradient there is now $(2x, 8z) = (4, 8)$, which no longer points directly away from the origin. Hence several iterations will be required for convergence.

[9]Equation (4-35) is derived from (4-34) by using the fact that the gradient of the quadratic form $f(\mathbf{x}) = \mathbf{x}^T\mathbf{H}\mathbf{x}$, where $\mathbf{H}$ is a symmetric n-by-n matrix, is $\nabla \mathbf{f}(\mathbf{x}) = 2\mathbf{H}\mathbf{x}$. This can be verified without great difficulty. Writing out the quadratic form as a sum of its scalar terms, we have

$$f(\mathbf{x}) = \mathbf{x}^T\mathbf{H}\mathbf{x} \equiv \sum_{i=1}^{n} h_{ii}x_i^2 + \sum_{i=1}^{n-1} \sum_{j=i+1}^{n} 2h_{ij}x_ix_j.$$

Note that for any given x_k the double summation contains exactly one term of the form $2h_{kj}x_kx_j$ or $2h_{jk}x_jx_k$ for every $j \neq k$. Inasmuch as $h_{jk} = h_{kj}$, it follows that for any x_k, $k = 1, \ldots, n$,

$$\frac{\delta f}{\delta x_k} = 2h_{kk}x_k + \sum_{j \neq k} 2h_{kj}x_j = 2\mathbf{h}_k \cdot \mathbf{x},$$

where $\mathbf{h}_k$ is the kth row of $\mathbf{H}$. But this equation is just the kth component of $\nabla \mathbf{f}(\mathbf{x}) = 2\mathbf{H}\mathbf{x}$, as was to be shown.

Lemma 3.1(a) guarantees that the Hessian is invertible, so the maximizing value of $\mathbf{x}$ is given by

$$(\mathbf{x} - \mathbf{x}_k) = -\mathbf{H}^{-1}(\mathbf{x}_k) \cdot \nabla \mathbf{f}(\mathbf{x}_k). \tag{4-36}$$

Equation (4-36) suggests using a direct climbing algorithm that proceeds from the current point $\mathbf{x}_k$ in the direction $-\mathbf{H}^{-1}(\mathbf{x}_k)\,\nabla\mathbf{f}(\mathbf{x}_k)$—that is, toward the point that maximizes the quadratic approximation of $f(\mathbf{x})$. This approach is known as *Newton's method.* Notice that because the gradient vanishes at a local optimum, so does the direction vector (4-36); this condition is, of course, the signal that convergence has occurred. Newton's method is fully as important, though perhaps not so widely used, as the method of steepest ascent (which is sometimes referred to as *Cauchy's method*); as we shall see, the two approaches can be used very effectively in combination.

We should mention here that in solving a *minimization* problem Newton's method uses precisely the same direction as in maximizing, rather than its negative. If $\mathbf{x}_k$ is in the vicinity of a local minimum, $\mathbf{H}(\mathbf{x}_k)$ will be positive definite, and the critical point of the approximation $f_Q(\mathbf{x})$ will be its *minimum* value. Thus it will again be desirable to move toward the critical point, that is, in the direction (4-36). Notice that, in fact, the direction vector (4-36) may point toward any type of critical point, depending on whether the Hessian is positive or negative (semi)definite or indefinite; this condition leads to difficulties in obtaining convergence, a topic we shall discuss later on.

It is interesting that Newton's method is closely related to the Newton-Raphson iterative procedure for solving simultaneous nonlinear equations, which was discussed in Section 4.3. Suppose the function $f(\mathbf{x})$ is being optimized via the classical indirect approach of setting its partial derivatives equal to zero and solving the resulting simultaneous equations by Newton-Raphson iteration. In that case the functions f_i and $\delta f_i/\delta x_j$ in the Newton-Raphson system (4-12) are, respectively, the first and second partial derivatives of $f(\mathbf{x})$, and the ith equation of the system becomes

$$\frac{\delta f(\mathbf{x}_0)}{\delta x_i} + \frac{\delta^2 f(\mathbf{x}_0)}{\delta x_i\,\delta x_1}\Delta x_1 + \cdots + \frac{\delta^2 f(\mathbf{x}_0)}{\delta x_i\,\delta x_n}\Delta x_n = 0, \qquad i = 1, \ldots, n, \tag{4-37}$$

where $\mathbf{x}_0$ denotes the starting or current point, and $\Delta x_j = x_j - x_{0j}$ is the adjustment to be made to the value of the jth variable. Representing (4-37) in vector notation, we have

$$\nabla\mathbf{f}(\mathbf{x}_0) + \mathbf{H}(\mathbf{x}_0) \cdot (\mathbf{x} - \mathbf{x}_0) = \mathbf{0},$$

and the new point $\mathbf{x}$ is computed via

$$(\mathbf{x} - \mathbf{x}_0) = -\mathbf{H}^{-1}(\mathbf{x}_0) \cdot \nabla\mathbf{f}(\mathbf{x}_0),$$

which is precisely Eq. (4-36). Thus from the current point $\mathbf{x}_0$ the Newton-Raphson procedure takes a step in the direction $-\mathbf{H}^{-1}(\mathbf{x}_0) \cdot \nabla \mathbf{f}(\mathbf{x}_0)$, covering a distance equal to the magnitude of that vector. From this point of view Newton-Raphson can be regarded as a special case of the Newton climbing method: Whereas the former always uses $\theta = 1$ in the recursion

$$\mathbf{x}_{k+1} = \mathbf{x}_k + \theta[-\mathbf{H}^{-1}(\mathbf{x}_k) \cdot \nabla \mathbf{f}(\mathbf{x}_k)], \tag{4-38}$$

the latter may choose any positive value for θ, depending on the tactics being employed.

One very obvious and effective choice for θ is the optimizing distance $\theta = \theta_m$, defined exactly as in the optimal gradient method. To illustrate how this step length is used in what might be called the "optimal Newton" method, we shall reconsider the example problem of Section 4.3:

$$\text{Min } f(x_1, x_2) = x_1^2 + x_1^2 x_2^2 + 3x_2^4. \tag{4-13}$$

In solving this problem via Newton-Raphson iteration, we began at $\mathbf{x}_0 = (1.0, 1.0)$, computed the first set of adjustments

$$\Delta x_1 = -\tfrac{12}{17} = -.706 \quad \text{and} \quad \Delta x_2 = -\tfrac{5}{17} = -.294,$$

and obtained the next point

$$\mathbf{x}_1 = (.294, .706), \qquad \text{with } f(\mathbf{x}_1) = .874. \tag{4-39}$$

Let us now return to the starting point $\mathbf{x}_0$ and work through an iteration of Newton's method. The direction of displacement is again $(-.706, -.294)$, but we now wish to compute the minimizing distance θ_m instead of simply using $\theta = 1$. Accordingly, we seek a point of the form

$$\mathbf{x}_1 = \begin{pmatrix} 1.0 \\ 1.0 \end{pmatrix} + \theta \begin{pmatrix} -.706 \\ -.294 \end{pmatrix},$$

whose objective value is

$$\begin{aligned} g(\theta) &= f(\mathbf{x}_1) = f(1.0 - .706\theta, 1.0 - .294\theta) \\ &= 5.0 - 6.941\theta + 3.470\theta^2 - .720\theta^3 + .066\theta^4. \end{aligned}$$

To minimize $g(\theta)$ we take its derivative:

$$\frac{dg}{d\theta} = -6.941 + 6.941\theta - 2.161\theta^2 + .262\theta^3 = 0.$$

The only real root is $\theta = \theta_m = 1.755$, which yields

(4-40) $$\mathbf{x}_1 = (-.239, .484), \qquad \text{with } f(\mathbf{x}_1) = .235.$$

Notice that this point is substantially closer to the optimum (i.e., the origin) and has a much smaller objective value than the Newton-Raphson point (4-39).

The major advantage of the Newton method over the Cauchy or steepest-ascent algorithm is that it uses a better direction of movement at each stage; thus it generally requires fewer iterations for convergence. However, Newton's method also has two important disadvantages:

1. It calls for a great deal of calculation at each stage, and
2. It may fail to converge.

The excessive calculation, most of which is devoted to inverting the Hessian matrix, can be reduced by using the same Hessian inverse for three or four consecutive iterations. Although this adjustment tends to increase the number of iterations, it reduces the average time required for each and may provide a significant increase in overall efficiency.

The second disadvantage of Newton's method—its possible failure to converge—causes more serious difficulties. In general it is true that convergence to an unconstrained local maximum will occur provided that at every iteration the Newton vector $-\mathbf{H}^{-1}(\mathbf{x}_k)\,\nabla\mathbf{f}(\mathbf{x}_k)$ points in a direction of increasing f (and provided also that the step length chosen does in fact produce an increase in f). This means that in order for convergence to be guaranteed the gradient vector $\nabla\mathbf{f}(\mathbf{x}_k)$ must at every iteration have a positive component in the "Newton direction":

(4-41) $$\nabla\mathbf{f}^T(\mathbf{x}_k) \cdot [-\mathbf{H}^{-1}(\mathbf{x}_k)\,\nabla\mathbf{f}(\mathbf{x}_k)] > 0$$

[the actual rate of change of f per unit of displacement from $\mathbf{x}_k$ can be obtained by dividing the left-hand side of (4-41) by the magnitude of the Newton vector]. Because, from Lemma 3.1(b), the inverse $\mathbf{H}^{-1}$ is negative definite if and only if $\mathbf{H}$ is negative definite, we conclude from (4-41) that *convergence to an unconstrained local maximum (minimum) is guaranteed if the Hessian matrix is negative (positive) definite* throughout some region R that includes both the starting point and the local optimum being sought.

In view of Theorem 3.10, we can also say that Newton's method must converge to a local maximum (minimum) of $f(\mathbf{x})$ provided that the function is concave (convex) throughout R. This implies that the Newton approach can be used with any starting point to obtain the unique unconstrained local optimum of any convex or concave function. For general functions, however, convergence can be guaranteed only from a starting point that is close to a local maximum (minimum), where concavity (convexity) is assured by Theorem 4.4. This observation suggests combining the method of steepest ascent with that of Newton, using the former initially and switching to the latter as the optimum is approached. This sort of "two-phase" strategy, which tries to exploit some of the strengths of both methods, is usually quite efficient and can always be relied upon to converge—provided that the switch to Newton's method is not premature. Increased computational efficiency can be sought by using any of the short-cuts and modifications discussed in the previous section.

EXERCISES

Section 4.2

4-1. Suppose an economist has conjectured that, on a per-capita basis, national consumption expenditure y is a linear function of disposable income x, and suppose further that he wishes to fit a linear regression model of the form

$$y = \alpha + \beta x + u,$$

where α and β are constant parameters and u is a random disturbance term with mean zero, to an m-year history of income/expenditure data. If X_i and Y_i are the per-capita disposable income and consumption expenditure in the ith year, $i = 1, \ldots, m$, what are the *least-squares* values of α and β? That is, what values of α and β minimize the sum of the squares of the residuals $e_i = Y_i - \alpha - \beta X_i$?

4-2. In the regression model of Exercise 4-1 the disturbance term u represents the aggregated effect of all variable factors other than disposable income which influence consumption expenditure during any given year; among these factors might be retail price levels, the previous year's consumption, and such psychological forces as the Dow-Jones average. Suppose the economist wishes to take some of these variables explicitly into account by fitting to m years of historical data the more sophisticated model

$$y = \beta_1 x_1 + \beta_2 x_2 + \cdots + \beta_n x_n + u.$$

(Note that we have omitted the constant term α; such a term can actually be included within the above model, say as β_k, provided that the fictitious variable x_k is automatically assigned a value of 1 at every "observation.") Define the column vector $\boldsymbol{\beta} \equiv (\beta_1, \ldots, \beta_n)$ and let the historical data be assembled into an m-component column vector $\mathbf{Y}$ and an m-by-n matrix $\mathbf{X}$, where Y_i is the consumption expenditure and X_{ij} is the value of the jth independent variable during the ith year, $i = 1, \ldots, m$. What are the least-squares values of $\beta_1, \ldots, \beta_n$?

4-3. For each of the following functions find all unconstrained local maxima and minima and the global maximum and minimum:

(a) $f(x_1, x_2) = (2x_1 - x_2)^2 + (x_2 - 1)^2$

(b) $f(x_1, x_2, x_3) = 2x_1 - 3x_2 + 4x_3$

(c) $f(x_1, x_2) = x_1^2 + 3x_1x_2 + x_2^2$
(d) $f(x_1, x_2) = x_1^4 - 4x_1^3 + 4x_1^2 + x_2^4 - x_2^3 - 6x_2^2$
(e) $f(x_1, x_2) = (x_1 - 1)^2(x_2^2 + x_2) + x_1^2$
(f) $f(x_1, x_2) = 2x_1^3 - x_1^2x_2 + x_2^3 + x_2^2 - 2x_2$
(g) $f(x_1, x_2) = (x_1 - 1)^2e^{x_2} + x_1$

4-4. If possible, determine a and b so that $f(x) = x^3 + ax^2 + bx$ has
(a) A local maximum at $x = -1$ and a local minimum at $x = +1$.
(b) Vanishing first and second derivatives at $x = 1$.
(c) A local maximum at $x = 1$ and a local minimum at $x = 0$.

4-5. Write the Hessian matrix of the function

$$f(x_1, x_2, x_3) = x_1^3 + x_1x_3 + x_2x_3^2.$$

For what values of the variables, if any, is the Hessian positive semidefinite? Negative semidefinite? Identify all local maxima and minima of f.

4-6. Complete the proof of Theorem 4.4; that is, show that if $\mathbf{x}_0$ in E^n is an unconstrained local maximum of $f(\mathbf{x})$, and $\mathbf{H}(\mathbf{x}_0)$ is negative semidefinite, then $f(\mathbf{x})$ is concave in a small region around $\mathbf{x}_0$.

Section 4.3

4-7. Apply the Newton-Raphson method to the simultaneous equation system

$$f_1(x_1, x_2) = x_1^3x_2 + 2x_2^2 - 20 = 0$$
$$f_2(x_1, x_2) = x_1^2x_2^3 - x_1x_2 - 20 = 0$$

Start at the point $\mathbf{x}_0 = (2, 2)$ and perform two iterations, maintaining three significant figures or slide-rule accuracy.

4-8. One thousand cubic feet of a certain precious liquid is to be sent from the Fountain of Youth to an eccentric millionaire in a specially constructed cubic container whose siding (all six sides) costs \$100 per square foot. The liquid is to be shipped in a number of stages, and the cost of each shipment, including return of the container to the fountain, is \$10. How long should each side of the container be if the millionaire wants the total shipping and container cost to be exactly \$4000? Assume the siding is of negligible thickness.

Section 4.7

4-9. Prove that no sequence of points $\{\mathbf{x}_i\}$ in E^n can have more than one limit point (as defined in Section 4.6.)

4-10. Is it possible for the optimal gradient method to converge to a stationary point that is not a local optimum after an infinite number of iterations? Prove or disprove.

4-11. Construct an example to show that in maximizing a function $f(\mathbf{x})$ the optimal gradient method can generate an infinite sequence of points $\{\mathbf{x}_i\}$ such that

$f(\mathbf{x}_i)$ is finite for any finite integer i but $f(\mathbf{x}_i) \longrightarrow \infty$ as $i \longrightarrow \infty$. That is, show that it is possible for the optimal gradient method to approach an unbounded optimal solution without ever generating an infinite step length θ_m. What computational problems does this possibility raise? What rules might be built into a computer code to enable this situation to be recognized?

Section 4.8

4-12. Suppose we wish to minimize $f(x_1, x_2) = (x_1 - 1)^2 + 2x_2^2$, using the optimal gradient method and obtaining θ_m by solving

$$\frac{dg(\theta)}{d\theta} \equiv \frac{d}{d\theta}[f(\mathbf{x} + \theta\, \nabla \mathbf{f}(\mathbf{x}))] = 0. \tag{4-25}$$

Derive a general expression for θ_m in terms of the coordinates (x_1, x_2) of the current point.

4-13. Starting at $\mathbf{x}_0 = (0, 0)$, take a single optimal gradient step toward the nearest local *minimum* of

$$f(x_1, x_2) = 2x_1^3 - x_1^2x_2 + x_2^3 + x_2^2 - x_1 - 2x_2,$$

computing θ_m via solution of (4-25). Verify that the new gradient vector is perpendicular to $\nabla\mathbf{f}(\mathbf{x}_0)$ and that the value of the objective has, in fact, decreased.

4-14. Find the point that maximizes $f(x_1, x_2) = 2x_1x_2 - 2x_1^2 - x_2^2 - x_2$ by setting its partial derivatives equal to zero and solving. Now seek the maximum of $f(x_1, x_2)$ by performing four iterations of the optimal gradient method, starting at the origin and using (4-25) to compute the exact value of θ_m at each stage. Make a simple diagram of the path generated by the algorithm.

4-15. Discuss some of the computational aspects of the "stepwise" trial-and-error method of obtaining an approximation to the optimal gradient value θ_m. What should the analyst consider in choosing ϵ and μ? Assuming that only the longer ϵ-steps are used during the early iterations, what sort of automatic rules might be included in a computer program to initiate and then regulate the "fine tuning" of the later stages, when steps of shorter and shorter length are being tried? How should the local optimum be identified—that is, what should be the stopping rule?

4-16. Express the numerator and denominator of (4-29) in vector/matrix notation. Under what circumstances is it possible for the value of θ_m computed via (4-29) to be negative? State a necessary condition for convergence of the optimal gradient method to a local maximum when θ_m is being computed via a second-order Taylor approximation.

4-17. Derive an approximation formula analogous to (4-29) for *minimization* search.

4-18. Repeat Exercise 4-14, but compute θ_m at each stage from the Taylor approximation (4-29) instead of via exact solution of (4-25). Why do the two different methods yield the same value of θ_m at each stage?

4-19. Suppose you were required to find a local maximum of

$$f(x_1, x_2) = \frac{x_1 x_2}{(x_1^2 + 1)(2x_2^2 + 1)}$$

via the optimal gradient method, starting at $\mathbf{x}_0 = (.5, .5)$. What method would you use for computing θ_m if you were working (a) with a hand calculator or (b) on a computer? Give reasons for your choices.

Section 4.10

4-20. Find the maximum of $f(x_1, x_2) = 2x_1x_2 - 2x_1^2 - x_2^2 - x_2$ by partial differentiation (this function was used in Exercise 4-14). Now, starting at $\mathbf{x}_0 = (1, 1)$, seek the maximum of f by performing two optimal gradient steps, followed by a "diagonal" step in which the direction of displacement is $\mathbf{x}_2 - \mathbf{x}_0$. At each stage compute the optimal step length by finding the exact solution of

$$\frac{dg(\theta)}{d\theta} \equiv \frac{d}{d\theta}[f(\mathbf{x}_k + \theta \mathbf{s}_k)] = 0,$$

where $\mathbf{s}_k$ is the appropriate direction of displacement. Make a diagram of the path generated thus far. Did the diagonal step accelerate convergence? Why, in general, should diagonal steps be so helpful when $f(\mathbf{x})$ is quadratic?

4-21. Use Newton's method with the optimal step length θ_m to find the maximum of $f(x_1, x_2) = 2x_1x_2 - 2x_1^2 - x_2^2 - x_2$, starting at (1, 1). Obtain the exact value of θ_m by solving $dg(\theta)/d\theta = 0$. How many iterations are required to locate the maximum? Why could this have been predicted in advance?

4-22. Starting at $\mathbf{x}_0 = (1, 0)$, perform a single iteration of Newton's method on the function

$$f(x_1, x_2) = 2x_1^3 - x_1^2x_2 + x_2^3 + x_2^2 - x_1 - 2x_2,$$

using the optimal step length θ_m and maintaining accuracy to two decimal places. Is the new direction vector perpendicular to the old? If the method were continued, would you expect it to converge to a local maximum, a local minimum, or neither?

4-23. Show that when Newton-Raphson iteration—i.e., Newton's climbing method with a step length of $\theta = 1$ at every stage—is applied to $f(x_1, x_2) = x_1^3 + x_2^3$, convergence to a critical point is achieved after an infinite number of iterations. What type of critical point is it? Would the same critical point have been obtained using the "optimal Newton" or the optimal gradient method? Why or why not?

5

Optimum Seeking by Experimentation

5.1 EXPERIMENTAL SEEKING

In this chapter we shall consider the general problem of optimizing a function of one or more variables which can be evaluated at any discrete point but which *cannot be explicitly represented in mathematical form*. Such a problem is called an *experimental seeking problem*. In most cases the function to be optimized is the performance or *response* of some real-world system, where the response is known to depend, perhaps exclusively, on the values of certain controllable parameters called *control variables*, but where the precise relationship between control variables and response is not completely known or cannot be quantified. To evaluate the response function at any specific point —that is, to measure the response for any set of values of the control variables —it is necessary to perform a costly and/or time-consuming *experiment*.

A classic example of an experimental seeking problem would be the determination of the optimal temperature and vapor pressure at which to conduct some industrial chemical reaction. In this problem the objective or response function would be the yield or percent completion of the reaction, and an experiment would consist of actually performing the reaction at some specific temperature and pressure. Each experiment would consume a certain amount of the reagents and a certain amount of time. If either or both of these resources were in limited supply, the analyst's aim would be not to find the optimum as efficiently as possible but rather to come as close to the optimum as he could while performing only a limited number of experiments. As a second example of experimental seeking, consider the problem of assigning a small number of satellite communication channels to a large number of users. Here, one possible objective would be to minimize the expected delay between the instant at which a message is generated by any user and the next instant at which his channel is free for transmission. The experiment necessary to determine the overall expected delay for any particular set of channel assignments might be a 6-hour computer simulation run.

Let us move on from these specific examples to a more general consideration of some of the characteristics and properties of experimental seeking problems. Their basic algebraic formulation is quite simple:

$$\text{(5-1)} \qquad \begin{cases} \text{Max } f(\mathbf{x}), & \mathbf{x} \text{ in } E^n, \\ \text{subject to } a_j \leqq x_j \leqq b_j, & j = 1, \ldots, n, \end{cases}$$

where the x_j are the control variables, $f(\mathbf{x})$ is the unknown response function, and a_j and b_j are fixed lower and upper bounds on x_j. In addition, one or more of the control variables may be restricted to integer values only or to some other sets of discrete values. It should be noted that the a_j and b_j need not be finite, although the experimenter will frequently impose bounds himself by choosing ranges of values within which he believes or assumes the optimum to lie. Because, as we shall see, the upper and lower bounds on the variables have no significant effect on the way in which experimental seeking problems are solved, and because the solution methods themselves are of venerable classical origin, it would not have been totally unreasonable to have included these problems in the previous chapter on "classical unconstrained optimization." Strictly speaking, however, problems with bounded variables and/or integrality restrictions cannot properly be described as unconstrained, and we have accordingly assigned them a special chapter of their own.

Apart from their "nearly unconstrained" structure, experimental seeking problems have other very important general properties that distinguish them within the overall family of optimization problems. First, the high cost—in dollars, elapsed time, or whatever—of performing an experiment plays a significant role in determining how the problem is to be solved. This is usually not true of the purely mathematical optimization techniques discussed elsewhere in this text. For example, when a single algebraic function $f(\mathbf{x})$ is to be maximized via direct climbing for some engineering analysis, it is probably not a matter of major concern whether, say, 5 or 10 hours of computer time are required. To be sure, the 5-hour difference may amount to a few hundred dollars in extra expenses, but this is likely to be a small fraction of the total cost of the analysis. When a great many computer hours are involved, so that their cost *is* significant, these factors are, of course, taken into account in the initial choice of a solution algorithm and in making such decisions as whether or not to pay the price of developing an especially suitable computer code. But computation time and cost would not normally influence the solution process to such an extent that the analyst might be forced to stop short of finding the optimum because he could not afford to continue the search.

This, however, is precisely what can happen in experimental seeking. In most cases the analyst has the resources available to perform only a limited number of experiments. After observing all the results, he is then expected either

1. To recommend a single solution (i.e., a value for each of the experimental variables) which he believes to be optimal or near-optimal, or
2. To state a set of ranges within which he believes the optimal values of the experimental variables must lie.

The conclusions he draws about the location of the optimum will frequently be based on some general assumption about the form of the response function; for example, in problems involving one variable a property known as *quasi-concavity* (see the next section) is usually assumed. Nevertheless, the final intervals within which he can place the optimal values of the variables may be uncomfortably broad. Moreover, because he will probably not have performed enough experiments to enable the optimal solution to be identified with certainty, he will usually not know exactly what the optimal value of the response function is (although he will be able to cite his best observed result as an inferior bound).

Another and more distinctive characteristic of the experimental seeking problem is that, because the response function is unknown, its partial derivatives are not available. It is therefore impossible in experimental seeking to make use of any solution method that relies on Taylor approximations, a restriction that immediately rules out such techniques as Newton-Raphson iteration. More importantly, the experimental measurement of the response at any single point can, by itself, yield no information about its relation to immediately adjacent points (i.e., no slopes or other derivatives, no sense of which way is "up"), and it follows that at any stage all deductions as to the direction in which the optimum lies must rest on the analyst's assumptions about the shape and character of the response function.

It should be clear from the foregoing that the optimal solution to an experimental seeking problem cannot readily be obtained by the methods of Chapter 4, that is, by climbing along a continuous and steadily rising path. Instead, it must be sought by drawing inferences from a set of judiciously placed discrete experiments, inferences based on mathematical properties that the response function is assumed to possess. The various algorithms or strategies for choosing the experiments are known as *experimental seeking methods* or *direct search methods*, although some writers prefer to broaden the term "direct search" to include both experimental seeking and direct climbing methods. Following a presentation of some important background material in the upcoming section, we shall devote the remainder of the chapter to a rather eclectic discussion of several of the most commonly used experimental seeking methods. More comprehensive coverage of this material can be found in the texts of Wilde [50] and Wilde and Beightler [51]; the former, in particular, is relatively unsophisticated and quite easy to read.

5.2 QUASI-CONCAVE AND QUASI-CONVEX FUNCTIONS

As will be seen later on, many experimental seeking algorithms rely on the response function's possessing a property known as *quasi-concavity*, which is defined as follows:

A scalar function $f(\mathbf{x})$ *is a* quasi-concave *function over a convex set* X *in* E^n *if for any two points* $\mathbf{x}_1$ *and* $\mathbf{x}_2$ *in* X,

(5-2) $$f(\lambda \mathbf{x}_1 + (1 - \lambda)\mathbf{x}_2) \geqq \min \{f(\mathbf{x}_1), f(\mathbf{x}_2)\}.$$

for all λ *such that* $0 \leqq \lambda \leqq 1$. *Similarly,* $f(\mathbf{x})$ *is a* quasi-convex *function over* X *if for any* $\mathbf{x}_1$ *and* $\mathbf{x}_2$ *in* X *and any* λ *satisfying* $0 \leqq \lambda \leqq 1$,

(5-3) $$f(\lambda \mathbf{x}_1 + (1 - \lambda)\mathbf{x}_2) \leqq \max \{f(\mathbf{x}_1), f(\mathbf{x}_2)\}.$$

The first definition implies that if $f(\mathbf{x})$ is a quasi-concave function, then it is impossible to find three collinear points such that the point in the middle has a smaller objective value than both of the other two. Thus the hypersurface $f(\mathbf{x})$ cannot have a "pit" or valley (in the ordinary sense of those terms), nor can it have two distinct peaks, although it may have one. For this reason quasi-concave functions are also called *unimodal*, or "one-humped."

Several quasi-concave functions of one variable are diagrammed in Fig. 5-1. A typical quasi-concave profile is that of the bell-shaped curve 5-1(a),

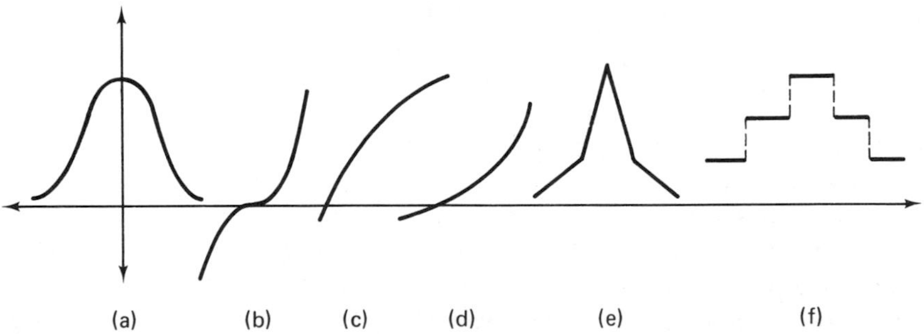

FIGURE 5-1

which represents the unit normal probability density function

$$f(x) = \frac{1}{\sqrt{2\pi}} e^{-x^2/2}.$$

Note that this function is (strictly) concave in the middle but convex on either side; thus a quasi-concave function is not necessarily concave. It is easily shown, however, that concavity does imply quasi-concavity; the proof is left as an exercise.

Another kind of quasi-concave shape is illustrated by the curves (b), (c), and (d) in Fig. 5-1, which have no humps, or at least not in the intervals shown. It should be clear from these curves that any monotonically nonincreasing or nondecreasing function is *both* quasi-concave *and* quasi-convex. Such a function may in addition be concave, as is (c), or convex, as is (d)—and, of course, any linear function would be concave, convex, quasi-concave, and quasi-convex.

The last two curves of Fig. 5-1 illustrate certain pathologies of quasi-concave functions. Although 5-1(e) does not possess a well-defined derivative at every point and 5-1(f) is not even continuous, both satisfy the definition (5-2).

Although quasi-concave and quasi-convex functions—those whose derivatives are continuous—have certain properties that are desirable in mathematical programming, they are not nearly so important or useful as concave and convex functions. One major difficulty in working with them is that the properties of quasi-concavity and quasi-convexity are not necessarily preserved under addition. For example, $f(x) = x^3$ and $g(x) = -3x$ are both quasi-concave for all x, but $h(x) \equiv f(x) + g(x) = x^3 - 3x$ is not, inasmuch as it has a "pit" or local minimum at $x = 1$. The quasi-concavity of a function may even be destroyed by the addition of an "independent" quasi-concave term in some other variable. Thus $f(x_1) = x_1^3$ is quasi-concave everywhere, but $g(x_1, x_2) = x_1^3 + x_2$ is not, as is demonstrated by the following three collinear points:

$$g(0, 0) = 0, \quad g(1, -3) = -2, \quad \text{and} \quad g(2, -6) = 2.$$

As this example might suggest, there are few quasi-concave functions of two or more variables that are not concave as well—hence the relative unimportance of quasi-concavity and quasi-convexity in mathematical programming.

However, quasi-concave functions of one variable do have an important property that can be exploited in experimental seeking. Suppose that a certain response function $f(x)$ is evaluated at two points, $x = p$ and $x = q$, with $p < q$, and suppose it is found that $f(p) < f(q)$, as shown in Fig. 5-2(a).

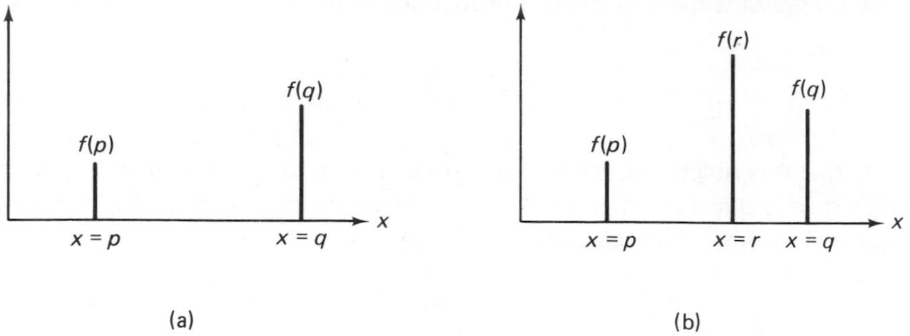

FIGURE 5-2

If $f(x)$ is known or assumed to be quasi-concave, then x^*, the value of x that maximizes $f(x)$, cannot be less than p—for if it were, the point $x = p$ would have a lesser response or objective value than points on either side of it, and the definiton of quasi-concavity would be violated. Similarly, if the response function is next evaluated at $x = r$ and it is found that $f(r) > f(q)$, as shown in Fig. 5-2(b), then x^* cannot exceed q and we can conclude that $p < x^* < q$.

By generalizing this line of reasoning, we can easily establish the following simple but useful result (the formal details of the proof are left for the reader to supply):

Theorem 5.1: Let the function $f(x)$ *be quasi-concave in the closed interval* $[x_0, x_{k+1}]$, *and let the maximum value of* $f(x)$ *over that interval occur at* x^*. *Suppose that* $f(x)$ *has been evaluated at the points* $x_1, \ldots, x_k$, *where*

$$x_0 < x_1 < \cdots < x_k < x_{k+1}.$$

If for some p, $1 \leqq p \leqq k$, *the value* $f(x_p)$ *is at least as great as all the other* $k - 1$ *values, then* $x_{p-1} \leqq x^* \leqq x_{p+1}$.

A similar result can be obtained for the minimum of a quasi-convex function. The interval $[x_{p-1}, x_{p+1}]$ within which x^* must lie is known as the *interval of uncertainty*. We shall see below how an experimental search for x^* can be conducted—that is, how the points $x_1, \ldots, x_k$ can be chosen—so that the interval of uncertainty is reduced in an optimal manner.

5.3 SIMULTANEOUS SEARCH STRATEGIES FOR ONE-VARIABLE PROBLEMS

Experimental seeking methods may be either *simultaneous* or *sequential*. A simultaneous search method is used when the analyst is given the resources for performing some fixed number of experiments but is required to plan them all before he has observed the results of any. A classic context of simultaneous search problems is agronomy: If an analyst is trying to find the best possible fertilizer mix or soil acidity for some new hybrid grain but must file his report after only one growing season, he must plan all his experiments simultaneously. In sequential search, on the other hand, the result of one experiment (i.e., the value of the response function at one point in variable space) can be used to help determine where to place the next. In other words, a sequential method is iterative. Sequential methods may be further subclassified according to whether the experiments are planned individually or several at a time; methods of the latter type are more difficult and will not be considered in any detail in this text.

We turn our attention first to experimental seeking problems that have only *one control variable*. If nothing at all is known or can be assumed about the form of the objective or response function, then there is no real basis on which to determine an optimal search strategy, whether simultaneous or sequential, and the placement of the experiments must essentially depend on the analyst's judgment. A typical approach in a one-variable, simultaneous search problem is to space the experiments equally across the range of possible values of the control variable, or across some subinterval in which the optimum is believed to lie. Alternatively, the experiments may be concentrated in the interval considered most likely to contain the optimum and scattered thinly over the less promising regions. In either case the experimental results are usually fitted to a polynomial of the appropriate degree, whose maximum (or minimum) can then be located by differentiation. This maximum is taken as an approximation to the optimal value of the control variable. It must be noted, however, that because nothing is assumed about the form of the unknown response function, the polynomial approximation can be no more than a piece of educated guesswork. In fact, regardless of the experimental results no statement whatever about the location of the optimum is rigorously justified—strictly speaking, the response function might rise to a very tall, thin peak at any of the untested points.

Suppose, however, that the response function whose maximum we are seeking is known or can be assumed to be quasi-concave (all results and search techniques to be described below will apply, with obvious modifications, to the minimization of quasi-convex functions). Quasi-concave functional behavior occurs with great frequency and in great variety throughout nature and in human affairs. For example, fully unimodal behavior, in which the response rises monotonically to a single peak and then falls away, is observed in such diverse phenomena as the following: per-acre yield of certain crops as a function of soil acidity, total sales income from a product as a function of its per-unit price, expected longevity as a function of daily calory intake, manufacturing efficiency or output per cost dollar as a function of production level, and total votes received by a politician as a function of his position on the liberal-to-conservative spectrum. To this list of examples might be added any phenomenon that obeys the familiar "law of diminishing returns." Other types of quasi-concave behavior that are often encountered naturally are illustrated by the monotonic curves (b), (c), and (d) in Fig. 5-1. Of course, a monotonically increasing or decreasing response function may also be "created" artificially in an experimental seeking problem when the response forms a legitimate peak but the control variable is restricted to a range of values that lies entirely on one side of it. In any case, it should be clear that an analyst may often assume quite reasonably that a given physical or economic function of one variable is quasi-concave, even though he may have no specific knowledge whatever about the system under study.

The assumption of quasi-concavity is very important because, for any given set of experimental results, it allows the analyst to identify, in accordance with Theorem 5.1, an interval of uncertainty within which the optimum x^* must lie. Consider first a *simultaneous search* problem in which just enough resources are available for, say, N experiments. The analyst's task is to plan those experiments in such a way as to reduce or minimize in some optimal manner the interval of uncertainty that will remain afterward. But inasmuch as he cannot know in advance which of the experimental points $x_1, x_2, \ldots, x_N$ will yield the greatest response, he has no way of predicting which of the N possible intervals of uncertainty will result. Therefore his safest strategy is to plan the experiments in such a way as to *minimize the largest of those N possible intervals.* This *minimax* criterion is rather conservative and pessimistic, in that it envisions the worst outcome and proceeds to guard against it, but it has the virtue of allowing an assured degree of success to be calculated, regardless of the experimental results.

One obvious and appealing search plan is to space the N simultaneous experiments equally across the permitted range of values of the control variable, where the range is assumed to be finite, with endpoints x_0 and x_{N+1} and length $L = x_{N+1} - x_0$. The equal spacing divides the original interval into $N + 1$ subintervals each of length $L/(N + 1)$; therefore, regardless of the experimental results, the final interval of uncertainty will have length $2L/(N + 1)$. It is not difficult to show that this common-sense search plan is in fact optimal, according to our minimax criterion, whenever N is *odd*. Let the experimental values selected for the control variable in any simultaneous search plan of size N (not necessarily an equally spaced plan) be $x_1, \ldots, x_N$, where the values have been ordered so that $x_1 < x_2 < \cdots < x_N$. Among the N possible intervals of uncertainty that may result are the following:

$$[x_0, x_2], [x_2, x_4], [x_4, x_6], \ldots, [x_{N-1}, x_{N+1}].$$

Inasmuch as the lengths of these $(N + 1)/2$ intervals must sum exactly to L, their average length is $2L/(N + 1)$ and at least one of them must equal or exceed that average. Thus, for any arrangement whatever of N simultaneous experiments, the maximum possible interval of uncertainty, which we are seeking to minimize, must be at least $2L/(N + 1)$. But this "minimal maximum" is achieved by the strategy of equal spacing, which must therefore be optimal. Q.E.D.

When N is *even*, however, a better strategy exists. Consider the following $(N/2) + 1$ intervals of uncertainty, any one of which might result from a simultaneous search plan involving N experiments:

$$[x_0, x_2], [x_2, x_4], \ldots, [x_{N-2}, x_N], [x_{N-1}, x_{N+1}].$$

These intervals cover the entire range of x, with an overlap between x_{N-1} and x_N; hence their total length is greater than L and their average length exceeds

$$\frac{L}{(N/2) + 1} = \frac{2L}{N + 2}. \tag{5-4}$$

It follows that the maximum possible interval of uncertainty in any search plan must also exceed $2L/(N + 2)$. This lower bound can be approached as nearly as physical circumstances permit by a strategy of arranging the experiments in equally spaced pairs, as shown in Fig. 5-3 for $N = 6$. The distance between the two experiments in any pair represents either the smallest incremental adjustment that can be made to the control variable or the smallest separation for which the difference in response can be measured accurately. Let this distance, known as the *experimental resolution*, be denoted by ϵ. Inasmuch as the $N/2$ pairs of experiments divide the original interval into $(N/2) + 1$ parts, it is easily seen that the final interval of uncertainty will be

$$\begin{cases} \dfrac{2L}{N+2} + \dfrac{\epsilon}{2} & \text{if } \max\limits_{1 \leq i \leq N}\{f(x_i)\} \text{ is } f(x_1) \text{ or } f(x_N), \\ \dfrac{2L}{N+2} & \text{otherwise.} \end{cases} \tag{5-5}$$

Thus the maximum possible interval of uncertainty is only $\epsilon/2$ greater than the theoretically calculated lower bound (5-4). Actually, as the reader might

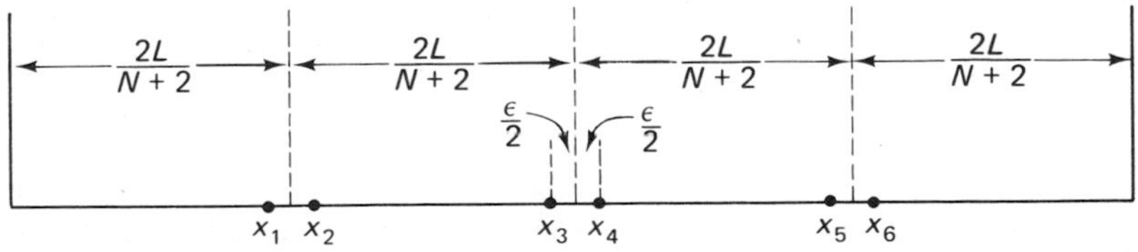

FIGURE 5-3

already suspect, the strategy of equally spaced pairs is not quite optimal: The experiments diagrammed in Fig. 5-3 can be displaced slightly to produce a very small decrease in the maximum interval of uncertainty (see Exercise 5-6).

Notice that for even values of N the largest interval in the set (5-5) is less than the largest that can result from the strategy of equally spaced experiments by an amount

$$\frac{2L}{N+1} - \left(\frac{2L}{N+2} + \frac{\epsilon}{2}\right) = \frac{2L}{(N+1)(N+2)} - \frac{\epsilon}{2}.$$

If N and ϵ are small, this can be a useful improvement. Note also that to increase the number of experiments from N (even) to $N + 1$ (odd) reduces the maximum possible interval of uncertainty by only

$$\left(\frac{2L}{N+2} + \frac{\epsilon}{2}\right) - \frac{2L}{(N+1)+1} = \frac{\epsilon}{2}.$$

Thus, unless ϵ is rather large, it is really uneconomical to use an odd number of simultaneous experiments.

5.4 SEQUENTIAL SEARCH STRATEGIES FOR ONE-VARIABLE PROBLEMS

Suppose now that our one-variable experimental seeking problem is to be solved by *sequential* rather than simultaneous search. In a sequential method the analyst plans his experiments one at a time, choosing each new test value of the control variable after he has observed the result of the previous experiment. We continue to assume that the objective function is quasi-concave, so that at any stage in the search there will remain, in accordance with Theorem 5.1, an interval of uncertainty within which the optimum must lie. Let I_k denote the interval of uncertainty remaining after the kth experiment, and let L_k denote the length of that interval. During the search each experiment will be placed inside the current interval of uncertainty (why would it be useless, under our assumptions, to place an experiment outside it?), and it is easily seen that I_{k+1} will be a subinterval of I_k for all $k \geqq 0$, where I_0 is the original finite range of values within which the control variable is assumed to be restricted. Note that Theorem 5.1 implies $I_1 = I_0$ and $L_1 = L_0$; thereafter, however, $L_{k+1} < L_k$ for all $k \geqq 1$. Thus every experiment except the first reduces the interval of uncertainty.

Having made all these preliminary remarks, we now turn our attention to the question of how to conduct an optimal sequential search. If we adopt the same minimax criterion that was used in simultaneous search, the question becomes, what strategy should the analyst use in planning his sequence of experiments so as to minimize the maximum interval of uncertainty that can possibly remain after N experiments? Let us suppose first that it is not known in advance exactly how many experiments are to be performed. Sequential search problems in which N is initially unknown are not at all uncommon. The uncertainty in N may be due to any of several factors:

1. The total amount of experimental resources to be made available may not be known until after experimenting has begun;
2. The resources required for an experiment may vary unpredictably, perhaps as a function of the control variable or of the response; or
3. The analyst may be free to stop experimenting whenever he feels he has learned enough about the response function and located the optimum with sufficient accuracy.

The optimal sequential search strategy for unknown N is also interesting because it is simpler and more elegant than the optimal strategy for known N—and, surprisingly, almost as efficient.

Assume, then, that N is unknown and consider the interval of uncertainty I_k that remains after the kth experiment, $k \geqq 1$. A typical interval is diagrammed in Fig. 5-4; its end points are denoted by A and C, while the experiment yielding the maximum response observed thus far is located at point B. Now, the essential nature of a minimax strategy is that it must be symmetric or balanced in such a way that the analyst is indifferent about the outcome of any particular experiment, i.e., indifferent as to whether or not the response for the current experiment exceeds the previous maximum. Any imbalance would create the possibility of an unfavorable outcome, which is precisely what the conservative minimax approach seeks to avoid. Hence, referring to Fig. 5-4, the next experiment must be placed at a point D that is symmetrically located with respect to point B within the interval $[A, C]$, so that $AD = BC$. Depending on whether or not the response at D exceeds the previously maximal response at B, the new interval of uncertainty I_{k+1} will be either $[A, D]$ or $[B, C]$, and in either case its length will be $L_{k+1} = BC$. Suppose, for preference, that the response at B remains maximal, so that I_{k+1} is $[A, D]$; then, continuing, the next or $(k + 2)$th experiment will be placed between A and B at a point E that satisfies $AB = ED$, and the next interval of uncertainty I_{k+2} will have length $L_{k+2} = AB$.

Based on the foregoing argument we can substitute into the identity

$$AC = AB + BC$$

to produce a very useful result about the lengths of successive intervals of uncertainty:

$$L_k = L_{k+1} + L_{k+2}, \qquad k \geqq 1, \tag{5-6}$$

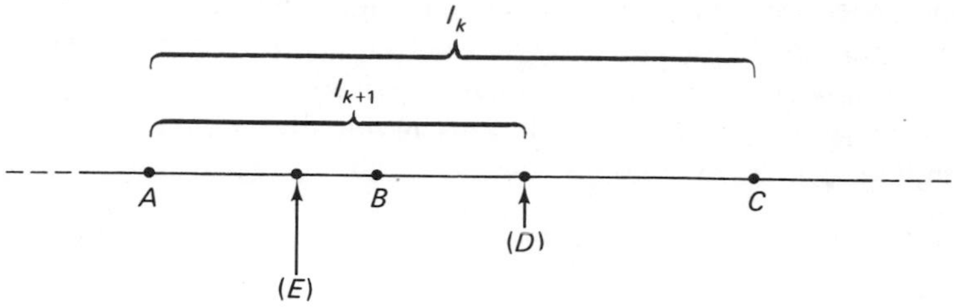

FIGURE 5-4

although, as noted earlier,

$$L_0 = L_1. \tag{5-7}$$

Let us now make the further observation that in the optimal search strategy the ratio of the lengths of any two consecutive intervals must be constant throughout:

$$\frac{L_{k+1}}{L_k} = r \qquad \text{for all } k \geqq 1. \tag{5-8}$$

This conjecture can be justified by arguing that, because the total number of experiments N is unknown, we cannot adopt a sophisticated strategy that carefully reduces the interval of uncertainty more slowly than necessary in the early stages in order to culminate triumphantly in an absolutely minimal interval precisely at the Nth stage. Instead, we must plan to do the best we can with each experiment, so that the interval of uncertainty is reduced at a steady rate r that can be maintained indefinitely. Substitution of (5-8) into (5-6) yields

$$L_k = rL_k + r^2 L_k$$

or

$$r^2 + r = 1; \tag{5-9}$$

the positive root of (5-9) is

$$r = \frac{-1 + \sqrt{5}}{2} \cong .618,$$

correct to three decimal places.

We can now specify the optimal sequential search strategy for unknown N. The placement of the first experiment is determined by the fact that it may become the end point of I_2 and must therefore be located at a distance L_2 from one end of the original interval. But $L_2 = rL_1 = rL_0$, so the first experiment should be placed a fraction r of the way from one end of the original interval to the other. Thereafter, as argued earlier, each new experiment must be placed symmetrically with respect to the "leading" experiment (i.e., the one whose response has been greatest thus far) within the current interval of uncertainty. It is easy to prove that if this strategy is employed, the following two conditions must hold after the completion of the kth experiment, for all $k \geqq 1$:

1. The length of the current interval of uncertainty I_k is $L_k = r^{k-1}L_0$, and
2. The current leading experiment is located at a distance rL_k from one end of I_k.

Clearly, both of these conditions hold after the first experiment E_1. The second experiment E_2 is then placed symmetrically with respect to E_1 within the current interval I_1, as shown in Fig. 5-5. Regardless of which of the first two experiments yields the greater response, it can be seen from Fig. 5-5 that

$$L_2 = rL_0$$

and that the leading experiment will be located at a distance

$$(1 - r)L_0 = \frac{(1 - r)L_2}{r} = rL_2 \tag{5-10}$$

from one end of I_2, where the second equality in (5-10) follows from (5-9). Thus "conditions 1 and 2" hold for $k = 2$ and, by induction, for all succeeding values of k. Q.E.D.

The procedure just described was originally suggested by Kiefer [26] and has come to be known as *golden section search*. The name traces back to Euclid's discovery that, using only a compass and straightedge, it is possible to divide any given line segment into two parts such that the ratio of the whole to the larger part equals the ratio of the larger part to the smaller. Taking the original line to be of unit length, the length λ of its larger part must obey the relation

$$\frac{1}{\lambda} = \frac{\lambda}{1 - \lambda}.$$

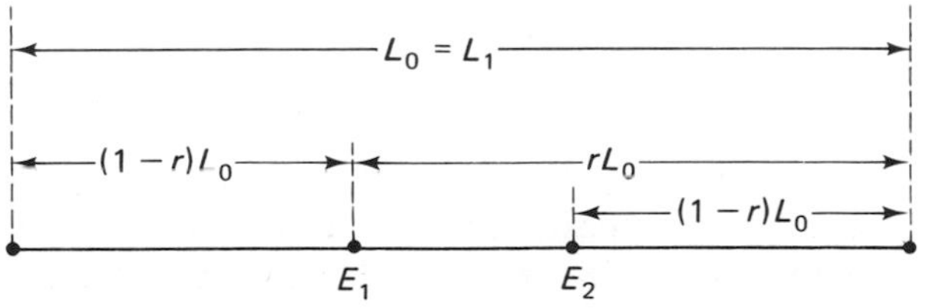

FIGURE 5-5

Cross multiplication leads to

$$\lambda^2 + \lambda = 1,$$

so that λ is precisely equal to our familiar ratio $r = (-1 + \sqrt{5})/2$. The division of a line in this manner came to be known as the *golden section,* both because it has several rather interesting geometric and numerical properties and because the proportions of the two parts seem quite harmonious and pleasing to the eye. Indeed, the golden section ratio has appeared and reappeared in art and architectural design since the time of the ancient Greeks.

Thus far in our discussion of one-variable sequential search, we have assumed that the precise number of experiments N was not known in advance. It is interesting that when N is known the minimax-optimal search strategy, proposed by Kiefer [26] and called *Fibonacci search,* is almost exactly the same as golden section search. Both methods operate automatically after the first experiment, with each succeeding experiment being placed symmetrically within the current interval of uncertainty, and in both methods the lengths of successive intervals satisfy the sequential relationship

(5-6) $$L_k = L_{k+1} + L_{k+2}, \qquad k \geqq 1.$$

In fact, it is this relationship that accounts for the name "Fibonacci search." Fibonacci was a thirteenth-century mathematician who studied the sequence of numbers 1, 1, 2, 3, 5, 8, 13, 21, 34, . . . , in which each term from the third on is the sum of its two predecessors. In a *generalized Fibonacci sequence* the first two terms may be any nonnegative numbers a and b, after which the sequence continues

(5-11) $$a, b, a + b, a + 2b, 2a + 3b, 3a + 5b, 5a + 8b, \ldots.$$

It is evident from (5-6) that the successive intervals of uncertainty in a Fibonacci or a golden section search, taken in reverse order, form a generalized Fibonacci sequence. Letting F_i denote the ith member of (5-11), the reader may be surprised to learn that, regardless of a and b, the limiting value of the fraction F_i/F_{i+1} as i approaches infinity is none other than $r = (-1 + \sqrt{5})/2$, the golden section ratio. A proof of this classical result for $a = b = 1$ is given in Wilde [50].

Returning to our comparison between Fibonacci and golden section search, the only difference between the two methods is due to a very slight disparity in the positioning of the first experiment. This disparity increases slowly from one stage to the next, until at the $(N - 1)$th stage of Fibonacci search the leading experiment lies exactly $\epsilon/2$ units away from the midpoint of the current interval of uncertainty I_{N-1}, where ϵ is the experimental resolution. Thus the last experiment can be symmetrically placed just on the other side of the midpoint, and the length of the final interval of uncertainty will be

$$L_N = \frac{L_{N-1}}{2} + \frac{\epsilon}{2}, \tag{5-12}$$

which represents a reduction of almost 50%. By contrast, the final reduction in an N-experiment golden section search is from L_{N-1} to $L_N = rL_{N-1}$, which amounts to only about 38%.

This 50% versus 38% difference on the final experiment actually overestimates the relative advantage of the Fibonacci over the golden section method. For any given starting interval and any $N \geqq 5$, the final interval left by a golden section search is about 17% longer than that left by a Fibonacci search. The efficiencies of Fibonacci, golden section, and simultaneous search are compared in Table 5-1 for various values of N; each entry is the appropriate *reduction ratio* L_N/L_0, that is, the ratio of the length of the final interval of uncertainty to that of the starting interval, with ϵ assumed to be completely negligible. The values of this ratio for $N = 20$ are, to seven decimal places, .0000914 for Fibonacci search and .0001070 for golden section. The latter exceeds the former by about 17%, as anticipated, but note that this difference would be reduced if the experimental resolution ϵ, which lengthens only the Fibonacci intervals, were included in the computations.

TABLE 5-1

REDUCTION RATIOS FOR ONE-VARIABLE SEARCH STRATEGIES (ϵ ASSUMED NEGLIGIBLE)

N	*Fibonacci*	*Golden Section*	*Simultaneous*
2	.500	.618	.500
3	.333	.382	.500
4	.200	.236	.333
5	.125	.146	.333
6	.077	.090	.250
8	.029	.034	.200
10	.011	.0132	.167
12	.0043	.0050	.143
14	.0016	.0019	.125
16	.0006	.0007	.111
20	.00009	.00011	.091

From a practical point of view, the very small advantage afforded by Fibonacci search seems scarcely enough to justify the effort of studying it in detail, although any reader wishing to do so may consult Wilde [50]. Far more important than the difference between Fibonacci and golden section is the great disparity between simultaneous and sequential search in general.

The superiority of the latter reaches a factor of 100 after only 15 experiments and a factor of 1000 after only 20, and it is clear that in most situations the analyst should be willing to pay a substantial premium for the capability of planning his experiments sequentially rather than simultaneously. If completely sequential experimentation is impossible, perhaps because not enough time is available, it would still be desirable to proceed "blockwise sequentially", that is, to perform several experiments at a time, assessing the results of each group or "block" before planning the next. The problems that arise in designing optimal blockwise search strategies under various constraints and assumptions can become extremely complicated, although the strategy of dividing the experiments evenly among the blocks and then arranging each block of experiments in equally spaced pairs seems to be optimal or near-optimal in a great many cases.

5.5 STRATEGIES FOR MULTIVARIABLE SEARCH

We conclude this chapter with a few brief remarks on experimental seeking problems having *two or more control variables.* The shift to higher dimensionality introduces certain difficulties that were not present in one-variable search problems, difficulties having to do with the fact that, even if the response function is assumed to be quasi-concave, segments of the feasible region cannot be eliminated so simply as in the one-variable case. For example, suppose that some experimental response function $f(x_1, x_2)$ has been evaluated at four points, with the results shown in Fig. 5-6(a). Assuming that f is quasi-concave, can it possibly take on its maximum value at the point $\hat{\mathbf{x}}$? Perhaps somewhat surprisingly, the answer is yes—the indifference curves or contours of f might be as shown in Fig. 5-6(b). As this example may suggest, the assumption of quasi-concavity is frequently not strong enough to allow useful conclusions about the location of the optimum to be

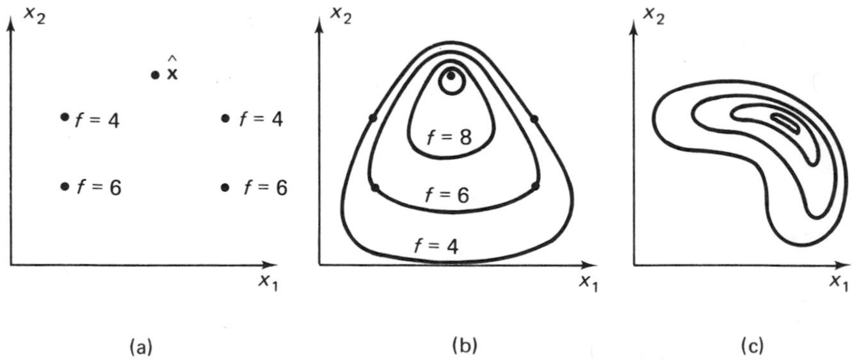

FIGURE 5-6

drawn from a given set of experiments. Moreover, the size and shape of the final region of uncertainty will inevitably depend on the actual results of the experiments—no guaranteed minimax-optimal search strategies, simultaneous or sequential, seem to exist for multivariable problems. And there is a further disturbing point: The assumption of quasi-concavity seems less natural and less persuasive than it did in the one-variable case. It is easy to imagine that there must be many experimental response functions with indifference contours similar, say, to those of Fig. 5-6(c), and one cannot help feeling that attention should focus primarily on those multivariable search strategies that can be applied to all types of experimental seeking problems.

The difficulties described above are felt most acutely in *simultaneous* search. If the analyst is unable or unwilling to postulate some specific algebraic form for the response function, he must plan his experiments with nothing but his intuition and common sense to guide him. One possible approach is to identify a discrete set of values for each variable and then simply test all possible combinations of those values (or almost all, depending on the permitted number of experiments N). Such a pattern of experiments is known as a *factorial* design. The placement of 20 experiments in a two-variable factorial design is schematically diagrammed in Fig. 5-7(a); note that the values chosen for a given control variable need not be equally spaced. The simple factorial approach is frequently rejected in favor of some other design that is more sophisticated and hopefully more efficient [such as the design of Fig. 5-7(b), which might be adopted if the response were known or believed to be more sensitive to the variable x_2 than to x_1]. In general, the proper choice of design depends on the experimenter's prior beliefs about how the control variables interact, i.e., about the form of the response function. This is a matter for probabilistic and statistical analysis; we shall not consider it further.

Multivariable *sequential* search is not quite so vague and amorphous as simultaneous search. Although it is again true that no search algorithms exist that are minimax-optimal in the sense defined earlier, we can at least devise sequential strategies that are guaranteed to converge toward or close

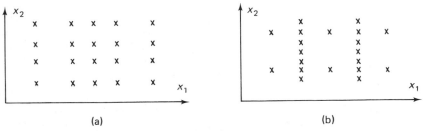

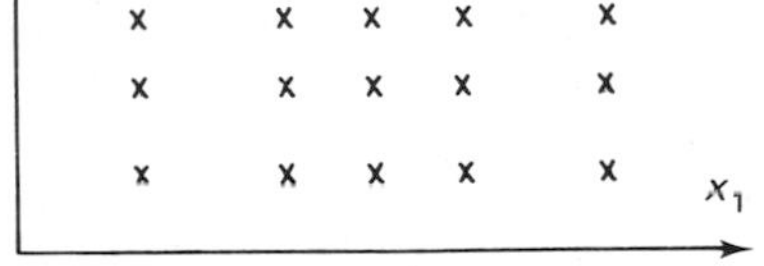

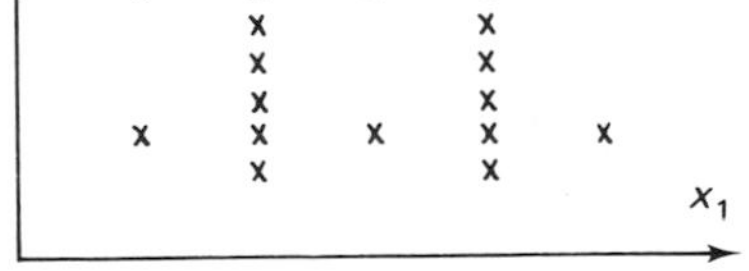

FIGURE 5-7

in on the optimum of any unimodal response function [in a multivariable context "unimodal" means "single-peaked" but not necessarily quasi-concave, as illustrated in Fig. 5-6(c)]. In all cases the rates at which these strategies proceed toward the optimum depend on the experimenter's luck and cannot be predicted in advance—by contrast, the rate at which a golden section search narrows the interval of uncertainty in a one-variable problem is completely predictable.

The simplest and most direct sequential search strategy is none other than the optimal gradient method, suitably modified as required to deal with the unknown experimental response function. Given some starting point $\mathbf{x}_0$ in E^n, the method begins with $n + 1$ simultaneous experiments at the points $\mathbf{x}_0, \mathbf{x}_0 + \epsilon_1\mathbf{e}_1, \mathbf{x}_0 + \epsilon_2\mathbf{e}_2, \ldots, \mathbf{x}_0 + \epsilon_n\mathbf{e}_n$, where the $\mathbf{e}_j$ are unit vectors in E^n and ϵ_j is the experimental resolution for the variable x_j (or some other small positive number). These experiments yield an approximation to the gradient vector $\nabla\mathbf{f}(\mathbf{x}_0)$, whose jth component is given by

$$\frac{\delta f(\mathbf{x}_0)}{\delta x_j} \approx \frac{f(\mathbf{x}_0 + \epsilon_j\mathbf{e}_j) - f(\mathbf{x}_0)}{\epsilon_j}. \tag{5-13}$$

Experiments are then performed sequentially "along the gradient," that is, at points of the form

$$\mathbf{x} = \mathbf{x}_0 + \theta\,\nabla\mathbf{f}(\mathbf{x}_0), \qquad \theta > 0,$$

exactly as in the direct climbing methods discussed in the previous chapter. The value of θ that maximizes

$$f(\mathbf{x}) = f(\mathbf{x}_0 + \theta\,\nabla\mathbf{f}(\mathbf{x}_0)) \equiv g(\theta)$$

can be sought via the following procedure:

1. Choose a positive step length μ and perform an experiment at $\mathbf{x} = \mathbf{x}_0 + \mu\,\nabla\mathbf{f}(\mathbf{x}_0)$.
2. If $f(\mathbf{x}_0 + \mu\,\nabla\mathbf{f}(\mathbf{x}_0))$ exceeds $f(\mathbf{x}_0)$, perform another experiment at $\mathbf{x} = \mathbf{x}_0 + 2\mu\,\nabla\mathbf{f}(\mathbf{x}_0)$ and continue taking steps of length μ until the response function shows a decrease, say, at the kth step.
3. The maximizing value θ_m must then lie between $(k - 2)\mu$ and $k\mu$ (or, if $k = 1$, between 0 and μ). Ignore the experiment at $\mathbf{x}_0 + (k - 1)\mu\,\nabla\mathbf{f}(\mathbf{x}_0)$ and perform a golden section search until the interval of uncertainty is as small as desired.

Using the point at which the eventual leading golden section experiment is located as the new starting point $\mathbf{x}_1$, a further set of n simultaneous experiments is performed to estimate the gradient $\nabla\mathbf{f}(\mathbf{x}_1)$, and the entire process is repeated. The method continues until all experimental resources have been exhausted or until a point $\mathbf{x}_p$ is found at which the gradient approximately vanishes.

In applying this optimal gradient method to an experimental seeking problem, the choice of μ at each stage, or for that matter at each step, must be governed by the analyst's judgment, as must the decision about how far to proceed with the golden section search. In the early stages it is usually best to omit the golden section phase entirely, using as the new starting point the last "μ-step" point that showed an increase, that is, using $\mathbf{x}_{i+1} = \mathbf{x}_i + (k - 1)\mu\, \nabla\mathbf{f}(\mathbf{x}_i)$. Later on, smaller steps are taken and, insofar as experimental resources permit, golden section is used to reduce the interval of uncertainty. Incidentally, golden section search can also be used as an aid in locating θ_m when the optimal gradient method is being applied to a *known* objective function in ordinary direct climbing.

Another rather straightforward type of multivariable sequential search strategy is the "contour tangent elimination" method described by Wilde [50]. This method is applicable whenever it can be assumed that the value of the response function increases monotonically along the straight line from any feasible point $\mathbf{x}_0$ to the maximum $\mathbf{x}^*$. An equivalent assumption would be that the gradient vector $\nabla\mathbf{f}(\mathbf{x}_0)$ at any feasible point $\mathbf{x}_0$ has a positive component in the direction of the straight line from $\mathbf{x}_0$ to $\mathbf{x}^*$. Quasi-concave response functions have this property, and so have certain other unimodal response functions, such as the one illustrated in Fig. 5-6(c).

To see how this property can be exploited in experimental search, suppose the gradient vector $\nabla\mathbf{f}$ has been evaluated at some feasible point $\mathbf{x}_0$, and consider the hyperplane that is perpendicular to $\nabla\mathbf{f}(\mathbf{x}_0)$ at $\mathbf{x}_0$ (and therefore tangent to the indifference contour through $\mathbf{x}_0$). If the response function $f(\mathbf{x})$ has the property described above, then its maximum $\mathbf{x}^*$ must lie on the "upward" side of that hyperplane, that is, on the side toward which $\nabla\mathbf{f}(\mathbf{x}_0)$ points. Thus every point on the "downward" side can be eliminated from consideration, reducing the region of uncertainty within which $\mathbf{x}^*$ must lie.

It should now be quite clear how contour tangent elimination search operates. At each stage in solving an n-variable problem, a cluster of $n + 1$ simultaneous experiments is performed to evaluate the gradient vector $\nabla\mathbf{f}$ at some suitably chosen feasible point; the hyperplane generated by that gradient vector then slices away a further portion of the region of uncertainty. In general, each feasible point selected for gradient evaluation should be roughly in the "middle" of the current region of uncertainty, so that the new region will be about half as large as the old. However, it must be emphasized that an irregular spatial region does not have an unambiguously defined midpoint, and that as a rule the fraction by which the region of uncertainty is reduced from one stage to the next depends significantly on the orientation of the gradient vector. Inasmuch as the orientation cannot be predicted in advance, there can be no "minimax-optimal" search strategies of the type designed for one-variable problems. Note that the contour tangent elimination procedure is the multivariable analogue of golden section and Fibonacci search; it is most suitable for two-variable problems because a simple planar map can be used to keep track of the region of uncertainty and to aid in choosing the search points.

Quite a few other multivariable sequential search methods have been developed and studied since experimental seeking began to be of interest in the early 1950s. They tend to be somewhat more sophisticated than the two discussed above and for the most part are designed to obtain more rapid convergence to the optimum for certain special classes of response functions. Among the more interesting are the *pattern search* technique of Hooke and Jeeves [24] and the *method of parallel tangents* ("partan") designed by Shah et al [41].

EXERCISES

Section 5.2

5-1. Sketch the indifference contours of a function of two variables that has only one "hump" or local maximum in some region R but is *not* quasi-concave throughout R. Must a function of one variable that has only a single constrained local maximum within some interval I necessarily be quasi-concave over I? Prove or disprove.

5-2. Prove or disprove:
(a) The negative of a quasi-concave function is quasi-convex.
(b) Any constrained local maximum of a quasi-concave function in a convex region R is a global maximum over R.
(c) The function $f(\mathbf{x})$ is quasi-concave if and only if the set of points $S \equiv \{\mathbf{x} \mid f(\mathbf{x}) \geqq b\}$ is convex for any real number b.

5-3. Let $f(x)$ be a continuous real-valued function of a single variable x, and suppose that $f(1) = 1$, $f(2) = 2$, $f(3) = 3$, $f(4) = 2$, and $f(5) = 1$. If $f(x)$ is assumed to be quasi-concave, what is the interval of uncertainty? What is it if $f(x)$ is *concave*? Answer the same questions if $f(3) = 2.5$ and the other four values remain as given above.

Section 5.3

5-4. Suggest some physical and/or economic contexts in which bimodal ("two-humped") or polymodal functions of one variable arise.

5-5. Suppose the value of x that maximizes the quasi-concave function $f(x)$ is known to lie in the interval [0, 1], and suppose it is to be sought by performing three simultaneous experiments having an experimental resolution of .01. In what sense would placing the experiments at .49, .50, and .51 be better than placing them at .25, .50, and .75?

5-6. Suppose the maximum of the quasi-concave function $f(x)$ is known to lie in the interval [0, 1], and suppose it is to be sought by performing N simultaneous experiments, where N is even, in such a way as to minimize the maximum interval of uncertainty that may result. Assuming that the experimental resolution is quite small but not negligible, exactly where should the N experiments be located and how long is the maximum possible interval of uncertainty? Why might you have guessed from (5-5) that the search strategy of equally spaced pairs could not be minimax-optimal?

Section 5.4

5-7. Using a golden section search, find within 5% the value of x within the interval [0, 1] that maximizes the function

$$f(x) \equiv \min\{x, 2x^2, 2 - 2x - x^2\}.$$

What assumption is needed in order to implement the search and how many experiments are required? Could Fibonacci search be used instead of golden section to solve the problem as posed? How? Assuming an infinitesimally small experimental resolution, how many experiments would be needed?

5-8. As a bit of recreation, determine the value of the fraction

$$\cfrac{1}{1 + \cfrac{1}{1 + \cfrac{1}{1 + \cdots}}}.$$

5-9. The terms $\{F_i\}$ in a *generalized* Fibonacci sequence can be specified as follows:

$$F_0 = a, \qquad F_1 = b, \quad \text{and} \quad F_i = F_{i-1} + F_{i-2} \qquad \text{for all } i \geqq 2.$$

Given that when $a = b = 1$

$$\lim_{i\to\infty} \frac{F_i}{F_{i+1}} = r = \frac{-1 + \sqrt{5}}{2},$$

prove that when a and b are *any* two positive numbers the limiting value of F_i/F_{i+1} is still r.

5-10. Using Eqs. (5-6) and (5-12), show that the fraction of the original interval that remains after an n-experiment Fibonacci search is

$$\frac{1}{F_n} + \frac{F_{n-2}\epsilon}{F_n},$$

where the Fibonacci numbers $\{F_i\}$ are given by

$$F_0 = F_1 = 1 \quad \text{and} \quad F_i = F_{i-1} + F_{i-2} \qquad \text{for all } i \geqq 2. \tag{5-14}$$

Then determine exactly where the first of the n experiments should be located.

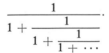

5-11. Given a quasi-concave response function of one variable defined over a finite interval, and assuming a negligible experimental resolution, how should eight experiments be performed in two blocks so that the maximum interval of uncertainty that may result is as small as possible? What is the minimax-optimal search plan for eight experiments in three blocks? For nine in two blocks? For nine in three blocks? Verify each answer by examining or logically eliminating all possible search schemes, and comment as seems appropriate on how search efficiency depends on the number of experiments and the number of blocks.

General

5-12. Suppose an experimental control variable x can take on any of K discrete values (not necessarily equally spaced), and suppose further that over some interval containing all these values the response function is known to be quasi-concave. Assuming a negligible experimental resolution, show that if $K = F_n - 1$, where F_n is the nth Fibonacci number as defined by (5-14), then the value of x for which the response function is greatest can be located with only $n - 1$ sequential experiments. What simple artifice can be used to guarantee that when $F_n - 1 < K < F_{n+1} - 1$ only n sequential experiments are required to locate the maximum? Discrete-value problems of this type are known as *lattice search problems*.

5-13. In conducting an experimental search, it is frequently necessary to consider the possibility of error in the observed values of the response function. Experimental error can arise either from limitations and inaccuracies in the instruments used in measuring or recording the response, or from an uncontrollable random component in the system (e.g., local rainfall or subsoil drainage in an agronomic experiment). Two simple approaches to dealing with such types of errors are as follows:

1. Include them within an enlarged experimental resolution ϵ, and
2. Define a *response resolution* δ such that two responses are considered to be equal if they differ by less than δ.

For which types of errors would each approach be the more suitable? Would the same ϵ or δ necessarily be used for all experiments? How might ϵ or δ in effect be reduced locally by repetition of an experiment or experiments? Indicate how your answers would vary according to whether the search is simultaneous or sequential.

6

Lagrange Multipliers and Kuhn-Tucker Theory

6.1 CONSTRAINED OPTIMIZATION THEORY

In Section 4.1 we distinguished three broad classes of mathematical programming problems according to the nature of their constraints:

1. Unconstrained problems,
2. Problems subject only to equality constraints, and
3. Problems subject to inequality constraints (with or without equality constraints as well).

Problems of the first type were given rather thorough coverage in Chapter 4, in which we discussed both the basic theory and the leading solution methods of unconstrained optimization. In the present chapter we shall establish the theoretical foundation for all constrained optimization, including problems of types 2 and 3. Solution algorithms for various subspecies of these problems will then be presented in Chapters 7, 8, and 9.

Equality-constrained optimization theory, to be covered in Sections 6.2 through 6.5, centers around the existence of certain numbers called *Lagrange multipliers*, named for the French mathematician Joseph-Louis Lagrange (1736–1813). This material, classical in origin, is relatively straightforward, and much of it will already be familiar to some readers. The remainder of the chapter will then be devoted to the more recently developed *Kuhn-Tucker theory*, which constitutes the cornerstone of mathematical programming and the heart of this text. The Kuhn-Tucker conditions and the various related results provide insight into inequality constraints and bounded feasible regions and give rise to computational algorithms for dealing with them; in addition, when applied to the special cases of equality-constrained problems and linear programs, Kuhn-Tucker theory reduces, respectively, to Lagrange multiplier theory and to the principles of duality.

But we are getting ahead of our story.

6.2 LAGRANGE MULTIPLIERS: SOME HEURISTIC GEOMETRY

Let us turn our attention to mathematical programming problems of the second type, those including one or more *equality* constraints; the general format is

$$\text{(6-1)} \qquad \begin{cases} \text{Max or Min } f(\mathbf{x}), & \mathbf{x} = (x_1, \ldots, x_n), \\ \text{subject to } g_i(\mathbf{x}) = b_i, & i = 1, \ldots, m. \end{cases}$$

It was noted in Section 4.1 that, in theory, we can reduce this problem to one of *unconstrained* optimization by using the constraints to eliminate variables, that is, by solving the constraints for m of the variables in terms of the others and then substituting into the objective function to eliminate those m variables. That idea is worth noting because it underscores the fact that the addition of equality constraints has relatively little theoretical impact. This is essentially because *equality constraints act only to reduce dimensionality; they do not establish boundaries.*

Consider for example, the following problem:

$$\text{Min } x_1^2 + x_2^2 + x_3^2$$
$$\text{subject to } x_1 + x_2 + x_3 = 10.$$

The feasible region is simply a plane in E^3, and we can move freely over it in any direction without ever running into a boundary. We are *constrained*, in that we cannot escape from the plane surface, but we are *not bounded.* From a different point of view, our example can also be represented as an unconstrained optimization problem in two variables:

$$\text{Min } x_1^2 + x_2^2 + (10 - x_1 - x_2)^2.$$

This can be solved by setting the two partial derivatives equal to zero and solving for x_1 and x_2, thereby determining x_3. If the optimal solution is finite, it must therefore occur at an unconstrained local optimum in E^2 (not E^3).

Generalizing this idea, the optimal solution to problem (6-1) must be an unconstrained local optimum on an $(n - m)$-dimensional hypersurface in E^n. We can imagine roaming freely over that hypersurface, just as we move freely through E^n during a gradient search, until we reach a point where the first-order partial derivatives of f with respect to displacement along the hypersurface are all equal to zero. Thus the theoretical similarity between unconstrained and equality-constrained optimization should be evident: We can solve a problem of either type by identifying all points at which the partial derivatives, suitably defined, are equal to zero and then choosing the best of them. In other words, for either type of problem the vanishing of the partial derivatives constitutes a set of necessary conditions which must obtain at the optimal solution (as well as at assorted saddle points, inflection points, and unconstrained local maxima and minima).

By contrast, however, problems with inequality constraints are basically quite different, because the inequality constraints establish "walls" or boundaries which cannot be crossed. If we now imagine moving through the feasible region, perhaps following the gradient and steadily improving the value of the objective, we can see that it would be quite possible to run abruptly into a boundary or a "corner," thereby establishing a constrained local optimum which might well be the optimal solution to the problem. *But at such a point the partial derivatives would not equal zero.* Hence we cannot, in general, solve a mathematical program with inequality constraints by merely examining all points at which the partial derivatives vanish; some other, more sophisticated set of necessary conditions (namely, the Kuhn-Tucker conditions, to be derived below) must be established which can identify constrained local optima as well. And this is what constitutes the essential theoretical difference between inequality-constrained problems on the one hand and unconstrained or equality-constrained problems on the other.

Returning to the main topic of this section, we have noted that problem (6-1) can in theory be reduced to an unconstrained problem having only $n - m$ variables. We now hasten to add that this procedure becomes extremely messy and laborious whenever there are more than two or three constraints. Even for small numbers of variables and constraints it is usually more efficient to use the method of *Lagrange multipliers* to solve problem (6-1). This procedure generates a system of equations representing necessary conditions which must hold at any local optimum. If all solutions to the system can be found, then the global optimum—assuming it is finite—must be among them.

To develop a basic understanding of the method of Lagrange multipliers, let us consider in detail the following example problem:

$$\text{Max } f(\mathbf{x}) \equiv f(x_1, x_2, x_3) \tag{6-2}$$

$$\text{subject to } g(x_1, x_2, x_3) = 0 \tag{6-3}$$

$$\text{and } h(x_1, x_2, x_3) = 0. \tag{6-4}$$

In what follows we shall provide a geometric explanation of why the method works; rigorous algebraic proofs will then be supplied in Section 6.3. The two constraints $g(\mathbf{x}) = 0$ and $h(\mathbf{x}) = 0$ in the problem above describe two surfaces in three-dimensional space; if there are to be any feasible solutions, these surfaces must intersect, forming a curve C. We can find the maximum value of the objective by examining the surfaces $f(\mathbf{x}) = k$ as the constant k is gradually increased. The intersection of any such surface $f = k_0$ with the curve C is a set of feasible points whose objective value is k_0, and any local maximum on C must occur at a point P where the surface $f = k_{\max}$ is *tangent* to C. This familiar geometrical result is illustrated in two dimensions by Fig. 6-1, where $k_1 < k_2 < \cdots < k_{\max}$.

Reverting to our three-dimensional problem, we recall from Section 4.4 that the gradient vector $\nabla\mathbf{f}$ is normal to the surface $f = k_{\max}$, since $k_{\max}$ is a constant. Thus $\nabla\mathbf{f}$ must also be normal to the curve C at the point P, where C is tangent to that surface. The other two gradients $\nabla\mathbf{g}$ and $\nabla\mathbf{h}$ are normal to the surfaces $g = 0$ and $h = 0$, respectively, so both of them are also normal to C. We have shown that the plane normal to C at P contains all three gradient vectors. It follows that they must be linearly dependent; that is, there must exist real numbers $\alpha_1, \alpha_2, \alpha_3$ not all zero such that

(6-5) $$\alpha_1 \nabla\mathbf{f} + \alpha_2 \nabla\mathbf{g} + \alpha_3 \nabla\mathbf{h} = \mathbf{0} \qquad \text{at } P.$$

However, because the two constraint surfaces intersect to form a curve, the vectors $\nabla\mathbf{g}$ and $\nabla\mathbf{h}$, which are normal to them, must also intersect.[1] This implies that no values for α_2 and α_3 other than $\alpha_2 = \alpha_3 = 0$ can be found to satisfy

(6-6) $$\alpha_2 \nabla\mathbf{g} + \alpha_3 \nabla\mathbf{h} = \mathbf{0}.$$

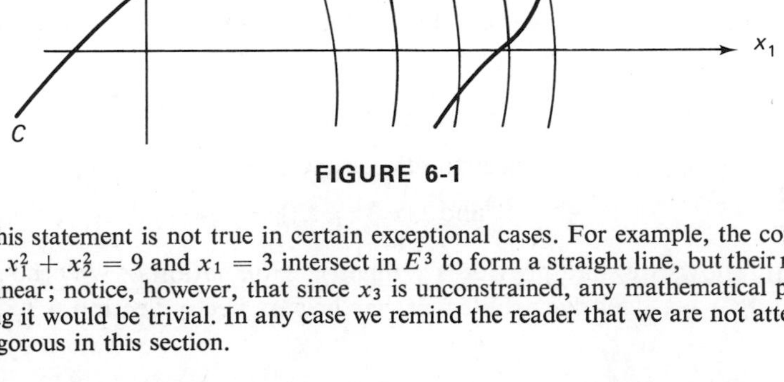

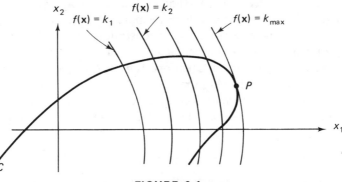

FIGURE 6-1

[1]This statement is not true in certain exceptional cases. For example, the constraint surfaces $x_1^2 + x_2^2 = 9$ and $x_1 = 3$ intersect in E^3 to form a straight line, but their normals are collinear; notice, however, that since x_3 is unconstrained, any mathematical program involving it would be trivial. In any case we remind the reader that we are not attempting to be rigorous in this section.

Hence α_1 cannot equal zero in (6-5), and we may write

$$\nabla \mathbf{f} - \lambda_1 \nabla \mathbf{g} - \lambda_2 \nabla \mathbf{h} = \mathbf{0} \quad \text{at } P, \tag{6-7}$$

where $\lambda_1 = -\alpha_2/\alpha_1$ and $\lambda_2 = -\alpha_3/\alpha_1$. The λ_i, whose definition may now seem rather arbitrary, are called *Lagrange multipliers.*

The three equations implied by (6-7) are

$$\begin{cases} \dfrac{\delta f}{\delta x_1} - \lambda_1 \dfrac{\delta g}{\delta x_1} - \lambda_2 \dfrac{\delta h}{\delta x_1} = 0, \\ \dfrac{\delta f}{\delta x_2} - \lambda_1 \dfrac{\delta g}{\delta x_2} - \lambda_2 \dfrac{\delta h}{\delta x_2} = 0, \\ \dfrac{\delta f}{\delta x_3} - \lambda_1 \dfrac{\delta g}{\delta x_3} - \lambda_2 \dfrac{\delta h}{\delta x_3} = 0. \end{cases} \tag{6-8}$$

These three, together with the two constraints

$$g(x_1, x_2, x_3) = 0 \tag{6-3}$$

and

$$h(x_1, x_2, x_3) = 0, \tag{6-4}$$

yield a total of five equations in the five variables x_1, x_2, x_3, λ_1, and λ_2. The solutions are critical points of various kinds: local maxima and minima, inflection points, and saddle points. Unless these five equations are linear there will usually be several critical points; the global maximum and global minimum, if they are finite, will be among them.

If the student does not care to memorize all those equations, he may instead simply remember the Lagrangian mnemonic function

$$F(\mathbf{x}, \boldsymbol{\lambda}) \equiv f(\mathbf{x}) - \lambda_1 g(\mathbf{x}) - \lambda_2 h(\mathbf{x}). \tag{6-9}$$

Setting the five partial derivatives of F equal to zero produces the five equations above. This function, known also as the *Lagrangian*, will play an important role when we consider mathematical programs with inequality constraints.

To summarize, the classical method of Lagrange multipliers provides a set of *necessary conditions which must obtain at all local optima* of an equality-constrained mathematical program. The geometric argument can, of course, be modified to fit any number of variables and equality constraints. If there were three variables and one constraint, for example, the constraint surface of feasible points $g(\mathbf{x}) = 0$ would be tangent to the optimal objective surface $f(\mathbf{x}) = k_{\max}$ at a single point P. The gradient vectors $\nabla \mathbf{f}$ and $\nabla \mathbf{g}$, being normal to these surfaces at P, would therefore be collinear, so we could write

$$\nabla \mathbf{f} - \lambda \nabla \mathbf{g} = \mathbf{0}. \tag{6-10}$$

The components of (6-10), along with $g(\mathbf{x}) = 0$, would provide four equations in the four variables x_1, x_2, x_3, and λ.

In the next section a generalized algebraic derivation of these necessary conditions will be presented. Students for whom the geometric argument has been sufficiently rigorous should at least take note of (6-22) and (6-12), which state the Lagrange multiplier results in general terms. Specific illustrations of the Lagrange techniques will then be found in Section 6.4.

6.3 LAGRANGE MULTIPLIERS: AN ALGEBRAIC DERIVATION

Consider the following equality-constrained mathematical program:

$$\text{(6-11)} \qquad \text{Max } f(\mathbf{x})$$

$$\text{(6-12)} \qquad \text{subject to } g_i(\mathbf{x}) = b_i, \qquad i = 1, \ldots, m,$$

where $\mathbf{x}$ is an n-component vector and $m < n$. Assuming that none of the constraints are redundant, this last requirement is necessary if the problem is to be interesting. When $m \geqq n$ there cannot be any continuous region over which $\mathbf{x}$ can vary while remaining feasible; thus there can be at most a countably infinite number of isolated feasible points.

We now postulate that a local maximum or minimum of $f(\mathbf{x})$ subject to (6-12) occurs[2] at some point $\mathbf{x}_0$ in E^n and proceed to derive certain interesting conditions which must hold at $\mathbf{x}_0$. Following the development of Hadley [22], one further assumption must be made: It must be possible to choose m of the variables, renumbering them $x_1, \ldots, x_m$ for convenience, such that the Jacobian matrix of first partial derivatives

$$\mathbf{J} \equiv \begin{bmatrix} \dfrac{\delta g_1}{\delta x_1} & \cdots & \dfrac{\delta g_1}{\delta x_m} \\ \cdot & & \cdot \\ \cdot & & \cdot \\ \cdot & & \cdot \\ \dfrac{\delta g_m}{\delta x_1} & \cdots & \dfrac{\delta g_m}{\delta x_m} \end{bmatrix}$$

has an inverse when evaluated at $\mathbf{x}_0$. This will be true in general provided that the constraint set is neither redundant nor inconsistent and that none of the variables is inessential. [A trivial example of an inessential variable would be

$$\text{Max } x_1 + x_2 + x_3$$
$$\text{subject to } (x_1 + x_2)^2 + x_3^2 = 4,$$

[2] A local optimum "subject to (6-12)" is, of course, a *constrained local optimum* in the sense defined in Section 3.2. However, we prefer to avoid using the term "constrained," which, as explained earlier, connotes boundary walls and nonvanishing partial derivatives —hence the circumlocution "subject to."

inasmuch as the three variables can be reduced to two by means of the substitution $x_4 = x_1 + x_2$.]

As long as all the $g_i(\mathbf{x})$ have continuous first derivatives and the Jacobian with respect to $x_1, \ldots, x_m$ has an inverse when evaluated at $\mathbf{x}_0$, we can in theory make use of the *implicit function theorem* to eliminate the constraints. This is done by solving them for the first m variables in terms of the $n - m$ others, thereby producing a new set of functions $h_1, \ldots, h_m$:

(6-13) $$x_i = h_i(\hat{\mathbf{x}}), \qquad i = 1, \ldots, m,$$

where $\hat{\mathbf{x}} \equiv (x_{m+1}, \ldots, x_n)$ is an $(n - m)$-component vector of variables.[3] After the substitution (6-13) the objective function $f(\mathbf{x})$ becomes

(6-14) $$\hat{f}(\hat{\mathbf{x}}) \equiv f[h_1(\hat{\mathbf{x}}), \ldots, h_m(\hat{\mathbf{x}}), \hat{\mathbf{x}}].$$

This function now has an unconstrained local maximum or minimum at the point $\hat{\mathbf{x}}_0 \equiv (x_{0,m+1}, \ldots, x_{0n})$ in E^{n-m}, so

(6-15) $$\frac{\delta \hat{f}(\hat{\mathbf{x}}_0)}{\delta x_j} = 0, \qquad j = m + 1, \ldots, n.$$

Applying the partial differentiation chain rule to (6-14),

(6-16) $$\frac{\delta \hat{f}}{\delta x_j} = \left(\sum_{k=1}^{m} \frac{\delta f}{\delta x_k} \cdot \frac{\delta h_k}{\delta x_j}\right) + \frac{\delta f}{\delta x_j} = 0 \qquad \text{at } \mathbf{x}_0$$

for $j = m + 1, \ldots, n$. Notice that we have used $\delta f/\delta x_k$ in the summation instead of $\delta f/\delta h_k$, which, in light of (6-13), is its equivalent.

Now, because the functions $h_i(\hat{\mathbf{x}})$ were obtained by solving the constraints $g_i(\mathbf{x}) = b_i$ for the first m variables, substitution of those functions back into the constraints will cause all the variable terms to cancel each other out, leaving

$$g_i[h_1(\hat{\mathbf{x}}), \ldots, h_m(\hat{\mathbf{x}}), \hat{\mathbf{x}}] \equiv b_i.$$

The left-hand side of this equation has the value b_i not just at $\hat{\mathbf{x}}_0$ but everywhere (in some neighborhood of $\hat{\mathbf{x}}_0$). Thus its derivative with respect to any of the remaining variables x_j must vanish:

(6-17) $$\frac{\delta g_i}{\delta x_j} + \sum_{k=1}^{m} \frac{\delta g_i}{\delta x_k} \cdot \frac{\delta h_k}{\delta x_j} = 0$$

for $i = 1, \ldots, m$ and $j = m + 1, \ldots, n$.

[3]The implicit function theorem is a result from mathematical analysis (see Rudin [39]) which guarantees that in the vicinity of $\mathbf{x}_0$ the functions $h_i(\hat{\mathbf{x}})$ exist and are differentiable.

One further preliminary step is needed. Let a set of real numbers $\lambda_1, \ldots, \lambda_m$ be defined by means of the following equations:

$$\sum_{i=1}^{m} \lambda_i \frac{\delta g_i(\mathbf{x}_0)}{\delta x_k} = \frac{\delta f(\mathbf{x}_0)}{\delta x_k} \qquad \text{for } k = 1, \ldots, m. \tag{6-18}$$

This is a system of m linear equations in m unknowns λ_i; since the Jacobian matrix $[\delta g_i(\mathbf{x}_0)/\delta x_k]$ was assumed to be invertible, (6-18) is sufficient to define unique real values for the λ_i.

The rest is simply manipulation. For any specific j, (6-17) represents m equations. Multiplying each of them by the respective λ_i and summing over i yields

$$\sum_{i=1}^{m} \lambda_i \frac{\delta g_i}{\delta x_j} + \sum_{k=1}^{m} \frac{\delta h_k}{\delta x_j}\left(\sum_{i=1}^{m} \lambda_i \frac{\delta g_i}{\delta x_k}\right) = 0 \tag{6-19}$$

for any j, where $j = m + 1, \ldots, n$. Evaluating (6-19) at $\mathbf{x}_0$ and subtracting it from (6-16) produces

$$\sum_{k=1}^{m} \left[\frac{\delta f(\mathbf{x}_0)}{\delta x_k} - \sum_{i=1}^{m} \lambda_i \frac{\delta g_i(\mathbf{x}_0)}{\delta x_k}\right] \frac{\delta h_k(\mathbf{x}_0)}{\delta x_j} + \frac{\delta f(\mathbf{x}_0)}{\delta x_j} - \sum_{i=1}^{m} \lambda_i \frac{\delta g_i(\mathbf{x}_0)}{\delta x_j} = 0 \tag{6-20}$$

for $j = m + 1, \ldots, n$. But, from (6-18), the term in brackets vanishes for every k, $k = 1, \ldots, m$, leaving

$$\frac{\delta f(\mathbf{x}_0)}{\delta x_j} = \sum_{i=1}^{m} \lambda_i \frac{\delta g_i(\mathbf{x}_0)}{\delta x_j}, \qquad j = m + 1, \ldots, n. \tag{6-21}$$

These combine neatly with (6-18), and we have established the following:

Theorem 6.1: Let $\mathbf{x}_0$ *in* E^n *be any local maximum or minimum of* $\mathrm{f}(\mathbf{x})$ *subject to the equality constraints* $\mathrm{g}_i(\mathbf{x}) = \mathrm{b}_i$, $\mathrm{i} = 1, \ldots, \mathrm{m}$, *where* $\mathrm{m} < \mathrm{n}$. *If it is possible to choose a set of* m *variables for which the Jacobian matrix* $[\delta \mathrm{g}_i(\mathbf{x}_0)/\delta \mathrm{x}_k]$ *has an inverse, then there exists a unique set of Lagrange multiplier values* $\lambda_1, \ldots, \lambda_m$ *satisfying*

$$\frac{\delta \mathrm{f}(\mathbf{x}_0)}{\delta \mathrm{x}_j} - \sum_{\mathrm{i}=1}^{\mathrm{m}} \lambda_{\mathrm{i}} \frac{\delta \mathrm{g}_i(\mathbf{x}_0)}{\delta \mathrm{x}_j} = 0, \qquad \mathrm{j} = 1, \ldots, \mathrm{n}. \tag{6-22}$$

If the constraints

$$g_i(\mathbf{x}_0) = b_i, \qquad i = 1, \ldots, m, \tag{6-12}$$

are included, (6-22) and (6-12) together form a system of $m + n$ equations in $m + n$ unknowns. Each local maximum and minimum, along with its associated Lagrange multiplier values, is among the solutions to this system, and therefore so is the global maximum (assuming it is finite).

Notice again that (6-22) and (6-12) can be obtained heuristically by equating to zero the partial derivatives of the mnemonic function

(6-23) $$F(\mathbf{x}, \boldsymbol{\lambda}) = f(\mathbf{x}) - \sum_{i=1}^{m} \lambda_i[g_i(\mathbf{x}) - b_i].$$

6.4 EXAMPLES AND COMPUTATIONAL REMARKS

Consider first the problem of finding the point on the plane

$$Ax_1 + Bx_2 + Cx_3 = D$$

that is closest to the origin in E^3. In the functional notation we have been using, the problem becomes

$$\text{Min } f(\mathbf{x}) = x_1^2 + x_2^2 + x_3^2$$
$$\text{subject to } g(\mathbf{x}) = Ax_1 + Bx_2 + Cx_3 = D.$$

From (6-22) we may immediately write

$$2x_1 - \lambda A = 0,$$
$$2x_2 - \lambda B = 0,$$

and

$$2x_3 - \lambda C = 0,$$

each of which holds at any local minimum. Substituting these into the constraint equation produces

$$\lambda\frac{A^2}{2} + \lambda\frac{B^2}{2} + \lambda\frac{C^2}{2} = D.$$

Thus $\lambda = 2D/(A^2 + B^2 + C^2)$, and we find that the only local optimum is

$$x_1 = \frac{AD}{A^2 + B^2 + C^2}, \qquad x_2 = \frac{BD}{A^2 + B^2 + C^2}, \qquad x_3 = \frac{CD}{A^2 + B^2 + C^2}.$$

Since the global minimum we are seeking is finite, it must occur at a local minimum; hence, the point we have found is the global minimum.

Let us now attack this problem in a different way: By solving the constraint for one of the variables, we can reduce the problem to one of unconstrained minimization. Thus

$$x_3 = \frac{D - Ax_1 - Bx_2}{C},$$

and the function to be minimized becomes

$$\hat{f}(\hat{\mathbf{x}}) = x_1^2 + x_2^2 + \frac{1}{C^2}(D - Ax_1 - Bx_2)^2.$$

We now take the two partial derivatives, equate them to zero, and manipulate, eventually arriving at the same solution.

Although this second approach may have been slightly more clumsy and time-consuming, it was nevertheless a reasonably satisfactory way of solving this particular problem. The method of Lagrange multipliers is indispensable, however, in dealing with larger problems in which the objective and constraint functions are more complex. A constraint equation such as

$$x_1x_2x_3x_4 + \log(x_1 + x_2) - e^{x_3x_4} = 10$$

simply cannot be solved conveniently for any one of the variables in terms of the others, and a problem involving several constraints as complicated as the one above would obviously prove intractable if the only solution method available were that of reduction by successive substitutions. Such problems can only be solved—albeit with substantial computational effort—by the Lagrange multiplier approach: Equations (6-22) are written directly and then solved, along with the constraints (6-12), by Newton-Raphson iteration or some similar technique. In general, of course, several applications of the procedure from different starting points might be needed before the optimal solution of a given equality-constrained mathematical program is found (and its global optimality verified). In any single application the solution obtained might be a local maximum or minimum or some other type of critical point; moreover, as was illustrated in Section 4.3, the Newton-Raphson technique might generate a sequence of points that fails to converge at all. Nevertheless, the method of Lagrange multipliers, using Newton-Raphson iteration, constitutes a straightforward procedure for solving equality-constrained problems that is well suited to machine computation: It is only necessary to code the various partial derivatives of (6-22) and (6-12), read in a starting point, and turn on the computer.

As a further illustration, we shall apply the technique of Lagrange multipliers to a special linear program P^*:

$$\text{Max } z = \mathbf{c}^T\mathbf{x}$$
$$\text{subject to } \mathbf{Ax} = \mathbf{b},$$
$$\text{with } \mathbf{x} \text{ unrestricted,}$$

where $\mathbf{A}$ is an m-by-n matrix. All constraints are equalities, so this problem fits the required format of (6-11) and (6-12). Assume that the objective is not unbounded (actually, most feasible problems in this format would have unbounded optimal solutions, and the rest would be trivial or uninteresting for other reasons). Then, since the partial derivatives are given by

$$\frac{\delta g_i}{\delta x_j} = a_{ij},$$

we know from Eq. (6-22) that at the (finite) maximum $\mathbf{x}_0$ there exist real numbers λ_i, $i = 1, \ldots, m$, such that

$$c_j - \sum_{i=1}^{m} \lambda_i a_{ij} = 0, \qquad j = 1, \ldots, n.$$

In matrix form this is

$$\mathbf{A}^T\boldsymbol{\lambda} = \mathbf{c},$$

where $\boldsymbol{\lambda}$ is an m-component column vector. Note that the optimal value of the objective function is

$$z_{\max} = \mathbf{c}^T\mathbf{x}_0 = \mathbf{x}_0^T\mathbf{c} = \mathbf{x}_0^T\mathbf{A}^T\boldsymbol{\lambda} = \mathbf{b}^T\boldsymbol{\lambda},$$

the last equality being due to the fact that $\mathbf{x}_0$ is a feasible solution.

By this time the student has realized that these Lagrange multipliers are none other than the dual variables. We have shown that if $\mathbf{x}_0$ is an optimal solution to P^*, then there exists a $\boldsymbol{\lambda}$ such that $\mathbf{A}^T\boldsymbol{\lambda} = \mathbf{c}$ and $\mathbf{c}^T\mathbf{x}_0 = \mathbf{b}^T\boldsymbol{\lambda}$. In other words, for problem P^*—which is by no means the general linear programming problem—we have made use of Lagrange multipliers to prove the duality theorem. To check this, recall from Section 2.5 that the dual of P^* is as follows:

$$\text{Min } z' = \mathbf{b}^T\mathbf{u}$$
$$\text{subject to } \mathbf{A}^T\mathbf{u} = \mathbf{c},$$
$$\text{with } \mathbf{u} \text{ unrestricted.}$$

Thus the Lagrange multipliers $\boldsymbol{\lambda}$ constitute a feasible solution to the dual.

This equivalence of Lagrange multipliers and dual variables is no coincidence; it will, in fact, be generalized in the next section to include all mathematical programs of the form (6-11) and (6-12). Furthermore, an even broader generalization will be developed from Kuhn-Tucker theory when we come to consider inequality constraints.

6.5 INTERPRETATION OF LAGRANGE MULTIPLIERS

Suppose that the *global optimum* of the general equality-constrained mathematical program

(6-11) $\qquad \text{Max } z = f(\mathbf{x}), \qquad \mathbf{x} = (x_1, \ldots, x_n),$

(6-12) $\qquad \text{subject to } g_i(\mathbf{x}) = b_i, \qquad i = 1, \ldots, m,$

occurs at $\mathbf{x}^* = (x_1^*, \ldots, x_n^*)$, with associated Lagrange multiplier values $\boldsymbol{\lambda}^* = (\lambda_1^*, \ldots, \lambda_m^*)$. If a different vector of values were now substituted for the right-hand side $\mathbf{b}$, the rest of the problem remaining unchanged, we would, in general, expect the maximum to occur at a different point. From this reasoning we may consider each x_j^* and λ_i^* to be a continuous function of $\mathbf{b}$. Again, the development follows [22].

The partial derivatives of the global optimum $z^* = f(\mathbf{x}^*)$ must be

$$\frac{\delta z^*}{\delta b_i} = \sum_{j=1}^{n} \frac{\delta f}{\delta x_j^*} \cdot \frac{\delta x_j^*}{\delta b_i}, \tag{6-24}$$

while the partial derivatives of the constraints $g_k(\mathbf{x}^*) = b_k$ become

$$\frac{\delta g_k}{\delta b_i} = \sum_{j=1}^{n} \frac{\delta g_k}{\delta x_j^*} \cdot \frac{\delta x_j^*}{\delta b_i} = \begin{cases} 1 & \text{if } k = i, \\ 0 & \text{if not.} \end{cases} \tag{6-25}$$

For any specific i, (6-25) represents m equations; multiplying each by the respective λ_k^*, $k = 1, \ldots, m$, and summing over k produces

$$\sum_{k=1}^{m} \lambda_k^* \sum_{j=1}^{n} \frac{\delta g_k}{\delta x_j^*} \cdot \frac{\delta x_j^*}{\delta b_i} = \sum_{j=1}^{n} \frac{\delta x_j^*}{\delta b_i} \sum_{k=1}^{m} \lambda_k^* \frac{\delta g_k}{\delta x_j^*} = \lambda_i^*. \tag{6-26}$$

The necessary condition (6-22), which is known to hold at $\mathbf{x}^*$, can now be substituted into (6-26), yielding

$$\sum_{j=1}^{n} \frac{\delta x_j^*}{\delta b_i} \cdot \frac{\delta f}{\delta x_j^*} = \lambda_i^*. \tag{6-27}$$

From (6-24) we conclude, finally, that

$$\lambda_i^* = \frac{\delta z^*}{\delta b_i}, \qquad i = 1, \ldots, m; \tag{6-28}$$

that is, *the ith Lagrange multiplier value associated with the global optimum* $\mathbf{x}^*$ *gives the rate of change of the optimal attainable value of the objective function with respect to change in* b_i.

Suppose we think of (6-11) and (6-12) in economic terms, precisely as we interpreted the linear program of Section 2.8. Let z represent the dollar income realized from selling various quantities $x_1, \ldots, x_n$ of goods produced by a factory, and let $g_i(\mathbf{x}) = b_i$ be a constraint ensuring that the amount of the ith input consumed when those quantities are produced is just equal to the limited amount b_i available. Then the λ_i^* give the rates of increase in sales income as the available quantities of inputs increase; that is, for a sufficiently small increment ϵ_i in b_i, an additional $\lambda_i^* \epsilon_i$ of income can be realized. Thus $\lambda_i^* \epsilon_i$ also represents the maximum price the manufacturer would just be willing to pay for that extra bit of input. Several alternative ways of referring to the Lagrange multipliers λ_i^*—as shadow prices, imputed values, or marginal values—are suggested by this economic framework.

The reader should recall that (6-28) is precisely the relationship we obtained during the proof of the duality theorem of linear programming,

$$u_{0i} = \frac{\delta f}{\delta b_i}, \tag{2-53}$$

except that we were then using the term "dual variable" rather than the more appropriate (as we shall see) phrase "generalized Lagrange multiplier." We still have not shown an exact equivalence between dual variables and Lagrange multipliers because we have not yet associated the latter with problems involving inequality constraints. That crucial relationship lies at the heart of the Kuhn-Tucker theory, to which we now turn our attention.

6.6 INEQUALITY CONSTRAINTS

The third and most difficult type of mathematical programming problem or *nonlinear program* takes the following general form:

$$\text{Max or Min } f(\mathbf{x}), \qquad \mathbf{x} = (x_1, \ldots, x_n), \tag{6-29}$$

$$\text{subject to } g_i(\mathbf{x}) \leqq b_i, \qquad i = 1, \ldots, r, \tag{6-30}$$

$$\text{and } h_i(\mathbf{x}) = b_i, \qquad i = 1, \ldots, s, \tag{6-31}$$

where constraints of the form $g_i(\mathbf{x}) \geqq b_i$ have been multiplied through by -1 to conform to (6-30). Bear in mind that the constraints (6-30) include any nonnegativity restrictions $-x_k \leqq 0$ that may be present. We continue to assume, as before, that the functions f, g_i, and h_i are continuous and have continuous first partial derivatives.

Most mathematical programs arising in the practice of operations research include inequality constraints and thus fit the general format given above. The most common type of inequality constraint is, of course, the nonnegativity restriction, which is often needed simply because negative quantities are not physically meaningful. In most other cases an inequality constraint can be interpreted in some way either as a resource limitation (the total amount of raw material, dollars, or time expended must not exceed the amount available) or as a minimum demand (the quantity of goods produced, manpower available, or riboflavin in the diet must be at least sufficient to satisfy customer demand, contract commitments, or minimum daily requirement). Constraints such as these are the natural expressions of physical and economic realities, and it is not surprising that they should occur with substantially greater frequency than equality constraints—how often, for example, is it *infeasible* for a plant to operate at less than full capacity, or for a department head to spend less than his budget permits? For these reasons inequality-constrained mathematical programs are in general more interesting and important to an operations research analyst, as well as being more varied and more difficult of solution, than problems having equality constraints only.

Quite apart from the above considerations, however, it will be convenient and economical in the discussion of the next few sections to omit equality constraints from consideration entirely. Theorems about the nonlinear program as formulated originally in (6-29) through (6-31) would be unpleasantly cumbersome, because each proof would require treatment of two objective and two constraint types. Accordingly, we define problem NLP ("nonlinear program"):

(6-32) $$\text{Max } f(\mathbf{x}), \qquad \mathbf{x} = (x_1, \ldots, x_n),$$

(6-33) $$\text{subject to } g_i(\mathbf{x}) \leqq b_i, \qquad i = 1, \ldots, m.$$

Observe that any mathematical program can be fitted to this format by means of simple transformations:

1. Minimizing $f(\mathbf{x})$ is the same as maximizing $-f(\mathbf{x})$,
2. Any constraint $h_i(\mathbf{x}) = b_i$ is equivalent to the pair of inequalities $h_i(\mathbf{x}) \leqq b_i$ and $h_i(\mathbf{x}) \geqq b_i$, and
3. Any constraint $h_i(\mathbf{x}) \geqq b_i$ can be converted to $-h_i(\mathbf{x}) \leqq -b_i$.

Thus problem NLP is sufficiently general to represent any mathematical program. Later on, the results obtained for it will be extended via a simple corollary to apply directly to mathematical programs of all types.

Classical mathematical theory provides two general approaches for solving problem NLP. One approach is to convert each inequality constraint $g_i(\mathbf{x}) \leqq b_i$ into an equation by adding a squared slack variable:

(6-34) $$g_i(\mathbf{x}) + s_i^2 = b_i, \qquad i = 1, \ldots, m.$$

Because s_i^2 is nonnegative for all real values of s_i, the new constraints are precisely equivalent to the old, and the problem can now be solved by the Lagrange multiplier technique. Note, however, that in order to solve problem NLP in this manner m extra variables must be introduced, an expedient that is both theoretically inelegant and computationally expensive.

There is a second classical approach to inequality-constrained optimization that can be used on problems of the form

(6-35) $$\begin{cases} \text{Max } f(\mathbf{x}), & \mathbf{x} = (x_1, \ldots, x_n), \\ \text{subject to } x_j \geqq 0, & j \text{ in } J, \end{cases}$$

where J is the set of subscripts of all variables required to be nonnegative. By means of partial differentiation and solution of the resulting systems of equations, unconstrained local optima are found and examined, first within the interior of the nonnegative orthant, then on all single boundaries (set some $x_\alpha = 0$, equate the remaining $n - 1$ partial derivatives to zero, and solve; repeat for all α in J), then on all double boundaries (set $x_\alpha = x_\beta = 0$ for all pairs α and β in J), and so on through the one k-tuple boundary, where k is the number of members of J. Provided that it is finite, the global maximum of f must be the largest of all those local maxima obtained at the various stages that satisfy $x_j \geqq 0$, $j \in J$. Note that this procedure entails finding unconstrained local optima for 2^k subproblems; obviously, as k increases beyond 2 or 3, the amount of computation required rapidly becomes prohibitive.

Both of these classical approaches use the indirect and inefficient tactic of converting inequality constraints into equations, with subsequent computations largely based on the vanishing of first partial derivatives. Not until the modern era of operations research, following World War II, did a body of mathematical theory evolve that dealt directly with inequality constraints and constrained local optima, thereby opening the way to the development of relatively efficient solution algorithms for many types of mathematical programs. The crucial step in that evolution was the derivation of a set of algebraic equations and inequalities called the *Kuhn-Tucker conditions* that necessarily hold at the optimal solution $\mathbf{x}^*$ of problem NLP. In fact, these conditions can be shown to hold at *any* constrained local maximum of $f(\mathbf{x})$ within the feasible region (6-33), and perhaps at certain other points as well. Provided that it is finite, the optimum $\mathbf{x}^*$ is known to be a constrained local maximum; this was established in Section 3.2. Therefore, to solve any mathematical programming problem—in theory, at least—one need only solve the set of nonlinear equations and inequalities represented by the Kuhn-Tucker necessary conditions: $\mathbf{x}^*$ will be among the solutions. In practice, of course, only the smallest and simplest problems will be amenable to algebraic manipulation of this sort. As we shall see, however, the Kuhn-Tucker conditions also provide the theoretical foundation for a great many automatic computational algorithms that can be applied to mathematical programs of various shapes and sizes.

6.7 HEURISTIC DESCRIPTION OF KUHN-TUCKER CONDITIONS

As an introduction to the serious mathematical development which lies ahead, we present in this section an entirely lighthearted and facile "derivation" of certain conditions which must obtain at the optimal solution $\mathbf{x}^*$ to problem NLP. This section is in no way intended to make a rigorous theoretical contribution—in fact it amounts to little more than a bit of intellectual legerdemain. However, by introducing some important concepts in a more or less persuasive manner, it will hopefully shed a useful light and provide motivation for the student.

Consider a function $f(\mathbf{x}) = f(x_1, x_2)$ with x_1 unrestricted and x_2 required to be nonnegative. At a point $\hat{\mathbf{x}} = (\hat{x}_1, \hat{x}_2)$ where $f(\mathbf{x})$ takes on a local maximum over the feasible region, it must be true that

$$\frac{\delta f(\hat{\mathbf{x}})}{\delta x_1} = 0. \tag{6-36}$$

This is the familiar classical property of an unconstrained (in the x_1-direction) optimum. Now, if $\hat{x}_2$ is different from zero, we effectively have a maximum unconstrained in the x_2-direction as well, so

$$\frac{\delta f(\hat{\mathbf{x}})}{\delta x_2} = 0 \qquad \text{if } \hat{x}_2 > 0. \tag{6-37}$$

But suppose $\hat{x}_2 = 0$; we can no longer be sure that the partial derivative is zero, because we might have the situation diagrammed (in two dimensions) in Fig. 6-2. Observe that at the optimal point $x_2 = \hat{x}_2 = 0$ the slope $\delta f(\hat{\mathbf{x}})/\delta x_2$ is negative. It could hardly be positive since if it were, a small increase in x_2 would increase the value of $f(\hat{x}_1, x_2)$ and contradict the stipulation that the maximum occurs at $x_2 = 0$. For the restricted variable x_2, then, we have shown that

$$\frac{\delta f(\hat{\mathbf{x}})}{\delta x_2} \leqq 0 \tag{6-38}$$

and

$$\hat{x}_2 \frac{\delta f(\hat{\mathbf{x}})}{\delta x_2} = 0; \tag{6-39}$$

Eq. (6-39) is derived from the fact that either $\hat{x}_2 = 0$ or (6-37) holds.

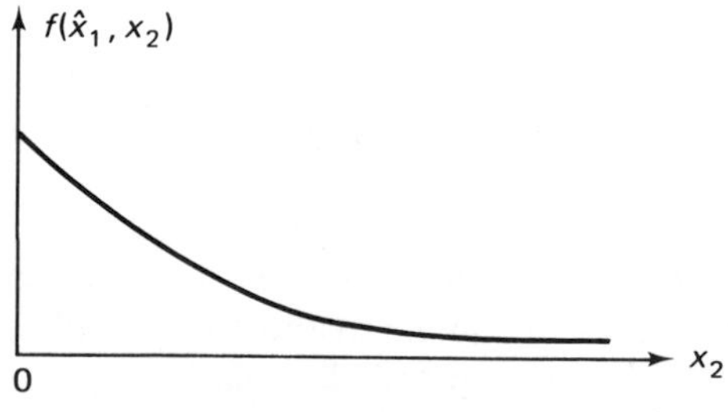

FIGURE 6-2

With these results in hand, let us return to the general inequality-constrained mathematical program NLP:

$$\text{(6-32)} \qquad \text{Max } f(\mathbf{x}), \qquad \mathbf{x} = (x_1, \ldots, x_n),$$

$$\text{(6-33)} \qquad \text{subject to } g_i(\mathbf{x}) \leqq b_i, \qquad i = 1, \ldots, m.$$

The constraints may be converted to equations by the addition of nonnegative slack variables s_i, $i = 1, \ldots, m$, producing the following problem:

$$\text{(6-32)} \qquad \text{Max } f(\mathbf{x})$$

$$\text{(6-40)} \qquad \text{subject to } g_i(\mathbf{x}) + s_i = b_i, \qquad i = 1, \ldots, m,$$

$$\text{(6-41)} \qquad \text{and } s_i \geqq 0, \qquad i = 1, \ldots, m.$$

The inequalities (6-33) are guaranteed in this new formulation by the nonnegativity of the slack variables.

Except for the special restrictions on s_i, we are now dealing with *equality constraints*. This suggests that we should be able to modify the development of Section 6.3, in which unconstrained local optima and their associated Lagrange multiplier values were found to satisfy certain relationships. It seems reasonable to expect that an analogous argument for our current problem would eventually lead to necessary conditions quite similar to (6-22) and (6-12), except that the provisions of (6-38) and (6-39) for the slack variables would have to be taken into account.

In line with these expectations, then, we write the Lagrangian mnemonic function implied by (6-32), (6-40), and (6-41):

$$\text{(6-42)} \qquad F(\mathbf{x}, \mathbf{s}, \boldsymbol{\lambda}) = f(\mathbf{x}) - \sum_{i=1}^{m} \lambda_i[g_i(\mathbf{x}) + s_i - b_i].$$

At any constrained local maximum the partial derivatives of the x_i and λ_i should equal zero as before, while for the slack variables, equations of the form (6-38) and (6-39) should hold. Since one of the constrained local maxima is global, the following relationships must all be satisfied by the global maximum $(\mathbf{x}^*, \mathbf{s}^*, \boldsymbol{\lambda}^*)$:

$$\text{(6-43)} \qquad \frac{\delta F}{\delta x_j} \equiv \frac{\delta f(\mathbf{x}^*)}{\delta x_j} - \sum_{i=1}^{m} \lambda_i^* \frac{\delta g_i(\mathbf{x}^*)}{\delta x_j} = 0, \qquad j = 1, \ldots, n;$$

$$\text{(6-44)} \qquad \frac{\delta F}{\delta \lambda_i} \equiv -[g_i(\mathbf{x}^*) + s_i^* - b_i] = 0, \qquad i = 1, \ldots, m;$$

but

$$\text{(6-45)} \qquad \frac{\delta F}{\delta s_i} \equiv -\lambda_i^* \leqq 0, \qquad i = 1, \ldots, m;$$

and

(6-46) $$s_i^* \frac{\delta F}{\delta s_i} \equiv -s_i^* \lambda_i^* = 0, \qquad i = 1, \ldots, m.$$

Of course, (6-44) are the original constraints. We can now remove the slack variables, which have served their purpose. Writing all of (6-43) as a single vector equation, the above reduce to

(6-47) $$\nabla \mathbf{f}(\mathbf{x}^*) - \sum_{i=1}^{m} \lambda_i^* \nabla \mathbf{g}_i(\mathbf{x}^*) = \mathbf{0};$$

(6-48) $$g_i(\mathbf{x}^*) \leqq b_i, \qquad i = 1, \ldots, m;$$

(6-49) $$\lambda_i^* \geqq 0, \qquad i = 1, \ldots, m;$$

and

(6-50) $$\lambda_i^*[b_i - g_i(\mathbf{x}^*)] = 0, \qquad i = 1, \ldots, m.$$

And, indeed, *these are precisely the Kuhn-Tucker necessary conditions.* That is, at the optimal solution $\mathbf{x}^*$ to problem NLP there must exist *generalized Lagrange multipliers* $\boldsymbol{\lambda}^* = (\lambda_1^*, \ldots, \lambda_m^*)$ such that (6-47) through (6-50) are satisfied.

These results, due to Kuhn and Tucker [28], will be properly and rigorously derived in Sections 6.8 and 6.9. In the meantime, we shall use them to solve an example (see Fig. 6-3):

$$\text{Max } f(\mathbf{x}) = 3x_1 + x_2$$
$$\text{subject to } g_1(\mathbf{x}) = x_1^2 + x_2^2 \leqq 5$$
$$\text{and } g_2(\mathbf{x}) = x_1 - x_2 \leqq 1.$$

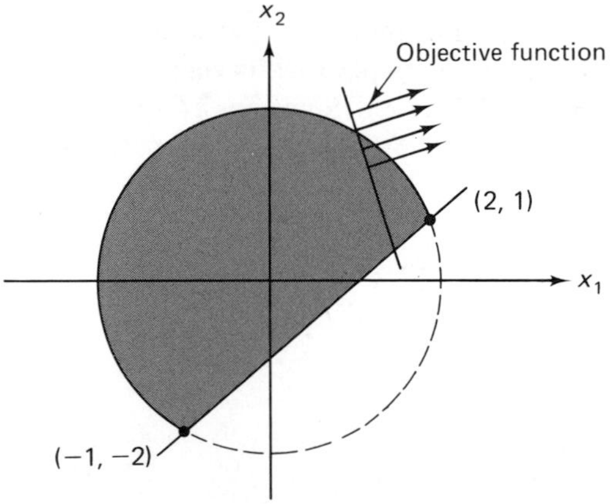

FIGURE 6-3

From the necessary conditions we have, omitting the asterisks,

$$\frac{\delta f}{\delta x_1} = 3 = \lambda_1(2x_1) + \lambda_2, \tag{6-51}$$

$$\frac{\delta f}{\delta x_2} = 1 = \lambda_1(2x_2) - \lambda_2, \tag{6-52}$$

$$\lambda_1(5 - x_1^2 - x_2^2) = 0, \tag{6-53}$$

$$\lambda_2(1 + x_2 - x_1) = 0. \tag{6-54}$$

We reason that if $\lambda_1 = 0$, two contradictory values for λ_2 arise from (6-51) and (6-52); therefore $\lambda_1 \neq 0$, and, from (6-53),

$$x_1^2 + x_2^2 = 5. \tag{6-55}$$

If $\lambda_2 = 0$, (6-51) and (6-52) yield

$$x_1 = \frac{3}{2\lambda_1} \quad \text{and} \quad x_2 = \frac{1}{2\lambda_1};$$

substituting these into (6-55) produces $\lambda_1 = \sqrt{2}/2$ (the negative root is forbidden by the nonnegativity of λ_1), so $x_1 = 3/\sqrt{2}$ and $x_2 = 1/\sqrt{2}$. But these do not satisfy the second constraint; therefore λ_2 must be different from zero, and (6-54) yields

$$x_1 - x_2 = 1. \tag{6-56}$$

There are two solutions to (6-55) and (6-56),

$$x_1 = 2, x_2 = 1 \quad \text{and} \quad x_1 = -1, x_2 = -2.$$

Substitution of the latter into (6-51) and (6-52) leads to $\lambda_1 = -\frac{2}{3}$, an impermissible value. Therefore, we choose $x_1 = 2$, $x_2 = 1$, which yields the acceptable values $\lambda_1 = \frac{2}{3}$, $\lambda_2 = \frac{1}{3}$; the maximum value of the objective function is then $f(\mathbf{x}) = 7$.

Using (6-47) through (6-50), the reader should now be able to show that the *minimum* value of $f(\mathbf{x}) = 3x_1 + x_2$, subject to the same constraints, is $-5\sqrt{2}$.

6.8 THE FEASIBLE REGION AND CONSTRAINT QUALIFICATION

Having reconnoitered the territory and identified the goal we are seeking, let us return to the starting point and begin a serious development of the mathematical results we shall need. First, we make a space-saving change in format: Subtract b_i from both sides of the constraint equation (6-33) and merge the $-b_i$ into the function $g_i(\mathbf{x})$. This produces the following, which we continue to call problem NLP:

$$\text{(6-57)} \qquad \text{Max } f(\mathbf{x}), \qquad \mathbf{x} = (x_1, \ldots, x_n),$$

$$\text{(6-58)} \qquad \text{subject to } g_i(\mathbf{x}) \leqq 0, \qquad i = 1, \ldots, m.$$

Again, bear in mind that these constraints are completely general: Any mathematical programming problem can be expressed in the form of (6-57) and (6-58).

Suppose now that $\mathbf{x}$ is a feasible solution to problem NLP—that is, the point $\mathbf{x}$ in E^n satisfies (6-58). Consider what may happen when we move away from $\mathbf{x}$ a very short distance in the direction $\mathbf{h}$, where $\mathbf{h}$ is a vector in E^n. If we remain in feasible territory throughout such a move, however small—that is, if $\mathbf{h}$ points in an *immediately feasible direction*—then for every i such that $g_i(\mathbf{x}) = 0$ it must be true that

$$\nabla \mathbf{g}_i^T(\mathbf{x}) \cdot \mathbf{h} \leqq 0$$

(throughout this text we are treating all gradients as column vectors). This holds because, if $\nabla \mathbf{g}_i(\mathbf{x})$ had a positive component in the direction $\mathbf{h}$, a small move in that direction would increase the value of $g_i(\mathbf{x})$ beyond its current value of zero, thereby violating (6-58). To express this idea more compactly we define

$$I(\mathbf{x}) \equiv \{i \mid g_i(\mathbf{x}) = 0\};$$

that is, $I(\mathbf{x})$ is the indexing set of subscript numbers of all constraints that are binding at the feasible point $\mathbf{x}$. Then if $\mathbf{h}$ points in a feasible direction, it follows that

$$\text{(6-59)} \qquad \nabla \mathbf{g}_i^T(\mathbf{x}) \cdot \mathbf{h} \leqq 0 \qquad \text{for all } i \text{ in } I(\mathbf{x}).$$

In general, however, the condition that (6-59) hold for all i such that $g_i(\mathbf{x}) = 0$ is *not sufficient* to guarantee that $\mathbf{h}$ points in a feasible direction. For example, Fig. 6-4 illustrates the feasible region in E^2 generated by the two constraints

$$g_1(\mathbf{x}) = x_1^3 - x_2 \leqq 0$$

and

$$g_2(\mathbf{x}) = x_1^3 + x_2 \leqq 0.$$

At the origin (0, 0) both of these are satisfied as equalities. Their gradients,

$$\nabla \mathbf{g}_1(\mathbf{x}) = (3x_1^2, -1)$$

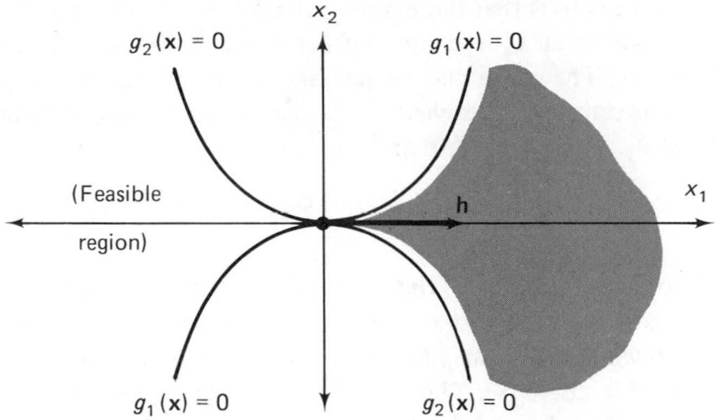

FIGURE 6-4

and

$$\nabla \mathbf{g}_2(\mathbf{x}) = (3x_1^2, 1),$$

when evaluated at the origin, both point directly along the x_2-axis. Thus for $\mathbf{h} = (1, 0)$,

$$\nabla \mathbf{g}_1^T(\mathbf{0}) \cdot \mathbf{h} = \nabla \mathbf{g}_2^T(\mathbf{0}) \cdot \mathbf{h} = 0,$$

and (6-59) is satisfied. But obviously **h** points in an infeasible direction: Any point on the x_1-axis to the right of the origin violates the constraints. [Notice that if the objective were to maximize $f(\mathbf{x}) = x_1$, the optimal solution, given the constraint set of Fig. 6-4, would be $\mathbf{x}^* = (0, 0)$, but there would be no associated values of λ_1^* and λ_2^* satisfying (6-47). The student should verify this.]

To rule out this type of pathology on the boundaries of the feasible region, Kuhn and Tucker excluded from their theoretical results all mathematical programs that do not satisfy a certain algebraic condition known as the *constraint qualification*. This condition would not be met, for example, by any problem having the feasible region shown in Fig. 6-4.

Constraint qualification: Let $\mathbf{x}_0$ *be any feasible solution to problem NLP, and let* **h** *be any vector originating at* $\mathbf{x}_0$ *which satisfies* (6-59). *Then, if the constraint qualification is to hold, there must exist an* n*-dimensional vector function* $\boldsymbol{\phi}(t)$ *of a nonnegative real variable* t *such that*

(a) $\boldsymbol{\phi}(0) = \mathbf{x}_0$;

(b) $\boldsymbol{\phi}(t) = \mathbf{x}$ *is a feasible point for all* $0 \leqq t \leqq T$, *where* T *is some positive number; and*

(c)

$$\lim_{t \to 0} \frac{\boldsymbol{\phi}(t) - \boldsymbol{\phi}(0)}{t} = \mathbf{h}. \tag{6-60}$$

What this amounts to is that there must exist some continuous, differentiable curve S originating at $\mathbf{x}_0$ and moving for some nonzero distance through feasible territory. The curve may be parameterized by defining $\mathbf{x} = \boldsymbol{\phi}(t)$, for $t \geqq 0$, to be the point on S reached by traveling a curvilinear distance t along S, starting at $\mathbf{x}_0$. Thus $\mathbf{x}_0 = \boldsymbol{\phi}(0)$ and

$$\lim_{t \to 0} \frac{\boldsymbol{\phi}(t) - \boldsymbol{\phi}(0)}{t}$$

is the slope of S at $\mathbf{x}_0$. Part (c) then provides that the vector $\mathbf{h}$ has the same slope as the feasible curve S at $\mathbf{x}_0$; hence, *if the constraint qualification is met, every* $\mathbf{h}$ *satisfying* (6-59) *must point in an immediately feasible direction* (i.e., along the feasible curve S). The situation is diagrammed in Fig. 6-5 for a point $\mathbf{x}_0$ on the boundary of the feasible region.

Notice that the provision (c) implies more than that $\mathbf{h}$ is tangent to the curve S. After all, we could take S in Fig. 6-4 to be the curve $x_1^3 = x_2$ extending from the origin down into the third quadrant. This curve is tangent to $\mathbf{h} = (1, 0)$ at the origin; however, at that point the limit in (6-60) would equal not $\mathbf{h}$ but $-\mathbf{h}$. Because, in fact, no curve $\boldsymbol{\phi}(t)$ proceeding from (0, 0) into the feasible region can satisfy (6-60), the constraint qualification is not satisfied in Fig. 6-4.

As this example illustrates, it is quite possible to construct mathematical programming problems whose feasible regions do not meet the constraint qualification. In the development to come, however, we shall need to make use of the function $\boldsymbol{\phi}(t)$ in order to prove a crucial theorem on which the Kuhn-Tucker necessary conditions depend. Accordingly, these conditions and their various related results will simply not apply to problems which fail to satisfy the constraint qualification. Fortunately, such "ill-behaved" problems are not very frequently encountered. They are characterized by the presence of one or more *cusps* at which the feasible region "comes to a point," as it does at the origin in Fig. 6-4.

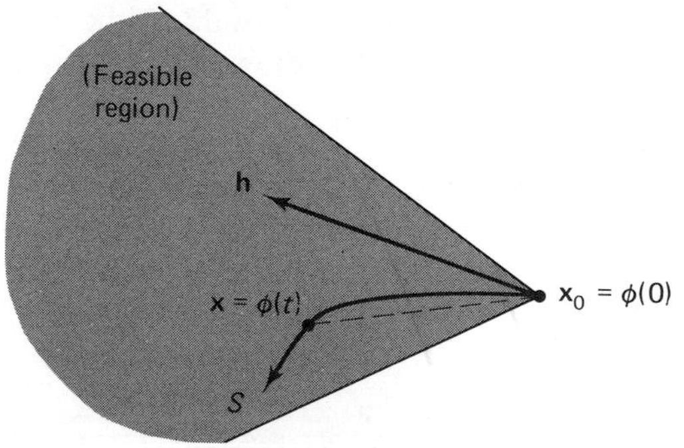

FIGURE 6-5

Returning now to problem NLP,

(6-57) $$\text{Max } f(\mathbf{x}), \qquad \mathbf{x} = (x_1, \ldots, x_n),$$

(6-58) $$\text{subject to } g_i(\mathbf{x}) \leqq 0, \qquad i = 1, \ldots, m,$$

suppose $\mathbf{x}^*$ is a constrained local maximum of $f(\mathbf{x})$ in the feasible region R defined by (6-58). From now on we shall simply write "local maximum" instead of "constrained local maximum"; all local maxima will be assumed to be of the constrained variety unless otherwise indicated. We first observe that because of our choice of $\mathbf{x}^*$ it must be impossible to achieve an *immediate* increase in the value of the objective function by moving away from $\mathbf{x}^*$ in any feasible direction. From a local maximum in the interior of R any direction is feasible; from a boundary point only those directions $\mathbf{h}$ that satisfy (6-59) can be feasible. We gather up these thoughts into the following proposition: *If* $\mathbf{x}^*$ *is a local maximum of problem NLP and the constraint qualification holds, then*

(6-61) $$\nabla \mathbf{f}^T(\mathbf{x}^*) \cdot \mathbf{h} \leqq 0$$

for all $\mathbf{h}$ *satisfying* (6-59). We remind the reader that all gradients are being treated as column vectors.

To prove (6-61), we begin with a Taylor expansion about $\mathbf{x}^*$:

(6-62) $$f(\mathbf{x}) = f(\mathbf{x}^*) + \nabla \mathbf{f}^T(\mathbf{x}^* + \theta(\mathbf{x} - \mathbf{x}^*)) \cdot (\mathbf{x} - \mathbf{x}^*)$$

for some θ, $0 \leqq \theta \leqq 1$. Since $\mathbf{x}^*$ is a local maximum, $f(\mathbf{x}) \leqq f(\mathbf{x}^*)$ for all feasible points $\mathbf{x}$ sufficiently close to $\mathbf{x}^*$, so

(6-63) $$0 \geqq \nabla \mathbf{f}^T(\mathbf{x}^* + \theta(\mathbf{x} - \mathbf{x}^*)) \cdot (\mathbf{x} - \mathbf{x}^*) \qquad \text{for feasible } \mathbf{x}.$$

Suppose some vector $\mathbf{h}$ satisfying (6-59) is given. We then make use of the constraint qualification, which guarantees the existence of a function $\boldsymbol{\phi}(t)$ such that (1) $\boldsymbol{\phi}(0) = \mathbf{x}^*$ and (2) the points $\mathbf{x} = \boldsymbol{\phi}(t)$ are feasible for small positive t. Making substitutions into (6-63) produces

(6-64) $$0 \geqq \nabla \mathbf{f}^T(\mathbf{x}^* + \theta[\boldsymbol{\phi}(t) - \boldsymbol{\phi}(0)]) \cdot \frac{\boldsymbol{\phi}(t) - \boldsymbol{\phi}(0)}{t};$$

we have divided both sides by $t > 0$, which preserves the inequality. If we now take the limit of (6-64) as t approaches zero, the constraint qualification gives us

$$0 \geqq \nabla \mathbf{f}^T(\mathbf{x}^*) \cdot \mathbf{h},$$

as was to be shown.

Let us aggregate the results of this section into a theorem:

Theorem 6.2: Let $\mathbf{x}^*$ *be a local maximum of* $f(\mathbf{x})$ *in the feasible region of the mathematical programming problem NLP. Let* $I(\mathbf{x}^*)$ *be the set of indices* i *for which*

$$g_i(\mathbf{x}^*) = 0.$$

Define a set H *of* n*-component vectors* **h** *as follows:* **h** *is in* H *if and only if*

$$\nabla \mathbf{g}_i^T(\mathbf{x}^*) \cdot \mathbf{h} \leqq 0 \qquad \text{for all i in } I(\mathbf{x}^*). \tag{6-59}$$

Then if **h** *is in* H *and problem NLP satisfies the constraint qualification,*

$$\nabla \mathbf{f}^T(\mathbf{x}^*) \cdot \mathbf{h} \leqq 0. \tag{6-61}$$

Note in passing that when $\mathbf{x}^*$ lies in the interior of the feasible domain, $I(\mathbf{x}^*)$ is empty. In this case *any n*-component vector **h** is a member of H, so from (6-61) the gradient vector $\nabla \mathbf{f}$ evaluated at $\mathbf{x}^*$ has no positive component in any direction. This implies

$$\nabla \mathbf{f}(\mathbf{x}^*) = \mathbf{0},$$

and $\mathbf{x}^*$ is evidently an unconstrained local maximum, as would have been expected.

6.9 THE KUHN-TUCKER NECESSARY CONDITIONS

We are now ready to obtain a result of fundamental importance in the theory of mathematical programming. The proof we shall use is essentially the same as the one originally presented by Kuhn and Tucker [28] in 1951.

Theorem 6.3 (Kuhn-Tucker conditions): Let $\mathbf{x}^*$ *be a constrained local maximum of* $f(\mathbf{x})$ *in the feasible region of the mathematical programming problem NLP:*

$$\text{Max } f(\mathbf{x}), \qquad \mathbf{x} = (x_1, \ldots, x_n), \tag{6-57}$$

$$\text{subject to } g_i(\mathbf{x}) \leqq 0, \qquad i = 1, \ldots, m, \tag{6-58}$$

where $f(\mathbf{x})$ *and all* $g_i(\mathbf{x})$ *are assumed to have continuous first derivatives. If the constraint qualification holds, then there must exist* $\boldsymbol{\lambda}^* = [\lambda_1^*, \ldots, \lambda_m^*]$ *such that*

$$\nabla \mathbf{f}(\mathbf{x}^*) - \sum_{i=1}^{m} \lambda_i^* \nabla \mathbf{g}_i(\mathbf{x}^*) = \mathbf{0}; \tag{6-65}$$

$$\lambda_i^* g_i(\mathbf{x}^*) = 0, \qquad i = 1, \ldots, m; \tag{6-66}$$

and

$$\lambda_i^* \geqq 0, \qquad i = 1, \ldots, m. \tag{6-67}$$

These Kuhn-Tucker conditions (6-65) through (6-67) must *necessarily* hold if $\mathbf{x}^*$ is to be a local maximum (the reader is reminded that all local optima are understood to be of the constrained variety unless stated otherwise). The λ_i^* are known as *generalized Lagrange multipliers*, and, as we shall see, they have many of the properties shared by dual variables and ordinary Lagrange multipliers. Notice, for example, that the condition (6-65) is exactly the same as the result (6-22) obtained for Lagrange multipliers. Moreover, (6-66) imposes complementary slackness on "primal" slacks and "dual" variables.

Like the duality theorem of linear programming and the Lagrange multiplier result of Section 6.3, this is an *existence theorem*: "There must exist $\boldsymbol{\lambda}^*$ such that, etc." By and large, theorems guaranteeing the existence of solutions (or dual variables, or whatever) tend to be the pivotal results in mathematics: Although they do not in general tell how to *find* solutions, they do establish the necessary theoretical foundations for problem-solving algorithms. Furthermore, from an esthetic point of view existence theorems are often the most harmonious and elegant, as the reader may discover for himself in studying the following proof.

As before, we define $I(\mathbf{x}^*)$ to be the set of indices i for which $g_i(\mathbf{x}^*) = 0$. The first step in the proof will be to establish the existence of a feasible solution to the following linear program in unknowns λ_i:

$$\text{(6-68)} \qquad \text{Min } z = \sum_{i \text{ in } I(\mathbf{x}^*)} 0 \cdot \lambda_i$$

$$\text{(6-69)} \qquad \text{subject to } \sum_{i \text{ in } I(\mathbf{x}^*)} \lambda_i \nabla \mathbf{g}_i(\mathbf{x}^*) = \nabla \mathbf{f}(\mathbf{x}^*),$$

$$\text{(6-70)} \qquad \lambda_i g_i(\mathbf{x}^*) = 0 \qquad \text{for all } i \text{ in } I(\mathbf{x}^*),$$

$$\text{(6-71)} \qquad \text{and} \qquad \lambda_i \geqq 0 \qquad \text{for all } i \text{ in } I(\mathbf{x}^*).$$

This linear program is different from the Kuhn-Tucker conditions themselves in that we are now considering only those constraints (and their associated multipliers λ_i) which are binding at $\mathbf{x}^*$.

The associated dual program may be written as follows:

$$\text{(6-72)} \qquad \text{Max } z' = \nabla \mathbf{f}^T(\mathbf{x}^*) \cdot \mathbf{h}$$

$$\text{(6-73)} \qquad \text{subject to } \nabla \mathbf{g}_i^T(\mathbf{x}^*) \cdot \mathbf{h} + w_i g_i(\mathbf{x}^*) \leqq 0 \qquad \text{for all } i \text{ in } I(\mathbf{x}^*),$$

with $\mathbf{h}$ and $\mathbf{w}$ unrestricted, where $\mathbf{h}$ is an n-component vector of dual variables associated with the primal constraints (6-69), and the $\mathbf{w}$ are the dual variables associated with (6-70). Obviously the dual program has at least one feasible solution, namely, $\mathbf{h} = \mathbf{0}$ and $\mathbf{w} = \mathbf{0}$.

Now, $g_i(\mathbf{x}^*) = 0$ for all i in $I(\mathbf{x}^*)$ by definition, so the dual constraints (6-73) only amount to

$$\text{(6-74)} \qquad \nabla \mathbf{g}_i^T(\mathbf{x}^*) \cdot \mathbf{h} \leqq 0 \qquad \text{for all } i \text{ in } I(\mathbf{x}^*).$$

But if the constraint qualification holds, then Theorem 6.2 requires that, for all $\mathbf{h}$ satisfying (6-74),

$$\nabla \mathbf{f}^T(\mathbf{x}^*) \cdot \mathbf{h} \leqq 0.$$

Thus the dual objective function cannot take on a value greater than zero for any feasible solution $\mathbf{h}$.

Since the dual problem is bounded but has a feasible solution, it must have an optimal solution (which happens to be $\mathbf{h} = \mathbf{0}$). We may therefore apply the duality theorem of linear programming (Theorem 2.4), which guarantees that *the primal problem* (6-68) *through* (6-71) *has a feasible solution*. Call this primal-feasible solution $\boldsymbol{\lambda}_I^*$, a vector with as many components as there are members of $I(\mathbf{x}^*)$. To this set of values of λ_i we add $\lambda_j = 0$ for all indices j *not* in $I(\mathbf{x}^*)$, that is, for all nonbinding constraints. These new $\lambda_j = 0$ expand $\boldsymbol{\lambda}_I^*$ into an m-component vector $\boldsymbol{\lambda}^*$ which fulfills the Kuhn-Tucker conditions, as desired. This is easily seen: $\boldsymbol{\lambda}_I^*$ satisfies (6-69) and the $\lambda_j = 0$ merely add a string of zeros to the left-hand side, so $\boldsymbol{\lambda}^*$ satisfies (6-65). Even more trivially, $\boldsymbol{\lambda}_I^*$ satisfying (6-70) and (6-71), combined with $\lambda_j = 0$, together establish (6-66) and (6-67) for the vector $\boldsymbol{\lambda}^*$. The proof is complete.

We have established the Kuhn-Tucker necessary conditions for problem NLP, whose constraints are all of the form $g_i(\mathbf{x}) \leqq 0$. As was mentioned earlier, problems with other types of constraints can be transformed so that Theorem 6.3 can be applied. It is more efficient, however, to make direct use of the following:

Theorem 6.4 (Kuhn-Tucker corollary): If $\mathbf{x}^*$ *is a local maximum for the mathematical programming problem*

$$\begin{aligned}
&\text{Max } f(\mathbf{x}), && \mathbf{x} = (x_1, \ldots, x_n),\\
&\text{subject to } g_i(\mathbf{x}) \leqq b_i, && i = 1, \ldots, u,\\
&\qquad g_j(\mathbf{x}) \geqq b_j, && j = u + 1, \ldots, v,\\
&\qquad g_k(\mathbf{x}) = b_k, && k = v + 1, \ldots, m,\\
&\text{and} \qquad \mathbf{x} \geqq \mathbf{0}, &&
\end{aligned}$$

and if the constraint qualification is satisfied, then there exists a vector of values $\boldsymbol{\lambda}^* = [\lambda_1^*, \ldots, \lambda_m^*]$ *such that*

$$\nabla \mathbf{f}(\mathbf{x}^*) - \sum_{i=1}^{m} \lambda_i^* \nabla \mathbf{g}_i(\mathbf{x}^*) \leqq \mathbf{0}; \tag{6-75}$$

$$x_j^* \left\{ \frac{\delta f(\mathbf{x}^*)}{\delta x_j} - \sum_{i=1}^{m} \lambda_i^* \frac{\delta g_i(\mathbf{x}^*)}{\delta x_j} \right\} = 0, \qquad j = 1, \ldots, n; \tag{6-76}$$

and

$$\lambda_i^*[b_i - g_i(\mathbf{x}^*)] = 0, \qquad i = 1, \ldots, m, \tag{6-77}$$

with

$$\lambda_i^* \geqq 0, \qquad i = 1, \ldots, u,$$
$$\lambda_j^* \leqq 0, \qquad j = u + 1, \ldots, v,$$

and

$$\lambda_k^* \text{ unrestricted}, \qquad k = v + 1, \ldots, m.$$

Notice that if the inequalities (6-75) are interpreted as dual constraints—this notion will be amplified in Section 9.9—then (6-76) and (6-77) are seen to be complementary slackness conditions: Either the dual (primal) constraint must be binding, or the primal (dual) variable must equal zero.

To prove the corollary, we rewrite the constraints in the form that allows us to apply the Kuhn-Tucker conditions directly. The symbols to be used for the associated multipliers are listed at the right:

$$\begin{array}{lll} \text{for} & g_i(\mathbf{x}) \leqq b_i, & \text{use } \lambda_i^*,\ i = 1, \ldots, u; \\ \text{for} & -g_j(\mathbf{x}) \leqq -b_j, & \text{use } \gamma_j^*,\ j = u + 1, \ldots, v; \\ \text{for} & g_k(\mathbf{x}) \leqq b_k, & \text{use } \alpha_k^*,\ k = v + 1, \ldots, m; \\ \text{for} & -g_k(\mathbf{x}) \leqq -b_k, & \text{use } \beta_k^*,\ k = v + 1, \ldots, m; \\ \text{and for} & -\mathbf{x} \leqq \mathbf{0}, & \text{use } \mu_j^*,\ j = 1, \ldots, n. \end{array}$$

Direct application of Theorem 6.3 shows that there must exist generalized Lagrange multipliers such that

$$\nabla \mathbf{f}(\mathbf{x}^*) - \sum_i \lambda_i^* \nabla \mathbf{g}_i(\mathbf{x}^*) + \sum_j \gamma_j^* \nabla \mathbf{g}_j(\mathbf{x}^*)$$
$$- \sum_k \alpha_k^* \nabla \mathbf{g}_k(\mathbf{x}^*) + \sum_k \beta_k^* \nabla \mathbf{g}_k(\mathbf{x}^*) + \boldsymbol{\mu}^* = \mathbf{0},$$
$$\lambda_i^*[g_i(\mathbf{x}^*) - b_i] = 0,$$
$$\gamma_j^*[b_j - g_j(\mathbf{x}^*)] = 0,$$
$$\alpha_k^*[g_k(\mathbf{x}^*) - b_k] = 0,$$
$$\beta_k^*[b_k - g_k(\mathbf{x}^*)] = 0,$$

and

$$\mu_j^*(-x_j^*) = 0,$$

with all λ_i^*, γ_j^*, α_k^*, β_k^*, and μ_j^* nonnegative.

Subscript limits for the multipliers correspond to the limits for the associated constraints. Now make the following changes:

$$\lambda_j^* = -\gamma_j^*, \qquad j = u + 1, \ldots, v,$$

and

$$\lambda_k^* = \alpha_k^* - \beta_k^*, \qquad k = v + 1, \ldots, m.$$

Notice that the multipliers λ_j^* are *nonpositive* and the λ_k^* are unrestricted. The reader will not find it difficult to complete the proof.

As a matter of terminology, any feasible solution $\mathbf{x}_0$ to a mathematical programming problem which, in conjunction with an associated set of values $\boldsymbol{\lambda}_0$, satisfies the Kuhn-Tucker conditions for that problem is known as a *constrained stationary point* or a *Kuhn-Tucker point*. It should be emphasized that a given problem may have constrained stationary points that are not constrained local optima of the objective function; in other words, the Kuhn-Tucker conditions are, in general, *not sufficient* to define and identify local optima. This will be an important consideration in our study of nonlinear programming algorithms, which can sometimes lead to Kuhn-Tucker points that are not locally optimal (just as direct climbing methods for unconstrained problems can end up at stationary points that are not unconstrained local optima). As an example of a nonoptimal Kuhn-Tucker point, consider the following problem:

$$\begin{aligned} \text{Max } f(x_1, x_2) &= x_2 \\ \text{subject to } x_1^2 + x_2^2 - 4 &\leqq 0 \\ \text{and } -x_1^2 + x_2 &\leqq 0. \end{aligned}$$

The feasible region is diagrammed in Fig. 6-6; it can easily be shown that the constraint qualification holds. The Kuhn-Tucker conditions are

$$\begin{aligned} 0 - \lambda_1(2x_1) - \lambda_2(-2x_1) &= 0, \\ 1 - \lambda_1(2x_2) - \lambda_2(1) &= 0, \\ \lambda_1(x_1^2 + x_2^2 - 4) &= 0, \\ \lambda_2(x_2 - x_1^2) &= 0, \end{aligned}$$

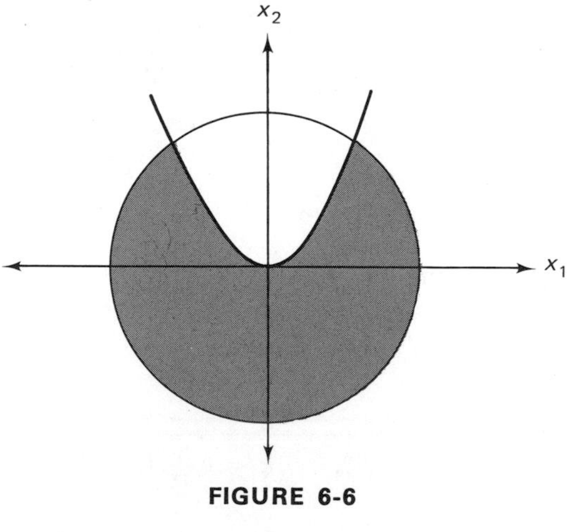

FIGURE 6-6

and

$$\lambda_1, \lambda_2 \geqq 0.$$

Note that the feasible point $x_1 = 0$, $x_2 = 0$, along with $\lambda_1 = 0$ and $\lambda_2 = 1$, satisfies all these equations. Yet the origin is obviously not a local maximum of $f(x_1, x_2)$ in the feasible region, since any neighboring point on the parabola $x_2 = x_1^2$ is feasible and yields a greater objective value.

Theorems 6.3 and 6.4 cannot be used directly in solving any but the smallest problems. After all, applying the Kuhn-Tucker conditions to problem NLP produces $m + n$ simultaneous nonlinear equations in $m + n$ unknowns, as well as nonnegativity restrictions on the λ_i^*. Even the example of Section 6.7, with two constraints in two variables, required a fair amount of effort. However, we shall see in the upcoming chapters that the Kuhn-Tucker results do provide the theoretical foundation for a great many computational algorithms. Most of these are designed specifically for problems with various types of special structure, and some are quite efficient—this is true, for example, of Wolfe's algorithm for solving quadratic programs.

In those cases where a problem is small enough to be attacked directly by writing out the conditions (6-65) through (6-67), the resulting system of equations may yield several solutions $(\mathbf{x}_k^*, \boldsymbol{\lambda}_k^*)$. Among these are all local maxima of $f(\mathbf{x})$ in the feasible region, and among the local maxima (unless the objective is unbounded) is the optimal solution to the given problem. This global maximum may thus be identified simply by evaluating the objective at each of the solution points $\mathbf{x}_k^*$.

6.10 APPLICATION TO LINEAR PROGRAMMING

As an example, let us derive the Kuhn-Tucker necessary conditions for one extremely important type of mathematical programming problem, the linear program in canonical form:

$$\text{(6-78)} \qquad \text{Max } f(\mathbf{x}) = \mathbf{c}^T\mathbf{x}, \qquad \mathbf{x} = (x_1, \ldots, x_n),$$

$$\text{(6-79)} \qquad \text{subject to } \mathbf{Ax} \leqq \mathbf{b}$$

$$\text{(6-80)} \qquad \text{and } \mathbf{x} \geqq \mathbf{0},$$

where $\mathbf{A}$ is an m-by-n matrix. For each constraint

$$g_i(\mathbf{x}) \equiv a_{i1}x_1 + \cdots + a_{in}x_n - b_i \leqq 0,$$

the gradient vector

$$\nabla \mathbf{g}_i(\mathbf{x}) = [a_{i1}, \ldots, a_{in}]$$

is perpendicular to the hyperplane $g_i(\mathbf{x}) = 0$ and points directly away from the feasible region. For any feasible point $\mathbf{x}_0$ on that hyperplane, the only n-component vectors $\mathbf{h}$ that satisfy

$$\nabla \mathbf{g}_i^T(\mathbf{x}_0) \cdot \mathbf{h} \leqq 0$$

are those that point either along the hyperplane or into the interior. Similarly, at an extreme point $\mathbf{x}_0$ where two or more hyperplanes intersect—refer to Fig. 6-7(a)—it is easily seen that again the only vectors $\mathbf{h}$ satisfying

$$\text{(6-59)} \qquad \nabla \mathbf{g}_i^T(\mathbf{x}_0) \cdot \mathbf{h} \leqq 0 \qquad \text{for all } i \text{ such that } g_i(\mathbf{x}_0) = 0$$

are those that point into (or along) the feasible region. We conclude from this that the constraint qualification is met by the linear feasible region (6-79) and (6-80). For a more rigorous demonstration, one could simply define $\boldsymbol{\phi}(t)$ to be the feasible point $\mathbf{x}$ on the line $\mathbf{h}$ a distance t from $\mathbf{x}_0$, which is tantamount to letting the curve S coincide with $\mathbf{h}$.

Before continuing with our discussion of the linear program, we pause to remark that the constraint qualification is also satisfied by any convex feasible region, that is, by any mathematical program whose feasible region is a convex set. The situation is illustrated in two dimensions in Fig. 6-7(b). Any vector $\mathbf{h}$ satisfying (6-59) must either point into the interior of the feasible region or must be tangent to one of the bounding hypersurfaces, say, $g_1(\mathbf{x}) = 0$, and directed inward from the other, as shown in the diagram. In the first case the curve S defined by the vector function $\boldsymbol{\phi}(t)$ can coincide with $\mathbf{h}$, while in the second it can coincide with that portion of the boundary $g_1(\mathbf{x}) = 0$ that lies to the left of $\mathbf{x}_0$ in the diagram. Although $\mathbf{h}$ is not an immediately feasible direction in the latter case, the provision of the constraint quealification that

$$\text{(6-60)} \qquad \lim_{t \to 0} \frac{\boldsymbol{\phi}(t) - \boldsymbol{\phi}(0)}{t} = \mathbf{h}$$

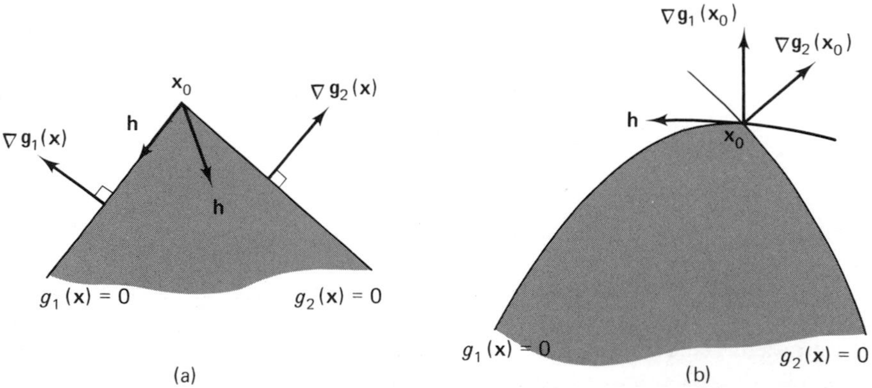

FIGURE 6-7

is satisfied because $\mathbf{h}$ is tangent to the boundary $g_1(\mathbf{x}) = 0$ at $\mathbf{x}_0$. The result we have obtained is important enough to be formalized:

Theorem 6.5: The constraint qualification is satisfied by any mathematical program having a convex feasible region, and therefore by any linear program.[4]

The argument leading up to this theorem was admittedly somewhat heuristic in character. The formal details that would be required for a rigorous proof are being omitted here, inasmuch as they are not particularly edifying.

Returning to the linear program (6-78) through (6-80), let us rewrite the constraints in the following form:

$$a_{i1}x_1 + \cdots + a_{in}x_n - b_i \leqq 0, \qquad i = 1, \ldots, m,$$

and

$$-x_j \leqq 0, \qquad j = 1, \ldots, n.$$

Assuming that the linear program is not unbounded, we may then apply the Kuhn-Tucker conditions as stated in Theorem 6.3: Associated with the (finite) optimal solution $\mathbf{x}^*$ is a set of values $\boldsymbol{\lambda}^* = (\lambda_1^*, \ldots, \lambda_m^*)$, $\mathbf{u}^* = (u_1^*, \ldots, u_n^*)$, such that

(6-81) $$\mathbf{c} - \mathbf{A}^T\boldsymbol{\lambda}^* + \mathbf{u}^* = \mathbf{0};$$

(6-82) $$\lambda_i^*(a_{i1}x_1^* + \cdots + a_{in}x_n^* - b_i) = 0, \qquad i = 1, \ldots, m;$$

(6-83) $$-u_j^*x_j^* = 0, \qquad j = 1, \ldots, n;$$

(6-84) $$\boldsymbol{\lambda}^* \geqq \mathbf{0};$$

and

(6-85) $$\mathbf{u}^* \geqq \mathbf{0}.$$

Since $\mathbf{u}^* = \mathbf{A}^T\boldsymbol{\lambda}^* - \mathbf{c} \geqq \mathbf{0}$, the first and last of these relations together imply

(6-86) $$\mathbf{A}^T\boldsymbol{\lambda}^* \geqq \mathbf{c},$$

and (6-83) may be written

(6-87) $$(a_{1j}\lambda_1^* + \cdots + a_{mj}\lambda_m^* - c_j)x_j^* = 0.$$

[4]It should be noted that the theorem as stated here is not quite accurate: Its conclusion is in fact applicable only to problems whose feasible regions include interior points. It is obvious, for example, that the constraint qualification cannot hold in the case of a feasible region consisting of only a single point, because then a feasible curve S of nonzero length could not possibly be defined.

For our first result, observe from (6-86) and (6-84) that the Kuhn-Tucker conditions guarantee the existence of a solution $\boldsymbol{\lambda} = \boldsymbol{\lambda}^*$ to the system

$$\mathbf{A}^T\boldsymbol{\lambda} \geqq \mathbf{c} \tag{6-88}$$

and

$$\boldsymbol{\lambda} \geqq \mathbf{0}. \tag{6-89}$$

These are the constraints of the dual problem of the given linear program. Thus we have shown that if the primal linear program has a finite optimal solution, then the dual has a feasible solution.

The dual objective function is

$$\text{Min } \mathbf{b}^T\boldsymbol{\lambda}, \tag{6-90}$$

and for *any* feasible solutions $\mathbf{x}$ and $\boldsymbol{\lambda}$ to the primal and dual problems, we may write

$$\begin{aligned} \mathbf{b}^T\boldsymbol{\lambda} \geqq (\mathbf{A}\mathbf{x})^T\boldsymbol{\lambda} \equiv \mathbf{x}^T\mathbf{A}^T\boldsymbol{\lambda} \qquad & \text{from (6-79)}, \\ \geqq \mathbf{x}^T\mathbf{c} \qquad & \text{from (6-88)}, \end{aligned}$$

making use of the fact that all feasible $\boldsymbol{\lambda}$ and $\mathbf{x}$ are nonnegative. Since $\mathbf{x}^*$ is one feasible solution to the primal,

$$\mathbf{b}^T\boldsymbol{\lambda} \geqq \mathbf{x}^{*T}\mathbf{c}, \tag{6-91}$$

so the dual problem is bounded from below. Knowing also that it has a feasible solution, we may conclude that it has a finite optimum.

What is the optimal solution to the dual problem? For every i, $i = 1, \ldots, m$, the complementary slackness condition (6-82) implies that either $\lambda_i^* = 0$ or else b_i equals the ith component of the vector $\mathbf{A}\mathbf{x}^*$; we may therefore write

$$\mathbf{b}^T\boldsymbol{\lambda}^* = (\mathbf{A}\mathbf{x}^*)^T\boldsymbol{\lambda}^* = \mathbf{x}^{*T}\mathbf{A}^T\boldsymbol{\lambda}^*. \tag{6-92}$$

Similarly, (6-87) implies

$$\mathbf{x}^{*T}\mathbf{A}^T\boldsymbol{\lambda}^* = \mathbf{x}^{*T}\mathbf{c}, \tag{6-93}$$

and (6-92) and (6-93) together yield

$$\mathbf{b}^T\boldsymbol{\lambda}^* = \mathbf{x}^{*T}\mathbf{c}. \tag{6-94}$$

But we know from (6-91) that no dual feasible solution can have an objective value less than $\mathbf{x}^{*T}\mathbf{c}$, so it follows that $\boldsymbol{\lambda}^*$ must be an optimal solution to the dual. We have shown that *if* $\mathbf{x}^*$ *is an optimal solution to the primal problem, then there exists an optimal solution* $\boldsymbol{\lambda}^*$ *to the dual problem such that* $\mathbf{c}^T\mathbf{x}^* = \mathbf{b}^T\boldsymbol{\lambda}^*$. This verifies the "only-if" half of the duality theorem of linear programming.[5]

6.11 SUFFICIENCY OF THE KUHN-TUCKER CONDITIONS

Consider the mathematical program

(6-95) $$\text{Max } f(\mathbf{x}), \qquad \mathbf{x} = (x_1, \ldots, x_n),$$

(6-96) $$\text{subject to } g_i(\mathbf{x}) \leqq 0, \qquad i = 1, \ldots, m,$$

where $f(\mathbf{x})$ is a concave function and all the $g_i(\mathbf{x})$ are convex. According to Theorem 3.6, the region defined by (6-96) is a convex set, so it follows from Theorem 3.7 that any constrained local maximum of $f(\mathbf{x})$ in the feasible region must be an optimal solution to the problem. Problems such as this one, in which a concave function is to be maximized or a convex function minimized over a convex set, are known as *convex programming problems.* They are characterized by having no suboptimal local maxima: Any local maximum is also global.

Can we conclude that any point $(\mathbf{x}_0, \boldsymbol{\lambda}_0)$ satisfying both (6-96) and the Kuhn-Tucker conditions must be an optimal solution? No, not yet, because we have no guarantee that $\mathbf{x}_0$ is a local maximum—recall the counterexample of Section 6.9. Nevertheless, that conclusion happens to be true: Any feasible solution to a convex program that satisfies (6-65) through (6-67) must indeed be a global maximum. In other words, for a convex program the Kuhn-Tucker conditions are *sufficient* to determine the optimal solution.

Theorem 6.6: Let $f(\mathbf{x})$ *be a concave function and all the* $g_i(\mathbf{x})$ *be convex in the mathematical program*

(6-95) $$\text{Max } f(\mathbf{x}), \qquad \mathbf{x} = (x_1, \ldots, x_n),$$

(6-96) $$\text{subject to } g_i(\mathbf{x}) \leqq 0, \qquad i = 1, \ldots, m.$$

If a feasible point $\mathbf{x}^*$ *and some* $\boldsymbol{\lambda}^* = (\lambda_1^*, \ldots, \lambda_m^*)$ *satisfy the Kuhn-Tucker conditions* (6-65) *through* (6-67), *then* $\mathbf{x}^*$ *is a global maximum. More generally, whenever a concave function is being maximized or a convex function minimized over a convex feasible region, any point at which the Kuhn-Tucker conditions are satisfied is a global optimum.*

[5] We cannot claim to have *proved* the duality theorem, because it was used in the derivation of the Kuhn-Tucker conditions.

We shall prove the result for the problem (6-95) and (6-96). Using the first Kuhn-Tucker condition (6-65),

$$\nabla \mathbf{f}^T(\mathbf{x}^*) \cdot (\mathbf{x} - \mathbf{x}^*) = \sum_{i=1}^{m} \lambda_i^* \nabla \mathbf{g}_i^T(\mathbf{x}^*) \cdot (\mathbf{x} - \mathbf{x}^*) \leqq \sum_{i=1}^{m} \lambda_i^*[g_i(\mathbf{x}) - g_i(\mathbf{x}^*)].$$

The inequality, based on Theorem 3.9, follows from the convexity of the functions $g_i(\mathbf{x})$ and the nonnegativity of the λ_i^*. In view of the second Kuhn-Tucker condition (6-66), the above reduces to

$$\nabla \mathbf{f}^T(\mathbf{x}^*) \cdot (\mathbf{x} - \mathbf{x}^*) \leqq \sum_{i=1}^{m} \lambda_i^* g_i(\mathbf{x}).$$

If $\mathbf{x}$ is a feasible solution to the mathematical programming problem, then $g_i(\mathbf{x})$ is nonpositive for all i, and since $\lambda_i^* \geqq 0$ by (6-67),

$$\nabla \mathbf{f}^T(\mathbf{x}^*) \cdot (\mathbf{x} - \mathbf{x}^*) \leqq 0$$

for all feasible $\mathbf{x}$. Finally, because $f(\mathbf{x})$ is concave, Theorem 3.9 implies that $f(\mathbf{x}) - f(\mathbf{x}^*)$ is also nonpositive, or

$$f(\mathbf{x}) \leqq f(\mathbf{x}^*)$$

for all feasible $\mathbf{x}$, as was to be shown. The proof can be extended in a straightforward manner to cover all convex programming problems.

6.12 THE LAGRANGIAN FUNCTION AND SADDLE POINTS

In the past few sections we have developed a method which allows us, *in theory*, to solve any "well-behaved" mathematical programming problem: Check to be sure the feasible region satisfies the constraint qualification, apply the Kuhn-Tucker conditions, and solve the resulting system of equations. If several solutions are found, the global optimum is among them and can be identified by evaluating the objective function at each.

In the remainder of this chapter a second theoretical method for solving mathematical programs will occupy our attention. It is even more awkward computationally than the application of Kuhn-Tucker conditions, but its value lies in the insights it affords into the properties and behavior of constrained optima. This second approach involves finding saddle points of a certain function which should look familiar to the reader:

Definition: Associated with any mathematical program

$$\text{Max } f(\mathbf{x}), \qquad \mathbf{x} = (x_1, \ldots, x_n),$$
$$\text{subject to } g_i(\mathbf{x}) \leqq 0, \qquad i = 1, \ldots, m,$$

is a generalized Lagrangian function

(6-97) $$\mathrm{F}(\mathbf{x}, \boldsymbol{\lambda}) \equiv \mathrm{f}(\mathbf{x}) - \sum_{i=1}^{m} \lambda_i g_i(\mathbf{x}),$$

where $\boldsymbol{\lambda} = (\lambda_1, \ldots, \lambda_m)$.

We assume that $f(\mathbf{x})$ and the $g_i(\mathbf{x})$ are continuous, so the Lagrangian function is continuous in $\mathbf{x}$ and $\boldsymbol{\lambda}$. Note that its values are real numbers.

To see how mathematical programs might be solved by finding saddle points of $F(\mathbf{x}, \boldsymbol{\lambda})$, we shall need some preliminary groundwork. Let us first specify exactly what we mean by a *saddle point* of the Lagrangian function. Ordinarily, a "local" saddle point for a function $T(\mathbf{x}, \boldsymbol{\lambda})$, where $\mathbf{x}$ is in E^n and $\boldsymbol{\lambda}$ is in E^m, is defined to be any point $(\mathbf{x}_0, \boldsymbol{\lambda}_0)$ in Euclidean $(n + m)$-space such that

(6-98) $$T(\mathbf{x}, \boldsymbol{\lambda}_0) \leqq T(\mathbf{x}_0, \boldsymbol{\lambda}_0) \leqq T(\mathbf{x}_0, \boldsymbol{\lambda})$$

for all $(\mathbf{x}, \boldsymbol{\lambda}_0)$ and all $(\mathbf{x}_0, \boldsymbol{\lambda})$ sufficiently close to $(\mathbf{x}_0, \boldsymbol{\lambda}_0)$. When (6-98) holds for all $(\mathbf{x}, \boldsymbol{\lambda}_0)$ and all $(\mathbf{x}_0, \boldsymbol{\lambda})$ in E^{n+m}, then $(\mathbf{x}_0, \boldsymbol{\lambda}_0)$ is known as a "global" saddle point.

A simple example of a global saddle point is $(x = 0, \lambda = 0)$ for the function $T(x, \lambda) = \lambda^2 - x^2$; this is easily verified from the definition. Similarly, the more complicated function

$$T(x, \lambda) = \lambda^2 - x^2(x - 1)^2$$

can be shown to have global saddle points at $(x = 0, \lambda = 0)$ and at $(x = 1, \lambda = 0)$.

For our purposes, however, we shall require a specialized version of this concept:

Definition: A Lagrangian function $\mathrm{F}(\mathbf{x}, \boldsymbol{\lambda})$ *will be said to have a Lagrangian global saddle point, or simply a global saddle point, at* $(\mathbf{x}^*, \boldsymbol{\lambda}^*)$ *if* $\boldsymbol{\lambda}^* \geqq \mathbf{0}$ *and*

(6-99) $$\mathrm{F}(\mathbf{x}, \boldsymbol{\lambda}^*) \leqq \mathrm{F}(\mathbf{x}^*, \boldsymbol{\lambda}^*) \leqq \mathrm{F}(\mathbf{x}^*, \boldsymbol{\lambda})$$

for all $\mathbf{x}$ *in* E^n *and all* $\boldsymbol{\lambda} \geqq \mathbf{0}$ *in* E^m.

These global saddle points can be identified via a *sequential optimization* approach that proceeds according to the following steps:

1. Holding $\boldsymbol{\lambda}$ fixed, find the maximum over all $\mathbf{x}$, given $\boldsymbol{\lambda}$, of $F(\mathbf{x}, \boldsymbol{\lambda})$; this is denoted

$$\max_{\mathbf{x}|\boldsymbol{\lambda}} F(\mathbf{x}, \boldsymbol{\lambda})$$

and is a function of $\boldsymbol{\lambda}$ alone.

2. Then find the minimum of this function over nonnegative $\boldsymbol{\lambda}$; that is, find

$$\min_{\boldsymbol{\lambda} \geqq \mathbf{0}}\{\max_{\mathbf{x}|\boldsymbol{\lambda}} F(\mathbf{x}, \boldsymbol{\lambda})\}.$$

Note that the inner symbol, "max," corresponds to the operation to be performed first.

To show how and why this procedure leads to saddle points, let us derive a rather simple result:

Theorem 6.7: If the Lagrangian function $F(\mathbf{x}, \boldsymbol{\lambda})$ *has a (Lagrangian) global saddle point at* $(\mathbf{x}^*, \boldsymbol{\lambda}^*)$, *then*

$$\min_{\boldsymbol{\lambda} \geqq \mathbf{0}}\{\max_{\mathbf{x}|\boldsymbol{\lambda}} F(\mathbf{x}, \boldsymbol{\lambda})\} = F(\mathbf{x}^*, \boldsymbol{\lambda}^*). \tag{6-100}$$

To expedite the proof we define

$$h(\boldsymbol{\lambda}) \equiv \max_{\mathbf{x}|\boldsymbol{\lambda}} F(\mathbf{x}, \boldsymbol{\lambda}). \tag{6-101}$$

From the fact that $(\mathbf{x}^*, \boldsymbol{\lambda}^*)$ is a global saddle point we have

$$F(\mathbf{x}^*, \boldsymbol{\lambda}) \geqq F(\mathbf{x}^*, \boldsymbol{\lambda}^*)$$

for all $\boldsymbol{\lambda} \geqq \mathbf{0}$. Clearly, then, for any such $\boldsymbol{\lambda}$

$$h(\boldsymbol{\lambda}) \equiv \max_{\mathbf{x}|\boldsymbol{\lambda}} F(\mathbf{x}, \boldsymbol{\lambda}) \geqq F(\mathbf{x}^*, \boldsymbol{\lambda}) \geqq F(\mathbf{x}^*, \boldsymbol{\lambda}^*) \equiv h(\boldsymbol{\lambda}^*); \tag{6-102}$$

the last equality follows from the definition of a global saddle point. But if (6-102) holds for all nonnegative $\boldsymbol{\lambda}$, then

$$\min_{\boldsymbol{\lambda} \geqq \mathbf{0}} h(\boldsymbol{\lambda}) = h(\boldsymbol{\lambda}^*) = F(\mathbf{x}^*, \boldsymbol{\lambda}^*),$$

which is the same as (6-100).

It can also be shown in an analogous way that if $(\mathbf{x}^*, \boldsymbol{\lambda}^*)$ is a global saddle point, then

$$\max_{\mathbf{x}}\{\min_{\boldsymbol{\lambda} \geqq \mathbf{0}|\mathbf{x}} F(\mathbf{x}, \boldsymbol{\lambda})\} = F(\mathbf{x}^*, \boldsymbol{\lambda}^*). \tag{6-103}$$

But both (6-100) and (6-103) are required to prove (6-99).

6.13 SADDLE POINTS AND OPTIMAL SOLUTIONS

In light of the foregoing discussion the reader will not be surprised to learn that there is a close relationship between global saddle points and optimal solutions to mathematical programming problems. In particular, in the case of convex programs the two are equivalent. We shall, however, prove the "if" and "only if" halves of this equivalence in two separate theorems. This is convenient because the equivalence does not extend to non-convex mathematical programming problems, for which the occurrence of a global saddle point at $\mathbf{x}^*$ is sufficient but not necessary to establish that $\mathbf{x}^*$ is an optimal solution.

Let us prove the sufficiency first:

Theorem 6.8: Let $(\mathbf{x}^*, \boldsymbol{\lambda}^*)$, *where* $\boldsymbol{\lambda}^* \geqq \mathbf{0}$, *constitute a (Lagrangian) global saddle point for the Lagrangian function associated with the mathematical program*

$$\text{Max } f(\mathbf{x}), \qquad \mathbf{x} = (x_1, \ldots, x_n),$$
$$\text{subject to } g_i(\mathbf{x}) \leqq 0, \qquad i = 1, \ldots, m.$$

Then $\mathbf{x}^*$ *is an optimal solution to this problem.*

Since $(\mathbf{x}^*, \boldsymbol{\lambda}^*)$ is a saddle point, $F(\mathbf{x}^*, \boldsymbol{\lambda}^*) \leqq F(\mathbf{x}^*, \boldsymbol{\lambda})$ for all $\boldsymbol{\lambda} \geqq \mathbf{0}$; from the definition of the Lagrangian function this implies

$$\sum_i \lambda_i g_i(\mathbf{x}^*) \leqq \sum_i \lambda_i^* g_i(\mathbf{x}^*) \tag{6-104}$$

for nonnegative $\boldsymbol{\lambda}$. Suppose some $g_k(\mathbf{x}^*) > 0$; then the vector $\boldsymbol{\lambda} = (\lambda_1^*, \lambda_2^*, \ldots, \lambda_k^* + 1, \ldots, \lambda_m^*) \geqq \mathbf{0}$, when substituted into the left-hand side of (6-104), would violate the inequality. Therefore all $g_i(\mathbf{x}^*) \leqq 0$ and $\mathbf{x}^*$ is a feasible solution to the problem.

Now suppose that some $g_k(\mathbf{x}^*) < 0$. If $\lambda_k^* > 0$, then the vector $\boldsymbol{\lambda} = (\lambda_1^*, \ldots, \lambda_{k-1}^*, 0, \lambda_{k+1}^*, \ldots, \lambda_m^*) \geqq \mathbf{0}$ would again violate (6-104). It follows that $g_k(\mathbf{x}^*) < 0$ implies $\lambda_k^* = 0$, or

$$\sum_{i=1}^{m} \lambda_i^* g_i(\mathbf{x}^*) = 0,$$

so that

$$F(\mathbf{x}^*, \boldsymbol{\lambda}^*) \equiv f(\mathbf{x}^*) - \sum_{i=1}^{m} \lambda_i^* g_i(\mathbf{x}^*) = f(\mathbf{x}^*). \tag{6-105}$$

Since $(\mathbf{x}^*, \boldsymbol{\lambda}^*)$ is a saddle point,

$$F(\mathbf{x}, \boldsymbol{\lambda}^*) \leqq F(\mathbf{x}^*, \boldsymbol{\lambda}^*)$$

for all $\mathbf{x}$, or, using (6-105),

$$f(\mathbf{x}) - \sum_{i=1}^{m} \lambda_i^* g_i(\mathbf{x}) \leqq f(\mathbf{x}^*). \tag{6-106}$$

But for any feasible $\mathbf{x}$ all the $g_i(\mathbf{x})$ are nonpositive, and, because each λ_i^* is nonnegative, (6-106) yields

$$f(\mathbf{x}) \leqq f(\mathbf{x}^*)$$

for all feasible $\mathbf{x}$. Therefore $\mathbf{x}^*$ is an optimal solution to the given problem, as was to be shown.

Having obtained this result, we can now attack any mathematical programming problem, convex or not, by searching for a global saddle point. If we find one, we have found an optimal solution to the problem: The saddle point is *sufficient*. But if we fail, we cannot conclude that no finite optimal solution exists—*unless the problem is a convex program*. That the optimal solution of a convex program must *necessarily* be a global saddle point is established by the following:

Theorem 6.9: Let $\mathbf{x}^*$ *be an optimal solution to the mathematical program*

$$\begin{aligned} &\text{Max } \mathrm{f}(\mathbf{x}), & \mathbf{x} = (\mathrm{x}_1, \ldots, \mathrm{x}_n), \\ &\text{subject to } \mathrm{g}_i(\mathbf{x}) \leqq 0, & i = 1, \ldots, m, \end{aligned}$$

where $\mathrm{f}(\mathbf{x})$ *is a concave function and all the* $\mathrm{g}_i(\mathbf{x})$ *are convex. Then there exists a nonnegative* $\boldsymbol{\lambda}^* = (\lambda_1^*, \ldots, \lambda_m^*)$ *such that* $(\mathbf{x}^*, \boldsymbol{\lambda}^*)$ *constitutes a global saddle point of the associated Lagrangian function* $\mathrm{F}(\mathbf{x}, \boldsymbol{\lambda})$.

Applying the Kuhn-Tucker conditions at $\mathbf{x}^*$, we know there exists some $\boldsymbol{\lambda}^* \geqq \mathbf{0}$ such that

(6-65) $$\nabla \mathbf{f}(\mathbf{x}^*) - \sum_{i=1}^{m} \lambda_i^* \nabla \mathbf{g}_i(\mathbf{x}^*) = \mathbf{0}$$

and

(6-66) $$\lambda_i^* g_i(\mathbf{x}^*) = 0, \qquad i = 1, \ldots, m.$$

We next observe that

$$F(\mathbf{x}, \boldsymbol{\lambda}^*) \equiv f(\mathbf{x}) - \sum_{i=1}^{m} \lambda_i^* g_i(\mathbf{x})$$

must be a concave function of $\mathbf{x}$. This is true because each of the functions $g_i(\mathbf{x})$ is convex by hypothesis, implying that

$\lambda_i^* g_i(\mathbf{x})$ is convex because $\lambda_i^* \geqq 0$ [see Exercise 3-8(b)];
$\sum_{i=1}^m \lambda_i^* g_i(\mathbf{x})$ is convex by Theorem 3.3;
$-\sum_{i=1}^m \lambda_i^* g_i(\mathbf{x})$ is concave by Theorem 3.2; and
$f(\mathbf{x}) - \sum_{i=1}^m \lambda_i^* g_i(\mathbf{x})$ is concave by Theorem 3.3.

Since $F(\mathbf{x}, \boldsymbol{\lambda}^*)$ is concave, Theorem 3.9 allows us to write

$$F(\mathbf{x}, \boldsymbol{\lambda}^*) \leqq F(\mathbf{x}^*, \boldsymbol{\lambda}^*) + \nabla_{\mathbf{x}} F^T(\mathbf{x}^*, \boldsymbol{\lambda}^*) \cdot (\mathbf{x} - \mathbf{x}^*), \tag{6-107}$$

where

$$\nabla_{\mathbf{x}} F \equiv \left(\frac{\delta F}{\delta x_1}, \ldots, \frac{\delta F}{\delta x_n}\right).$$

But

$$\nabla_{\mathbf{x}} F(\mathbf{x}^*, \boldsymbol{\lambda}^*) \equiv \nabla \mathbf{f}(\mathbf{x}^*) - \sum_{i=1}^{m} \lambda_i^* \nabla \mathbf{g}_i(\mathbf{x}^*) = \mathbf{0}$$

because of (6-65). Hence (6-107) yields

$$F(\mathbf{x}, \boldsymbol{\lambda}^*) \leqq F(\mathbf{x}^*, \boldsymbol{\lambda}^*). \tag{6-108}$$

Now we may also check by simple substitution that for any $\boldsymbol{\lambda}$

$$F(\mathbf{x}^*, \boldsymbol{\lambda}) = F(\mathbf{x}^*, \boldsymbol{\lambda}^*) + \nabla_{\lambda} F^T(\mathbf{x}^*, \boldsymbol{\lambda}^*) \cdot (\boldsymbol{\lambda} - \boldsymbol{\lambda}^*), \tag{6-109}$$

where

$$\nabla_{\lambda} F(\mathbf{x}^*, \boldsymbol{\lambda}^*) = (-g_1(\mathbf{x}^*), \ldots, -g_m(\mathbf{x}^*)).$$

The second Kuhn-Tucker condition (6-66) implies

$$\nabla_{\lambda} F^T(\mathbf{x}^*, \boldsymbol{\lambda}^*) \cdot \boldsymbol{\lambda}^* = 0,$$

and (6-109) therefore yields

$$F(\mathbf{x}^*, \boldsymbol{\lambda}) = F(\mathbf{x}^*, \boldsymbol{\lambda}^*) - \sum_{i=1}^{m} \lambda_i g_i(\mathbf{x}^*).$$

But each $g_i(\mathbf{x}^*)$ is nonpositive, so for any $\boldsymbol{\lambda} \geqq \mathbf{0}$

$$F(\mathbf{x}^*, \boldsymbol{\lambda}) \geqq F(\mathbf{x}^*, \boldsymbol{\lambda}^*). \tag{6-110}$$

And (6-108) and (6-110) together establish the theorem.

Theorems 6.7, 6.8, and 6.9 together imply that the two-step sequential optimization procedure described in Section 6.12 can, in theory at least, be used to obtain optimal solutions to mathematical programming problems. We shall now demonstrate this approach on the example problem of Section 6.7. Even though it is practically impossible to construct a simpler nonlinear program, we shall see that substantial amounts of logic and calculation are needed to find the global saddle point.

$$\text{Max } f(\mathbf{x}) = 3x_1 + x_2$$
$$\text{subject to } g_1(\mathbf{x}) = x_1^2 + x_2^2 - 5 \leqq 0,$$
$$\text{and } g_2(\mathbf{x}) = x_1 - x_2 - 1 \leqq 0.$$

Notice that this is a convex programming problem. Forming the Lagrangian,

$$F(\mathbf{x}, \boldsymbol{\lambda}) = 3x_1 + x_2 - \lambda_1(x_1^2 + x_2^2 - 5) - \lambda_2(x_1 - x_2 - 1),$$

we hold $\boldsymbol{\lambda}$ fixed and maximize $F(\mathbf{x}, \boldsymbol{\lambda})$ over $\mathbf{x}$:

$$\text{(6-111)} \qquad \begin{cases} \dfrac{\delta F}{\delta x_1} = 3 - 2x_1\lambda_1 - \lambda_2 = 0, \\ \dfrac{\delta F}{\delta x_2} = 1 - 2x_2\lambda_1 + \lambda_2 = 0. \end{cases}$$

We now remark that it is impossible to have a global saddle point with $\lambda_1^* = 0$: This would imply that

$$F(\mathbf{x}, \boldsymbol{\lambda}^*) = (3 - \lambda_2^*)x_1 + (1 + \lambda_2^*)x_2 + \lambda_2^*,$$

and, regardless of the value of λ_2^*, no finite $\mathbf{x}^*$ could then exist to make

$$F(\mathbf{x}, \boldsymbol{\lambda}^*) \leqq F(\mathbf{x}^*, \boldsymbol{\lambda}^*)$$

for all $\mathbf{x}$ in E^n. Hence we need be concerned only with positive λ_1. The equations (6-111) yield

$$\text{(6-112)} \qquad x_1 = \frac{3 - \lambda_2}{2\lambda_1} \quad \text{and} \quad x_2 = \frac{1 + \lambda_2}{2\lambda_1},$$

and the second derivative test confirms that these are indeed the maximizing values.

Substitution of (6-112) into the Lagrangian produces

$$\text{(6-113)} \qquad \max_{\mathbf{x} \mid \boldsymbol{\lambda}} F(\mathbf{x}, \boldsymbol{\lambda}) \equiv h(\boldsymbol{\lambda}) = \frac{10\lambda_1^2 + 2\lambda_1\lambda_2 + \lambda_2^2 - 2\lambda_2 + 5}{2\lambda_1},$$

which we now must minimize over $\lambda_1 > 0$, $\lambda_2 \geqq 0$. The partial derivatives of (6-113) are

$$\frac{\delta h}{\delta \lambda_1} = \frac{10\lambda_1^2 - \lambda_2^2 + 2\lambda_2 - 5}{2\lambda_1^2} = 0$$

and

$$\frac{\delta h}{\delta \lambda_2} = 1 + \frac{\lambda_2}{\lambda_1} - \frac{1}{\lambda_1} = 0.$$

The latter yields $\lambda_1 = 1 - \lambda_2$, which may be substituted into the former to produce

$$(3\lambda_2 - 5)(3\lambda_2 - 1) = 0.$$

If $\lambda_2 = \frac{5}{3}$, then $\lambda_1 = -\frac{2}{3}$, which is forbidden. Therefore we take

$$\lambda_2 = \tfrac{1}{3}, \quad \lambda_1 = \tfrac{2}{3}, \quad \text{and} \quad h(\boldsymbol{\lambda}) = 7; \tag{6-114}$$

for these values both second derivatives of $h(\boldsymbol{\lambda})$ are positive, as they should be at a minimum.

We must still check the boundaries: $\lambda_1 = 0$ is impossible, but what about $\lambda_2 = 0$? We find that

$$h(\lambda_1, 0) = \frac{10\lambda_1^2 + 5}{2\lambda_1},$$

and the minimum occurs at $\lambda_1 = \frac{1}{2}\sqrt{2}$; this yields

$$h\left(\frac{1}{2}\sqrt{2}, 0\right) = \frac{10}{\sqrt{2}} \cong 7.07 > 7,$$

so the minimum of $h(\boldsymbol{\lambda})$ over nonnegative $\boldsymbol{\lambda}$ is given by (6-114). Substituting into (6-112), we find that the global saddle point occurs at $x_1 = 2$, $x_2 = 1$, which is precisely the solution we obtained by applying the Kuhn-Tucker conditions in Section 6.7.

If he has worked his way through this example, the reader will not be surprised to hear that Lagrangian saddle-point theory is of rather limited practical utility in mathematical programming. As a general rule, it is much easier to work with the Kuhn-Tucker conditions, which are a relatively simple set of equations and inequalities, than with saddle-point theory, which involves optimization of the Lagrangian function. In fact, not one of the solution algorithms to be presented in the remainder of this text will be based on the material discussed in the past two sections. There are, however, a few applications of saddle-point theory in mathematical programming. The most important at present are to be found in a newly developing area known as "large-scale mathematical programming," which is concerned with solving certain types of very large problems using computational methods that exploit their special structural characteristics. One recommended reference in this area is Lasdon [29].

EXERCISES

Section 6.3

6-1. Solve the following equality-constrained mathematical programs by the method of Lagrange multipliers.

(a) Min $z = 3x_1^2 + 4x_2^2 + 5x_3^2$
subject to $x_1 + x_2 + x_3 = 10$.

(b) Max $z = x_1 - x_2 + 2x_3$
subject to $x_1^2 + x_2^2 = 4$
and $x_1 + x_2 + x_3 = 5$.

(c) Min $z = x_1 - x_2 + x_3^2$
subject to $x_1^2 + x_2^2 = 4$
and $x_1 + x_2 + x_3 = 5$.

6-2. Solve the following problem using the Lagrange multiplier technique:

$$\text{Min } z = x_1^2 + (x_2 - 1)^2$$
$$\text{subject to } -2x_1^2 + x_2 = 4.$$

Now solve by eliminating the variable x_1. Why is the wrong answer obtained?

6-3. Suppose that $g_{m-1}(\mathbf{x}) \equiv g_m(\mathbf{x})$ in the problem

$$\text{Max } f(\mathbf{x})$$
$$\text{subject to } g_i(\mathbf{x}) = 0, \qquad i = 1, \ldots, m.$$

That is, suppose that a certain constraint has inadvertently been included twice. Does Theorem 6.1 apply to this problem? If not, show where it breaks down. Would your answers be the same if $g_m(\mathbf{x}) \equiv g_1(\mathbf{x}) + g_2(\mathbf{x}) + \cdots + g_{m-1}(\mathbf{x})$?

6-4. Determine whether or not the following problem has any feasible solutions:

$$\text{Min } x_1 + 2x_2 + 3x_3 + 3x_4^2$$
$$\text{subject to } x_1 + x_2 + x_3 + x_4 = 4,$$
$$x_2 + x_3 + x_4^2 = 3,$$
$$\text{and } x_3 + 2x_4^2 = 3.$$

Now attempt to solve it via the method of Lagrange multipliers. What happens? Does Theorem 6.1 apply to this problem? Find the optimal solution.

6-5. Consider the following equality-constrained mathematical program:

$$\text{Min } x_1^2 + (x_4 - 1)^2$$
$$\text{subject to } x_1 + 2x_2 + x_4 = 3,$$
$$2x_1 - x_2 + x_3 - x_4^2 = -1,$$
$$\text{and } 5x_2 - x_3 + 2x_4^2 = 6.$$

(a) Can the constraints be solved for the other three variables in terms of x_4? In terms of x_3? Of x_2? Of x_1? Does the implicit function theorem apply to the constraints?
(b) Solve the problem using Lagrange multipliers. Does Theorem 6.1 apply? Discuss.

Section 6.5

6-6. (a) Find the dimensions of the rectangle that encloses the largest possible area within a perimeter of P feet. Show that the value of the Lagrange multiplier equals the rate of change of the maximum attainable area with respect to change in the perimeter.
(b) What are the dimensions of the (three-dimensional) box that encloses the great-

est possible volume within a total surface area of S square feet? Interpret the Lagrange multiplier.

6-7. In using Lagrange multipliers to solve the linear program

$$\text{Max } z = \mathbf{c}^T\mathbf{x}$$
$$\text{subject to } \mathbf{Ax} = \mathbf{b}, \qquad \text{with } \mathbf{x} \text{ unrestricted,}$$

how would (a) unboundedness and (b) infeasibility be discovered?

6-8. If the Lagrange multiplier associated with the kth constraint $g_k(\mathbf{x}) = b_k$ of an equality-constrained problem has a value of zero, is it necessarily true that the optimal objective value would be unaffected by small changes in the value of b_k? Suggest two reasons why a Lagrange multiplier might have a value of zero, and devise a problem in which a zero-valued Lagrange multiplier actually occurs.

6-9. Determine by inspection the optimal solution to the following problem:

$$\text{Min } x_1^2 + x_2^2$$
$$\text{subject to } x_1x_2 + x_3 = 1$$
$$\text{and } x_1 + x_2x_3 = 0.$$

Now use the necessary conditions (6-22) to compute the associated values of the Lagrange multipliers. Are they in fact equal to the rates of change of the optimal objective value with respect to changes in the right-hand-side constants? If not, why not?

Section 6.6

6-10. Solve the following problem by the method of Lagrange multipliers, using squared slack variables to transform the constraints into equalities:

$$\text{Max } 3x_1 + x_2$$
$$\text{subject to } x_1^2 + x_2^2 \leqq 5$$
$$\text{and } x_1 - x_2 \leqq 1.$$

6-11. Use the classical procedure of finding unconstrained local optima within the feasible region and on all possible boundaries to solve the following inequality-constrained problems.

(a) Min $x_1^2 + 3x_1x_2 + x_2^2 - 4x_1 - x_2$
subject to $x_1, x_2 \geqq 0$.

(b) Min $(x_1^4 - 14x_1^2 + 24x_1 + 1)(x_2 + 1)$
subject to $x_1, x_2 \geqq 0$.

(c) Min $x_1^2 + 2x_2^2 + 3x_3^2 + 4x_1x_2 - x_1x_3 - 2x_2x_3 - 5x_1 - 6x_2 - 3x_3$
subject to $x_1, x_2, x_3 \geqq 0$.

(d) Max $(x_1 - 1)^2 + (x_2 - 1)^2$
subject to $x_1 + 2x_2 \leqq 5$
and $x_1, x_2 \geqq 0$.

Section 6.7

Exercises 6-12 through 6-15: Solve each of the following mathematical programming problems by applying the Kuhn-Tucker necessary conditions (6-47) through (6-50).

6-12. Min $3x_1 + x_2$
subject to $x_1^2 + x_2^2 \leqq 5$
and $x_1 - x_2 \leqq 1.$

6-13. Max $x_1 - x_2$
subject to $x_1^2 + x_2^2 \leqq 5$
and $x_1 - x_2 \leqq 1.$

6-14. Max $x_1 + x_2$
subject to $x_1^2 + x_2^2 \leqq 5$
and $-(x_1 - 3)^2 + x_2 \leqq 3.$

6-15. Max $x_1^2 - x_2$
subject to $x_1^2 + x_2^2 + x_3^2 \leqq 9$
and $3x_1 - x_2 + x_3 \leqq 6.$

Section 6.9

Exercises 6-16 and 6-17: Solve each of the following problems, finding all points that satisfy the Kuhn-Tucker conditions.

6-16. Max $x_1^2 + 3x_2^2$
subject to $x_1^2 + x_2^2 = 5$
and $2x_1 - 2x_2 \leqq 1.$

6-17. Max $x_1^2 + 2x_2^2$
subject to $x_1^2 + x_2^2 \leqq 5$
and $2x_1 - 2x_2 = 1.$

6-18. Attempt to solve the following two problems by applying the Kuhn-Tucker conditions:

(a) Max x_1
subject to $x_1^2 - x_2 \leqq 0$
and $x_1 - x_2 \leqq -1.$

(b) Max $x_1^2 - x_2$
subject to $x_1^2 - x_2 \leqq 0$
and $x_1 - x_2 \leqq -1.$

Based on your results, discuss the relationships between the Kuhn-Tucker conditions and unboundedness.

6-19. For each of the following problems, solve graphically and then attempt to solve via the Kuhn-Tucker conditions. Show that the constraint qualification does *not* hold at the optimal solution $\mathbf{x}^*$ by finding some vector $\mathbf{h}$ which satisfies

$$\nabla \mathbf{g}_i^T(\mathbf{x}^*) \cdot \mathbf{h} \leqq 0 \qquad \text{for all } i \text{ in } I(\mathbf{x}^*)$$

but which does not point in an immediately feasible direction.

(a) Max x_1
subject to $(x_1 - 1)^5 + x_2 \leqq 0$
and $x_1, x_2 \geqq 0.$

(b) Max x_1
subject to $x_1^4 + x_1^3 - 3x_1^2 - 5x_1 - x_2 \geqq 2,$
$x_1 \leqq 1,$
and $x_2 \geqq 0.$

6-20. (a) Formally derive the Kuhn-Tucker conditions for the following general problem:

$$\begin{aligned} &\text{Max } f(\mathbf{x}), && \mathbf{x} = (x_1, \ldots, x_n) \\ &\text{subject to } g_i(\mathbf{x}) \leqq b_i && i = 1, \ldots, u, \\ &g_j(\mathbf{x}) \geqq b_j, && j = u + 1, \ldots, v, \\ \text{and } &g_k(\mathbf{x}) = b_k, && k = v + 1, \ldots, m. \end{aligned}$$

This is the problem of Theorem 6.4 *without* the nonnegativity restriction on **x**.
(b) Use the result of part (a) to establish the Kuhn-Tucker conditions for the problem

$$\begin{aligned} &\text{Max } f(\mathbf{x}), && \mathbf{x} = (x_1, \ldots, x_n), \\ &\text{subject to } g_i(\mathbf{x}) - b_i, && i = 1, \ldots, m. \end{aligned}$$

They should, of course, be the Lagrange multiplier conditions (6-22).

6-21. Use the Kuhn-Tucker conditions to solve the mathematical programming problem

$$\begin{aligned} &\text{Max } x_1 \\ &\text{subject to } x_1^2 + x_2^2 \leqq k \\ &\text{and } -x_1 + x_2 \leqq 0, \end{aligned}$$

where k is any positive constant. Now show that the optimal value of the multiplier associated with the first constraint equals the rate of change of the optimal objective value with respect to change in k. Repeat for the multiplier associated with the second constraint. A diagram of the feasible region will be helpful.

6-22. In the problem of Exercise 6-21 we saw that the optimal values of the multipliers equaled the rates at which the optimal objective value would increase if the constraints were relaxed. This same relationship has been shown to hold for dual variables in linear programming (Section 2.8) and for ordinary Lagrange multipliers (Section 6.5). Use the proof of the Kuhn-Tucker conditions to argue informally that this relationship also holds for mathematical programming problems in general.

Section 6.11

6-23. Prove that no optimal solution to the linear program

$$\begin{aligned} &\text{Max } z = \mathbf{c}^T\mathbf{x} \\ &\text{subject to } \mathbf{Ax} \leqq \mathbf{b} \text{ and } \mathbf{x} \geqq \mathbf{0}, \end{aligned}$$

where $\mathbf{c} \neq \mathbf{0}$, can occur at an interior point (as defined in Section 3.2) of the feasible region. Could a point at which exactly one constraint is binding be optimal? What would this require?

Exercises 6-24 through 6-26: Solve each of the following mathematical programs *either* by finding all feasible points that satisfy the Kuhn-Tucker conditions *or* by finding one such point and demonstrating that the problem is a convex program.

6-24. Min $x_1^2 + x_2^2 - 4x_1$
subject to $x_1 x_2 \geqq 1$
and $x_1^2 + x_2 \leqq 4.5$.

6-25. Max $x_1 - e^{-x_2}$
subject to $-\sin x_1 + x_2 \leqq 0$
and $x_1 \leqq 3$

6-26. Max $x_1 - 2x_2$
subject to $\max(0, x_1) - x_2 \leqq 0$
and $x_1^2 + x_2 \leqq 4$.

6-27. The peace-loving nation of Pacifica has three cities which must be protected by antiballistic-missile missiles; each ABM can defend *only* the city at which it has been deployed. Assuming perfect ABM performance, an enemy nation must send in $k + 1$ missiles to destroy a city protected by k ABMs; if the "value" of the city is V, then the enemy has achieved a payoff of $V/(k + 1)$ value units per attacking missile. Pacifica therefore wishes to deploy its N ABMs "evenly," in order to minimize the maximum payoff an enemy can achieve by destroying any one city. Formulate Pacifica's deployment problem as an ordinary mathematical program (that is, ignore the integrality constraints) and show that it is a *convex program.* Now suggest a common-sense, short-cut method for solving the problem when the variables are required to take on integer values.

6-28. A *quadratic program* calls for the maximization of a quadratic objective function subject to linear constraints; in general it may be represented as follows:

$$\text{Max } \mathbf{c}^T\mathbf{x} + \mathbf{x}^T\mathbf{D}\mathbf{x}$$
$$\text{subject to } \mathbf{A}\mathbf{x} = \mathbf{b}$$
$$\text{and } \mathbf{x} \geqq \mathbf{0},$$

where $\mathbf{D}$ is a symmetric n-by-n matrix and $\mathbf{A}$ is m-by-n. Under what circumstances is any local optimum also a global optimum? Write the Kuhn-Tucker conditions for this problem [recalling that if $f(\mathbf{x}) = \mathbf{x}^T\mathbf{D}\mathbf{x}$, then $\nabla \mathbf{f}(\mathbf{x}) = 2\mathbf{D}\mathbf{x}$]. Show how it would be possible to add artificial variables and solve via the simplex method, *provided* that a certain additional restriction could legitimately be placed upon the choice of the entering variable.

Section 6.13

6-29. Prove or disprove each of the following, restricting your attention throughout to linear programs of the form

$$\text{Max } z = \mathbf{c}^T\mathbf{x}$$
$$\text{subject to } \mathbf{A}\mathbf{x} \leqq \mathbf{b} \text{ and } \mathbf{x} \geqq \mathbf{0}.$$

(a) Every linear program (LP) is a convex program.
(b) Every LP with a finite optimal solution has a global saddle point.
(c) No infeasible LP can have a global saddle point.
(d) No unbounded LP can have a global saddle point.

6-30. Give an example of a mathematical programming problem with more than one global saddle point.

6-31. (a) Show that the Lagrangian function associated with the problem

$$\begin{aligned} &\text{Max } x^2 \\ &\text{subject to } x \leqq 1 \\ &\text{and } -x \leqq 0 \end{aligned}$$

has no global saddle point.
(b) Prove that the problem

$$\begin{aligned} &\text{Min } x^2 \\ &\text{subject to } x \leqq 1 \\ &\text{and } -x \leqq 0 \end{aligned}$$

must have a global saddle point $(x^*, \boldsymbol{\lambda}^*)$, and then proceed to find it.

6-32. Given a pair of primal and dual linear programs,

$$\begin{aligned} &\text{Max } z = \mathbf{c}^T\mathbf{x} \\ &\text{subject to } \mathbf{A}\mathbf{x} \leqq \mathbf{b} \\ &\text{and } \mathbf{x} \geqq \mathbf{0} \end{aligned}$$

and

$$\begin{aligned} &\text{Min } z' = \mathbf{b}^T\mathbf{u} \\ &\text{subject to } \mathbf{A}^T\mathbf{u} \geqq \mathbf{c} \\ &\text{and } \mathbf{u} \geqq \mathbf{0}, \end{aligned}$$

define $L(\mathbf{x}, \mathbf{u}) \equiv \mathbf{c}^T\mathbf{x} + \mathbf{b}^T\mathbf{u} - \mathbf{u}^T\mathbf{A}\mathbf{x}$. Let $\mathbf{x}^*$ and $\mathbf{u}^*$ be the optimal solutions to the two programs and show that for all *feasible* solutions $\mathbf{x}$ and $\mathbf{u}$

$$L(\mathbf{x}, \mathbf{u}^*) \leqq L(\mathbf{x}^*, \mathbf{u}^*) \leqq L(\mathbf{x}^*, \mathbf{u}).$$

6-33. (a) Write the Lagrangian function for the equality-constrained mathematical program

$$\begin{aligned} &\text{Max } f(\mathbf{x}), \qquad && \mathbf{x} = (x_1, \ldots, x_n), \\ &\text{subject to } g_i(\mathbf{x}) = 0, \qquad && i = 1, \ldots, m, \end{aligned}$$

and show that $(\mathbf{x}^*, \boldsymbol{\lambda}^*)$ is a global saddle point if

$$F(\mathbf{x}, \boldsymbol{\lambda}^*) \leqq F(\mathbf{x}^*, \boldsymbol{\lambda}^*) \leqq F(\mathbf{x}^*, \boldsymbol{\lambda})$$

for all $\mathbf{x}$ in E^n and all $\boldsymbol{\lambda}$ in E^m (note that the nonnegativity restriction on $\boldsymbol{\lambda}$ is absent here). Bear in mind that the definition of the Lagrangian function was stated in terms of an inequality-constrained problem.

(b) Find the optimal solution $\mathbf{x}^*$ to the problem

$$\text{Min } x_1^2 + x_2^2 + \cdots + x_n^2$$
$$\text{subject to } x_1 + x_2 + \cdots + x_n = 1$$

and show that there exists a value λ^* such that $(\mathbf{x}^*, \lambda^*)$ is a global saddle point.

7

Quadratic Programming

7.1 LINEARLY CONSTRAINED PROBLEMS AND QUADRATIC OBJECTIVES

We have spoken of mathematical programming problems as being partitioned into three broad classes: those without constraints, those having equality constraints only, and those whose constraints include at least one inequality or nonnegativity restriction. These three classes of problems are here arranged in order of increasing theoretical sophistication and computational difficulty, as reflected in the increasing complexity of the three crucial theorems that establish necessary conditions for their optimal solutions: Theorems 4.1 (vanishing of partial derivatives at an unconstrained optimum), 6.1 (existence of Lagrange multipliers for equality-constrained problems), and 6.3 (Kuhn-Tucker conditions). The rest of this textbook will be devoted to algorithms for solving problems in the third category, algorithms which will all depend in some way on the Kuhn-Tucker conditions.

Mathematical programs in the third, or inequality-constrained, class, however, should really be further divided into two subclasses, according to whether or not all their constraints are linear. Problems with nonlinear constraints are substantially more difficult to work with than those having linear constraints only, essentially because the latter can be solved with much greater efficiency by means of what are called "methods of feasible directions." These are direct climbing methods, of the type studied in Chapter 4, with the added proviso that displacement at any stage along the current direction vector must be stopped *either* when the value of the objective function stops improving *or* when a boundary of the feasible region is encountered. One major advantage in dealing with linear constraints via these methods is that it is possible to move in a fixed straight-line direction along a boundary of the feasible region, that is, along an edge or along a straight path within a hyperplane. This ability to search along the boundary or constraint surface, using the relatively simple computational methods of Chapter 4, is important because the constraints that will prove to be binding on the optimal solution (i.e., the portion of the boundary that contains the optimal point) are usually identified before the optimum itself has been found. By contrast, searching along a curved path embedded in a nonlinear constraint surface would require parameterizing the variables and would thus be much more difficult computationally. Yet the alternative—moving in simple straight-line steps through the interior of the feasible region—effectively rules out any seeking along a nonlinear boundary and thus hampers the analyst in his efforts to narrow the area of search.

Another important advantage afforded by linear constraints is that, in using a "feasible-directions" method, it is possible to determine with very simple computations whether or not the current direction vector points toward any given nonbinding constraint hyperplane, and if so, exactly how much displacement is required to reach it. Suppose the current point and direction vector are $\mathbf{x}_k$ and $\mathbf{s}_k$; then the new point $\mathbf{x}_{k+1}$ will be of the form $\mathbf{x}_k + \theta\mathbf{s}_k$ for some positive step length θ. To determine the value of θ at which the vector $\mathbf{x}_k + \theta\mathbf{s}_k$ just reaches the constraint hyperplane

$$a_{i1}x_1 + a_{i2}x_2 + \cdots + a_{in}x_n \equiv \boldsymbol{\alpha}_i \cdot \mathbf{x} = b_i$$

it is only necessary to solve the equation

$$\boldsymbol{\alpha}_i \cdot (\mathbf{x}_k + \theta\mathbf{s}_k) = b_i,$$

which yields

$$\theta = \frac{b_i - \boldsymbol{\alpha}_i \cdot \mathbf{x}_k}{\boldsymbol{\alpha}_i \cdot \mathbf{s}_k}.$$

Notice, however, how much more difficult the same computations would be if the constraints were nonlinear: To obtain the value of θ at which the direction vector $\mathbf{x}_k + \theta\mathbf{s}_k$ intersects the constraint hypersurface $g(\mathbf{x}) = 0$, it is necessary to solve the nonlinar equation

$$g(\mathbf{x}_k + \theta\mathbf{s}_k) = 0,$$

which might well be very complicated indeed.

Because of these difficulties, problems with nonlinear constraints are not usually solved by feasible-directions methods—other strategies, more complicated but somewhat more efficient, are employed instead. It is therefore natural and convenient for us to treat separately the two subclasses of inequality-constrained mathematical programs: Those with linear constraints only will be considered in this chapter and the next, while those with nonlinear constraints will be covered in Chapter 9.

Having explained our reasons for dividing inequality-constrained problems into two subclasses, we shall now attempt to justify a further partitioning of one of them. Problems having *linear constraints only* vary significantly in mathematical sophistication and computational difficulty according to whether the objective function is (1) linear, (2) quadratic, or (3) anything else. Problems of subtype (1) are, of course, linear programs and can be solved via the supremely efficient simplex methods, with which the student is already familiar. It is interesting, however, that quadratic objective functions, while not so easy to work with as linear objectives, are nevertheless significantly more tractable than those of any other nonlinear type. The basic reason for this is that the gradient of a quadratic function is entirely linear. As a result, the Kuhn-Tucker conditions for a *quadratic program*—that is, a mathematical program with linear constraints (presumed to include some inequalities) and a quadratic objective—assume a very simple form that can be exploited by relatively efficient solution algorithms. Accordingly, we shall devote the present chapter to a discussion of the quadratic program and several algorithms for solving it; solution methods applicable to all linearly constrained problems, regardless of objective function, will then be presented in Chapter 8.

7.2 THE QUADRATIC PROGRAM: EXAMPLES AND APPLICATIONS

The quadratic programming problem (QPP) can be represented in the following general form:

$$
\text{(7-1)} \qquad \begin{cases} \text{Max } z = \sum_{j=1}^{n} c_j x_j + \sum_{j=1}^{n} \sum_{k=j}^{n} c_{jk} x_j x_k \\ \text{subject to } \sum_{j=1}^{n} a_{ij} x_j = b_i, \qquad i = 1, \ldots, m, \\ \text{and} \qquad x_j \geqq 0, \qquad j = 1, \ldots, n, \end{cases}
$$

where the a_{ij}, b_i, c_j, and c_{jk} are any real numbers. Note that the objective function includes all possible quadratic terms, that is, terms in x_j^2 and $x_j x_k$ for all j and k. The formulation (7-1) is sufficiently general to accommodate all the usual variations:

1. A quadratic objective z can be minimized by maximizing its negative $-z$;
2. Inequality constraints can be converted to equations by the addition of nonnegative slack or surplus variables; and

3. A set of p unrestricted variables x'_j, $j \in J$, can be transformed into $p + 1$ nonnegative variables via the following substitutions:

$$x'_j = x_j - x_0, \qquad j \in J,$$

where $x_j \geqq 0$, $j \in J$, and $x_0 \geqq 0$.

We do not at this point rule out the possibility that some of the constraints may be redundant.

The quadratic program (7-1) can be written more compactly using matrix/vector notation:

$$\text{(7-2)} \qquad \begin{cases} \text{Max } z = \mathbf{c}^T\mathbf{x} + \mathbf{x}^T\mathbf{D}\mathbf{x} \\ \quad \text{subject to } \mathbf{A}\mathbf{x} = \mathbf{b} \\ \qquad\quad \text{and } \quad \mathbf{x} \geqq \mathbf{0}, \end{cases}$$

where $\mathbf{x}$ and $\mathbf{c}$ are n-component column vectors, $\mathbf{A}$ is m-by-n, $\mathbf{b}$ is m-by-1, and $\mathbf{D}$ is a *symmetric* n-by-n matrix whose components are defined as follows:

$$d_{jj} = c_{jj} \qquad \text{for all } j$$

$$d_{jk} = d_{kj} = \frac{c_{jk}}{2} \qquad \text{for all } j < k.$$

The scalar term $\mathbf{x}^T\mathbf{D}\mathbf{x}$ is a *quadratic form*; the reader should recall from Section 3.3 that even when a quadratic form is given initially as

$$q(\mathbf{x}) = \sum_{j=1}^{n} \sum_{k=1}^{n} c_{jk} x_j x_k$$

[which, because of the duplicated terms, is less well defined than the expression appearing in (7-1) above], the symmetric matrix $\mathbf{D}$ that satisfies $q(\mathbf{x}) \equiv \mathbf{x}^T\mathbf{D}\mathbf{x}$ is unique. The quadratic programming problem (7-2)—or some equivalent representation of it—will occupy our attention throughout this chapter.

Before beginning our presentation of solution methods for the quadratic program, however, we shall attempt in the rest of this section to indicate its importance in the practice of operations research. Although *linear* functions are the type most useful and most widely used in the modeling of mathematical optimization problems, *quadratic* functions are quite firmly established in second place. This is due both to the fact that quadratic functions and problems are the least difficult to handle computationally of all nonlinear types, and to the fact that a fair number of the functional relationships occurring in the real world either are truly quadratic or may be so approximated. The surface area, for example, of a circle, cube, or other regular figure is proportional to the square of its characteristic linear dimension. The kinetic energy carried by a rocket or atomic particle is proportional to the square of its velocity, while the potential energy of a rigid standing wall or dam is a quadratic function of its height. The sales revenue of a monopolistic firm that sells x_1 units of some product at a price of x_2 apiece is x_1x_2, which is a quadratic term in the variables x_1 and x_2. In statistics, the variance of a given sample of observations is a quadratic function of the values that constitute the sample. Many additional examples could be cited. Furthermore, countless other nonlinear relationships occurring in nature and in human affairs are capable of being adequately *approximated* by quadratic functions, at least over some ranges of the variables involved. Such approximations are welcomed eagerly by engineers and operations research analysts because of the relative computational advantages afforded by quadratic functions.

In the formulation of a quadratic programming problem, of course, only the objective function may include quadratic terms. One very common example of a QPP arises when statistical data are to be fitted to a mathematical model by the method of *least squares.* In Section 4.2 we presented an example of an unconstrained least-squares problem in which the object was to determine the values of the unknown parameters $\beta_0, \beta_1, \ldots, \beta_n$ in the model

$$y = \beta_0 + \beta_1 x_1 + \cdots + \beta_n x_n + u$$

that would best fit a given set of joint observations $[Y_i, X_{i1}, \ldots, X_{in}]$, $i = 1, \ldots, m$, where u is a random disturbance term with mean zero. The measure of the closeness of fit was taken, somewhat arbitrarily but in accordance with long-standing tradition, to be the sum of the squares of the residuals

$$e_i = Y_i - \beta_0 - \beta_1 X_{i1} - \cdots - \beta_n X_{in}, \qquad i = 1, \ldots, m,$$

where each residual e_i is the algebraic difference between the value of y actually observed and the value that would be expected if the parameter values were $\beta_0, \beta_1, \ldots, \beta_n$. The objective function then took the form

$$\text{(7-3)} \qquad \text{Min} \sum_{i=1}^{m} e_i^2 = \sum_{i=1}^{m} (Y_i - \beta_0 - \beta_1 X_{i1} - \cdots - \beta_n X_{in})^2,$$

which is in fact a quadratic function in the variables $\beta_0, \ldots, \beta_n$.

It is, however, quite common in econometrics and in other applications of statistical analysis for constraints to be placed on the unknown parameter values. Suppose an economist theorizes that the fraction β_j of the American public's total consumption expenditure that is allocated to goods and services in the jth class, $j = 1, \ldots, n$, is constant from year to year, regardless of the overall level of consumption (assume that for purposes of analysis all types of goods and services have been partitioned into n mutually exclusive classes). He has available as data a detailed breakdown of all consumption expenditures over an m-year period and wishes to use these statistics to form least-squares estimates of the unknown parameters β_j. Let C_i be the total consumption expenditure and let X_{ij} be the amount spent on goods and services in the jth class, $j = 1, \ldots, n$, during the ith year, $i = 1, \ldots, m$, where

$$C_i = \sum_{j=1}^{n} X_{ij}, \qquad i = 1, \ldots, m.$$

Then the residual or error associated with X_{ij} is

$$e_{ij} = X_{ij} - \beta_j C_i,$$

and the economist's problem can be formulated as follows:

$$\text{Min } z = \sum_{i=1}^{m} \sum_{j=1}^{n} (X_{ij} - \beta_j C_i)^2$$

$$\text{subject to } \sum_{j=1}^{n} \beta_j = 1$$

$$\text{and } 0 \leqq \beta_j \leqq 1, \qquad j = 1, \ldots, n.$$

This is evidently a quadratic programming problem; we could, if we wished, convert it to the form (7-1) by reversing the objective function and adding slack variables to the upper bound constraints $\beta_j = 1$.

Another quadratic program known as the *portfolio problem* arises (theoretically, anyway) in the planning of investments. Suppose a prosperous third baseman wishes to invest a total of \$$B$ in n different stocks and bonds. The annual return on a \$1 investment in the jth security is a random variable whose expected value and variance, based on historical data, are r_j and s_{jj}; the covariance of the ith and jth returns is $s_{ij} = s_{ji}$. If the third baseman wished only to maximize his expected overall return, regardless of risk, he would simply invest all his money in the security offering the greatest r_j. If, however, he were to some extent risk-averse, he might prefer to invest in such a way as to minimize the overall variance of his portfolio, subject to the condition that his expected total return be at least some specified amount R per year. The pursuit of the latter objective is what constitutes the "portfolio problem."

Before proceeding with the formulation, we recall from elementary probability theory that if u_1 and u_2 are random variables with expected values r_1 and r_2, variances s_{11} and s_{22}, and covariance s_{12}, then

1. The expected value and variance of the random variable $v = xu_1$, where x is a constant, are xr_1 and $x^2 s_{11}$, and

2. The expected value and variance of the random variable $w = u_1 + u_2$ are $r_1 + r_2$ and $s_{11} + s_{22} + 2s_{12}$.

By extending these results one can derive expressions for the overall mean and variance of any investment plan or portfolio. Let x_j be the dollar amount invested in the jth security, $j = 1, \ldots, n$. Then the total annual return is a random variable with expected value

$$x_1 r_1 + \cdots + x_n r_n$$

and variance

$$\sum_{j=1}^{n} x_j^2 s_{jj} + \sum_{i=1}^{n-1} \sum_{j=i+1}^{n} 2x_i x_j s_{ij} = \sum_{i=1}^{n} \sum_{j=1}^{n} x_i x_j s_{ij} = \mathbf{x}^T\mathbf{S}\mathbf{x},$$

where $\mathbf{x} = (x_1, \ldots, x_n)$ and $\mathbf{S}$ is a symmetric n-by-n matrix with elements s_{ij}. The portfolio problem is then formulated as follows: Find the values of $x_1, \ldots, x_n$ that

$$\text{Min } z = \mathbf{x}^T\mathbf{S}\mathbf{x}$$

$$\text{subject to } \sum_{j=1}^{n} r_j x_j \geqq R,$$

$$\sum_{j=1}^{n} x_j = B,$$

$$\text{and} \qquad x_j \geqq 0, \qquad j = 1, \ldots, n.$$

As expected, this is a quadratic program.

An interesting parametric programming problem arises when the risk-averse investment strategy is generalized. The notion of minimizing the overall variance of the portfolio, conditional on guaranteeing at least a minimum expected return R, appears rather arbitrary and even to some extent irrational. Would an investor really prefer the "minimum-variance-given-R" portfolio over one which offered a substantially higher expected return with only a slightly greater variance? Perhaps he would, but in any case his answer to this question would surely depend on the amounts by which the expected returns and variances differed and on his personality and financial circumstances. It can, however, be stated that, given two portfolios with exactly the same overall variance, any rational investor would choose the one offering the higher expected return; similarly, given two portfolios with the same average return, an investor would normally prefer the one having the lesser variance. This suggests defining a class of *efficient* portfolios: A portfolio or investment plan will be said to be efficient if no other plan exists that has *either* as high a return and a lesser variance *or* a higher return and no greater variance. Thus the class of efficient portfolios must include every plan that could be considered optimal by any rational (according to the criteria above) investor.

It might therefore be desirable to have a computational procedure for generating efficient portfolios. Such a procedure is provided by the following parametric quadratic programming problem:

$$\text{(7-4)} \qquad \begin{cases} \text{Min } z = \mathbf{x}^T\mathbf{S}\mathbf{x} - \theta\mathbf{r}^T\mathbf{x} \\ \text{subject to } \sum_{j=1}^{n} x_j = B \\ \text{and} \quad x_j \geqq 0, \quad j = 1, \ldots, n, \end{cases}$$

where $\mathbf{r} = (r_1, \ldots, r_n)$ and θ is a nonnegative scalar parameter. We have seen that the overall return for any portfolio $\mathbf{x}$ has expected value $\mathbf{r}^T\mathbf{x}$ and variance $\mathbf{x}^T\mathbf{S}\mathbf{x}$. It follows that for any given nonnegative value of θ the optimal solution $\hat{\mathbf{x}}$ to (7-4) must be an efficient portfolio: If, for example, there existed some feasible $\mathbf{x}$ satisfying

$$\mathbf{r}^T\mathbf{x} > \mathbf{r}^T\hat{\mathbf{x}} \quad \text{and} \quad \mathbf{x}^T\mathbf{S}\mathbf{x} \leqq \hat{\mathbf{x}}^T\mathbf{S}\hat{\mathbf{x}},$$

then $\hat{\mathbf{x}}$ would have a greater objective value than $\mathbf{x}$ and could not be optimal. We leave as an exercise the question of whether or not *all* efficient portfolios are generated by (7-4) as θ is varied continuously from 0 to $+\infty$.

7.3 WOLFE'S ALGORITHM

One of the earliest and simplest methods for solving quadratic programs was that proposed by Philip Wolfe [52] in 1959; despite its rather venerable history, it is still quite commonly used today. In theory, Wolfe's algorithm can be applied directly to any quadratic programming problem of the form

$$\text{(7-2)} \qquad \begin{cases} \text{Max } z = \mathbf{c}^T\mathbf{x} + \mathbf{x}^T\mathbf{D}\mathbf{x} \\ \text{subject to } \mathbf{A}\mathbf{x} = \mathbf{b} \\ \text{and} \quad \mathbf{x} \geqq \mathbf{0}, \end{cases}$$

where $\mathbf{x}$ and $\mathbf{c}$ are n-component column vectors, $\mathbf{A}$ is an m-by-n matrix, $\mathbf{b}$ is m-by-1, and $\mathbf{D}$ is n-by-n and symmetric. The basic approach of the algorithm is to generate, by means of a modified simplex pivoting procedure, a sequence of feasible points that terminates at a solution point $\mathbf{x}^*$ where the Kuhn-Tucker conditions are satisfied. Note that the feasible region of (7-2), which is bounded entirely by hyperplanes, is a convex set. Therefore, provided that the objective function is concave—that is, provided that the matrix $\mathbf{D}$ is negative definite or semidefinite—the point $\mathbf{x}^*$ will of necessity be an optimal solution. This follows from the sufficiency of the Kuhn-Tucker conditions for convex programming problems (Theorem 6.6).

As we shall see in the next section, when $\mathbf{D}$ is negative definite Wolfe's algorithm can be shown to converge to the optimal solution, or to demonstrate infeasibility, within a finite number of iterations, provided that the possibility of infinite cycling due to degeneracy is excluded. When **D** is negative *semi*definite, finite convergence to the optimum can be guaranteed only by perturbing **D** very slightly so as to make it negative definite; note, however, that if the algorithm is applied without perturbing **D**, the optimal solution will usually be obtained anyway. Finally, when **D** is indefinite or positive (semi)definite, convergence may fail entirely, or if it occurs, the solution obtained may not even be a local optimum. In view of these pathologies, Wolfe's algorithm tends to be employed in practice only when the matrix **D** is known or can be assumed to be at least negative semidefinite; other methods are usually preferred when **D** is indefinite.

A remark or two would be in order at this point on the subject of how it is determined whether or not the n-by-n matrix **D** is negative definite or semidefinite. When n is small, or when all or most of the off-diagonal elements d_{ij} are zero, **D** can be characterized (as negative definite or whatever) simply by inspection or by hand calculation of a few determinants, according to the rule given at the end of Section 3.3. When n is fairly large, however, the characterization of **D** is an extremely tedious computational problem that can require substantial amounts of running time on a digital computer. Fortunately, in many quadratic programs arising in operations research it can be deduced from physical, economic, or other considerations that **D** must be negative definite or, in the case of a minimization problem, positive definite. For example, we saw in the previous section that the overall variance of an investment portfolio **x** can be represented by $\mathbf{x}^T\mathbf{S}\mathbf{x}$, where the (i, j)th element of **S** gives the covariance of the annual returns of the ith and jth securities. But because the overall variance must be positive for all $\mathbf{x} \neq \mathbf{0}$, the covariance matrix **S** must be positive definite. To offer another example, in any least-squares linear regression problem the total sum of the squared residuals, which is to be minimized, must be positive or zero for any set of parameter values, and it can easily be shown that the quadratic-form component of such an objective function must be positive semidefinite.

We begin our development of Wolfe's algorithm by writing the Kuhn-Tucker conditions for the quadratic program (7-2). Observe first that because all the constraints are linear, the constraint qualification is satisfied. Therefore, from Theorem 6.4 (the Kuhn-Tucker corollary), if $\mathbf{x} = (x_1, \ldots, x_n)$ is the optimal solution to (7-2), then there must exist generalized Lagrange multiplier values $\boldsymbol{\lambda} = (\lambda_1, \ldots, \lambda_m)$ such that

$$\text{(7-5)} \qquad \mathbf{c} + 2\mathbf{D}\mathbf{x} \leqq \sum_{i=1}^{m} \lambda_i \boldsymbol{\alpha}_i = \mathbf{A}^T\boldsymbol{\lambda}$$

and

$$\text{(7-6)} \qquad x_j\left\{c_j + 2\boldsymbol{\delta}_j \cdot \mathbf{x} - \sum_{i=1}^{m} \lambda_i a_{ij}\right\} = 0, \qquad j = 1, \ldots, n,$$

$$\text{with } \lambda_i \text{ unrestricted,} \qquad i = 1, \ldots, m,$$

where $\boldsymbol{\alpha}_i$ is the ith row of $\mathbf{A}$ and $\boldsymbol{\delta}_j$ is the jth row of $\mathbf{D}$ and where we have used the fact that the gradient of $\mathbf{x}^T\mathbf{D}\mathbf{x}$ is $2\mathbf{D}\mathbf{x}$ when $\mathbf{D}$ is symmetric. Note that we have abandoned the asterisk or starred notation used in Theorem 6.4; we shall be treating $\mathbf{x}$ and $\boldsymbol{\lambda}$ as ordinary variables and seeking values of them that satisfy both the Kuhn-Tucker conditions and the original constraints of (7-2). Continuing the development, we define a set of slack variables for the inequalities (7-5):

$$\text{(7-7)} \qquad \mathbf{v} = \mathbf{A}^T\boldsymbol{\lambda} - \mathbf{c} - 2\mathbf{D}\mathbf{x} \geqq \mathbf{0}.$$

The jth component of (7-7) is

$$v_j = \sum_{i=1}^{m} a_{ij}\lambda_i - c_j - 2\boldsymbol{\delta}_j \cdot \mathbf{x}, \qquad j = 1, \ldots, n,$$

and the Kuhn-Tucker conditions (7-5) and (7-6) become

$$\text{(7-8)} \qquad 2\mathbf{D}\mathbf{x} - \mathbf{A}^T\boldsymbol{\lambda} + \mathbf{v} = -\mathbf{c},$$

$$\text{(7-9)} \qquad \mathbf{v} \geqq \mathbf{0},$$

and

$$\text{(7-10)} \qquad x_j v_j = 0, \qquad j = 1, \ldots, n,$$

dropping a minus sign from the left-hand side of (7-10).

As stated earlier, Wolfe's algorithm aims at finding values of $\mathbf{x}$, $\mathbf{v}$, and $\boldsymbol{\lambda}$ that satisfy the Kuhn-Tucker conditions and the original constraints

$$\text{(7-11)} \qquad \mathbf{A}\mathbf{x} = \mathbf{b}, \qquad \mathbf{x} \geqq \mathbf{0}.$$

If it were not for the complementary slackness (primal variables, dual slacks) relations (7-10), this would be a very simple matter: The constraints (7-8), (7-9), and (7-11) are linear, and a feasible solution to them could be found via a phase I simplex approach. Fortunately, the nonlinear constraints (7-10) are of a very special type, and Wolfe was able to modify the simplex method in a way that enables it to solve quadratic programs by finding feasible points that satisfy the Kuhn-Tucker conditions. We remark again that any such point must be optimal if the matrix $\mathbf{D}$ is negative definite or semidefinite.

Before the modified simplex method can be applied, the constraints (7-8), (7-9), and (7-11) must be converted to standard form, as defined in Section 2.4. This requires only that the m unrestricted variables λ_i be replaced by $m + 1$ nonnegative variables, as follows:

$$\text{(7-12)} \qquad \lambda_i = \lambda_i' - \lambda_0', \qquad i = 1, \ldots, m, \qquad \text{with } \lambda_i', \lambda_0' \geqq 0.$$

In vector form (7-12) becomes

$$\boldsymbol{\lambda} = \boldsymbol{\lambda}' - \lambda_0' \mathbf{1}, \tag{7-13}$$

where **1** is an m-component column vector consisting entirely of 1s. After making the substitution (7-13) and dropping the primes, we arrive at the following system of equations:

$$\begin{cases} \mathbf{Ax} = \mathbf{b}, \\ 2\mathbf{Dx} - \mathbf{A}^T\boldsymbol{\lambda} + \mathbf{A}^T\mathbf{1}\lambda_0 + \mathbf{v} = -\mathbf{c} \\ \mathbf{x}, \boldsymbol{\lambda}, \lambda_0, \mathbf{v} \geqq \mathbf{0}, \\ x_j v_j = 0, \qquad j = 1, \ldots, n, \end{cases} \tag{7-14}$$

Our goal, then, is to obtain any set of values of $\mathbf{x}$, $\boldsymbol{\lambda}$, λ_0, and $\mathbf{v}$ that satisfy (7-14)—that is, any feasible solution to (7-14)—or to determine that none exists. At this point we make the crucial observation that if a feasible solution to (7-14) *does* exist, then so does a *basic* feasible solution to

$$\begin{cases} \mathbf{Ax} = \mathbf{b}, \\ 2\mathbf{Dx} - \mathbf{A}^T\boldsymbol{\lambda} + \mathbf{A}^T\mathbf{1}\lambda_0 + \mathbf{v} = -\mathbf{c}, \\ \mathbf{x}, \boldsymbol{\lambda}, \lambda_0, \mathbf{v} \geqq \mathbf{0}. \end{cases} \tag{7-15}$$

For suppose that $[\hat{\mathbf{x}}, \hat{\boldsymbol{\lambda}}, \hat{\lambda}_0, \hat{\mathbf{v}}]$ satisfies (7-14); then let $\lambda^* = \min\{\hat{\lambda}_0, \hat{\lambda}_1, \ldots, \hat{\lambda}_m\}$ and consider the alternative solution

$$\begin{bmatrix} \mathbf{x} \\ \boldsymbol{\lambda} \\ \lambda_0 \\ \mathbf{v} \end{bmatrix} = \begin{bmatrix} \hat{\mathbf{x}} \\ \hat{\boldsymbol{\lambda}} - \lambda^*\mathbf{1} \\ \hat{\lambda}_0 - \lambda^* \\ \hat{\mathbf{v}} \end{bmatrix}. \tag{7-16}$$

This solution must also satisfy (7-14) because the changes in the values of $\boldsymbol{\lambda}$ and λ_0 (which remain nonnegative) cancel each other out:

$$\begin{aligned} -\mathbf{A}^T\hat{\boldsymbol{\lambda}} + \mathbf{A}^T\mathbf{1}\hat{\lambda}_0 &= -\mathbf{A}^T(\hat{\boldsymbol{\lambda}} - \hat{\lambda}_0\mathbf{1}) \\ &= -\mathbf{A}^T(\hat{\boldsymbol{\lambda}} - \lambda^*\mathbf{1} + \lambda^*\mathbf{1} - \hat{\lambda}_0\mathbf{1}) \\ &= -\mathbf{A}^T(\hat{\boldsymbol{\lambda}} - \lambda^*\mathbf{1}) + \mathbf{A}^T\mathbf{1}(\hat{\lambda}_0 - \lambda^*). \end{aligned}$$

Now, by the definition of λ^*, at least one of the variables $\lambda_0, \lambda_1, \ldots, \lambda_m$ must have a zero value in (7-16), so no more than m of those values can be different from zero. But because $\hat{x}_j\hat{v}_j = 0$, $j = 1, \ldots, n$, at most n of the components of $\hat{\mathbf{x}}$ and $\hat{\mathbf{v}}$ combined can differ from zero. It follows that the feasible solution (7-16) has no more than $n + m$ nonzero components; it must therefore be a basic feasible solution to the system (7-15), which has $m + n$ constraints.

We conclude from the foregoing argument that we should be able to obtain a feasible solution to (7-14), if one exists, by searching over the basic feasible solutions to (7-15). This is precisely the strategy of *Wolfe's algorithm*, which proceeds according to the following steps:

1. Use an ordinary phase I simplex approach to determine a basic feasible solution $\mathbf{x_B} \geqq \mathbf{0}$ to the original constraints $\mathbf{Ax} = \mathbf{b}$, where $\mathbf{x_B}$ contains the values of the m basic variables. Note that if no such $\mathbf{x_B}$ exists, the quadratic program (7-2) has no feasible solution. Let $\mathbf{B}$ denote the m-by-m basis matrix associated with $\mathbf{x_B}$, so that $\mathbf{Bx_B} = \mathbf{b}$, and construct the n-by-m matrix $\mathbf{D_B}$ by collecting those columns of $\mathbf{D}$ that correspond, in order, to the columns of $\mathbf{A}$ in $\mathbf{B}$.

2. Compute the n-by-1 vector $\hat{\mathbf{u}} = -\mathbf{c} - 2\mathbf{D_B x_B}$ and add the artificial variable u_j, with scalar coefficient Δ_j, to the $(m+j)$th constraint equation in (7-15), $j = 1, \ldots, n$, where

$$\Delta_j = \begin{cases} +1 & \text{if } \hat{u}_j \geqq 0, \\ -1 & \text{if } \hat{u}_j < 0. \end{cases} \tag{7-17}$$

If we define $\mathbf{u} = (u_1, \ldots, u_n)$ and let $\boldsymbol{\Delta}$ denote an n-by-n diagonal matrix whose jth diagonal element is Δ_j, the constraints (7-15) then take the form

$$\begin{cases} \mathbf{Ax} = \mathbf{b}, \\ 2\mathbf{Dx} - \mathbf{A}^T\boldsymbol{\lambda} + \mathbf{A}^T\mathbf{1}\lambda_0 + \mathbf{v} + \boldsymbol{\Delta}\mathbf{u} = -\mathbf{c}, \\ \mathbf{x}, \boldsymbol{\lambda}, \lambda_0, \mathbf{v}, \mathbf{u} \geqq \mathbf{0}. \end{cases} \tag{7-18}$$

3. Letting the variables $\mathbf{x}$ retain the values computed for them in step 1, assign to the remaining variables the following initial values:

$$\boldsymbol{\lambda} = \mathbf{0}, \quad \lambda_0 = 0, \quad \mathbf{v} = \mathbf{0}, \quad \text{and} \quad u_j = \Delta_j \hat{u}_j \geqq 0, \qquad j = 1, \ldots, n.$$

It is easily verified that these values, of which at most $m + n$ can be positive (namely, $\mathbf{x_B}$ and $\mathbf{u}$), constitute a feasible solution to the $m + n$ constraint equations (7-18). Moreover, $(\mathbf{x_B}, \mathbf{u})$ is a *basic* feasible solution because its associated basis matrix

$$\begin{bmatrix} \mathbf{B} & \mathbf{0} \\ 2\mathbf{D_B} & \boldsymbol{\Delta} \end{bmatrix}$$

has an inverse

$$\begin{bmatrix} \mathbf{B}^{-1} & \mathbf{0} \\ -2\boldsymbol{\Delta}\mathbf{D_B}\mathbf{B}^{-1} & \boldsymbol{\Delta} \end{bmatrix}.$$

In checking that these matrices are indeed the inverses of each other, note that $\mathbf{\Delta\Delta} = \mathbf{I}_n$, the identity matrix of order n.

4. Starting with the initial basic feasible solution constructed in step 3, use the "Wolfe-simplex" method to minimize the sum of the artificial variables (that is, to maximize $z' = -u_1 - \cdots - u_n$) subject to the constraints (7-18). The Wolfe-simplex method differs from the ordinary simplex method of linear programming in one respect only: The variable v_j cannot be chosen to enter the basis if x_j is already a basic variable (unless x_j will be held at zero or driven to zero by the ensuing pivot), and vice versa. Thus the complementary slackness conditions $x_j v_j = 0, j = 1, \ldots, n$, are enforced at every pivot; note that they are also satisfied by the initial basic solution, in which $\mathbf{v}$ was set equal to $\mathbf{0}$. This special pivoting rule, whereby certain otherwise eligible variables are prohibited from being brought into the basis, is an example of what is known as *restricted basis entry*; other types of basis entry restrictions are used in other simplex-related algorithms, most notably in separable programming (see Chapter 8). The Wolfe-simplex method terminates when no variable having a negative reduced cost—that is, no variable eligible for basis entry according to the usual simplex criterion—can be pivoted into the basis without violating the conditions $x_j v_j = 0, j = 1, \ldots, n$. At that point the $\mathbf{x}$-component of the current solution is taken as the solution to the original quadratic program; we shall prove in the next section that when $\mathbf{D}$ is negative definite the solution obtained by Wolfe's algorithm must in fact be optimal.

7.4 CONVERGENCE OF WOLFE'S ALGORITHM IN THE NEGATIVE DEFINITE CASE

Having dealt with the case of infeasibility in step 1 above, we shall restrict the present discussion to quadratic programming problems having feasible solutions. Before attempting to prove that whenever $\mathbf{D}$ is negative definite Wolfe's algorithm will eventually find a feasible point satisfying the Kuhn-Tucker conditions, we must first demonstrate that such a point actually exists. To obtain this result we need only show that when $\mathbf{D}$ is negative definite the problem (7-2) cannot have an unbounded maximum. It will then follow that some finite feasible point must be a global maximum and therefore a constrained local maximum, at which the Kuhn-Tucker conditions necessarily hold.

Accordingly, let us consider what happens to the value of the objective function $z = \mathbf{c}^T\mathbf{x} + \mathbf{x}^T\mathbf{D}\mathbf{x}$, where $\mathbf{D}$ is negative definite, as $\mathbf{x}$ is displaced from some fixed point $\mathbf{x}_0$ in any specified direction $\mathbf{y}_0 \neq \mathbf{0}$. The points thus generated are of the form

$$\mathbf{x} = \mathbf{x}_0 + \theta \mathbf{y}_0,$$

where θ is a nonnegative scalar parameter; the direction $\mathbf{y}_0$ might, for example, represent an extreme ray of the feasible region emanating from the extreme point $\mathbf{x}_0$. The objective value at any point $\mathbf{x}$, expressed as a function of θ, is then

$$z(\theta) = \mathbf{c}^T\mathbf{x}_0 + \theta\mathbf{c}^T\mathbf{y}_0 + \mathbf{x}_0^T\mathbf{D}\mathbf{x}_0 + 2\theta\mathbf{x}_0^T\mathbf{D}\mathbf{y}_0 + \theta^2\mathbf{y}_0^T\mathbf{D}\mathbf{y}_0.$$

But because $\mathbf{y}_0 \neq \mathbf{0}$, the quadratic form $\mathbf{y}_0^T\mathbf{D}\mathbf{y}_0$ has a negative value; therefore, as θ increases, the term $\theta^2\mathbf{y}_0^T\mathbf{D}\mathbf{y}_0 < 0$ eventually becomes large enough in magnitude to dominate the sum and cause $z(\theta)$ to begin decreasing, regardless of the values of $\mathbf{c}$, $\mathbf{D}$, $\mathbf{x}_0$, and $\mathbf{y}_0$. We conclude that z cannot become infinitely large; in fact, as θ increases without limit z must approach $-\infty$.

Notice that our argument does not hold when $\mathbf{D}$ is negative *semi*definite. A simple illustration of unboundedness in this case is provided by the following quadratic program:

$$\begin{aligned} \text{Max } z = x_1 + x_2 - (2x_1 - x_2)^2 \\ \text{subject to } x_1 - x_2 \leqq 1 \\ \text{and} \quad x_1, x_2 \geqq 0, \end{aligned}$$

where the quadratic form is negative semidefinite. Any solution of the form

$$x_1 = \theta, \qquad x_2 = 2\theta, \qquad z = 3\theta,$$

where θ is a nonnegative parameter, is feasible, and as θ is increased without limit z becomes infinitely large. As an exercise, the student might wish to try solving this example via Wolfe's algorithm, using the origin as the initial solution; it is, of course, necessary to begin by adding a slack variable to the constraint. We shall return to the negative semidefinite case at the end of this section.

It has thus far been established that when $\mathbf{D}$ is negative definite there exists at least one point in the feasible region of the quadratic program (7-2) that satisfies the Kuhn-Tucker conditions—which is to say, there exists at least one feasible solution to (7-14). Wolfe's algorithm seeks that feasible solution by starting with an initial set of values $\mathbf{x}_\mathbf{B}$ that satisfies some of the constraints of (7-14), adding nonnegative artificial variables u_j to satisfy the others, and then using a modified simplex pivoting procedure to maximize $z' = -\sum_j u_j$ subject to the augmented constraints (7-18), with $x_j v_j = 0$ for all j being held in force throughout. The variable chosen to enter the basis at each pivot will have a negative reduced cost, precisely as the standard simplex method requires (for a maximization problem); the objective value will therefore increase or remain the same at each iteration. Assuming that an infinitely repeating sequence of degenerate bases is not encountered,[1] the procedure must eventually terminate at a solution having the property that no nonbasic variable with a negative reduced cost can be pivoted into the basis without violating the condition $x_j v_j = 0$ for some j. We wish to show that when this point is reached the value of every artificial variable will be zero; it will then follow that the values of $\mathbf{x}$, $\boldsymbol{\lambda}$, λ_0, and $\mathbf{v}$ satisfy (7-14), as desired.

Accordingly, suppose Wolfe's algorithm has terminated at a solution having the property just described. Let W denote this solution and let the vector of variables $\mathbf{x}$ be partitioned into three subvectors, as follows:

$$x_j \text{ is included in } \begin{cases} \mathbf{x}_1 & \text{if } x_j > 0 \text{ and } v_j = 0 \text{ in } W, \\ \mathbf{x}_2 & \text{if } x_j = 0 \text{ and } v_j > 0 \text{ in } W, \\ \mathbf{x}_3 & \text{if } x_j = 0 \text{ and } v_j = 0 \text{ in } W. \end{cases} \tag{7-19}$$

Note that every component of $\mathbf{x}$ will be assigned by the partition (7-19) to one and only one of the subvectors: the fourth combination, $x_j > 0$ and $v_j > 0$, is prohibited because the condition $x_j v_j = 0$ is known to obtain. Similarly, we partition $\mathbf{v}$ in precisely the same way—that is, v_j is in $\mathbf{v}_1$ (or $\mathbf{v}_2$ or $\mathbf{v}_3$) if and only if x_j is in $\mathbf{x}_1$ (or $\mathbf{x}_2$ or $\mathbf{x}_3$). Observe that $\mathbf{x}_2$, $\mathbf{x}_3$, $\mathbf{v}_1$, and $\mathbf{v}_3$ all have zero values in the solution W.

Consider now the linear program

$$\begin{cases} \text{Max } z' = -u_1 - u_2 - \cdots - u_n \\ \text{subject to} \qquad \mathbf{A}\mathbf{x} = \mathbf{b}, \\ \qquad 2\mathbf{D}\mathbf{x} - \mathbf{A}^T\boldsymbol{\lambda} + \mathbf{A}^T\mathbf{1}\lambda_0 + \mathbf{v} + \boldsymbol{\Delta}\mathbf{u} = -\mathbf{c}, \\ \qquad \mathbf{x}_2 = \mathbf{0}, \\ \qquad \mathbf{v}_1 = \mathbf{0}, \\ \text{and} \qquad \mathbf{x}, \boldsymbol{\lambda}, \lambda_0, \mathbf{v}, \mathbf{u} \geqq \mathbf{0}, \end{cases} \tag{7-20}$$

which differs from the problem originally constructed in having the constraints $\mathbf{x}_2 = \mathbf{0}$ and $\mathbf{v}_1 = \mathbf{0}$ in place of the conditions $x_j v_j = 0, j = 1, \ldots, n$.

[1]This assumption is justified by a quarter-century of practical experience with the simplex and simplex-related methods, during which time *cycling*, or the infinite repetition of a sequence of basic solutions, has never (to the author's knowledge) been encountered in an actual problem. To be sure, whenever the simplex method is applied to a newly formulated linear program, there is always the theoretical possibility that cycling might occur. If it were judged desirable, this remote possibility could in fact be eliminated entirely by using a somewhat more complicated basis entry criterion or by performing certain auxiliary calculations (see Chapter 6 of [21] for details], either of which modifications would permit the finiteness of the simplex method to be proved rigorously. But these anti-cycling precautions require extra computer time and storage, and because they do not appear to be necessary in practice, they are simply not used.

It is not difficult to see that the optimal solution to (7-20) must be precisely the solution W obtained by Wolfe's algorithm for the original problem [i.e., the solution obtained when $z' = -\sum_j u_j$ is maximized subject to (7-18)]. To verify this assertion, imagine deleting the variables $\mathbf{x}_2$ and $\mathbf{v}_1$ and the constraints $\mathbf{x}_2 = \mathbf{0}$ and $\mathbf{v}_1 = \mathbf{0}$ from the problem (7-20) and solving what is left of it via the standard simplex method, using as the initial basic solution the basic portion of W. The initial simplex tableau must then be the final tableau generated by Wolfe's algorithm.[2] But note that none of the nonbasic variables —that is, none of the components of $\mathbf{x}_3$ or $\mathbf{v}_3$ that are nonbasic—can possibly have a negative reduced cost: If this were not true, Wolfe's algorithm would not have terminated at W. We conclude, therefore, that W is an optimal solution to (7-20).

The next step in the proof is to write the dual of the linear program (7-20). Before doing so, however, we shall make several changes in the formulation. We begin by reversing the transformation

$$\text{(7-13)} \qquad \boldsymbol{\lambda} = \boldsymbol{\lambda}' - \lambda_0'\mathbf{1},$$

a maneuver that can be accomplished simply by dropping λ_0 from the problem and treating $\boldsymbol{\lambda}$ as a vector of unrestricted rather than nonnegative variables. We next reorder the components of $\mathbf{x}$ and $\mathbf{v}$ so that $\mathbf{x} \equiv (\mathbf{x}_1, \mathbf{x}_2, \mathbf{x}_3)$ and $\mathbf{v} \equiv (\mathbf{v}_1, \mathbf{v}_2, \mathbf{v}_3)$, and reorder and partition the columns of $\mathbf{A}$ and $\mathbf{D}$ in exactly the same way:

$$\mathbf{A} \equiv [\mathbf{A}_1 \,\vdots\, \mathbf{A}_2 \,\vdots\, \mathbf{A}_3] \quad \text{and} \quad \mathbf{D} \equiv [\mathbf{D}_1 \,\vdots\, \mathbf{D}_2 \,\vdots\, \mathbf{D}_3].$$

At this point the constraints of (7-20) can be represented as

$$\begin{aligned} \mathbf{A}_1\mathbf{x}_1 + \mathbf{A}_2\mathbf{x}_2 + \mathbf{A}_3\mathbf{x}_3 \qquad\qquad\qquad\quad &= \mathbf{b}, \\ 2\mathbf{D}_1\mathbf{x}_1 + 2\mathbf{D}_2\mathbf{x}_2 + 2\mathbf{D}_3\mathbf{x}_3 - \mathbf{A}^T\boldsymbol{\lambda} + \mathbf{v} + \boldsymbol{\Delta}\mathbf{u} &= -\mathbf{c}, \\ \mathbf{x}_2 &= \mathbf{0}, \\ \mathbf{v}_1 &= \mathbf{0}, \end{aligned}$$

and

$$\mathbf{x}_1, \mathbf{x}_2, \mathbf{x}_3, \mathbf{v}, \mathbf{u} \geqq \mathbf{0}.$$

We then delete $\mathbf{x}_2$ and $\mathbf{v}_1$ from the formulation and partition the second set of constraints into three groups that conform with the partitions of $\mathbf{x}$ and $\mathbf{v}$. The

[2] Strictly speaking, our argument is valid only if no component of $\mathbf{x}_2$ or $\mathbf{v}_1$ appears in the basic portion of W. This condition, however, can always be attained: If any of the components of $\mathbf{x}_2$ or $\mathbf{v}_1$ are included in the final (degenerate) basis obtained by Wolfe's algorithm, they can subsequently be removed and replaced by components of $\mathbf{x}_3$ or $\mathbf{v}_3$ in "zero-for-zero" pivots. Such pivots do not, of course, alter the values of any of the variables.

problem (7-20) has now been converted into the following equivalent form:

$$
(7\text{-}21)\quad
\begin{cases}
\text{Max } z' = -u_1 - u_2 - \cdots - u_n \\
\text{subject to } \mathbf{A}_1\mathbf{x}_1 + \mathbf{A}_3\mathbf{x}_3 = \mathbf{b}, \\
2\mathbf{D}_{11}\mathbf{x}_1 + 2\mathbf{D}_{13}\mathbf{x}_3 - \mathbf{A}_1^T\boldsymbol{\lambda} + \boldsymbol{\Delta}_1\mathbf{u} = -\mathbf{c}_1, \\
2\mathbf{D}_{21}\mathbf{x}_1 + 2\mathbf{D}_{23}\mathbf{x}_3 - \mathbf{A}_2^T\boldsymbol{\lambda} + \mathbf{v}_2 + \boldsymbol{\Delta}_2\mathbf{u} = -\mathbf{c}_2, \\
2\mathbf{D}_{31}\mathbf{x}_1 + 2\mathbf{D}_{33}\mathbf{x}_3 - \mathbf{A}_3^T\boldsymbol{\lambda} + \mathbf{v}_3 + \boldsymbol{\Delta}_3\mathbf{u} = -\mathbf{c}_3, \\
\text{and } \quad \mathbf{x}_1, \mathbf{x}_3, \mathbf{v}_2, \mathbf{v}_3, \mathbf{u} \geqq \mathbf{0},
\end{cases}
$$

where the rows of $\mathbf{D}_1$, $\mathbf{D}_3$, and $\boldsymbol{\Delta}$ and the components of $\mathbf{c}$ have been partitioned as required—that is,

$$
\mathbf{D}_1 \equiv \begin{bmatrix} \mathbf{D}_{11} \\ \mathbf{D}_{21} \\ \mathbf{D}_{31} \end{bmatrix}
$$

and so on. Finally, we write the dual of the linear program (7-21), letting $\mathbf{p}$, $\mathbf{q}_1$, $\mathbf{q}_2$, and $\mathbf{q}_3$ be the four sets of dual variables associated with the four groups of primal constraints:

$$
(7\text{-}22)\quad
\begin{cases}
\text{Min } z'' = \mathbf{b}^T\mathbf{p} - \mathbf{c}_1^T\mathbf{q}_1 - \mathbf{c}_2^T\mathbf{q}_2 - \mathbf{c}_3^T\mathbf{q}_3 \\
\text{subject to } \mathbf{A}_1^T\mathbf{p} + 2\mathbf{D}_{11}^T\mathbf{q}_1 + 2\mathbf{D}_{21}^T\mathbf{q}_2 + 2\mathbf{D}_{31}^T\mathbf{q}_3 \geqq \mathbf{0}, \\
\mathbf{A}_3^T\mathbf{p} + 2\mathbf{D}_{13}^T\mathbf{q}_1 + 2\mathbf{D}_{23}^T\mathbf{q}_2 + 2\mathbf{D}_{33}^T\mathbf{q}_3 \geqq \mathbf{0}, \\
- \mathbf{A}_1\mathbf{q}_1 - \mathbf{A}_2\mathbf{q}_2 - \mathbf{A}_3\mathbf{q}_3 = \mathbf{0}, \\
\mathbf{q}_2 \geqq \mathbf{0}, \\
\mathbf{q}_3 \geqq \mathbf{0}, \\
\text{and } \quad \boldsymbol{\Delta}_1^T\mathbf{q}_1 + \boldsymbol{\Delta}_2^T\mathbf{q}_2 + \boldsymbol{\Delta}_3^T\mathbf{q}_3 \geqq -\mathbf{1},
\end{cases}
$$

with all dual variables unrestricted (except as provided by the dual constraints in the case of $\mathbf{q}_2$ and $\mathbf{q}_3$). The student can verify by reviewing Section 2.5 that (7-22) is indeed the dual of (7-21) and therefore of (7-20). Note that $\mathbf{1}$ is an n-component column vector of 1s.

We are now in position to complete our proof, that every artificial variable u_j has a zero value in the optimal primal solution W, by exploiting some of the joint properties of optimal solutions to primal and dual linear programs. To avoid any further proliferation of notation, let $(\mathbf{p}, \mathbf{q}_1, \mathbf{q}_2, \mathbf{q}_3)$ henceforth represent an optimal solution to (7-22)—that is, a set of optimal values of the dual variables—rather than the variables themselves. Then, inasmuch as all components of $\mathbf{x}_1$ and $\mathbf{v}_2$ are known to be positive in the optimal primal solution W, the principle of complementary slackness (recall Theorem 2.6) dictates that the first and fourth groups of constraints in (7-22) must be satisfied as equalities by the optimal dual solution. In particular, this implies $\mathbf{q}_2 = \mathbf{0}$; thus the optimal value of the dual objective is

$$z''_{\text{opt}} = \mathbf{b}^T\mathbf{p} - \mathbf{c}_1^T\mathbf{q}_1 - \mathbf{c}_3^T\mathbf{q}_3, \tag{7-23}$$

and the optimal dual solution must satisfy the following constraint relationships:

$$\begin{cases} \mathbf{A}_1^T\mathbf{p} + 2\mathbf{D}_{11}^T\mathbf{q}_1 + 2\mathbf{D}_{31}^T\mathbf{q}_3 = \mathbf{0}, \\ \mathbf{A}_3^T\mathbf{p} + 2\mathbf{D}_{13}^T\mathbf{q}_1 + 2\mathbf{D}_{33}^T\mathbf{q}_3 \geqq \mathbf{0}, \\ \qquad \mathbf{A}_1\mathbf{q}_1 + \quad \mathbf{A}_3\mathbf{q}_3 = \mathbf{0}, \\ \qquad\qquad\qquad \mathbf{q}_3 \geqq \mathbf{0}, \\ \qquad \mathbf{\Delta}_1^T\mathbf{q}_1 + \quad \mathbf{\Delta}_3^T\mathbf{q}_3 \geqq -\mathbf{1}. \end{cases} \tag{7-24}$$

We now premultiply the first set of constraints in (7-24) by $\mathbf{q}_1^T$ and the second by $\mathbf{q}_3^T \geqq \mathbf{0}$ and take the transposes of the products, obtaining

$$\mathbf{p}^T\mathbf{A}_1\mathbf{q}_1 + 2\mathbf{q}_1^T\mathbf{D}_{11}\mathbf{q}_1 + 2\mathbf{q}_3^T\mathbf{D}_{31}\mathbf{q}_1 = 0$$
$$\mathbf{p}^T\mathbf{A}_3\mathbf{q}_3 + 2\mathbf{q}_1^T\mathbf{D}_{13}\mathbf{q}_3 + 2\mathbf{q}_3^T\mathbf{D}_{33}\mathbf{q}_3 \geqq 0.$$

Adding these two relations and noting in (7-24) that $\mathbf{A}_1\mathbf{q}_1 + \mathbf{A}_3\mathbf{q}_3 = \mathbf{0}$, we can write, in matrix notation,

$$2[\mathbf{q}_1^T, \mathbf{q}_3^T]\begin{bmatrix} \mathbf{D}_{11} & \mathbf{D}_{13} \\ \mathbf{D}_{31} & \mathbf{D}_{33} \end{bmatrix}\begin{bmatrix} \mathbf{q}_1 \\ \mathbf{q}_3 \end{bmatrix} \geqq 0 \tag{7-25}$$

But (7-25) is equivalent to

$$2[\mathbf{q}_1^T, \mathbf{0}, \mathbf{q}_3^T]\begin{bmatrix} \mathbf{D}_{11} & \mathbf{D}_{12} & \mathbf{D}_{13} \\ \mathbf{D}_{21} & \mathbf{D}_{22} & \mathbf{D}_{23} \\ \mathbf{D}_{31} & \mathbf{D}_{32} & \mathbf{D}_{33} \end{bmatrix}\begin{bmatrix} \mathbf{q}_1 \\ \mathbf{0} \\ \mathbf{q}_3 \end{bmatrix} \equiv 2[\mathbf{q}_1^T, \mathbf{0}, \mathbf{q}_3^T]\mathbf{D}\begin{bmatrix} \mathbf{q}_1 \\ \mathbf{0} \\ \mathbf{q}_3 \end{bmatrix} \geqq 0,$$

as is easily verified, and because $\mathbf{D}$ is *negative definite*, (7-25) can hold only if $\mathbf{q}_1 = \mathbf{0}, \mathbf{q}_3 = \mathbf{0}$. This is the crucial step in the proof; observe that the argument would break down here if $\mathbf{D}$ were only negative *semi*definite, because in that case there would exist values of $\mathbf{q}_1$ and $\mathbf{q}_3$ other than $\mathbf{0}$ that would satisfy (7-25).

Having established that $\mathbf{q}_1 = \mathbf{0}$ and $\mathbf{q}_3 = \mathbf{0}$, we can deduce from the first set of constraints in (7-24) that $\mathbf{A}_1^T\mathbf{p} = \mathbf{0}$ or $\mathbf{p}^T\mathbf{A}_1 = \mathbf{0}$, which implies

$$\mathbf{p}^T\mathbf{A}_1\mathbf{x}_1 = \mathbf{0} \cdot \mathbf{x}_1 = 0 \tag{7-26}$$

as well. Now, the primal constraints (7-21) require $\mathbf{A}_1\mathbf{x}_1 + \mathbf{A}_3\mathbf{x}_3 = \mathbf{b}$. But

$\mathbf{x}_3 = \mathbf{0}$ in the optimal solution W; therefore $\mathbf{A}_1\mathbf{x}_1 = \mathbf{b}$, and from (7-26) we have

$$\mathbf{p}^T\mathbf{b} = \mathbf{b}^T\mathbf{p} = 0.$$

It is then evident from (7-23) that the optimal value of the dual objective function is zero, from which we conclude that the primal objective function

$$z' = -u_1 - u_2 - \cdots - u_n$$

must also have a value of zero at the optimal solution W. This implies $u_1 = u_2 = \cdots = u_n = 0$, as was to be shown.

What we have proved in this section is that when Wolfe's algorithm is applied to a quadratic programming problem of the form (7-2), with **D** *negative definite*, it will obtain within a finite number of iterations (barring degeneracy) a feasible solution $(\mathbf{x}^*, \boldsymbol{\lambda}^*, \lambda_0^*, \mathbf{v}^*)$ to (7-14), that is, a feasible solution $\mathbf{x}^*$ to (7-2) at which the Kuhn-Tucker conditions are satisfied. Because **D** is negative definite, (7-2) is a convex program, and the sufficiency of the Kuhn-Tucker conditions thus guarantees that $\mathbf{x}^*$ is an optimal solution.

When **D** is only negative *semi*definite, however, two theoretical difficulties arise. First, the problem (7-2) might have an unbounded optimum; there would then be no constrained local maximum at all and therefore no feasible solution whatever to (7-14). Second, even when a bounded optimum exists it is possible for Wolfe's algorithm to terminate before all the artificial variables have been driven to zero, as was noted in the proof just above. In such a case the final set of values $(\mathbf{x}^*, \boldsymbol{\lambda}^*, \lambda_0^*, \mathbf{v}^*)$ will not satisfy (7-14) and $\mathbf{x}^*$ will not, in general, be a constrained local maximum.

In practice there are two ways of dealing with the negative semidefinite case, assuming it is known that the matrix **D** is indeed negative semidefinite. The more rigorous and, from a computational point of view, safer procedure is to perturb the diagonal elements of **D** very slightly so that the matrix becomes negative *definite*. This is accomplished by using in place of **D** the matrix

$$\hat{\mathbf{D}} = \mathbf{D} - \epsilon\mathbf{I}_n,$$

where $\mathbf{I}_n$ is an n-by-n identity matrix and ϵ is a very small positive number. Observe that the quadratic form

$$\mathbf{x}^T\hat{\mathbf{D}}\mathbf{x} \equiv \mathbf{x}^T\mathbf{D}\mathbf{x} - \epsilon\mathbf{x}^T\mathbf{I}_n\mathbf{x} = \mathbf{x}^T\mathbf{D}\mathbf{x} - \epsilon\sum_{j=1}^{n} x_j^2$$

must be negative definite: If $\mathbf{x} \neq \mathbf{0}$, $\mathbf{x}^T\hat{\mathbf{D}}\mathbf{x}$ is less than $\mathbf{x}^T\mathbf{D}\mathbf{x}$, which is at most zero. A typical value for ϵ might be .0001 times the magnitude of the smallest diagonal element; the resulting change in **D** would be so slight as to have only a negligible effect on the location of the optimal solution.

The other way of approaching the quadratic program (7-2) when **D** is negative semidefinite is to skip the perturbation and simply apply Wolfe's algorithm directly. In most cases a feasible solution to (7-14) will be obtained in routine fashion; because the sufficiency of the Kuhn-Tucker conditions still applies, the **x**-component of any such solution must be an optimal solution to (7-2). This does not imply, however, that Wolfe's method can be used successfully when **D** is *indefinite*. Although for any given problem there is a reasonable chance that a solution to (7-14) will be obtained, its **x**-component will not necessarily be a constrained local optimum of the quadratic program —recall in this connection that in general the Kuhn-Tucker conditions may be satisfied at points other than local optima. Because of this additional difficulty, it is better to use some other solution method, such as that of Beale (see Section 7.6), when **D** is indefinite.

7.5 AN EXAMPLE

We shall now illustrate Wolfe's algorithm by applying it to the following example problem:

$$\begin{aligned} \text{Max } z = 3x_1 + 4x_2 - x_1^2 + 2x_1x_2 - 2x_2^2 \\ \text{subject to} \quad x_1 + 2x_2 \leqq 7, \\ -x_1 + 2x_2 \leqq 4, \\ \text{and} \qquad x_1, x_2 \geqq 0. \end{aligned}$$

Observe that the quadratic form

$$-x_1^2 + 2x_1x_2 - 2x_2^2 = -(x_1 - x_2)^2 - x_2^2$$

is negative definite. The constraints may be converted to the standard form (7-2) by the addition of slack variables:

$$\begin{aligned} x_1 + 2x_2 + x_3 \qquad &= 7, \\ -x_1 + 2x_2 \qquad + x_4 &= 4, \end{aligned}$$

and

$$x_1, x_2, x_3, x_4 \geqq 0.$$

In the notation of this chapter

$$\mathbf{A} = \begin{bmatrix} 1 & 2 & 1 & 0 \\ -1 & 2 & 0 & 1 \end{bmatrix} \quad \text{and} \quad \mathbf{D} = \begin{bmatrix} -1 & 1 & 0 & 0 \\ 1 & -2 & 0 & 0 \\ 0 & 0 & 0 & 0 \\ 0 & 0 & 0 & 0 \end{bmatrix},$$

and the system of constraint equations (7-14) is written as follows:

$$
\begin{array}{llllllll}
x_1 + 2x_2 + x_3 & & & & & & & = 7, \\
-x_1 + 2x_2 & + x_4 & & & & & & = 4, \\
-2x_1 + 2x_2 & & - \lambda_1 + \lambda_2 & + v_1 & & & & = -3, \\
2x_1 - 4x_2 & & - 2\lambda_1 - 2\lambda_2 + 4\lambda_0 & & + v_2 & & & = -4, \\
& & - \lambda_1 \qquad + \lambda_0 & & & + v_3 & & = 0, \\
& & \qquad - \lambda_2 + \lambda_0 & & & & + v_4 & = 0,
\end{array}
$$

with

$$\mathbf{x}, \boldsymbol{\lambda}, \lambda_0, \mathbf{v} \geqq \mathbf{0}$$

and

$$x_j v_j = 0, \qquad j = 1, \ldots, 4.$$

It is convenient to use as a starting point the origin $x_1 = x_2 = 0$; the associated basic solution to $\mathbf{Ax} = \mathbf{b}$ is then $\mathbf{x_B} = (x_3, x_4) = (7, 4)$. We now compute

$$\hat{\mathbf{u}} = -\mathbf{c} - 2\mathbf{D_B x_B} = (-3, -4, 0, 0)$$

and add artificial variables to the constraints above in accordance with (7-17):

$$
(7\text{-}27)\quad \left\{
\begin{array}{llllllllllll}
x_1+2x_2+x_3 & & & & & & & & & & & =7, \\
-x_1+2x_2 & +x_4 & & & & & & & & & & =4, \\
-2x_1+2x_2 & & -\lambda_1+\lambda_2 & +v_1 & & & & -u_1 & & & & =-3, \\
2x_1-4x_2 & & -2\lambda_1-2\lambda_2+4\lambda_0 & & +v_2 & & & & -u_2 & & & =-4, \\
& & -\lambda_1 \qquad +\lambda_0 & & & +v_3 & & & & +u_3 & & =0, \\
& & \qquad -\lambda_2+\lambda_0 & & & & +v_4 & & & & +u_4 & =0, \\
& & \text{with } \mathbf{x}, \boldsymbol{\lambda}, \lambda_0, \mathbf{v}, \mathbf{u} \geqq \mathbf{0}. & & & & & & & & &
\end{array}
\right.
$$

The initial basic solution to this set of constraints is then

$$
\begin{bmatrix} x_3 \\ x_4 \\ u_1 \\ u_2 \\ u_3 \\ u_4 \end{bmatrix} = \begin{bmatrix} 7 \\ 4 \\ 3 \\ 4 \\ 0 \\ 0 \end{bmatrix}, \quad \text{with basis matrix} \begin{bmatrix} 1 & 0 & 0 & 0 & 0 & 0 \\ 0 & 1 & 0 & 0 & 0 & 0 \\ 0 & 0 & -1 & 0 & 0 & 0 \\ 0 & 0 & 0 & -1 & 0 & 0 \\ 0 & 0 & 0 & 0 & 1 & 0 \\ 0 & 0 & 0 & 0 & 0 & 1 \end{bmatrix},
$$

and our task now is to use the simplex method with restricted basis entry to maximize $z' = -u_1 - u_2 - u_3 - u_4$ subject to the constraints (7-27) and to the complementary slackness conditions $x_j v_j = 0$, $j = 1, \ldots, 4$. Before beginning the computations, however, we note that two of the artificial basic variables, u_3 and u_4, have zero values. We can therefore attempt to simplify matters somewhat by dropping them from the problem entirely and replacing them in the basis with v_3 and v_4, whose columns of constraint coefficients in (7-27) are, respectively, identical. This maneuver is permissible in that it does not violate the conditions $x_j v_j = 0$. However, it does have the disadvantage of introducing two complementary pairs of variables into the basis simultaneously, which means that in subsequent pivots we shall have to keep an eye on the values of all basic variables in order to be sure that complementary slackness is preserved.

We can now proceed with the computations. The problem will be solved by the revised simplex method, as described in Chapter 5 of [42], with the usual simplex notation being adopted. Thus the initial basic variables are $\mathbf{x_B} = (x_3, x_4, u_1, u_2, v_3, v_4) = (7, 4, 3, 4, 0, 0)$, with cost coefficients $\mathbf{c_B} = (0, 0, -1, -1, 0, 0)$, and the basis matrix $\mathbf{B}$ is as given above. Clearly, $\mathbf{B}^{-1} = \mathbf{B}$, so the auxiliary vector for computing reduced costs is $\mathbf{c_B^T B}^{-1} = [0, 0, 1, 1, 0, 0]$, and the augmented matrix inverse used by the revised simplex method is

$$(\mathbf{B}^*)^{-1} = \begin{bmatrix} \mathbf{B}^{-1} & \mathbf{0} \\ \mathbf{c_B^T B}^{-1} & 1 \end{bmatrix} = \begin{bmatrix} 1 & 0 & 0 & 0 & 0 & 0 & 0 \\ 0 & 1 & 0 & 0 & 0 & 0 & 0 \\ 0 & 0 & -1 & 0 & 0 & 0 & 0 \\ 0 & 0 & 0 & -1 & 0 & 0 & 0 \\ 0 & 0 & 0 & 0 & 1 & 0 & 0 \\ 0 & 0 & 0 & 0 & 0 & 1 & 0 \\ 0 & 0 & 1 & 1 & 0 & 0 & 1 \end{bmatrix}.$$

In general, the reduced cost of any nonbasic variable x_j is obtained by multiplying the last row of $(\mathbf{B}^*)^{-1}$ by the augmented "activity column" $\mathbf{a}_j^* \equiv (\mathbf{a}_j, -c_j)$, where $\mathbf{a}_j$ is the column of constraint coefficients associated with x_j. In the case of x_1, for example, the reduced cost is

$$\begin{aligned} z_1 - c_1 &= [\mathbf{c_B^T B}^{-1}, 1] \cdot (\mathbf{a}_1, -c_1) \\ &= [0, 0, 1, 1, 0, 0, 1] \cdot (1, -1, -2, 2, 0, 0, 0) = 0, \end{aligned}$$

and we find that the reduced costs for the other nonbasic variables x_2, λ_1, λ_2, λ_0, v_1, and v_2 are, respectively, -2, -3, -1, 4, 1, and 1. Because we are maximizing, we prefer to choose the variable having the most negative reduced cost, in this case λ_1, to enter the basis; note, however, that we shall have to reject this choice if it leads to a violation of complementary slackness.

To determine which basic variable will be replaced, we compute

$$\mathbf{y}_5^* = (\mathbf{B}^*)^{-1}\mathbf{a}_5^* = \begin{bmatrix} 1 & 0 & 0 & 0 & 0 & 0 & 0 \\ 0 & 1 & 0 & 0 & 0 & 0 & 0 \\ 0 & 0 & -1 & 0 & 0 & 0 & 0 \\ 0 & 0 & 0 & -1 & 0 & 0 & 0 \\ 0 & 0 & 0 & 0 & 1 & 0 & 0 \\ 0 & 0 & 0 & 0 & 0 & 1 & 0 \\ 0 & 0 & 1 & 1 & 0 & 0 & 1 \end{bmatrix} \begin{bmatrix} 0 \\ 0 \\ -1 \\ -2 \\ -1 \\ 0 \\ 0 \end{bmatrix}$$

$$= \begin{bmatrix} 0 \\ 0 \\ 1 \\ 2 \\ -1 \\ 0 \\ -3 \end{bmatrix} = \begin{bmatrix} \left.\begin{matrix} \\ \\ \\ \\ \\ \end{matrix}\right\} \mathbf{y}_5 \\ z_5 - c_5 \end{bmatrix},$$

where the subscript 5 is used because λ_1 is the fifth variable, reading from left to right, in the constraint set (7-27). Applying the usual simplex exit criterion (2-28), we find that

$$\min_i \left(\frac{x_{Bi}}{y_{i5}}, y_{i5} > 0\right) = \min\left(\frac{3}{1}, \frac{4}{2}\right) = \frac{4}{2} = \frac{x_{B4}}{y_{45}},$$

so $x_{B4} = u_2 = 4$ leaves the basis. The columns of $(\mathbf{B}^*)^{-1}$ and the vector of values $\mathbf{x}_B^* = (\mathbf{x}_B, z')$ are updated in exactly the same way as the columns of the simplex tableau in a standard simplex pivot. In the present case the computational vector (2-30) is

$$\begin{aligned} \mathbf{\Phi} &= \left(\frac{-y_{15}}{y_{45}}, \frac{-y_{25}}{y_{45}}, \frac{-y_{35}}{y_{45}}, \frac{1}{y_{45}} - 1, \frac{-y_{55}}{y_{45}}, \frac{-y_{65}}{y_{45}}, \frac{-(z_5 - c_5)}{y_{45}}\right) \\ &= (0, 0, -.5, -.5, .5, 0, 1.5), \end{aligned}$$

and the new values of the basic variables turn out to be

$$\text{(7-28)} \qquad \mathbf{x}_B^* = (x_3, x_4, u_1, \lambda_1, v_3, v_4, z') = (7, 4, 1, 2, 2, 0, -1).$$

But observe that $x_3 v_3 \neq 0$ in this solution. Complementary slackness has evidently been violated, and we conclude that some variable other than λ_1 must be chosen to enter the basis. Accordingly, we return to the initial basic solution and select x_2, which has the second most favorable reduced cost, to be the entering variable. We now find that $\mathbf{y}_2^* = (2, 2, -2, 4, 0, 0, -2)$, the exiting variable is $x_{\mathbf{B}4} = u_2 = 4$, the computational vector $\mathbf{\Phi} = (-.5, -.5, .5, -.75, 0, 0, .5)$, and the new values are

$$(\mathbf{B}^*)^{-1} = \begin{bmatrix} 1 & 0 & 0 & .5 & 0 & 0 & 0 \\ 0 & 1 & 0 & .5 & 0 & 0 & 0 \\ 0 & 0 & -1 & -.5 & 0 & 0 & 0 \\ 0 & 0 & 0 & -.25 & 0 & 0 & 0 \\ 0 & 0 & 0 & 0 & 1 & 0 & 0 \\ 0 & 0 & 0 & 0 & 0 & 1 & 0 \\ 0 & 0 & 1 & .5 & 0 & 0 & 1 \end{bmatrix} \quad \text{and} \quad \begin{bmatrix} x_3 \\ x_4 \\ u_1 \\ x_2 \\ v_3 \\ v_4 \\ z' \end{bmatrix} = \begin{bmatrix} 5 \\ 2 \\ 5 \\ 1 \\ 0 \\ 0 \\ -5 \end{bmatrix}.$$

Complementary slackness has been preserved, and the pivot is accepted.

Moving on to the next iteration, we determine that the reduced costs of the nonbasic variables x_1, λ_1, λ_2, λ_0, v_1, and v_2 are -1, -2, 0, 2, 1, and .5. Our first choice for basis entry is therefore the fifth variable λ_1. However, we find that $\mathbf{y}_5^* = (-1, -1, 2, 0.5, -1, 0, -2)$ and

$$\min_i \left(\frac{x_{\mathbf{B}i}}{y_{i5}}, y_{i5} > 0\right) = \min\left(\frac{5}{2}, \frac{1}{.5}\right) = \frac{1}{.5} = \frac{x_{\mathbf{B}4}}{y_{45}},$$

so that if λ_1 were to enter the basis the variable ejected would be $x_{\mathbf{B}4} = x_2 = 1$. This would again produce the basic solution (7-28), which is already known to be unacceptable. We therefore pass on to our second choice and try to bring x_1 into the basis. The computations proceed as follows:

$$\mathbf{y}_1^* = (2, 0, 1, -0.5, 0, 0, -1),$$

$$\min_i\left(\frac{x_{\mathbf{B}i}}{y_{i1}}, y_{i1} > 0\right) = \min\left(\frac{5}{2}, \frac{5}{1}\right) = \frac{5}{2} = \frac{x_{\mathbf{B}1}}{y_{11}},$$

$$\mathbf{\Phi} = (-.5, 0, -.5, .25, 0, 0, .5)$$

and

$$(\mathbf{B}^*)^{-1} = \begin{bmatrix} .5 & 0 & 0 & .25 & 0 & 0 & 0 \\ 0 & 1 & 0 & .5 & 0 & 0 & 0 \\ -.5 & 0 & -1 & -.75 & 0 & 0 & 0 \\ .25 & 0 & 0 & -.125 & 0 & 0 & 0 \\ 0 & 0 & 0 & 0 & 1 & 0 & 0 \\ 0 & 0 & 0 & 0 & 0 & 1 & 0 \\ .5 & 0 & 1 & .75 & 0 & 0 & 1 \end{bmatrix} \quad \text{and} \quad \begin{bmatrix} x_1 \\ x_4 \\ u_1 \\ x_2 \\ v_3 \\ v_4 \\ z' \end{bmatrix} = \begin{bmatrix} 2.5 \\ 2 \\ 2.5 \\ 2.25 \\ 0 \\ 0 \\ -2.5 \end{bmatrix}.$$

The complementary slackness conditions are satisfied, and the new solution is accepted.

In recomputing the reduced costs for the third iteration we again find that λ_1 is the preferred candidate for basis entry. This time, however, the pivot is successful:

$$\mathbf{y}_5^* = (-.5, -1, 2.5, .25, -1, 0, -2.5),$$

$$\min_i\left(\frac{x_{\mathbf{B}i}}{y_{i5}}, y_{i5} > 0\right) = \min\left(\frac{2.5}{2.5}, \frac{2.25}{.25}\right) = \frac{2.5}{2.5} = \frac{x_{\mathbf{B}3}}{y_{35}},$$

$$\mathbf{\Phi} = (.2, .4, -.6, -.1, .4, 0, 1),$$

and

$$\mathbf{x}_\mathbf{B}^* = (x_1, x_4, \lambda_1, x_2, v_3, v_4, z')$$
$$= (3, 3, 1, 2, 1, 0, 0).$$

Complementary slackness remains satisfied, and the last artificial variable has been driven out of the basis. The current basic solution is therefore optimal, which means that

$$(x_1, x_2, x_3, x_4) = (3, 2, 0, 3)$$

constitutes an optimal solution to the original quadratic program. The value of the objective at this point is $z = 12$.

The feasible region of the original problem and the sequence of solution points generated by Wolfe's algorithm are plotted in Fig. 7-1. Several indifference curves of the objective function are also shown. Notice that the solutions generated at the various stages do not always coincide with extreme points. In fact, in the general case they need not even be boundary points:

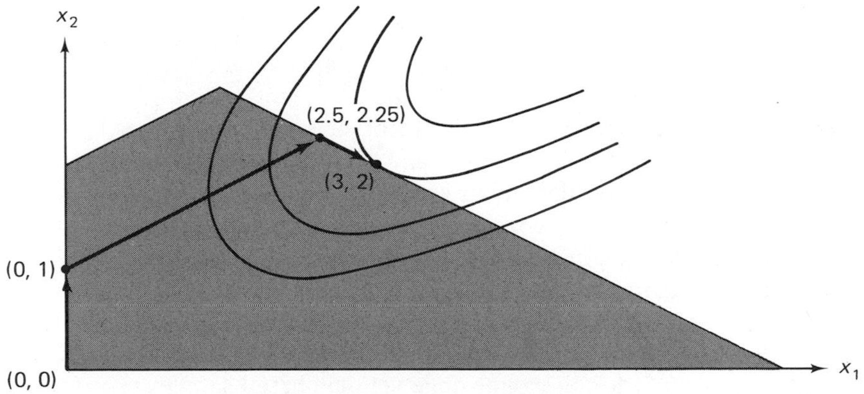

FIGURE 7-1

An example of this would be provided by any quadratic program whose optimal solution is itself in the interior of the feasible region.

One further computational remark should be made. In solving our example problem we were twice obliged to perform most of the calculations necessary for a pivot (all but the updating of the augmented matrix inverse) *before* discovering that the pivot in question would lead to a violation of complementary slackness and would therefore be unacceptable. This difficulty was caused by our decision to insert v_3 and v_4 into the initial basic solution at a zero level, despite the fact that the initial basis already included x_3 and x_4. The potential for future violations of complementary slackness was thereby introduced, and it is questionable whether an overall saving in computational effort was actually achieved. Wolfe's algorithm works best when no complementary pair of variables x_k and v_k are in the basis together, because then, by simply choosing an entering variable whose complementary partner is not already in the basis, one can guarantee that the ensuing pivot will not violate complementary slackness.

7.6 THE METHOD OF BEALE

In 1955 E. M. L. Beale published a paper in a British journal describing a method for solving quadratic programs that was based on arguments from classical calculus rather than on the Kuhn-Tucker conditions. The paper initially attracted little attention, especially in the United States, so an amplified version of it was published in this country in 1959 [4]. This second effort was more favorably received, and Beale's method has since become quite well known, although certainly not so widely used as that of Wolfe.

Beale's method, like Wolfe's, is directly applicable to any quadratic program of the following form:

$$(7\text{-}2) \qquad \begin{cases} \text{Max } z = \mathbf{c}^T\mathbf{x} + \mathbf{x}^T\mathbf{D}\mathbf{x} \\ \quad \text{subject to } \mathbf{A}\mathbf{x} = \mathbf{b} \\ \qquad\qquad \text{and} \quad \mathbf{x} \geqq \mathbf{0}, \end{cases}$$

where $\mathbf{x}$ is an n-component column vector of variables, $\mathbf{A}$ is m-by-n, $\mathbf{b}$ is m-by-1, $\mathbf{c}$ is n-by-1, and $\mathbf{D}$ is n-by-n and symmetric. We do *not*, however, require $\mathbf{D}$ to be negative definite or semidefinite—that is, we do not require the objective function to be concave. Thus the algorithm will in general yield local optima, although, of course, when the objective function is concave the first solution obtained will be the global optimum.

To develop Beale's method, we begin by assuming that we have eliminated all redundant constraints and determined an initial basic feasible solution $\mathbf{x_B}$ to $\mathbf{A}\mathbf{x} = \mathbf{b}$; this can be done via the phase I simplex procedure, if necessary. By reordering the columns of $\mathbf{A}$, the elements of $\mathbf{c}$, and the rows *and* columns of $\mathbf{D}$ as necessary, we can partition the vector of variables into two pieces,

$$\mathbf{x} \equiv (\mathbf{x_B}, \mathbf{x_R}),$$

where $\mathbf{x_R}$ is an $(n - m)$-component vector containing the nonbasic variables. Similarly, the constraint matrix $\mathbf{A}$ can be partitioned into $[\mathbf{B} \quad \mathbf{R}]$, where $\mathbf{B}$ is the basis matrix associated with $\mathbf{x_B}$ and $\mathbf{R}$ is the m-by-$(n - m)$ matrix of nonbasic columns; the constraints $\mathbf{Ax} = \mathbf{b}$ can then be rewritten as

$$\mathbf{Bx_B} + \mathbf{Rx_R} = \mathbf{b}. \tag{7-29}$$

Note that the values of the variables in the initial solution are

$$\mathbf{x_B} = \mathbf{B}^{-1}\mathbf{b} \quad \text{and} \quad \mathbf{x_R} = \mathbf{0} \tag{7-30}$$

and that these values satisfy the constraints (7-29). Finally, we partition $\mathbf{c}$ and $\mathbf{D}$ to conform with $\mathbf{x} = (\mathbf{x_B}, \mathbf{x_R})$:

$$\mathbf{c}^T \equiv [\mathbf{c}_\mathbf{B}^T, \mathbf{c}_\mathbf{R}^T] \quad \text{and} \quad \mathbf{D} \equiv \begin{bmatrix} \mathbf{D}_{11} & \mathbf{D}_{12} \\ \mathbf{D}_{21} & \mathbf{D}_{22} \end{bmatrix}.$$

Here, $\mathbf{D}_{11}$ is m-by-m, $\mathbf{D}_{12}$ is m-by-$(n - m)$, and so on.

Although we know of one feasible solution to the problem (7-2), namely (7-30), suppose we consider other values of $\mathbf{x_B}$ and $\mathbf{x_R}$ that satisfy (7-29)—we do not restrict ourselves to basic solutions. From the constraints (7-29) we have

$$\mathbf{x_B} = \mathbf{B}^{-1}\mathbf{b} - \mathbf{B}^{-1}\mathbf{Rx_R}, \tag{7-31}$$

where (7-31) must hold for *any* feasible solution to (7-2), not just for the one that happens to correspond to the above partition. Substitution of (7-31) into the partitioned objective function produces the following:

$$\begin{aligned} z &= \mathbf{c}_\mathbf{B}^T\mathbf{x_B} + \mathbf{c}_\mathbf{R}^T\mathbf{x_R} + [\mathbf{x}_\mathbf{B}^T \quad \mathbf{x}_\mathbf{R}^T]\begin{bmatrix} \mathbf{D}_{11} & \mathbf{D}_{12} \\ \mathbf{D}_{21} & \mathbf{D}_{22} \end{bmatrix}\begin{bmatrix} \mathbf{x_B} \\ \mathbf{x_R} \end{bmatrix} \\ &= \cdots \\ &= \{(\mathbf{c}_\mathbf{B}^T + (\mathbf{B}^{-1}\mathbf{b})^T\mathbf{D}_{11})\mathbf{B}^{-1}\mathbf{b}\} \\ &\quad + \{2(\mathbf{B}^{-1}\mathbf{b})^T\mathbf{D}_{12} - 2(\mathbf{B}^{-1}\mathbf{b})^T\mathbf{D}_{11}\mathbf{B}^{-1}\mathbf{R} + \mathbf{c}_\mathbf{R}^T - \mathbf{c}_\mathbf{B}^T\mathbf{B}^{-1}\mathbf{R}\}\mathbf{x_R} \\ &\quad + \mathbf{x}_\mathbf{R}^T\{\mathbf{D}_{22} - 2\mathbf{D}_{21}\mathbf{B}^{-1}\mathbf{R} + (\mathbf{B}^{-1}\mathbf{R})^T\mathbf{D}_{11}\mathbf{B}^{-1}\mathbf{R}\}\mathbf{x_R}. \end{aligned} \tag{7-32}$$

Let the three expressions within braces be denoted, respectively, by the scalar z_0, the $(n - m)$-component row vector $\mathbf{p}$, and the $(n - m)$-by-$(n - m)$ matrix

Q; we then have

(7-33) $$z = z_0 + \mathbf{p} \cdot \mathbf{x_R} + \mathbf{x_R^T Q x_R}.$$

We shall assume henceforth that the off-diagonal elements of **Q** have been adjusted, if necessary, in order to make **Q** symmetric; this can, of course, be done without changing the quadratic form $\mathbf{x_R^T Q x_R}$ in any way.

Equation (7-33) gives the value of the objective z as a function of the variables $\mathbf{x_R} \equiv (x_{\mathbf{R}1}, \ldots, x_{\mathbf{R},n-m})$. For the initial solution we have $\mathbf{x_R} = \mathbf{0}$ and $z = z_0$, but suppose we consider the effect on z of changing the value of each of the nonbasic variables in $\mathbf{x_R}$. The first partial derivative of (7-33) with respect to $x_{\mathbf{R}j}$ is

(7-34) $$\frac{\delta z}{\delta x_{\mathbf{R}j}} = p_j + 2 \sum_{k=1}^{n-m} q_{jk} x_{\mathbf{R}k}, \qquad j = 1, \ldots, n - m.$$

Initially every nonbasic variable equals zero, so

$$\frac{\delta z}{\delta x_{\mathbf{R}j}} = p_j,$$

and the objective function can evidently be increased by increasing any nonbasic variable $x_{\mathbf{R}j}$ for which p_j is positive. Inasmuch as we would like to obtain the optimal solution as rapidly as possible, let us bring into the basis the variable that promises the greatest initial rate of improvement; that is, let $x_{\mathbf{R}h}$ enter the basis, where $p_h = \max_j(p_j)$.

As $x_{\mathbf{R}h}$ begins to increase, the values of the variables $x_{\mathbf{B}1}, \ldots, x_{\mathbf{B}m}$ begin to change in accordance with (7-31). If we were solving a linear program via the simplex method, this procedure would, of course, continue, with the value of the objective increasing at the same constant rate, until one of the basic variables reached zero, thereby completing the pivot. When the objective is quadratic, however, it is evident from (7-34) that the partial derivative $\delta z/\delta x_{\mathbf{R}h}$ changes as a function of $x_{\mathbf{R}h}$; in particular, it is possible for $\delta z/\delta x_{\mathbf{R}h}$ to fall to zero during a pivot (i.e., while all the $x_{\mathbf{B}i}$ are still positive). If this should happen, any further increase in $x_{\mathbf{R}h}$ would worsen rather than improve the value of the objective function and would therefore not be desirable. Accordingly, in the method of Beale we allow the entering variable $x_{\mathbf{R}h}$ to increase only until *either*

1. One of the basic variables is driven to zero, *or*
2. The partial derivative $\delta z/\delta x_{\mathbf{R}h}$ vanishes.

The critical values of $x_{\mathbf{R}h}$ at which conditions 1 and 2 occur are not at all difficult to compute. The basic variable $x_{\mathbf{B}r}$ that will reach zero first is determined by computing the **y**-column of the entering variable (call it $\mathbf{y}_h$, although in general the hth nonbasic variable will not be x_h) and applying the usual simplex basis exit criterion: The value of $x_{\mathbf{R}h}$ at which $x_{\mathbf{B}r}$ vanishes is then

$$x_{\mathbf{R}h}^{(1)} = \frac{x_{\mathbf{B}r}}{y_{rh}} = \min_i \left(\frac{x_{\mathbf{B}i}}{y_{ih}}, y_{ih} > 0\right). \tag{7-35}$$

To compute the second critical value, we need only observe that, because all the other nonbasic variables $x_{\mathbf{R}j}, j \neq h$, are being held at zero throughout, Eq. (7-34) reduces to

$$\frac{\delta z}{\delta x_{\mathbf{R}h}} = p_h + 2q_{hh}x_{\mathbf{R}h}. \tag{7-36}$$

Thus condition 2 can arise only when $q_{hh} < 0$, in which case the partial derivative vanishes when $x_{\mathbf{R}h}$ reaches the critical value

$$x_{\mathbf{R}h}^{(2)} = \frac{-p_h}{2q_{hh}}. \tag{7-37}$$

The new value attained by $x_{\mathbf{R}h}$ in a "Beale-simplex" pivot is then

$$\min\{x_{\mathbf{R}h}^{(1)}, x_{\mathbf{R}h}^{(2)}\},$$

where we set $x_{\mathbf{R}h}^{(1)} = \infty$ if no basic variable is driven toward zero as $x_{\mathbf{R}h}$ is increased, and $x_{\mathbf{R}h}^{(2)} = \infty$ if $q_{hh} \geqq 0$. Note that if $x_{\mathbf{R}h}^{(1)} = x_{\mathbf{R}h}^{(2)} = \infty$—that is, if $x_{\mathbf{R}h}$ can be increased indefinitely without causing either of the conditions 1 or 2 to occur—then the quadratic program must have an unbounded optimal solution.

Suppose now that when $x_{\mathbf{R}h}$ is allowed to increase case 1 occurs—that is, the objective function continues to increase until some basic variable $x_{\mathbf{B}r}$ is driven to zero. We can then bring $x_{\mathbf{R}h}$ into the basis to replace $x_{\mathbf{B}r}$ by means of a simplex pivot, using virtually the same tableau and computational scheme as in linear programming. Forming the $\mathbf{\Phi}$-vector from the column $\mathbf{y}_h$, we update the various **y**-columns (or, if the revised simplex approach has been adopted, the basis matrix inverse) in the usual manner, *except* that the reduced costs $z_j - c_j$ of linear programming are no longer present. Instead, we recompute the vector **p** and matrix **Q**, after which we are ready for the next pivot step. In solving very small problems it is reasonable to compute the new values of **p** and **Q** directly from their definitions at each iteration. The same is true for much larger problems in which only a few of the variables are involved in the quadratic portion of the objective function; in such problems most of the elements of **D** are zero, and the direct computation of **p** and **Q** via (7-32) is greatly simplified. For other quadratic programs, however, it is more efficient to update the current values of $\mathbf{p}$ and $\mathbf{Q}$ by means of certain transformation formulas; these formulas, which would be incorporated into any computer-coded version of Beale's method, are unfortunately rather messy and will not be inflicted on the reader.

Note also in connection with case 1 that if a tie occurs in computing the critical value $x_{\mathrm{R}h}^{(1)}$, the introduction of the variable $x_{\mathrm{R}h}$ into the basis will drive two basic variables to zero and therefore give rise to degeneracy. Then, when some new variable $x_{\mathrm{R}k}$ is chosen to enter the basis at the next iteration, it might turn out that $x_{\mathrm{R}k}^{(1)} = 0$; in that case it would not be possible to increase $x_{\mathrm{R}k}$ at all and a degenerate or "zero-for-zero" pivot would take place. Moreover, the same thing could also happen at the next iteration, and the next, and it is clear that, as with the simplex method of linear programming, there is at least a theoretical possibility of cycling through an endlessly repeating sequence of degenerate bases. However—again by analogy with the simplex method—the danger that infinite cycling might occur in an actual problem appears extremely remote and can thus quite reasonably be disregarded in practice. We shall have more to say on the general subject of degeneracy in Section 8.8.

Returning to the pivot we have been considering, let us turn our attention to case 2. Suppose that before any basic variable has been driven to zero the entering variable $x_{\mathrm{R}h}$ reaches its second critical value $x_{\mathrm{R}h}^{(2)} > 0$. At this point $m + 1$ variables are positive, barring degeneracy, and we no longer have a basic solution to the constraints $\mathbf{Ax} = \mathbf{b}$. However, we do have $\delta z/\delta x_{\mathrm{R}h} = 0$, and the $m + 1$ positive variables can be considered a basic solution to the $m + 1$ constraints

$$\mathbf{Ax} = \mathbf{b}$$

and

$$u_h \equiv \frac{\delta z}{\delta x_{\mathrm{R}h}} = p_h + 2 \sum_{k=1}^{n-m} q_{hk} x_{\mathrm{R}k}. \tag{7-38}$$

The new "constraint," which can be rewritten in standard form as

$$u_h - 2 \sum_{k=1}^{n-m} q_{hk} x_{\mathrm{R}k} = p_h, \tag{7-39}$$

is evidently satisfied by the set of values

$$u_h = 0, \quad x_{\mathrm{R}h} = x_{\mathrm{R}h}^{(2)} = \frac{-p_h}{2q_{hh}}, \quad \text{and} \quad x_{\mathrm{R}k} = 0, \qquad k \neq h.$$

It must be noted, however, that u_h is merely an auxiliary variable introduced for computational purposes: It is *not* required to take on a nonnegative value, just as the expression

$$p_h + 2\sum_{k=1}^{n-m} q_{hk}x_{\mathbf{R}k}$$

is not required to be nonnegative (if it were, that would constitute a real constraint over and above the original set $\mathbf{Ax} = \mathbf{b}$). For this reason u_h is called a *free variable*. The subscript h is assigned to it in order to simplify the notation, although it would be more correct to use the subscript of the variable that happens to be $x_{\mathbf{R}h}$; earlier, we used $\mathbf{y}_h$ to denote the $\mathbf{y}$-column of $x_{\mathbf{R}h}$ for the same reason.

To continue performing the computations within a simplex framework, it is necessary to add (7-39) to the set of constraints, which means adding a row to the simplex tableau. Before the pivot begins, u_h has the positive value p_h and must therefore be the $(m + 1)$th basic variable. Because none of the current basic variables appear in the new constraint, the new basis matrix $\hat{\mathbf{B}}$ takes the form

$$\hat{\mathbf{B}} = \begin{bmatrix} \mathbf{B} & \mathbf{0} \\ \mathbf{0} & 1 \end{bmatrix}, \quad \text{with } \hat{\mathbf{B}}^{-1} = \begin{bmatrix} \mathbf{B}^{-1} & \mathbf{0} \\ \mathbf{0} & 1 \end{bmatrix},$$

where $\mathbf{B}$ is the old m-by-m basis. The new nonbasic columns must be of the form $\hat{\mathbf{a}}_k \equiv (\mathbf{a}_k, -2q_{hk})$; this follows from (7-39), with the same informal convention of identical subscripts again being adopted. The new $\mathbf{y}$-columns are

$$\hat{\mathbf{y}}_k = \hat{\mathbf{B}}^{-1}\hat{\mathbf{a}}_k = \begin{bmatrix} \mathbf{B}^{-1} & \mathbf{0} \\ \mathbf{0} & 1 \end{bmatrix}\begin{bmatrix} \mathbf{a}_k \\ -2q_{hk} \end{bmatrix} = \begin{bmatrix} \mathbf{y}_k \\ -2q_{hk} \end{bmatrix},$$

and the new values of the basic variables are

$$\hat{\mathbf{x}}_{\mathbf{B}} = \hat{\mathbf{B}}^{-1}\hat{\mathbf{b}} = \begin{bmatrix} \mathbf{B}^{-1} & \mathbf{0} \\ \mathbf{0} & 1 \end{bmatrix}\begin{bmatrix} \mathbf{b} \\ p_h \end{bmatrix} = \begin{bmatrix} \mathbf{B}^{-1}\mathbf{b} \\ p_h \end{bmatrix},$$

so the constraint (7-39) can be added immediately to the old simplex tableau without altering any of the values already present and without performing a single computation. The nonbasic variable $x_{\mathbf{R}h}$ can now be brought into the basis, as planned. Because $x_{\mathbf{R}h}^{(2)} = p_h/(-2q_{hh}) > 0$ is less than $x_{\mathbf{R}h}^{(1)}$—this was the reason for defining the free variable u_h in the first place—the variable chosen by the simplex exit criterion must necessarily be u_h, and the pivoting computions then proceed in the usual manner. Afterward, $\mathbf{p}$ and $\mathbf{Q}$ are recomputed and the next iteration begins.

In future iterations the free variable u_h may be brought back into the basis if $\delta z/\delta u_h$ is large enough *in magnitude*, either positive or negative, to justify its selection; in those cases where $\delta z/\delta u_h$ is negative, u_h will be *decreased* from its nonbasic value of zero to a negative level, thereby causing z to increase, as desired. If and when u_h does reenter the basis, whether with a positive or a negative value, there is no longer any need to keep track of it or of the constraint (7-39) associated with it. This follows from the fact that we do not really care whether the free variable u_h is greater or less than zero; it will never be chosen to leave the basis, nor figure in any way in the selection of the exiting variable, because there is no need to prevent its becoming negative (it is the nonnegativity requirement that dictates the choice of the exiting variable in the ordinary simplex method of linear programming). Thus the row containing the basic variable u_h can simply be dropped from the tableau—the missing values will never be needed in updating the other tableau entries. The reader may imagine, if he prefers, that the variable u_h and the missing tableau row still exist but that we are no longer bothering to calculate and update their values.

Pivoting continues as described above, with the size of the basis increasing and decreasing at irregular intervals, until a point is reached from which it is impossible to improve the objective value by making any permitted change in a nonbasic variable. Permitted changes include increases in all variables and decreases in free variables; thus the condition for termination is

$$(7\text{-}40) \qquad \frac{\delta z}{\delta x_{\mathbf{R}j}} \begin{cases} \leq 0 & \text{if } x_{\mathbf{R}j} \text{ is a restricted (i.e., nonnegative) variable,} \\ = 0 & \text{if } x_{\mathbf{R}j} \text{ is a free variable.} \end{cases}$$

It is not difficult to see that (7-40) constitutes a necessary condition for optimality.[3] When we manipulated the constraints $\mathbf{Ax} = \mathbf{b}$ to produce

$$(7\text{-}31) \qquad \mathbf{x_B} = \mathbf{B}^{-1}\mathbf{b} - \mathbf{B}^{-1}\mathbf{R}\mathbf{x_R}$$

and then substituted for $\mathbf{x_B}$ in the objective function to obtain

$$(7\text{-}33) \qquad z = z_0 + \mathbf{p} \cdot \mathbf{x_R} + \mathbf{x_R}^T\mathbf{Q}\mathbf{x_R},$$

we in effect transformed the original quadratic program into a problem in the nonbasic variables only. At any current basic solution each of the nonbasic variables has a value of zero, so we can think of such a solution as being

[3] We have slightly oversimplified the argument here. It is more correct to say that Beale's algorithm terminates when a solution has been obtained having the property that any small increase or decrease in a single nonbasic variable that improves the value of the objective function must immediately cause the solution to become infeasible. This condition admits the possibility that $\delta z/\delta x_{\mathbf{R}j}$ may be positive for some nonbasic variable $x_{\mathbf{R}j}$ but that any increase in $x_{\mathbf{R}j}$ would immediately cause a zero-valued (restricted) basic variable to decrease. Thus (7-40) is a correct statement of the termination condition of Beale's algorithm and of the necessary condition for optimality if and only if the possibility of a degenerate basic solution is barred.

located at the origin in *nonbasic space*. Suppose now that the current set of values $\mathbf{x_B}$ is a locally optimal solution to the quadratic program, which means that the origin is a constrained local maximum of the objective function z in nonbasic space. Then no small displacement away from the origin in a permitted, or feasible, direction can possibly increase the value of z. But, because a nonnegative variable can only be permitted to increase, while a free variable can be perturbed in either direction, it follows immediately that the condition (7-40) must obtain at the origin. Q.E.D.

Although (7-40) is a *necessary* condition for optimality, it is, of course, *not sufficient* in the general case. When the matrix $\mathbf{D}$ is indefinite it is quite possible for (7-40) to hold at a constrained stationary point or even at a local minimum; an example will be given in the next section. If $\mathbf{D}$ is negative definite or semidefinite, however, the objective function is concave, the problem is a convex program, and the necessary condition for optimality is also sufficient. Proof of the last assertion within the context of nonbasic space is quite straightforward.

It should be noted that although the size of the basis changes from time to time as free variables are defined and then eliminated from the problem, the number of nonbasic variables remains constant throughout at $n - m$. Inasmuch as free variables need not be carried along in the basis, the greatest possible number of basic variables at any stage is just n, the number of variables x_j in the original problem. Thus, in the course of applying Beale's algorithm to a quadratic program having n nonnegative variables and m constraints, the total number of variables can range from n to $2n - m$, while the number of constraints and basic variables may vary from m to n. By comparison, Wolfe's algorithm requires a total of $m + n$ constraints and basic variables at every stage. A more extensive comparison of the theoretical and computational aspects of various quadratic programming algorithms will appear in a later section.

7.7 AN EXAMPLE

Let us apply Beale's algorithm to the following quadratic programming problem:

$$\begin{aligned} \text{Max } z &= 18x_1 - x_1^2 + x_1x_2 - x_2^2 \\ \text{subject to} \quad & x_1 + x_2 \leqq 12, \\ & -x_1 + x_2 \leqq 6, \\ \text{and} \quad & x_1, x_2 \geqq 0. \end{aligned}$$

We observe that the quadratic form

$$-x_1^2 + x_1x_2 - x_2^2 = -(x_1 - \tfrac{1}{2}x_2)^2 - \tfrac{3}{4}x_2^2$$

is negative definite, so the objective function is strictly concave. With the addition of slack variables the constraints become

$$x_1 + x_2 + x_3 \qquad = 12$$

and

$$-x_1 + x_2 \qquad + x_4 = 6.$$

It is immediately obvious that $\mathbf{x_B} = (x_3, x_4) = (12, 6)$ will serve as an initial basic feasible solution, and we begin with the following standard simplex tableau:

	x_1	x_2	x_3	x_4
$x_{B1} = x_3 = 12$	1	1	1	0
$x_{B2} = x_4 = 6$	−1	1	0	1

The initial value of the objective function is zero. Note that the row of reduced costs required in linear programming is not present in the tableau above; the computations needed to evaluate the partial derivatives $\delta z/\delta x_{Rj}$ will be carried out "on the side."

In the initial basic solution the variables arrange themselves as follows:

$$\text{(7-41)} \qquad \mathbf{x} = (\mathbf{x_B}, \mathbf{x_R}) = (x_3, x_4 \,\vdots\, x_1, x_2) = (12, 6 \,\vdots\, 0, 0).$$

The vector $\mathbf{c}$ and the matrices $\mathbf{A}$ and $\mathbf{D}$ are then partitioned in accordance with (7-14):

$$\mathbf{c}^T = [\mathbf{c}_\mathbf{B}^T, \mathbf{c}_\mathbf{R}^T] = [0, 0 \,\vdots\, 18, 0]$$

$$\mathbf{D} = \begin{bmatrix} \mathbf{D}_{11} & \mathbf{D}_{12} \\ \mathbf{D}_{21} & \mathbf{D}_{22} \end{bmatrix} = \left[\begin{array}{cc:cc} 0 & 0 & 0 & 0 \\ 0 & 0 & 0 & 0 \\ \hdashline 0 & 0 & -1 & .5 \\ 0 & 0 & .5 & -1 \end{array}\right]$$

$$\mathbf{A} = [\mathbf{B} \quad \mathbf{R}] = \left[\begin{array}{cc:cc} 1 & 0 & 1 & 1 \\ 0 & 1 & -1 & 1 \end{array}\right]$$

It should be clear that although we have reordered the variables and the elements of $\mathbf{c}$, $\mathbf{A}$, and $\mathbf{D}$, the problem itself has not been changed: For example, the constraints are still

$$\mathbf{Ax} = [\mathbf{B} \quad \mathbf{R}]\begin{bmatrix} \mathbf{x_B} \\ \mathbf{x_R} \end{bmatrix} = \mathbf{Bx_B} + \mathbf{Rx_R} = \begin{bmatrix} 1 & 0 \\ 0 & 1 \end{bmatrix}\begin{bmatrix} x_3 \\ x_4 \end{bmatrix} + \begin{bmatrix} 1 & 1 \\ -1 & 1 \end{bmatrix}\begin{bmatrix} x_1 \\ x_2 \end{bmatrix}$$

$$= \begin{bmatrix} x_3 + x_1 + x_2 \\ x_4 - x_1 + x_2 \end{bmatrix} = \mathbf{b} = \begin{bmatrix} 12 \\ 6 \end{bmatrix},$$

exactly as in the original problem. We can now substitute into (7-32) to compute the values of z_0, $\mathbf{p}$, and $\mathbf{Q}$:

$$z_0 = [0, 0]\cdot(12, 6) = 0,$$
$$\mathbf{p} = 2\cdot[0, 0] - 2\cdot[0, 0] + [18, 0] - [0, 0] = [18, 0],$$

and

$$\mathbf{Q} = \begin{bmatrix} -1 & .5 \\ .5 & -1 \end{bmatrix} - 2\cdot\begin{bmatrix} 0 & 0 \\ 0 & 0 \end{bmatrix} + \begin{bmatrix} 0 & 0 \\ 0 & 0 \end{bmatrix} = \begin{bmatrix} -1 & .5 \\ .5 & -1 \end{bmatrix}.$$

Inasmuch as $\mathbf{Q}$ is already symmetric, it need not be adjusted at all, and (7-33) becomes

(7-42) $$z = z_0 + \mathbf{p}\cdot\mathbf{x_R} + \mathbf{x_R^T Q x_R} = 18x_1 - x_1^2 + x_1x_2 - x_2^2.$$

In this equation the objective is expressed as a function of the nonbasic variables, which currently happen to be the original variables x_1 and x_2.

To determine the entering variable we simply examine the vector $\mathbf{p}$:

$$\frac{\delta z}{\delta x_{\mathbf{R}1}} = \frac{\delta z}{\delta x_1} = p_1 = 18 \quad \text{and} \quad \frac{\delta z}{\delta x_{\mathbf{R}2}} = \frac{\delta z}{\delta x_2} = p_2 = 0.$$

We could also have obtained these results by taking the first partial derivatives of (7-42) and evaluating them at $x_1 = x_2 = 0$. Because $\delta z/\delta x_1$ is positive, the current solution is not optimal; $x_{\mathbf{R}1} = x_1$ is chosen to enter the basis, and the first critical value is found to be

$$x_{\mathbf{R}1}^{(1)} = \min_i \left(\frac{x_{\mathbf{B}i}}{y_{i1}}, y_{i1} > 0\right) = \frac{12}{1} = 12.$$

The second critical value, however, is

$$x_{\mathbf{R}1}^{(2)} = \frac{-p_1}{2q_{11}} = \frac{-18}{2(-1)} = 9,$$

so x_1 will be increased only to 9, and neither of the current basic variables will be driven to zero. We must therefore add to the problem a new free variable u_1 and a new constraint of the form (7-38),

$$u_1 = \frac{\delta z}{\delta x_{\mathbf{R}1}} = p_1 + 2\sum_k q_{1k}x_{\mathbf{R}k}$$
$$= 18 - 2x_1 + x_2,$$

which is equivalent to

(7-43) $$u_1 + 2x_1 - x_2 = 18.$$

Initially, before x_1 is increased, we have $u_1 = 18$; thus u_1 must be added to the basis, and the current tableau is augmented as follows:

	x_1	x_2	x_3	x_4	u_1
$x_{B1} = x_3 = 12$	1	1	1	0	0
$x_{B2} = x_4 = 6$	−1	1	0	1	0
$x_{B3} = u_1 = 18$	2	−1	0	0	1

We are finally ready to bring x_1 into the basis; applying the exit criterion,

$$\min_i \left(\frac{x_{Bi}}{y_{i1}}, y_{i1} > 0\right) = \min\left(\frac{12}{1}, \frac{18}{2}\right) = \frac{18}{2} = \frac{x_{B3}}{y_{31}},$$

so u_1, as expected, is driven to zero and out of the basis. The computational vector $\mathbf{\Phi}$ is

$$\mathbf{\Phi} = \left(\frac{-y_{11}}{y_{31}}, \frac{-y_{21}}{y_{31}}, \frac{1}{y_{31}} - 1\right) = (-.5, .5, -.5),$$

and the new tableau becomes

	x_1	x_2	x_3	x_4	u_1
$x_{B1} = x_3 = 3$	0	1.5	1	0	−.5
$x_{B2} = x_4 = 15$	0	.5	0	1	.5
$x_{B3} = x_1 = 9$	1	−.5	0	0	.5

Because this solution has three variables with positive values, it is not an extreme point of the original set of constraints—see Fig. 7-2.

Having generated a new basic feasible solution, we must now repartition the problem:

$$\mathbf{x} = (\mathbf{x_B}, \mathbf{x_R}) = (x_3, x_4, x_1 \,\vdots\, x_2, u_1),$$

$$\mathbf{c}^T = [\mathbf{c}_\mathbf{B}^T, \mathbf{c}_\mathbf{R}^T] = (0, 0, 18 \,\vdots\, 0, 0),$$

$$\mathbf{A} = [\mathbf{B} \quad \mathbf{R}] = \left[\begin{array}{ccc:cc} 1 & 0 & 1 & 1 & 0 \\ 0 & 1 & -1 & 1 & 0 \\ 0 & 0 & 2 & -1 & 1 \end{array}\right],$$

and

$$\mathbf{D} = \begin{bmatrix} \mathbf{D}_{11} & \mathbf{D}_{12} \\ \mathbf{D}_{21} & \mathbf{D}_{22} \end{bmatrix} = \left[\begin{array}{ccc:cc} 0 & 0 & 0 & 0 & 0 \\ 0 & 0 & 0 & 0 & 0 \\ 0 & 0 & -1 & .5 & 0 \\ \hdashline 0 & 0 & .5 & -1 & 0 \\ 0 & 0 & 0 & 0 & 0 \end{array}\right].$$

Note that because u_1 does not appear in the objective function, the row and column of **D** corresponding to u_1 contain only zeros. We can now compute

$$z_0 = \left([0, 0, 18] + [3, 15, 9]\begin{bmatrix} 0 & 0 & 0 \\ 0 & 0 & 0 \\ 0 & 0 & -1 \end{bmatrix}\right)\cdot(3, 15, 9) = 81,$$

$$\mathbf{p} = 2\cdot[3, 15, 9]\begin{bmatrix} 0 & 0 \\ 0 & 0 \\ .5 & 0 \end{bmatrix} - 2[3, 15, 9]\begin{bmatrix} 0 & 0 & 0 \\ 0 & 0 & 0 \\ 0 & 0 & -1 \end{bmatrix}\begin{bmatrix} 1.5 & -.5 \\ .5 & .5 \\ -.5 & .5 \end{bmatrix}$$

$$+ [0, 0] - [0, 0, 18]\begin{bmatrix} 1.5 & -.5 \\ .5 & .5 \\ -.5 & .5 \end{bmatrix} = [9, 0],$$

and

$$\mathbf{Q} = \begin{bmatrix} -1 & 0 \\ 0 & 0 \end{bmatrix} - 2\begin{bmatrix} 0 & 0 & .5 \\ 0 & 0 & 0 \end{bmatrix}\begin{bmatrix} 1.5 & -.5 \\ .5 & .5 \\ -.5 & .5 \end{bmatrix}$$

$$+ \begin{bmatrix} 1.5 & .5 & -.5 \\ -.5 & .5 & .5 \end{bmatrix}\begin{bmatrix} 0 & 0 & 0 \\ 0 & 0 & 0 \\ 0 & 0 & -1 \end{bmatrix}\begin{bmatrix} 1.5 & -.5 \\ .5 & .5 \\ -.5 & .5 \end{bmatrix}$$

$$= \begin{bmatrix} -.75 & -.25 \\ .25 & -.25 \end{bmatrix},$$

which we adjust to

$$\mathbf{Q} = \begin{bmatrix} -.75 & 0 \\ 0 & -.25 \end{bmatrix}.$$

The objective, expressed as a function of the nonbasic variables, is then

$$z = z_0 + \mathbf{p} \cdot \mathbf{x_R} + \mathbf{x_R^T Q x_R} = 81 + 9x_2 - .75x_2^2 - .25u_1^2.$$

Examining the vector **p**, we see that the next variable to enter the basis will be $x_{\mathbf{R}1} = x_2$. The critical points are

$$x_{\mathbf{R}1}^{(1)} = \min_i \left(\frac{x_{\mathbf{B}i}}{y_{i2}}, y_{i2} > 0\right) = \min\left(\frac{3}{1.5}, \frac{15}{.5}\right) = \frac{3}{1.5} = 2 = \frac{x_{\mathbf{B}1}}{y_{12}}$$

and

$$x_{\mathbf{R}1}^{(2)} = \frac{-p_1}{2q_{11}} = \frac{-9}{2(-.75)} = 6,$$

so x_2 replaces $x_{B1} = x_3$ in the basis in a normal simplex pivot. The new tableau is as follows:

	x_1	x_2	x_3	x_4	u_1
$x_{B1} = x_2 = 2$	0	1	2/3	0	−1/3
$x_{B2} = x_4 = 14$	0	0	−1/3	1	2/3
$x_{B3} = x_1 = 10$	1	0	1/3	0	1/3

Partitioning the problem again, we find

$$\mathbf{x} = (\mathbf{x_B}, \mathbf{x_R}) = (x_2, x_4, x_1 \mid x_3, u_1),$$

$$\mathbf{c}^T = [\mathbf{c}_\mathbf{B}^T, \mathbf{c}_\mathbf{R}^T] = [0, 0, 18 \mid 0, 0],$$

$$\mathbf{A} = [\mathbf{B} \quad \mathbf{R}] = \left[\begin{array}{rrr|rr} 1 & 0 & 1 & 1 & 0 \\ 1 & 1 & -1 & 0 & 0 \\ -1 & 0 & 2 & 0 & 1 \end{array}\right],$$

and

$$\mathbf{D} = \begin{bmatrix} \mathbf{D}_{11} & \mathbf{D}_{12} \\ \mathbf{D}_{21} & \mathbf{D}_{22} \end{bmatrix} = \left[\begin{array}{rrr|rr} -1 & 0 & .5 & 0 & 0 \\ 0 & 0 & 0 & 0 & 0 \\ .5 & 0 & -1 & 0 & 0 \\ \hline 0 & 0 & 0 & 0 & 0 \\ 0 & 0 & 0 & 0 & 0 \end{array}\right]$$

which yield

$$z_0 = 96, \quad \mathbf{p} = [-4, 2], \quad \text{and} \quad \mathbf{Q} = \begin{bmatrix} -1/3 & 1/6 \\ 1/6 & -1/3 \end{bmatrix}.$$

Because $p_2 = 2$ is positive, the current solution is still not optimal, and $x_{R2} = u_1$ enters the basis. Note, incidentally, that if p_2 had been negative, the solution still would not have been optimal, for we could then have increased the objective function by *decreasing* the free variable u_1. The critical values of the entering variable u_1 are

$$x_{\mathbf{R}2}^{(1)} = \min\left(\frac{14}{2/3}, \frac{10}{1/3}\right) = 21$$

and

$$x_{\mathbf{R}2}^{(2)} = \frac{-p_2}{2q_{22}} = 3,$$

so another free variable must be defined:

$$u_2 = \frac{\delta z}{\delta x_{R2}} = p_2 + 2 \sum_k q_{2k} x_{Rk}$$
$$= 2 + \frac{1}{3} x_3 - \frac{2}{3} u_1.$$

Before the pivot begins we have $u_2 = 2$; thus u_2 becomes the fourth basic variable and the following row is added to the current tableau:

$x_{B4} = u_2 = 2$	0	0	−1/3	0	2/3	1

The element 1 at the far right is associated with the new variable u_2; all other elements in the tableau column of u_2 are zero. We can now bring u_1 into the basis; as expected, the exit criterion selects $x_{B4} = u_2$, and the new tableau turns out to be

	x_1	x_2	x_3	x_4	u_1	u_2
$x_{B1} = x_2 = 3$	0	1	1/2	0	0	1/2
$x_{B2} = x_4 = 12$	0	0	0	1	0	−1
$x_{B3} = x_1 = 9$	1	0	1/2	0	0	−1/2
$x_{B4} = u_1 = 3$	0	0	−1/2	0	1	3/2

Because the new basis entrant $x_{B4} = u_1$ is a free variable, not restricted to nonnegative values, we can drop it and the constraint (7-43) from the above tableau and from the entire problem. Thus in repartitioning the problem we have only five variables and three constraints to deal with:

$$\mathbf{x} = (\mathbf{x_B}, \mathbf{x_R}) = (x_2, x_4, x_1 \,\vdots\, x_3, u_2),$$
$$\mathbf{c}^T = [\mathbf{c}_B^T, \mathbf{c}_R^T] = [0, 0, 18 \,\vdots\, 0, 0],$$
$$\mathbf{A} = [\mathbf{B} \quad \mathbf{R}] = \left[\begin{array}{ccc|cc} 1 & 0 & 1 & 1 & 0 \\ 1 & 1 & -1 & 0 & 0 \\ 0 & 0 & 0 & -1/3 & 1 \end{array}\right],$$

and

$$\mathbf{D} = \begin{bmatrix} \mathbf{D}_{11} & \mathbf{D}_{12} \\ \mathbf{D}_{21} & \mathbf{D}_{22} \end{bmatrix} - \left[\begin{array}{ccc|cc} -1 & 0 & .5 & 0 & 0 \\ 0 & 0 & 0 & 0 & 0 \\ .5 & 0 & -1 & 0 & 0 \\ \hline 0 & 0 & 0 & 0 & 0 \\ 0 & 0 & 0 & 0 & 0 \end{array}\right].$$

These values yield

$$z_0 = 99, \quad \mathbf{p} = [-3, 0], \quad \text{and} \quad \mathbf{Q} = \begin{bmatrix} -.25 & 0 \\ 0 & -.75 \end{bmatrix}.$$

Examining the components of **p**, we see that the criterion for termination of Beale's algorithm has been satisfied:

$$\frac{\delta z}{\delta x_{\mathbf{R}1}} = p_1 = -3 \qquad \text{for the restricted variable } x_{\mathbf{R}1} = x_3$$

and

$$\frac{\delta z}{\delta x_{\mathbf{R}2}} = p_2 = 0 \qquad \text{for the free variable } x_{\mathbf{R}2} = u_2.$$

Because the objective function is concave, the set of values

$$x_1 = 9, \qquad x_2 = 3, \qquad x_3 = 0, \qquad x_4 = 12, \qquad z = 99$$

must constitute an optimal solution to the quadratic program. This conclusion can be verified, in part, by computing the second derivative $\delta^2 z/\delta u_2^2$, which is obtained by differentiation of (7-34):

$$\frac{\delta^2 z}{\delta u_2^2} = \frac{\delta^2 z}{\delta x_{\mathbf{R}2}^2} = \frac{\delta}{\delta x_{\mathbf{R}2}}\left(p_2 + 2\sum_{k=1}^{n-m} q_{2k}x_{\mathbf{R}k}\right) = 2q_{22} = -1.5.$$

The second derivative with respect to the free variable u_2 is negative at the origin (in nonbasic space), so the origin is indeed an unconstrained local maximum with respect to displacement in the u_2-direction.

The sequence of points generated in solving this problem is diagrammed in Fig. 7-2. Obviously, Beale's method, like that of Wolfe, does not restrict itself to extreme points, at least not in the space of the original set of variables. Notice, however, that every pivot step does terminate either when a boundary of the feasible region is encountered (step 2) or at a point where the direction of displacement is tangent to an indifference contour (steps 1 and 3). Because of this characteristic, Beale's method can be viewed as a classical hill-climbing procedure, in which the strategy is always to continue climbing in the chosen direction until either the objective stops improving or a boundary of the feasible region is reached. In Chapter 8 we shall study several other methods of this type.

It was mentioned in Section 7.6 that when **D** is neither negative definite nor semidefinite, Beale's algorithm can terminate at a point that is not a local maximum. As an example, consider the following problem, in which the

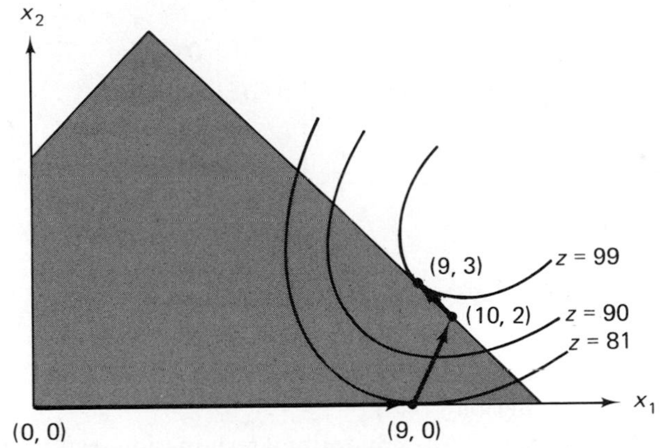

FIGURE 7-2

objective function is *positive* definite:

$$\text{Max } z = -12x_2 + x_1^2 + x_2^2$$

$$\begin{aligned} \text{subject to} \quad & x_1 + x_2 + x_3 = 12, \\ & -x_1 + x_2 + x_4 = 6, \\ \text{and} \quad & x_j \geqq 0, \qquad j = 1, \ldots, 4. \end{aligned}$$

Here the feasible region is the same as in the previous example, while the objective is to maximize the distance from the point (0, 6). Suppose we begin at $x_1 = 0$, $x_2 = 6$, which is obviously a local *minimum*; the corresponding basic solution is then $x_2 = 6$, $x_3 = 6$, with basis matrix and inverse

$$\mathbf{B} = \begin{bmatrix} 1 & 1 \\ 1 & 0 \end{bmatrix} \quad \text{and} \quad \mathbf{B}^{-1} = \begin{bmatrix} 0 & 1 \\ 1 & -1 \end{bmatrix},$$

and the initial tableau is as follows:

	x_1	x_2	x_3	x_4
$x_{B1} = x_2 = 6$	−1	1	0	1
$x_{B2} = x_3 = 6$	2	0	1	−1

Partitioning the problem as usual,

$$\mathbf{x} = (\mathbf{x_B}, \mathbf{x_R}) = (x_2, x_3 \,\vdots\, x_1, x_4),$$

$$\mathbf{c}^T = [\mathbf{c}_\mathbf{B}^T, \mathbf{c}_\mathbf{R}^T] = [-12, 0 \,\vdots\, 0, 0],$$

$$\mathbf{A} = [\mathbf{B} \quad \mathbf{R}] = \left[\begin{array}{cc:cc} 1 & 1 & 1 & 0 \\ 1 & 0 & -1 & 1 \end{array}\right],$$

and

$$\mathbf{D} = \begin{bmatrix} \mathbf{D}_{11} & \mathbf{D}_{12} \\ \mathbf{D}_{21} & \mathbf{D}_{22} \end{bmatrix} = \left[\begin{array}{cc:cc} 1 & 0 & 0 & 0 \\ 0 & 0 & 0 & 0 \\ \hdashline 0 & 0 & 1 & 0 \\ 0 & 0 & 0 & 0 \end{array}\right],$$

and we find that

$$z_0 = -36, \quad \mathbf{p} = [0, 0], \quad \text{and} \quad \mathbf{Q} = \begin{bmatrix} 2 & -1 \\ -1 & 1 \end{bmatrix}.$$

The components of $\mathbf{p}$ are both zero, so our initial solution immediately satisfies the termination criterion (7-40).

It is clear, however, that this nonoptimal termination resulted solely from our "unlucky" choice of a starting point at which the partial derivatives $\delta z/\delta x_{\mathbf{R}j}$ happened to vanish. Fortunately, when Beale's algorithm is being used there is a very simple way of recovering or "coming unstuck" from a false optimum in such cases as this: It is only necessary to examine the second partial derivatives $\delta^2 z/\delta x_{\mathbf{R}j}^2$. If the second derivative is positive for some $x_{\mathbf{R}h}$, then the current solution must be a minimum, not a maximum, with respect to displacement in the $x_{\mathbf{R}h}$-direction (in nonbasic space), and the value of the objective function can be increased by bringing $x_{\mathbf{R}h}$ into the basis. In the example above, both q_{11} and q_{22} are positive, with $q_{11} > q_{22}$; accordingly, $x_{\mathbf{R}1} = x_1$ should be chosen to enter the basis. The reader can verify that the new basic solution will be $x_1 = 3$, $x_2 = 9$, with an improved objective value of $z = -18$.

7.8 LEMKE'S ALGORITHM

We shall now consider a third quadratic programming algorithm, that of Lemke [30], which is rather more sophisticated and elegant than either of the two we have studied thus far. The methods of Wolfe, Beale, and Lemke are three of the best known and most widely used quadratic programming algorithms in existence; in both respects they probably should be ranked in the order given. All of them, incidentally, are of surprisingly advanced age in this hypermodern field of nonlinear programming: Lemke's method, the last of the three to be developed, appeared originally in 1962.

The algorithm of Lemke is applicable to any quadratic programming problem of the following form:

$$\text{(7-44)} \qquad \begin{cases} \text{Max } z = \mathbf{c}^T\mathbf{x} - \frac{1}{2}\mathbf{x}^T\mathbf{D}\mathbf{x} \\ \text{subject to } \mathbf{A}\mathbf{x} \leqq \mathbf{b}, \end{cases}$$

where $\mathbf{x}$ and $\mathbf{c}$ are n-component column vectors, $\mathbf{b}$ is m-by-1, $\mathbf{A}$ is m-by-n, and

$\mathbf{D}$ is n-by-n, symmetric, and *positive definite*. This last requirement is crucial and will be used at several points in the discussion below. Note that the constraint set $\mathbf{Ax} \leqq \mathbf{b}$ includes any nonnegativity restrictions that may be present. The formulation in terms of inequality constraints rather than equations is necessary because the Kuhn-Tucker conditions then assume a particularly simple form, which Lemke was able to exploit quite profitably in developing his algorithm. It is strictly for purposes of convenience, however, that the objective function has been left as it appeared in Lemke's original article; other formulations, such as our more familiar $z = \mathbf{c}^T\mathbf{x} + \mathbf{x}^T\mathbf{Dx}$, could be made to serve just as well.

If there were no constraints in the problem (7-44), the objective would take on its maximum value at a point $\mathbf{x}_0$ at which the gradient $\nabla \mathbf{z}$ vanishes; this follows from the fact that z cannot have an unbounded maximum (recall the argument at the beginning of Section 7.4). Inasmuch as the gradient of z at any point $\mathbf{x}$ is

$$\nabla \mathbf{z} = \mathbf{c} - \mathbf{Dx},$$

the unconstrained optimum must lie at the unique point

$$\mathbf{x}_0 = \mathbf{D}^{-1}\mathbf{c}. \tag{7-45}$$

Recall from Lemma 3.1 that the inverse of a positive definite matrix is known to exist. If $\mathbf{x}_0$ happens to satisfy the constraints $\mathbf{Ax} \leqq \mathbf{b}$, we can conclude immediately that it must be the optimal solution to the quadratic program (7-44). But if it does not, as will usually be the case, then the optimum will lie somewhere on the boundary of the feasible region, and it will be necessary to use some type of iterative computational procedure in order to find it.

We begin our presentation of Lemke's algorithm by writing the Kuhn-Tucker conditions for the problem (7-44): Applying Theorem 6.3 directly, if $\mathbf{x}$ is the optimal solution to (7-44), then there exist values $\mathbf{y} = (y_1, \ldots, y_m)$ such that

$$\mathbf{c} - \mathbf{Dx} = \sum_{i=1}^{m} y_i \boldsymbol{\alpha}_i^T = \mathbf{A}^T\mathbf{y}, \tag{7-46}$$

$$y_i(\boldsymbol{\alpha}_i\mathbf{x} - b_i) = 0, \qquad i = 1, \ldots, m, \tag{7-47}$$

and

$$\mathbf{y} \geqq \mathbf{0}, \tag{7-48}$$

where $\boldsymbol{\alpha}_i$ is the ith row of $\mathbf{A}$ and where (7-47) can be rewritten in matrix form as

$$\mathbf{y}^T(\mathbf{Ax} - \mathbf{b}) = 0. \tag{7-49}$$

Note that, because $\mathbf{y}$ is nonnegative and $\mathbf{Ax} - \mathbf{b}$ is nonpositive, (7-49) implies each of the conditions (7-47). We now observe that, inasmuch as $\mathbf{D}$ is positive definite, the objective function of the quadratic program (7-44) is strictly concave, so the Kuhn-Tucker conditions must be sufficient: If we can find a set of values $\mathbf{x} = \mathbf{x}^*$ and $\mathbf{y} = \mathbf{y}^*$ that satisfy

$$\text{(7-50)} \qquad \begin{cases} \mathbf{Ax} \leqq \mathbf{b}, \\ \mathbf{c} - \mathbf{Dx} = \mathbf{A}^T\mathbf{y}, \\ \mathbf{y}^T(\mathbf{Ax} - \mathbf{b}) = 0, \\ \mathbf{y} \geqq \mathbf{0}, \end{cases}$$

then $\mathbf{x}^*$ must be the optimal solution to (7-44). This set of relations can be reduced by manipulating its second member so as to eliminate the variables $\mathbf{x}$; we find that

$$\text{(7-51)} \qquad \mathbf{x} = \mathbf{D}^{-1}(\mathbf{c} - \mathbf{A}^T\mathbf{y}),$$

so (7-50) reduces to

$$\text{(7-52)} \qquad \begin{cases} \mathbf{AD}^{-1}\mathbf{c} - \mathbf{AD}^{-1}\mathbf{A}^T\mathbf{y} \leqq \mathbf{b}, \\ \mathbf{y}^T(\mathbf{AD}^{-1}\mathbf{c} - \mathbf{AD}^{-1}\mathbf{A}^T\mathbf{y} - \mathbf{b}) = 0, \\ \mathbf{y} \geqq \mathbf{0}. \end{cases}$$

If we now define

$$\text{(7-53)} \qquad \mathbf{Q} = \mathbf{AD}^{-1}\mathbf{A}^T \quad \text{and} \quad \mathbf{p} = \mathbf{b} - \mathbf{AD}^{-1}\mathbf{c},$$

where $\mathbf{Q}$ is m-by-m and $\mathbf{p}$ is m-by-1, (7-52) becomes

$$\text{(7-54)} \qquad \begin{cases} -\mathbf{Qy} \leqq \mathbf{p}, \\ \mathbf{y}^T(-\mathbf{Qy} - \mathbf{p}) = 0, \\ \mathbf{y} \geqq \mathbf{0}. \end{cases}$$

Finally, we introduce a new set of variables

$$\text{(7-55)} \qquad \boldsymbol{\lambda} \equiv \mathbf{p} + \mathbf{Qy},$$

where $\boldsymbol{\lambda}$ is m-by-1, and substitute into (7-54), producing

$$\text{(7-56)} \qquad \mathbf{0} \leqq \boldsymbol{\lambda}, \quad \mathbf{y}^T(-\boldsymbol{\lambda}) = 0, \quad \text{and} \quad \mathbf{y} \geqq \mathbf{0}.$$

The set of relations (7-55) and (7-56) are jointly equivalent to the set (7-50); that is, if we can find values $\mathbf{y} = \mathbf{y}^*$ and $\boldsymbol{\lambda} = \boldsymbol{\lambda}^*$ that satisfy

$$\text{(7-57)} \qquad \boldsymbol{\lambda} = \mathbf{p} + \mathbf{Qy}, \quad \boldsymbol{\lambda}^T\mathbf{y} = 0, \quad \text{and} \quad \mathbf{y}, \boldsymbol{\lambda} \geqq \mathbf{0},$$

then, from (7-51), $\mathbf{x}^* = \mathbf{D}^{-1}(\mathbf{c} - \mathbf{A}^T\mathbf{y}^*)$ must be the optimal solution to the original problem.

Consider now a second quadratic program

$$\text{(7-58)} \qquad \begin{cases} \text{Min } f(\mathbf{y}) = \mathbf{p}^T\mathbf{y} + \tfrac{1}{2}\mathbf{y}^T\mathbf{Q}\mathbf{y} \\ \qquad \text{subject to } \mathbf{y} \geqq \mathbf{0}, \end{cases}$$

where $\mathbf{y}$ is m-by-1 and $\mathbf{p}$ and $\mathbf{Q}$ are as given by (7-53). Note that $\mathbf{Q}$ must be symmetric,

$$\mathbf{Q}^T = (\mathbf{A}\mathbf{D}^{-1}\mathbf{A}^T)^T = (\mathbf{A}^T)^T(\mathbf{D}^{-1})^T\mathbf{A}^T = \mathbf{A}\mathbf{D}^{-1}\mathbf{A}^T = \mathbf{Q},$$

using the fact that the inverse of a symmetric matrix is also symmetric. The Kuhn-Tucker conditions for the problem (7-58) are as follows: If $\mathbf{y}$ is the optimal solution to (7-58), then there exist multipliers $\boldsymbol{\lambda} = (\lambda_1, \ldots, \lambda_m)$ such that

$$\text{(7-59)} \qquad \begin{cases} -\mathbf{p} - \mathbf{Q}\mathbf{y} = \sum_{i=1}^{m} \lambda_i(-\mathbf{e}_i) = -\boldsymbol{\lambda}, \\ \boldsymbol{\lambda}^T\mathbf{y} = 0, \\ \boldsymbol{\lambda} \geqq \mathbf{0}, \end{cases}$$

where $\mathbf{e}_i$ is an m-component unit vector with a 1 in the ith position and 0s elsewhere. Because $\mathbf{D}$ is positive definite, we know from Lemma 3.1 that $\mathbf{D}^{-1}$ is also positive definite and that $\mathbf{Q} = \mathbf{A}\mathbf{D}^{-1}\mathbf{A}^T$ is positive semidefinite. Therefore, the objective function of (7-58) is convex and the Kuhn-Tucker conditions (7-59) are sufficient: If we can find values $\mathbf{y} = \mathbf{y}^*$ and $\boldsymbol{\lambda} = \boldsymbol{\lambda}^*$ satisfying $\mathbf{y} \geqq \mathbf{0}$ and (7-59), then $\mathbf{y}^*$ is the optimal solution to (7-58). But observe that *this set of conditions is precisely the same as the set* (7-57) *that was derived from the original quadratic program.* What we propose to do, therefore, is to solve the problem (7-58); when we have obtained its optimal solution $\mathbf{y}^*$, we will know that the values $\mathbf{y}^*$ and $\boldsymbol{\lambda}^* = \mathbf{p} + \mathbf{Q}\mathbf{y}^*$ satisfy (7-57), and it will then follow that $\mathbf{x}^* = \mathbf{D}^{-1}(\mathbf{c} - \mathbf{A}^T\mathbf{y}^*)$ is the optimal solution to the original problem.

Although it would be possible to solve the problem (7-58) via any number of different general methods, including Wolfe's algorithm and several of the methods to be presented in Chapter 8, Lemke devised a rather efficient special procedure which takes advantage of the very simple nature of the constraints. Notice that the gradient of the objective function is

$$\nabla \mathbf{f} = \mathbf{p} + \mathbf{Q}\mathbf{y} = \boldsymbol{\lambda}.$$

Basically, Lemke's strategy is to generate a sequence of feasible points, each having a lesser objective value than its predecessor, until a point $\mathbf{y}^*$ is reached at which the gradient $\boldsymbol{\lambda}^*$ satisfies both $\boldsymbol{\lambda}^* \geqq \mathbf{0}$ and $\boldsymbol{\lambda}^{*T}\mathbf{y}^* = 0$. The point $\mathbf{y}^*$ will then be the optimal solution to (7-58). At each stage the new point is obtained by proceeding from the old in a certain straight-line direction, until either the objective $f(\mathbf{y})$ stops decreasing or the value of one of the variables falls to zero. Thus in its underlying strategy Lemke's algorithm is quite similar to that of Beale.

The computational procedure begins with the feasible solution $\mathbf{y} = \mathbf{0}$. It is clear that this initial solution would immediately be optimal if $\mathbf{p}$ were nonnegative, for then $\mathbf{y} = \mathbf{0}$, $\boldsymbol{\lambda} = \mathbf{p}$ would satisfy the Kuhn-Tucker conditions (7-59). In that case the optimal solution to the original quadratic program would be $\mathbf{x}^* = \mathbf{D}^{-1}(\mathbf{c} - \mathbf{A}^T \cdot \mathbf{0}) = \mathbf{D}^{-1}\mathbf{c}$, which is just the unconstrained maximum of the original objective function. In general, however, $\mathbf{p} = \mathbf{b} - \mathbf{A}\mathbf{D}^{-1}\mathbf{c}$ will not be nonnegative, and neither $\mathbf{y} = \mathbf{0}$ nor $\mathbf{x} = \mathbf{D}^{-1}\mathbf{c}$ will be the optimal solutions to their respective problems. It will then be necessary to apply Lemke's algorithm, which we now proceed to develop.

Accordingly, suppose $\mathbf{y} = (y_1, \ldots, y_m)$ is any current feasible solution, and let K be the set of subscripts of all variables y_k whose current values are zero:

$$k \text{ is in } K \text{ if and only if } y_k = 0, \qquad k = 1, \ldots, m.$$

Further, let $\mathbf{B}$ be a nonsingular m-by-m matrix whose columns $\mathbf{b}_1, \ldots, \mathbf{b}_m$ will normally include, but need not be restricted to, a unit vector $\mathbf{e}_k$ for every k in K. That is, for each k in K we stipulate that, except in certain "degenerate" situations (see below), the vector $\mathbf{e}_k$ must be one of the columns $\mathbf{b}_j$ of $\mathbf{B}$; we call these $\mathbf{b}_j$ the *restricted columns* of $\mathbf{B}$. Inasmuch as we do not intend to require that $\mathbf{e}_k$ be the kth column of $\mathbf{B}$, we use the symbol J to denote the set of subscripts of the restricted columns. All other $\mathbf{b}_j$, where j is not in J, will be known as *free columns*; they will be generated by the algorithm as described below. Note that for the initial solution $\mathbf{y} = \mathbf{0}$, all the subscripts $k = 1, \ldots, m$ are included in K and all the unit vectors $\mathbf{e}_1, \ldots, \mathbf{e}_m$ are columns of $\mathbf{B}$; for convenience these columns are usually arranged in their natural order, so at the initial stage $\mathbf{B} = \mathbf{B}^{-1} = \mathbf{I}_m$, the identity matrix of order m.

We now define an m-component vector

$$\mathbf{w} = \mathbf{B}^{-1}(\mathbf{p} + \mathbf{Q}\mathbf{y}), \tag{7-60}$$

which implies

$$\boldsymbol{\lambda} = \mathbf{p} + \mathbf{Q}\mathbf{y} = \sum_{j=1}^{m} w_j \mathbf{b}_j, \tag{7-61}$$

where $\mathbf{b}_j$ is the jth column of $\mathbf{B}$. Note that $\mathbf{w}$ is not necessarily nonnegative.

Next, we derive the following important result:

Lemma 7.1: If $\mathbf{w} \geqq \mathbf{0}$ *and* $w_j = 0$ *for all* j *not in* J, *then the current feasible point* $\mathbf{y}$ *is an optimal solution to* (7-58).

The hypothesis of the lemma reduces (7-61) to

$$\boldsymbol{\lambda} = \mathbf{p} + \mathbf{Q}\mathbf{y} = \sum_{j \text{ in } J} w_j \mathbf{b}_j,$$

where every $\mathbf{b}_j$, j in J, is some unit vector. Then for each component of $\boldsymbol{\lambda}$ we have

$$\lambda_k = \begin{cases} \text{some } w_j \geqq 0 & \text{if } k \text{ is in } K \text{ and } \mathbf{e}_k = \mathbf{b}_j \text{ is in } \mathbf{B}, \\ 0 & \text{otherwise.} \end{cases}$$

That is, $\boldsymbol{\lambda}$ is nonnegative, and for every k, $k = 1, \ldots, m$, either k is in K, implying $y_k = 0$, or $\lambda_k = 0$ (or possibly both). It follows that the current $\mathbf{y}$ and $\boldsymbol{\lambda}$ satisfy the Kuhn-Tucker conditions, so $\mathbf{y}$ is optimal. Q.E.D.

We conclude from Lemma 7.1 that if the current solution $\mathbf{y}$ is not optimal, then either

1. $w_j \neq 0$ for at least one j not in J, or
2. $w_j = 0$ for all j not in J but $w_j < 0$ for some j in J.

In general there may be several "candidate" components $w_\alpha, w_\beta, \ldots$ that have nonzero values in case 1 or negative values in case 2. We shall begin the next iteration by removing the column $\mathbf{b}_r$ from the matrix $\mathbf{B}$, where w_r is the largest *in magnitude* of the candidate values; that is,

$$|w_r| = \max\{|w_\alpha|, |w_\beta|, \ldots\}.$$

As we shall see, $\mathbf{b}_r$ is not necessarily a restricted column.

Consider now the points $\hat{\mathbf{y}}$ on the straight line

$$\hat{\mathbf{y}} = \mathbf{y} - \theta \boldsymbol{\beta}_r, \tag{7-62}$$

where $\mathbf{y}$ is the current solution point, $\boldsymbol{\beta}_r^T$ is the rth *row* of $\mathbf{B}^{-1}$, and θ is any real number. We wish to find the value of θ that minimizes $f(\hat{\mathbf{y}})$, conditional on $\hat{\mathbf{y}} \geqq \mathbf{0}$. Observe that

$$\begin{aligned} f(\mathbf{y} - \theta\boldsymbol{\beta}_r) &= \mathbf{p}^T\mathbf{y} - \theta\mathbf{p}^T\boldsymbol{\beta}_r + \tfrac{1}{2}(\mathbf{y} - \theta\boldsymbol{\beta}_r)^T\mathbf{Q}(\mathbf{y} - \theta\boldsymbol{\beta}_r) \\ &= f(\mathbf{y}) - \theta\mathbf{p}^T\boldsymbol{\beta}_r - \theta\boldsymbol{\beta}_r^T\mathbf{Q}\mathbf{y} + \tfrac{1}{2}\theta^2\boldsymbol{\beta}_r^T\mathbf{Q}\boldsymbol{\beta}_r \\ &= f(\mathbf{y}) + \tfrac{1}{2}\theta^2\boldsymbol{\beta}_r^T\mathbf{Q}\boldsymbol{\beta}_r - \theta\boldsymbol{\beta}_r^T(\mathbf{p} + \mathbf{Q}\mathbf{y}), \end{aligned} \tag{7-63}$$

using the fact that $\mathbf{Q} = \mathbf{Q}^T$. But from (7-60) we have

$$\boldsymbol{\beta}_r^T(\mathbf{p} + \mathbf{Q}\mathbf{y}) = w_r,$$

so (7-63) leads to

$$f(\mathbf{y} - \theta\boldsymbol{\beta}_r) = f(\mathbf{y}) + \tfrac{1}{2}\theta^2\boldsymbol{\beta}_r^T\mathbf{Q}\boldsymbol{\beta}_r - \theta w_r. \tag{7-64}$$

Note that, because $\mathbf{Q}$ is positive semidefinite, the quadratic form $\boldsymbol{\beta}_r^T\mathbf{Q}\boldsymbol{\beta}_r$ must be either positive or zero; these alternatives give rise to two different procedures for determining the value of θ that minimizes (7-64) over the feasible region.

Case A. $\boldsymbol{\beta}_r^T\mathbf{Q}\boldsymbol{\beta}_r$ *is positive.* In this case $f(\mathbf{y} - \theta\boldsymbol{\beta}_r)$ is a quadratic function whose minimizing value θ^* is obtained by setting the first derivative $df/d\theta$ equal to zero, which yields

$$\theta^* = \frac{w_r}{\boldsymbol{\beta}_r^T\mathbf{Q}\boldsymbol{\beta}_r}. \tag{7-65}$$

We know that $\theta = \theta^*$ is a minimum, not a maximum, because the second derivative $d^2f/d\theta^2 = \boldsymbol{\beta}_r^T\mathbf{Q}\boldsymbol{\beta}_r$ is positive. We must now determine whether $\mathbf{y} - \theta^*\boldsymbol{\beta}_r$ is nonnegative, that is, whether it is possible to proceed from $\mathbf{y}$ to $\mathbf{y} - \theta^*\boldsymbol{\beta}_r$ without leaving the feasible region $\mathbf{y} \geqq \mathbf{0}$. Accordingly, we compute

$$t_0 = \min_i \left\{\frac{y_i}{\theta^*(\boldsymbol{\beta}_r)_i},\ \theta^*(\boldsymbol{\beta}_r)_i > 0\right\}, \tag{7-66}$$

where $(\boldsymbol{\beta}_r)_i$ is the ith component of $\boldsymbol{\beta}_r$; if there is no i such that $\theta^*(\boldsymbol{\beta}_r)_i$ is positive, we set $t_0 = \infty$.

In general, the value of t_0 computed via (7-66) will be positive. This conclusion proceeds from the fact that for every $y_i = 0$ the unit vector $\mathbf{e}_i$ will normally be a column $\mathbf{b}_j$ of $\mathbf{B}$. If $\mathbf{e}_i = \mathbf{b}_r$—that is, if $\mathbf{e}_i$ is the vector being removed from $\mathbf{B}$—the subscript r must be in J; then, from the way in which $\mathbf{b}_r$ was chosen, w_r must be negative, and (7-65) implies $\theta^* < 0$. But, because the rth diagonal element of $\mathbf{B}^{-1}\mathbf{B} = \mathbf{I}$ must be 1, we also have

$$1 = (\mathbf{B}^{-1}\mathbf{B})_{rr} = \boldsymbol{\beta}_r^T\mathbf{b}_r = \boldsymbol{\beta}_r^T\mathbf{e}_i = (\boldsymbol{\beta}_r)_i.$$

Therefore $\theta^*(\boldsymbol{\beta}_r)_i$ is negative and is ineligible for consideration in (7-66). On the other hand, for all other unit vectors $\mathbf{e}_i = \mathbf{b}_j \neq \mathbf{b}_r$, we have

$$0 = (\mathbf{B}^{-1}\mathbf{B})_{rj} = \boldsymbol{\beta}_r^T\mathbf{b}_j = \boldsymbol{\beta}_r^T\mathbf{e}_i = (\boldsymbol{\beta}_r)_i,$$

so $\theta^*(\boldsymbol{\beta}_r)_i = 0$ is again ineligible in (7-66). We have thus established that under normal circumstances, i.e., when $\mathbf{e}_i$ is in $\mathbf{B}$ for every $y_i = 0$, the minimizing quotient in (7-66) will have a positive numerator. It then follows that the value of t_0 will be positive, which implies that at least some nonzero displacement along the line from $\mathbf{y}$ to $\mathbf{y} - \theta^*\boldsymbol{\beta}_r$ will be possible. Whether or not the full distance can be covered will depend on whether or not t_0 is greater than 1.

Subcase A1. If $t_0 > 1$, it follows that $y_i \geqq \theta^*(\boldsymbol{\beta}_r)_i$ for all i; thus $\mathbf{y} - \theta^*\boldsymbol{\beta}_r \geqq \mathbf{0}$, and we can proceed all the way to the minimizing point $\theta = \theta^*$. The new column that replaces $\mathbf{b}_r$ in $\mathbf{B}$ is then taken to be $\mathbf{Q}\boldsymbol{\beta}_r$; we know that the resulting set of columns $\mathbf{b}_1, \ldots, \mathbf{b}_{r-1}, \mathbf{Q}\boldsymbol{\beta}_r, \mathbf{b}_{r+1}, \ldots, \mathbf{b}_m$ forms a legitimate basis because $\boldsymbol{\beta}_r^T\mathbf{Q}\boldsymbol{\beta}_r$, the rth component of $\mathbf{B}^{-1}(\mathbf{Q}\boldsymbol{\beta}_r)$, is positive.[4] The columns of the Lemke inverse $\mathbf{B}^{-1}$ are updated in exactly the same way as are the tableau columns (or the basis inverse) in the standard (or revised) simplex method. Note that the new entrant $\mathbf{Q}\boldsymbol{\beta}_r$ will be a free column of $\mathbf{B}$—that is, r will not be in J. Our choice of the vector $\mathbf{Q}\boldsymbol{\beta}_r$ to replace $\mathbf{b}_r$ not only ensures that the new set of columns will again form a basis but also guarantees that the algorithm will converge to a solution within a finite number of iterations. The convergence proof is quite difficult, however, and will not be given here.

Subcase A2. If $t_0 \leqq 1$, the straight-line path from $\mathbf{y}$ to $\mathbf{y} - \theta^*\boldsymbol{\beta}_r$ must cross through or at least arrive at one of the bounding hyperplanes of the feasible region. More specifically, let y_k be the component of $\mathbf{y}$ that determines t_0 in (7-66); that is,

$$t_0 = \frac{y_k}{\theta^*(\boldsymbol{\beta}_r)_k} \leqq 1. \tag{7-67}$$

Then if $\theta = \theta^*$ were used the new value of the kth variable would be

$$\hat{y}_k = y_k - \theta^*(\boldsymbol{\beta}_r)_k \leqq 0.$$

Unless $t_0 = 1$ this value is negative, and it is clear that in order to decrease $f(\hat{\mathbf{y}})$ as much as possible while preserving feasibility we should proceed only until the boundary $y_k = 0$ is just reached. This occurs at

$$\theta = \frac{y_k}{(\boldsymbol{\beta}_r)_k} = \theta^* t_0,$$

[4]Here we are using a theorem from linear algebra that may be stated as follows. Let the m-component vectors $\mathbf{b}_1, \mathbf{b}_2, \ldots, \mathbf{b}_m$ constitute a basis and suppose that

$$\mathbf{a} = \mu_1\mathbf{b}_1 + \mu_2\mathbf{b}_2 + \cdots + \mu_m\mathbf{b}_m \equiv \mathbf{B}\boldsymbol{\mu},$$

where the μ_i are scalar coefficients. If any vector $\mathbf{b}_r$ for which μ_r, the rth component of $\mathbf{B}^{-1}\mathbf{a}$, is nonzero is removed from $\mathbf{B}$ and replaced by $\mathbf{a}$, then the resulting set of vectors also constitutes a basis. The proof may be found in [42] or [20].

at which point the new values of the variables are given by

$$\hat{\mathbf{y}} = \mathbf{y} - \theta^* t_0 \boldsymbol{\beta}_r.$$

Inasmuch as $\hat{y}_k = 0$, it follows from our original definition of $\mathbf{B}$ that the new column replacing $\mathbf{b}_r$ in $\mathbf{B}$ should be $\mathbf{e}_k$, which will of course be a restricted column (our stipulation that $\mathbf{B}$ will normally include a column $\mathbf{e}_k$ for every $y_k = 0$ is precisely the reason the outcome $t_0 = 1$ was placed in subcase A2 rather than A1). Notice as before that because the rth component of $\mathbf{B}^{-1}\mathbf{e}_k$ is $\boldsymbol{\beta}_r^T\mathbf{e}_k = (\boldsymbol{\beta}_r)_k \neq 0$, the new set of columns of $\mathbf{B}$ will form a legitimate basis. Thus $\mathbf{B}$ can again be updated via the usual simplex transformation formulas.

The reader may already have observed that a sort of degeneracy can arise when a tie occurs in the determination of t_0 via (7-66). This creates a situation in which the values of two variables fall simultaneously to zero, but only one of their associated unit vectors can enter $\mathbf{B}$ as a restricted column. Because we shall then have a variable $y_k = 0$ for which $\mathbf{e}_k$ is *not* a column of $\mathbf{B}$, it may happen in some future iteration that the value of t_0 computed via (7-66) will be zero. In such a case the new point $\mathbf{y} - \theta^* t_0 \boldsymbol{\beta}_r$ generated by the iteration will be the same as the old, giving rise to the theoretical possibility of infinite cycling. Fortunately, Lemke was able to prove that a given point $\mathbf{y}$ can persist through only a finite number of iterations, after which $f(\mathbf{y})$ must decrease; finite convergence of the algorithm is therefore assured.

Case B. $\boldsymbol{\beta}_r^T\mathbf{Q}\boldsymbol{\beta}_r$ *equals zero.* In this case (7-64) reduces to

$$f(\hat{\mathbf{y}}) \equiv f(\mathbf{y} - \theta\boldsymbol{\beta}_r) = f(\mathbf{y}) - \theta w_r,$$

and the new point $\hat{\mathbf{y}}$ is obviously obtained by allowing θ to increase or decrease, depending on the sign of w_r, until a boundary of the feasible region is encountered. Thus if $w_r > 0$, so that a positive value of θ is desired, we compute

$$\theta = \min_i \left\{ \frac{y_i}{(\boldsymbol{\beta}_r)_i}, \ (\boldsymbol{\beta}_r)_i > 0 \right\}, \tag{7-68}$$

whereas if $w_r < 0$, we compute

$$\theta = \max_i \left\{ \frac{y_i}{(\boldsymbol{\beta}_r)_i}, \ (\boldsymbol{\beta}_r)_i < 0 \right\}. \tag{7-69}$$

In either case let y_k be the component of $\mathbf{y}$ that determines θ:

$$\theta = \frac{y_k}{(\boldsymbol{\beta}_r)_k}. \tag{7-70}$$

Then $\hat{y}_k = y_k - \theta(\boldsymbol{\beta}_r)_k = 0$, and the vector $\mathbf{e}_k$ is chosen to replace $\mathbf{b}_r$ in $\mathbf{B}$. As in subcase A2, a tie can occur in computing the value of θ via (7-68) or (7-69). When this happens degeneracy is again introduced and with it the possibility that in some future iteration a value of zero may be computed for t_0 or θ.

Another and more important possibility is that none of the components $(\boldsymbol{\beta}_r)_i$ may have the proper sign required to make it eligible for consideration in (7-68) or (7-69). In this event we can proceed infinitely far along the line $\hat{\mathbf{y}} = \mathbf{y} - \theta\boldsymbol{\beta}_r$ without ever encountering a boundary of the feasible region, so that $f(\mathbf{y})$ can be made to approach $-\infty$. Inasmuch as the objective function $f(\mathbf{y}) = \mathbf{p}^T\mathbf{y} + \frac{1}{2}\mathbf{y}^T\mathbf{Q}\mathbf{y}$ is convex, it must then have no local minimum within the feasible region $\mathbf{y} \geqq \mathbf{0}$. This implies that there can exist no set of values $(\mathbf{y}^*, \boldsymbol{\lambda}^*)$ satisfying the conditions (7-57) and therefore no locally optimal solution to the original quadratic program (7-44). But it has been established that (7-44) cannot have an unbounded optimal solution, so we must conclude that *when $f(\mathbf{y})$ is unbounded the original problem has no feasible solution at all.*

Regardless of whether subcase A1 or A2 or case B prevails, unless unboundedness is discovered a new column is chosen to replace $\mathbf{b}_r$ in $\mathbf{B}$, and the iteration ends with the inverse $\mathbf{B}^{-1}$ being updated via the usual simplex transformation formula. To begin the next iteration, we use the new $\mathbf{B}^{-1}$ to recompute $\mathbf{w} = \mathbf{B}^{-1}(\mathbf{p} + \mathbf{Q}\mathbf{y})$ and then apply the test of Lemma 7.1 to determine whether or not optimality has been achieved. The process continues until the optimal solution $\mathbf{y}^*$ is obtained, after which $\mathbf{x}^* = \mathbf{D}^{-1}(\mathbf{c} - \mathbf{A}^T\mathbf{y}^*)$, the optimal solution to the original quadratic program, can be computed.

7.9 AN EXAMPLE

To illustrate Lemke's algorithm, let us apply it to the quadratic program already solved via Beale's method in Section 7.7:

$$\begin{aligned}
\text{Max } z &= 18x_1 - x_1^2 + x_1x_2 - x_2^2 \\
&= 18x_1 - \tfrac{1}{2}(2x_1^2 - 2x_1x_2 + 2x_2^2) \\
\text{subject to} \quad & x_1 + x_2 \leqq 12, \\
& -x_1 + x_2 \leqq 6, \\
& -x_1 \qquad \leqq 0, \\
\text{and} \quad & \qquad -x_2 \leqq 0.
\end{aligned}$$

The data for this problem include

$$\mathbf{c} = \begin{bmatrix} 18 \\ 0 \end{bmatrix}, \quad \mathbf{D} = \begin{bmatrix} 2 & -1 \\ -1 & 2 \end{bmatrix}, \quad \mathbf{A} = \begin{bmatrix} 1 & 1 \\ -1 & 1 \\ -1 & 0 \\ 0 & -1 \end{bmatrix}, \quad \text{and} \quad \mathbf{b} = \begin{bmatrix} 12 \\ 6 \\ 0 \\ 0 \end{bmatrix},$$

and from it we can compute

$$\mathbf{D}^{-1} = \begin{bmatrix} 2/3 & 1/3 \\ 1/3 & 2/3 \end{bmatrix}, \qquad \mathbf{p} = \mathbf{b} - \mathbf{A}\mathbf{D}^{-1}\mathbf{c} = (-6, 12, 12, 6),$$

and

$$\mathbf{Q} = \mathbf{A}\mathbf{D}^{-1}\mathbf{A}^T = \begin{bmatrix} 2 & 0 & -1 & -1 \\ 0 & 2/3 & 1/3 & -1/3 \\ -1 & 1/3 & 2/3 & 1/3 \\ -1 & -1/3 & 1/3 & 2/3 \end{bmatrix}.$$

We then wish to solve the following problem via Lemke's algorithm:

$$\text{(7-58)} \qquad \begin{cases} \text{Min } f(\mathbf{y}) = \mathbf{p}^T\mathbf{y} + \tfrac{1}{2}\mathbf{y}^T\mathbf{Q}\mathbf{y} \\ \quad \text{subject to } \mathbf{y} \geqq \mathbf{0}, \end{cases}$$

beginning at $\mathbf{y} = \mathbf{0}$, $f(\mathbf{y}) = 0$. Note that the associated solution to the original quadratic program is the unconstrained maximum

$$\mathbf{x} = \mathbf{D}^{-1}(\mathbf{c} - \mathbf{A}^T\mathbf{y}) = \mathbf{D}^{-1}\mathbf{c} = (12, 6),$$

which is not a feasible point.

Initially the Lemke matrix and inverse are $\mathbf{B} = \mathbf{B}^{-1} = \mathbf{I}_4$, and all four columns $\mathbf{b}_1, \ldots, \mathbf{b}_4$ are restricted. To test the current solution $\mathbf{y} = \mathbf{0}$ for optimality, we compute

$$\mathbf{w} = \mathbf{B}^{-1}(\mathbf{p} + \mathbf{Q}\mathbf{y}) = \mathbf{p} = (-6, 12, 12, 6).$$

Inasmuch as all the subscripts $1, \ldots, 4$ are in J, the only component of $\mathbf{w}$ that violates the optimality condition is $w_1 = -6$; therefore the column $\mathbf{b}_r = \mathbf{b}_1$ will be removed from the matrix $\mathbf{B}$. The rth row of $\mathbf{B}^{-1}$ is $\boldsymbol{\beta}_r^T = \boldsymbol{\beta}_1^T = [1, 0, 0, 0]$, so we wish to consider points on the line

$$\hat{\mathbf{y}} = \mathbf{y} - \theta\boldsymbol{\beta}_r = (-\theta, 0, 0, 0).$$

We find that $\boldsymbol{\beta}_r^T\mathbf{Q}\boldsymbol{\beta}_r = 2$, which is positive, so $f(\hat{\mathbf{y}}) = f(\mathbf{y} - \theta\boldsymbol{\beta}_r)$ is minimized at

$$\theta^* = \frac{w_r}{\boldsymbol{\beta}_r^T\mathbf{Q}\boldsymbol{\beta}_r} = \frac{-6}{2} = -3.$$

To determine whether it is possible to proceed from $\mathbf{y}$ all the way to $\mathbf{y} - \theta^*\boldsymbol{\beta}_r$, we attempt to compute

$$t_0 = \min_i \left\{ \frac{y_i}{\theta^*(\boldsymbol{\beta}_r)_i}, \ \theta^*(\boldsymbol{\beta}_r)_i > 0 \right\}.$$

We find, however, that $\theta^*(\boldsymbol{\beta}_r)_i$ is nonpositive for all i; therefore, in effect, $t_0 = \infty$, which means that we can proceed as far as we like in the direction $-\theta^*\boldsymbol{\beta}_r = (3, 0, 0, 0)$ without ever encountering a boundary of the feasible region. We are thus within the domain of subcase A1, and we have

$$\theta = \theta^* = -3,$$
$$\hat{\mathbf{y}} = \mathbf{y} - \theta^*\boldsymbol{\beta}_r = (3, 0, 0, 0),$$

and

$$f(\hat{\mathbf{y}}) = \mathbf{p}^T\hat{\mathbf{y}} + \tfrac{1}{2}\hat{\mathbf{y}}^T\mathbf{Q}\hat{\mathbf{y}} = -9.$$

The associated solution to the original problem is now

$$\mathbf{x} = \mathbf{D}^{-1}(\mathbf{c} - \mathbf{A}^T\hat{\mathbf{y}}) = (9, 3), \qquad \text{with } z = 99.$$

Because the new point $\mathbf{y}$ was generated by stopping at a "directional minimum" of $f(\mathbf{y})$, rather than by running into a bounding hyperplane $y_k = 0$, the new vector chosen to replace $\mathbf{b}_r = \mathbf{b}_1$ in $\mathbf{B}$ is the free column $\mathbf{Q}\boldsymbol{\beta}_r = \mathbf{Q}\boldsymbol{\beta}_1 = (2, 0, -1, -1)$. To update $\mathbf{B}^{-1}$, we compute $\boldsymbol{\Phi} = (-1/2, 0, 1/2, 1/2)$, which leads to the new matrix inverse

$$\mathbf{B}^{-1} = \begin{bmatrix} 1/2 & 0 & 0 & 0 \\ 0 & 1 & 0 & 0 \\ 1/2 & 0 & 1 & 0 \\ 1/2 & 0 & 0 & 1 \end{bmatrix}.$$

Note that the subscripts 2, 3, and 4 are now in J but that 1 is not.

We begin the second iteration by computing

$$\mathbf{w} = \mathbf{B}^{-1}(\mathbf{p} + \mathbf{Q}\mathbf{y}) = (0, 12, 9, 3).$$

This time $\mathbf{w}$ is nonnegative and $w_j = 0$ for all j not in J, so the current solutions $\mathbf{y} = (3, 0, 0, 0)$ and $\mathbf{x} = (9, 3)$ must be optimal for their respective problems. We have thus solved the given quadratic program in a single iteration of Lemke's algorithm, whereas it may be recalled that Beale's method required three iterations.

7.10 COMPUTATIONAL CONSIDERATIONS IN QUADRATIC PROGRAMMING

For any given quadratic programming problem the major factor limiting the choice among solution methods is the character of $\mathbf{D}$ in the objective function

$$\text{(7-71)} \qquad \text{Max } z = \mathbf{c}^T\mathbf{x} + \mathbf{x}^T\mathbf{D}\mathbf{x}$$

(this objective can be represented equivalently as

$$\text{Max } z = \mathbf{c}^T\mathbf{x} - \tfrac{1}{2}\mathbf{x}^T\hat{\mathbf{D}}\mathbf{x},$$

where $\hat{\mathbf{D}} = -2\mathbf{D}$ has the opposite character of **D**). If **D** is or might be indefinite or positive (semi)definite, convergence cannot be guaranteed for either Wolfe's or Lemke's algorithm, and it becomes necessary to use some type of constrained hill-climbing approach to find the various local optima that may exist. This strategy is embodied in the quadratic programming algorithm of Beale, as well as in many other more general methods that can be applied to all linearly constrained problems, regardless of the nature of their objective functions. Methods of the latter type, which will be studied in Chapter 8, tend as a group to be somewhat less efficient in solving quadratic programs than the algorithms discussed in this chapter, but—perhaps surprisingly—some of them appear to be quite competitive.[5] The reader may also wish to take note that Eaves [12] succeeded in developing an extended version of Lemke's algorithm that can in theory be applied to any quadratic program, no matter what the character of **D** may be. The Eaves approach, however, does not appear to promise great increases in efficiency and is thus primarily of theoretical interest.

When **D** is negative *semi*definite in (7-71), the quadratic programming problem can also be solved by the algorithm of Wolfe (as well as by the hill-climbing methods of Beale and others). It will be recalled from Section 7.4 that, barring degeneracy, the finite convergence of Wolfe's algorithm to a virtually optimal solution can be guaranteed by perturbing the diagonal elements of **D** very slightly, so that **D** becomes negative definite. Moreover, finite termination of the algorithm is assured even without the perturbation, although there is then some chance that the final set of values obtained will not constitute an optimal solution.

[5]It is difficult to substantiate statements about the relative efficiencies of nonlinear programming algorithms, because very little rigorous testing of them has been conducted. The major difficulty in this regard is that the performance of a nonlinear programming algorithm inevitably varies greatly from one problem to the next, even when the test problems are structurally very similar. Therefore, to develop a reasonable estimate of the solution time of, say, Lemke's algorithm for quadratic programs having certain dimensions and certain densities of nonzero elements in the matrices **A** and **D**, it would be necessary to solve a rather large number of carefully constructed problems. The next steps would presumably be to repeat the process for problems of slightly different size or densities, and to continue in this fashion until a rule of thumb for predicting solution time could be formulated. It is not really surprising that there has been little enthusiasm for such research: It is tedious, requires many expensive hours of computer time, and might not even be particularly helpful to an operations research analyst confronted by a specific, possibly atypical problem. Moreover—and perhaps conclusively—the evaluation of currently existing algorithms has always been, from the academic point of view, a less virtuous pursuit than the inventing of new ones.

Finally, when **D** is negative *definite*, the inverse $\mathbf{D}^{-1}$ is known to exist, and it also becomes possible to apply such methods as that of Lemke. Another algorithm that makes use of $\mathbf{D}^{-1}$ in its computations was published by Theil and van de Panne [44] in 1960. Their method begins with the unconstrained optimum

$$\mathbf{x}_0 = \mathbf{D}^{-1}\mathbf{c}, \tag{7-45}$$

which will in general fail to satisfy one or more of the constraints. A sequence of subproblems is then solved in which the objective function is maximized subject to each of the violated constraints in turn, then to each pair of violated constraints, and so on, until a solution $\mathbf{x}^*$ is found that satisfies every constraint. Under certain conditions $\mathbf{x}^*$ will be the optimal solution to the original problem. The constraints in each subproblem are assumed to hold as equalities, so at each stage the method of Lagrange multipliers can be used. The overall procedure depends for its efficiency on being able to arrive at the optimal solution without being forced to deal with subproblems having large numbers of constraints. Thus it enjoys its greatest advantage, relative to other algorithms, when very few of the constraints are binding at the optimum.

In choosing among the various quadratic programming algorithms when several may be used, there are a number of different considerations to bear in mind. For example, suppose an analyst wishes to solve a quadratic programming problem with a concave objective function as quickly or as cheaply as possible, and suppose further that he has no computer-coded algorithm readily available, so that he is forced to modify an existing linear programming code. In such a case his primary consideration will be that of expedience, and it is clear that he should plan to use Wolfe's algorithm: Conversion from the simplex to the "Wolfe-simplex" method requires little more than modifying the logic by which the entering variable is chosen, so that the complementary slackness conditions $x_j v_j = 0, j = 1, \ldots, n$, are preserved. This may be a decisive advantage for users with limited budgets, because proprietary computer codes of quadratic programming algorithms are less generally available and rather more expensive than LP codes.

Apart from these considerations, Wolfe's algorithm appears to perform at least as well as any other when applied over a broad range of quadratic programming problems. Once a basic feasible solution to the constraints $\mathbf{Ax} = \mathbf{b}$ has been found, the time required to solve a quadratic program having m constraints and n variables is, on the average, not much greater than the solution time of the simplex method for a linear program with $m + n$ constraints.

For certain specific and identifiable groups of quadratic programs, however, other algorithms can be expected to be somewhat more efficient than that of Wolfe. In particular, for *weakly quadratic* problems—problems in which only a few of the variables appear in the quadratic portion of the objective function—the method of Beale is likely to yield a better performance. It is true that in using Beale's algorithm we must compute $\mathbf{p}$ and $\mathbf{Q}$ in

$$z = z_0 + \mathbf{p} \cdot \mathbf{x_R} + \mathbf{x_R^T Q x_R} \tag{7-33}$$

before updating the tableau at each iteration. However, Beale requires a considerably smaller tableau than Wolfe: The number of rows in the Beale tableau ranges between m and n, whereas the Wolfe tableau has $m + n$ rows throughout. Thus, on an average basis, the Beale tableau has only about half as many rows as Wolfe; similar considerations show that it also has roughly half as many columns. This advantage of the Beale algorithm is increased in the case of a weakly quadratic problem, because then most of the pivots will carry all the way to a boundary of the feasible region (why?), generating fewer free variables and keeping the tableau smaller. Another factor tending to favor Beale for weakly quadratic problems is that when most of the elements of

$$\mathbf{D} = \begin{bmatrix} \mathbf{D}_{11} & \mathbf{D}_{12} \\ \mathbf{D}_{21} & \mathbf{D}_{22} \end{bmatrix}$$

are zero—that is, when very few variables have quadratic terms—it becomes relatively easy to recompute $\mathbf{p}$ and $\mathbf{Q}$ directly from their definitions (7-32) at each iteration.

In contrast to Beale's method, the Lemke and Theil algorithms are at their best when very few constraints are binding on the optimal solution, that is, when the optimum is "internal" with respect to most of the constraints. In the extreme case where the unconstrained maximum $\mathbf{x}_0 = \mathbf{D}^{-1}\mathbf{c}$ lies inside the feasible region and is therefore itself the optimal solution, the quadratic program is solved by Lemke or Theil in one iteration. By extension, whenever the analyst knows or suspects that only a few constraints lie between $\mathbf{x}_0$ and the optimal solution he is seeking, he should prefer to use one of these procedures.

It should be noted that the Lemke algorithm has the advantage, with respect to Wolfe and Beale, of not requiring the precomputation of a starting basic feasible solution to $\mathbf{Ax} = \mathbf{b}$. On the other hand, it also has the marginal disadvantage of not operating within the feasible region: If computation must be interrupted for any reason before the problem has been completely solved, no feasible solution whatever will have been generated by the algorithm. This disadvantage can be described as "marginal" because in almost every real problem-solving environment the computer is allowed to run until the optimal solution is obtained, or until the analyst can conclude with reasonable confidence that convergence will not occur. In view of all the costs incurred in planning and performing an operations research analysis, it would be most unusual for computation time to be so expensive that the supervisor or decision maker would settle for a suboptimal feasible solution in preference to allowing the computer to run for a few more hours.

EXERCISES

Section 7.2

7-1. It was noted in the text that mathematical programs with nonlinear objective functions and linear constraints are appreciably more sophisticated theoretically and more difficult to solve than linear programs. Would problems having both nonlinear objectives and nonlinear constraints similarly be expected, as a class, to be substantially more difficult than problems with linear objectives and nonlinear constraints? Why or why not?

7-2. Consider the parametric linear programming problem

$$\text{Max } z = \theta \mathbf{c}^T\mathbf{x} + \mathbf{d}^T\mathbf{x}$$
$$\text{subject to } \mathbf{Ax} = \mathbf{b} \text{ and } \mathbf{x} \geqq \mathbf{0},$$

where $\mathbf{c}$ and $\mathbf{d}$ are nonvanishing n-component vectors, with $\mathbf{c} \neq \mathbf{d}$, and θ is a nonnegative scalar parameter. In general, θ may have several discrete critical values $\theta_j, j = 1, 2, \ldots,$ such that as θ is increased from just below any θ_j to just above it the optimal solution to the linear program changes discontinuously from one extreme point to another (with the two extreme points being co-optimal when $\theta = \theta_j$ exactly). Does this sort of behavior *characteristically* occur as the parameter θ is varied in the quadratic program

$$\text{Max } z = \theta \mathbf{c}^T\mathbf{x} + \mathbf{x}^T\mathbf{D}\mathbf{x}$$
$$\text{subject to } \mathbf{Ax} = \mathbf{b} \text{ and } \mathbf{x} \geqq \mathbf{0}$$

where $\mathbf{c} \neq \mathbf{0}$ and $\mathbf{D} \neq \mathbf{0}$? Under what *special* set of circumstances might the optimal solution to this quadratic program change discontinuously as θ is varied? Illustrate with an actual numerical example.

Exercises 7-3 and 7-4: Determine whether or not each of the following problems can be modeled as a quadratic program.

7-3. Consider a factory producing n different items that is located in a country having a centrally planned economy. The per-unit price at which items of type i can be sold is

$$p_i = r_i - s_{i1}x_1 - s_{i2}x_2 - \cdots - s_{in}x_n, \qquad i = 1, \ldots, n,$$

where the variable x_j is the number of type j items produced per year, and r_i and the s_{ij} are real constants. Price relationships of the above form might obtain if the various items were similar and could to some extent be substituted for each other. The production levels x_i for the various items must satisfy a set of resource constraints that can be represented in matrix form as

$$\mathbf{Ax} \leqq \mathbf{b}, \qquad \mathbf{x} \geqq \mathbf{0}.$$

Each year the manager of the factory is paid a government bonus equal to the total sales revenue (number of items sold times price per item) of that item that generates the *least* amount of revenue. How many items of each type should the factory manager plan to produce during the year in order to maximize his bonus?

7-4. Suppose an economist theorizes that each year the American public allocates its total consumption expenditure C in the following manner:

1. A fixed minimum expenditure of x_j dollars is first budgeted for goods and services in the jth class, $j = 1, \ldots, n$;
2. Then, in addition, a fixed fraction β_j of the remaining uncommitted funds

$$C - \sum_{j=1}^{n} x_j$$

is allocated to the jth class, where

$$\sum_{j=1}^{n} \beta_j = 1 \quad \text{and} \quad \beta_j \geqq 0, j = 1, \ldots, n.$$

The economist has detailed data available from which he can compute the total consumption expenditure in each class in each of m different years. How can he use this data to derive least-squares estimates of the unknown parameters x_j and β_j?

Section 7.5

7-5. If the starting basic feasible solution to the quadratic program

$$\text{Max } z = \mathbf{c}^T\mathbf{x} + \mathbf{x}^T\mathbf{D}\mathbf{x}$$
$$\text{subject to } \mathbf{Ax} = \mathbf{b} \text{ and } \mathbf{x} \geqq \mathbf{0},$$

with $\mathbf{D}$ negative definite, is also the optimal solution, will Wolfe's algorithm in general terminate immediately at the first iteration? If not, what are the maximum and minimum numbers of iterations that might be required, assuming the dimensions of $\mathbf{A}$ are m-by-n?

7-6. Prove or disprove: The unconstrained optimization problem

$$\text{Max } z = \mathbf{c}^T\mathbf{x} + \mathbf{x}^T\mathbf{D}\mathbf{x},$$

where all components of $\mathbf{c}$ and $\mathbf{D}$ are nonzero, must have an unbounded optimal solution if $\mathbf{D}$ is negative semidefinite.

7-7. Use Wolfe's algorithm to minimize or maximize each of the objective functions below, subject to the following set of constraints:

$$x_1 + 2x_2 \leqq 7,$$
$$-x_1 + 2x_2 \leqq 4,$$

and

$$x_1, x_2 \geqq 0.$$

In each case take the origin as the initial basic solution to the constraints.
(a) Max $z = x_1 + 6x_2 - 2x_1^2 + 2x_1x_2 - x_2^2$.
(b) Min $z = (x_1 - 1)^2 + (x_2 - 1)^2$.
(c) Max $z = 4x_1 + 9x_2 - 3x_1^2 + 2x_1x_2 - 2x_2^2$.
(d) Max $z = 2x_1 - 2x_2 - x_1^2 + 2x_1x_2 - x_2^2$.

7-8. It was noted in Section 7.4 that the following quadratic program has an unbounded optimal solution:

$$\text{Max } z = x_1 + x_2 - (2x_1 - x_2)^2$$
$$\text{subject to } x_1 - x_2 \leqq 1$$
$$\text{and} \quad x_1, x_2 \geqq 0.$$

Apply Wolfe's algorithm to this problem, starting with the basic feasible solution $\mathbf{x_B} = x_1 = 1$, and comment on the result. What would have happened if we had initially perturbed the quadratic form to make it negative definite, as described in the text?

7-9. Describe in detail how a computer code for solving linear programs via the standard (primal) simplex method must be modified in order to convert it to a "Wolfe-simplex" code.

7-10. Suppose we wish to apply Wolfe's algorithm directly to problems of the following form:

$$\text{Min } z = \mathbf{c}^T\mathbf{x} + \mathbf{x}^T\mathbf{D}\mathbf{x}$$
$$\text{subject to } \mathbf{A}\mathbf{x} = \mathbf{b} \text{ and } \mathbf{x} \geqq \mathbf{0},$$

where $\mathbf{D}$ is positive definite. Write the Kuhn-Tucker conditions for this problem and derive the linear program that must be solved via Wolfe-simplex pivoting.

7-11. Repeat Exercise 7-10 for the quadratic program

$$\text{Max } z = \mathbf{c}^T\mathbf{x} + \mathbf{x}^T\mathbf{D}\mathbf{x}$$
$$\text{subject to } \mathbf{A}\mathbf{x} = \mathbf{b},$$

where $\mathbf{D}$ is negative definite.

7-12. (a) Using Theorem 3.9, prove that if $\mathbf{x}_0$ and $\hat{\mathbf{x}}$ are any two feasible solutions to the quadratic program

$$\text{Max } f(\mathbf{x}) = \mathbf{c}^T\mathbf{x} + \mathbf{x}^T\mathbf{D}\mathbf{x}$$
$$\text{subject to } \mathbf{A}\mathbf{x} = \mathbf{b} \text{ and } \mathbf{x} \geqq \mathbf{0},$$

where $\mathbf{D}$ is negative definite or semidefinite, then

$$f(\hat{\mathbf{x}}) \leqq \mathbf{c}^T\hat{\mathbf{x}} + 2\mathbf{x}_0^T\mathbf{D}\hat{\mathbf{x}} - \mathbf{x}_0^T\mathbf{D}\mathbf{x}_0.$$

(b) Use the result of part (a) to prove that if there exists a set of values $(\mathbf{x}_0, \boldsymbol{\lambda}_0, \mathbf{v}_0)$ satisfying

$$\mathbf{A}\mathbf{x} = \mathbf{b},$$
$$2\mathbf{D}\mathbf{x} - \mathbf{A}^T\boldsymbol{\lambda} + \mathbf{v} = -\mathbf{c},$$
$$\text{and } \mathbf{x} \geqq \mathbf{0}, \mathbf{v} \geqq \mathbf{0}$$

(but not necessarily $x_jv_j = 0, j = 1, \ldots, n$), then the quadratic program given above cannot have an unbounded optimal solution.

Section 7.7

7-13. Use Beale's method to obtain a single local optimum of each of the objective functions below, subject to the following set of constraints:

$$x_1 + x_2 \leqq 12,$$
$$-x_1 + x_2 \leqq 6,$$

and

$$x_1, x_2 \geqq 0.$$

In each case take the origin as the starting basic feasible solution.
(a) Max $z = 6x_1 + 8x_2 - x_1^2 + 2x_1x_2 - 2x_2^2$.
(b) Max $z = x_2 + x_1^2 + x_1x_2 + 2x_2^2$.
(c) Max $z = (x_1 - 4)^2 + (x_2 - 2)^2$.
(d) Min $z = (x_1 - 4)^2 + (x_2 - 2)^2$.
(e) Max $z = -2x_1 + 16x_2 - x_1^2 + x_1x_2 - x_2^2$.

7-14. Continue with the application of Beale's method to the example problem at the end of Section 7.7 until a local optimal solution is obtained.

7-15. Solve the following quadratic program via the method of Beale:

$$\text{Max } z = x_1 - x_2 - x_1^2 + 4x_1x_2 - 4x_2^2$$
$$\text{subject to } \quad x_1 - 3x_2 \leqq 3,$$
$$-x_1 + x_2 \leqq 1,$$
$$\text{and } \quad x_1, x_2 \geqq 0.$$

7-16. Can a restricted variable x_j that enters the basis at one iteration of Beale's method be removed at the next? Can a restricted variable that is removed from the basis at one iteration be reinserted at the next?

7-17. How many multiplications and divisions are required to perform a single complete iteration of Beale's method (in which one restricted variable replaces another in the basis) if the original quadratic program had m constraints and n variables and if the number of basic variables is now k, where $m \leqq k \leqq n$? Assume that all elements of $\mathbf{A}$, $\mathbf{b}$, $\mathbf{c}$, and $\mathbf{D}$ are nonzero and that standard (i.e., not revised) simplex pivoting is being used. Comment on how this total *operation count* depends on k and compare it to the operation count for a single standard pivot of the Wolfe-simplex method.

Section 7.9

7-18. Answer the following questions about what may happen during a single iteration of Lemke's algorithm in each of the three (sub)cases A1, A2, and B.
(a) Must $f(\mathbf{y})$ decrease by a (possibly infinite) positive amount?
(b) May unboundedness of $f(\mathbf{y})$ be discovered?
(c) May a finite optimal solution be obtained?

7-19. Use Lemke's algorithm to optimize each of the following objective functions, subject to the constraints of Exercise 7-13:
(a) Max $z = 6x_1 + 8x_2 - x_1^2 + 2x_1x_2 - 2x_2^2$.
(b) Min $z = (x_1 - 4)^2 + (x_2 - 2)^2$.
(c) Max $z = -2x_1 + 16x_2 - x_1^2 + x_1x_2 - x_2^2$.

7-20. Solve the following problem via Lemke's algorithm:

$$\begin{aligned} \text{Max } z &= x_1 + x_2 - x_1^2 - 2x_2^2 \\ \text{subject to } \quad 2x_1 - 2x_2 &\leqq -1, \\ x_1 \qquad &\leqq \ \ 1, \\ \text{and } \ -x_1 + 2x_2 &\leqq \ \ 0. \end{aligned}$$

7-21. Can the optimal solution to the "equivalent" quadratic program

$$\begin{cases} \text{Min } f(\mathbf{y}) = \mathbf{p}^T\mathbf{y} + \frac{1}{2}\mathbf{y}^T\mathbf{Q}\mathbf{y} \\ \quad \text{subject to } \mathbf{y} \geqq \mathbf{0} \end{cases} \tag{7-58}$$

be obtained via Wolfe's algorithm? If so, derive the linear program that would have to be solved.

7-22. Are there any problems of the form

$$\begin{aligned} \text{Max } z &= \mathbf{c}^T\mathbf{x} - \tfrac{1}{2}\mathbf{x}^T\mathbf{D}\mathbf{x} \\ \text{subject to } &\mathbf{A}\mathbf{x} \leqq \mathbf{b}, \end{aligned}$$

where $\mathbf{D}$ is positive *semi*definite, to which Lemke's algorithm can be applied? [*Hint:* First prove that if $\mathbf{x}^T\mathbf{D}\mathbf{x} = 0$, then $\mathbf{D}\mathbf{x} = \mathbf{0}$.]

7-23. Prove or disprove: If a point $\hat{\mathbf{y}} \geqq \mathbf{0}$ generated by the Lemke algorithm is not an optimal solution to (7-58), then $\hat{\mathbf{x}} = \mathbf{D}^{-1}(\mathbf{c} - \mathbf{A}^T\hat{\mathbf{y}})$ cannot be a feasible solution to (7-44).

General

7-24. Another interesting quadratic programming algorithm is that of Frank and Wolfe [16], which is applicable to the problem

$$\text{Max } z = \mathbf{c}^T\mathbf{x} + \mathbf{x}^T\mathbf{D}\mathbf{x}$$
$$\text{subject to } \mathbf{A}\mathbf{x} = \mathbf{b} \text{ and } \mathbf{x} \geqq \mathbf{0},$$

where $\mathbf{D}$ is negative definite or semidefinite. The algorithm begins by using the phase I simplex technique to seek a feasible solution to the following system of linear equations in the unknowns $\mathbf{x}$, $\boldsymbol{\lambda}$, and $\mathbf{v}$:

$$\text{(7-72)} \qquad \begin{cases} \mathbf{A}\mathbf{x} = \mathbf{b}, \\ 2\mathbf{D}\mathbf{x} - \mathbf{A}^T\boldsymbol{\lambda} + \mathbf{v} = -\mathbf{c}, \\ \mathbf{x} \geqq \mathbf{0}, \quad \mathbf{v} \geqq \mathbf{0} \end{cases}$$

but *not* necessarily $\mathbf{v}^T\mathbf{x} = 0$ (the unrestricted variables $\boldsymbol{\lambda}$ are first transformed into nonnegative variables in the usual manner). Assuming a feasible solution is found—call it $(\mathbf{x}_0, \boldsymbol{\lambda}_0, \mathbf{v}_0)$—we then begin to solve the following linear program via the simplex method:

$$\text{Max } z = -\mathbf{v}_0^T\mathbf{x} - \mathbf{x}_0^T\mathbf{v}$$
$$\text{subject to the constraints (7-72).}$$

The notation is simplified if we define a column vector $\boldsymbol{\alpha} \equiv (\mathbf{x}, \mathbf{v})$ and a row vector $\boldsymbol{\beta} \equiv [\mathbf{v}^T, \mathbf{x}^T]$, each having $2n$ components; we also analogously define $\boldsymbol{\alpha}_0 \equiv (\mathbf{x}_0, \mathbf{v}_0)$, and so on. Thus the objective function above becomes

$$\text{Max } z = -\boldsymbol{\beta}_0 \cdot \boldsymbol{\alpha}.$$

Simplex pivoting continues until a solution $\hat{\boldsymbol{\alpha}}$ is found such that (1) $\hat{\boldsymbol{\beta}} \cdot \hat{\boldsymbol{\alpha}} = 0$ or (2) $-\boldsymbol{\beta}_0 \cdot \hat{\boldsymbol{\alpha}} \geqq -\frac{1}{2}\boldsymbol{\beta}_0 \cdot \boldsymbol{\alpha}_0$. In case 1 the algorithm terminates. In case 2 we compute

$$\mu_0 = \min\left\{1, \frac{\boldsymbol{\beta}_0 \cdot [\boldsymbol{\alpha}_0 - \hat{\boldsymbol{\alpha}}]}{[\hat{\boldsymbol{\beta}} - \boldsymbol{\beta}_0] \cdot [\hat{\boldsymbol{\alpha}} - \boldsymbol{\alpha}_0]}\right\},$$
$$\boldsymbol{\alpha}_1 = \mu_0\hat{\boldsymbol{\alpha}} + (1 - \mu_0)\boldsymbol{\alpha}_0,$$

and

$$\boldsymbol{\beta}_1 = \mu_0\hat{\boldsymbol{\beta}} + (1 - \mu_0)\boldsymbol{\beta}_0.$$

This completes the first iteration. In the second iteration we begin solving the linear program

$$\text{Max } z = -\boldsymbol{\beta}_1 \cdot \boldsymbol{\alpha}$$
$$\text{subject to the constraints (7-72).}$$

As before, pivoting continues until a new solution $\hat{\boldsymbol{\alpha}}$ is found such that either $\hat{\boldsymbol{\beta}} \cdot \hat{\boldsymbol{\alpha}} = 0$ or $-\boldsymbol{\beta}_1 \cdot \hat{\boldsymbol{\alpha}} \geqq -\frac{1}{2}\boldsymbol{\beta}_1 \cdot \boldsymbol{\alpha}_1$. In the former case—case 1—the algorithm terminates; in the latter, a new value μ_1 is computed exactly as above (with 1s replacing 0s in the subscripts), leading to $\boldsymbol{\alpha}_2$ and $\boldsymbol{\beta}_2$, and a further iteration follows. The procedure continues until termination in case 1 is achieved.

(a) How are infeasibility and unboundedness of the original quadratic program identified?

(b) Prove that when case 1 occurs an optimal solution to the original problem has been found.

(c) Let $\boldsymbol{\alpha}^* \equiv (\mathbf{x}^*, \mathbf{v}^*)$ denote the optimal solution to the quadratic program [i.e., the feasible solution to (7-72) that also satisfies $\mathbf{v}^T\mathbf{x} = 0$]; given that (7-72) has a feasible solution, we know from Exercise 7-12 that the quadratic program cannot have an unbounded optimum, so $\boldsymbol{\alpha}^*$ must exist. Verify the relations

$$\boldsymbol{\beta}_k \cdot \boldsymbol{\alpha}_k - 2\boldsymbol{\beta}_k \cdot \boldsymbol{\alpha}^* = (\boldsymbol{\beta}_k - \boldsymbol{\beta}^*) \cdot (\boldsymbol{\alpha}_k - \boldsymbol{\alpha}^*) \geqq 0,$$

where $\boldsymbol{\alpha}_k$ and $\boldsymbol{\beta}_k$ are the vectors computed at the end of the kth iteration of Frank and Wolfe's algorithm, and use these relations to prove that at each iteration, barring infinite cycling, case 2 will eventually be attained if case 1 is not.

(d) Noting that case 2 termination implies

$$\boldsymbol{\beta}_k \cdot [\boldsymbol{\alpha}_k - \hat{\boldsymbol{\alpha}}] \geqq \tfrac{1}{2}\boldsymbol{\beta}_k \cdot \boldsymbol{\alpha}_k > 0,$$

prove that $0 < \mu_k \leqq 1$.

(e) Prove that

$$\boldsymbol{\beta}_{k+1} \cdot \boldsymbol{\alpha}_{k+1} \leqq (1 - \tfrac{1}{2}\mu_k)\boldsymbol{\beta}_k \cdot \boldsymbol{\alpha}_k < \boldsymbol{\beta}_k \cdot \boldsymbol{\alpha}_k.$$

(f) Finally, show that the objective value $-\boldsymbol{\beta}_k \cdot \hat{\boldsymbol{\alpha}}$ that obtains when case 2 is reached at the end of the $(k + 1)$th iteration of the algorithm is greater (closer to zero) than the objective value attained at the kth iteration. Is this proof sufficient to support the conclusion that the algorithm must converge to an optimal solution within a finite number of iterations?

7-25. Consider the parametric quadratic programming problem

$$\text{Max } z = \theta\mathbf{c}^T\mathbf{x} + \mathbf{x}^T\mathbf{D}\mathbf{x}$$
$$\text{subject to } \mathbf{A}\mathbf{x} = \mathbf{b} \text{ and } \mathbf{x} \geqq \mathbf{0},$$

where it is desired to obtain the optimal solution for every nonnegative value of the parameter θ. Recall from the end of Section 7.2 that this problem can arise when an investor wishes to generate a family of efficient portfolios having different degrees of risk. Briefly outline two methods for solving the problem, one based on Wolfe's algorithm and one on Beale's. Which of the two seems superior? Compare and contrast each with a parametric *linear* programming procedure that is (more or less) similar to it.

8

Algorithms for Linearly Constrained Problems

8.1 INTRODUCTION

Throughout this chapter we shall be concerned with mathematical programming problems of the following form:

$$\text{(8-1)} \qquad \text{Max } z = f(\mathbf{x})$$

$$\text{(8-2)} \quad \text{subject to } a_{i1}x_1 + a_{i2}x_2 + \cdots + a_{in}x_n \begin{Bmatrix} \leqq \\ = \\ \geqq \end{Bmatrix} b_i, \qquad i = 1, \ldots, m,$$

where $\mathbf{x} = (x_1, \ldots, x_n)$ is an n-component column vector of unknowns, $f(\mathbf{x})$ is a scalar-valued function having continuous first partial derivatives, and the a_{ij} and b_i are real numbers. The linear constraints (8-2) can be taken to include whatever nonnegativity restrictions $x_j \geqq 0$ may be present, although for many of the algorithms to be covered below we shall impose nonnegativity on all variables and then single out those restrictions for special treatment, as is done for the simplex method. As usual, no generality is lost by representing the objective as a maximization.

In Section 7.1 we noted that linearly constrained mathematical programs are easier to deal with than those having general (i.e., nonlinear) constraints, because in the former case the so-called "feasible-directions" strategy can be used much more efficiently. This strategy, which we shall discuss in some detail later on, is a generalization of the direct climbing approach of Chapter 4. It operates by generating a sequence of feasible points $\mathbf{x}_0$, $\mathbf{x}_1$, $\mathbf{x}_2, \ldots$, each having a better objective value than its predecessor, with the sequence eventually converging to a local optimal solution or constrained stationary point. At each stage the new point $\mathbf{x}_{k+1}$ is obtained by selecting or computing a favorable direction of displacement from $\mathbf{x}_k$ and then proceeding in that direction either until the objective value stops improving or until a boundary of the feasible region is encountered. For unconstrained problems this strategy obviously reduces to ordinary direct climbing. One algorithm embodying the feasible-directions approach that we have studied already is the method of Beale.[1]

Almost all the well-known and useful algorithms for solving linearly constrained mathematical programs can be viewed as being of the feasible-directions type. However, we shall begin this chapter with one that is not normally so regarded, namely, the separable programming technique, which is based on a modification of the simplex method. In fact, when the nonlinear objective function $f(\mathbf{x})$ satisfies certain concavity conditions, separable programming reduces completely to linear programming.

8.2 SEPARABLE PROGRAMMING: THE APPROXIMATING PROBLEM

In the next few sections we shall present a simplex-based algorithm for obtaining an approximate solution to any linearly constrained mathematical program whose objective function $f(x_1, \ldots, x_n)$ can be represented as a summation of the form

$$f(x_1, \ldots, x_n) \equiv \sum_{j=1}^{n} f_j(x_j), \tag{8-3}$$

where each of the f_j is a continuous function of a single variable x_j. Functions that can be broken into single-variable components satisfying (8-3) are said to be *separable* or to possess the property of *separability*. For example, any linear function

$$f(x_1, \ldots, x_n) = c_1x_1 + \cdots + c_nx_n$$

[1]It is interesting to note that the simplex method, as applied to linear programs, can also be viewed as a special case of the feasible-directions strategy. The direction of displacement chosen at each iteration is along that edge of the feasible region that offers the greatest rate of improvement in the objective value z. Because a linear objective function has constant first partial derivatives, z can never "stop improving" during any straight-line displacement; it follows that, unless unboundedness is discovered, each pivot must terminate when a new boundary of the feasible region is encountered, i.e., at an adjacent extreme point.

is obviously separable, with each of the component functions being $f_j(x_j) = c_j x_j$. So is any quadratic form that lacks cross-product terms:

$$f(x_1, \ldots, x_n) = c_1 x_1^2 + \cdots + c_n x_n^2 \equiv \sum_{j=1}^{n} c_j x_j^2.$$

Optimization problems with nonlinear separable objective functions are not uncommon in operations research. For example, any objective that calls for minimizing the total variance of a number of *independent* random variables (such as investments) would possess the separability property, as would the objective of minimizing raw-material costs at a plant that orders each raw material from a different supplier. In other cases, objective functions that are not immediately separable can be made so by certain transformations of variables. The simplest and most useful transform is applicable to the "nonseparated" term $x_i x_j$ and requires defining two new variables

$$y_i = \tfrac{1}{2}(x_i + x_j) \quad \text{and} \quad y_j = \tfrac{1}{2}(x_i - x_j); \tag{8-4}$$

it then follows that

$$x_i x_j = y_i^2 - y_j^2,$$

as the reader can easily verify. Thus, wherever the term $x_i x_j$ appears in the objective, it is replaced by the separable expression $y_i^2 - y_j^2$, and the two linear equations (8-4) are added to the set of constraints. Other transformations for achieving separability are rather more complicated and require nonlinear substitutions, leading to nonlinear constraints, which are not within the province of this chapter (but see Chapter 9). Still others that do preserve linearity in the constraints may perhaps be more accurately regarded as "tricky" rather than complicated. For example, the nonseparated term in the objective function

$$\text{Max } z = \sum_{j=1}^{n} f_j(x_j) + \min\{x_1, x_2, \ldots, x_n\}$$

can be replaced by the single new variable

$$y = \min\{x_1, x_2, \ldots, x_n\}$$

provided that the linear inequalities

$$\begin{gathered} y - x_1 \leqq 0, \\ \vdots \\ y - x_n \leqq 0 \end{gathered}$$

are added to the constraint set—the new constraints ensure that y cannot be greater than $\min\{x_1, \ldots, x_n\}$, and the objective of maximizing ensures that it will not be less.

Consider, then, the *separable programming problem*

$$\text{(8-5)} \qquad \begin{cases} \text{Max } z = \sum_{j=1}^{n} f_j(x_j) \\ \text{subject to } \mathbf{Ax} = \mathbf{b} \\ \text{and} \quad \mathbf{x} \geqq \mathbf{0}, \end{cases}$$

where $\mathbf{x} = (x_1, \ldots, x_n)$, $\mathbf{A}$ is m-by-n, $\mathbf{b}$ is m-by-1, and the f_j are continuous scalar-valued functions. The linear constraints have been written in standard form, without loss of generality, because the solution technique to be presented for this problem is based on the simplex method. We begin by approximating each of the functions $f_j(x_j)$ as closely as is desired by a piecewise linear function $\hat{f}_j(x_j)$, as shown in Fig. 8-1. This is accomplished by determining a lower bound L_j and upper bound M_j on the value of x_j; choosing $r_j + 1$ *break points* or values of x_j, denoted $x_{j0}, x_{j1}, \ldots, x_{jr_j}$, where

$$x_{j0} = L_j < x_{j1} < x_{j2} < \cdots < x_{jr_j} = M_j;$$

and computing for each of these values the ordinate

$$f_{jk} = f_j(x_{jk}), \qquad k = 0, 1, \ldots, r_j.$$

The function $\hat{f}_j(x_j)$ is then the piecewise linear curve that is produced by joining the points $(x_{j0}, f_{j0}), (x_{j1}, f_{j1}), \ldots, (x_{jr_j}, f_{jr_j})$ with r_j successive straight-line segments. If the bounds L_j and M_j cannot be established log-

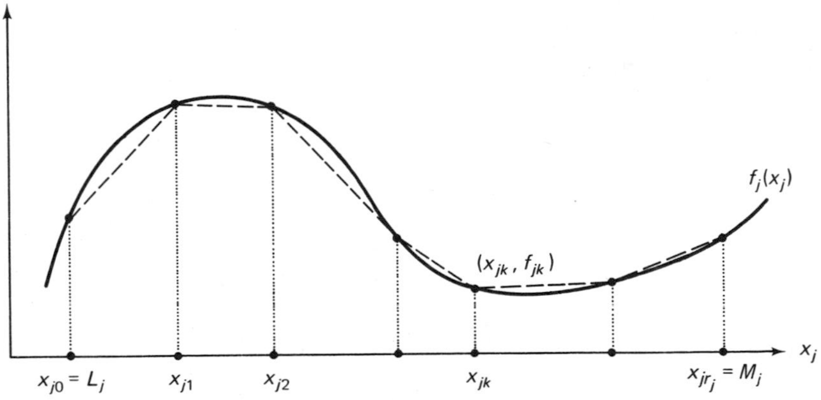

FIGURE 8-1

ically from the analyst's knowledge of the problem, they must be set at "safe" levels, that is, at levels low and high enough to guarantee that the optimal value of x_j will fall between them. Frequently the lower bound L_j will be zero, but sometimes the analyst will be able to assign it a positive value. As we shall see, it will be to his advantage to do so whenever possible: By narrowing the range defined for x_j, he can decrease the number of linear segments in the approximation $\hat{f}_j(x_j)$ and thereby reduce the computational effort required to solve the problem.

To represent the function $\hat{f}_j$ algebraically, we must deal with it on a piecewise basis. Any general point x_j in the interval $x_{jk} \leqq x_j \leqq x_{j,k+1}$ can be expressed as a *unique* convex combination of the two end points:

$$x_j = \lambda_{jk} x_{jk} + \lambda_{j,k+1} x_{j,k+1}, \tag{8-6}$$

where

$$\lambda_{jk} + \lambda_{j,k+1} = 1 \quad \text{and} \quad \lambda_{jk}, \lambda_{j,k+1} \geqq 0. \tag{8-7}$$

The approximated objective value of x_j is then

$$\hat{f}_j(x_j) = \lambda_{jk} f_{jk} + \lambda_{j,k+1} f_{j,k+1}, \tag{8-8}$$

where λ_{jk} and $\lambda_{j,k+1}$ have the same values as in (8-6). Thus, by means of the transformation (8-6), any value of x_j within the given interval can be expressed in terms of the new variables λ_{jk} and $\lambda_{j,k+1}$; moreover, the approximated objective value $\hat{f}_j(x_j)$ as given by (8-8) is a linear function of those variables.

To represent the *entire* piecewise linear function $\hat{f}_j$, however, it is necessary to use logical restrictions in addition to algebraic relations. Any x_j in the entire range $L_j \leqq x_j \leqq M_j$, together with its approximate objective value, can be expressed *uniquely* in terms of the variables $\lambda_{j0}, \lambda_{j1}, \ldots, \lambda_{jr_j}$ as follows:

$$x_j = \lambda_{j0} x_{j0} + \lambda_{j1} x_{j1} + \cdots + \lambda_{jr_j} x_{jr_j} = \sum_{k=0}^{r_j} \lambda_{jk} x_{jk} \tag{8-9}$$

and

$$\hat{f}_j(x_j) = \sum_{k=0}^{r_j} \lambda_{jk} f_{jk}, \tag{8-10}$$

where

$$\sum_{k=0}^{r_j} \lambda_{jk} = 1 \quad \text{and} \quad \lambda_{jk} \geqq 0, \qquad k = 0, 1, \ldots, r_j, \tag{8-11}$$

provided it is also required that

1. At most two of the λ_{jk} can be positive, and
2. If two are positive, they must be adjacent (i.e., if λ_{js} and λ_{jt} are positive, then either $t = s + 1$ or $s = t + 1$).

The logical restrictions serve to define a single linear segment—or a breakpoint if only one λ_{jk} is positive—and the relations (8-9) through (8-11) then reduce to (8-6), (8-8), and (8-7) on that segment.

We are now ready to transform the given separable programming problem (8-5) into an *approximating problem*, which we shall then be able to solve via a simplex approach. The approximating problem is constructed by choosing points that define a piecewise linear approximation for each $f_j(x_j)$ and then making substitutions of the form (8-9) through (8-11) for each variable x_j. The ith constraint

$$a_{i1}x_1 + \cdots + a_{in}x_n = \sum_{j=1}^{n} a_{ij}x_j = b_i$$

becomes

$$\sum_{j=1}^{n} a_{ij} \sum_{k=0}^{r_j} \lambda_{jk}x_{jk} \equiv \sum_{j=1}^{n} \sum_{k=0}^{r_j} a_{ijk}\lambda_{jk} = b_i, \tag{8-12}$$

where $a_{ijk} = a_{ij}x_{jk}$, and the approximating problem is

$$(8\text{-}13)\quad \begin{cases} \text{Max } z = \sum_{j=1}^{n} \hat{f}_j(x_j) \equiv \sum_{j=1}^{n} \sum_{k=0}^{r_j} f_{jk}\lambda_{jk} \\ \text{subject to } \sum_{j=1}^{n} \sum_{k=0}^{r_j} a_{ijk}\lambda_{jk} = b_i, \qquad i = 1, \ldots, m, \\ \qquad\qquad \sum_{k=0}^{r_j} \lambda_{jk} = 1, \qquad j = 1, \ldots, n, \\ \quad \text{and} \qquad \lambda_{jk} \geqq 0 \qquad \text{for all } j \text{ and } k, \\ \textit{and to restrictions } 1 \textit{ and } 2 \textit{ for each } j, \qquad j = 1, \ldots, n. \end{cases}$$

This is identical to the problem

$$(8\text{-}14)\quad \begin{cases} \text{Max } z = \sum_{j=1}^{n} \hat{f}_j(x_j) \\ \text{subject to } \mathbf{Ax} = \mathbf{b} \\ \qquad \text{and } L_j \leqq x_j \leqq M_j, \qquad j = 1, \ldots, n, \end{cases}$$

and *nearly* identical to the original separable program (8-5), assuming that the upper and lower bounds M_j and L_j do not encroach on the feasible region. The closer the approximations $\hat{f}_j$ are to the given functions f_j, the closer (8-13) is to (8-5). Notice that if it were not for restrictions 1 and 2, the problem (8-13) would be a linear program in the variables λ_{jk}, and we could solve it via the simplex method.

It is worth noting that in forming the approximating problem (8-13) there is no need to construct a piecewise linear approximation to any function $f_j(x_j)$ that is already linear. In such cases we can dispense with the transformations (8-9) through (8-11) and simply use the original variable x_j in the formulation of (8-13). This observation is important because problems in which only a few of the variables are involved nonlinearly are not at all uncommon in the practice of operations research.

As an example, let us construct an approximating problem for the following separable program:

$$\begin{aligned} \text{Max } z &= 2\sqrt{x_1} - x_2^2 \\ \text{subject to } x_1 + 2x_2 &\leqq 6, \\ x_1 - x_2 &\leqq 0, \\ \text{and} \qquad x_1, x_2 &\geqq 0. \end{aligned}$$

The lower bound for each variable is evidently zero, and a pair of upper bounds can be determined from consideration of the first constraint (in light of the nonnegativity restrictions): $x_1 \leqq 6$ and $x_2 \leqq 3$. Although it might be possible to establish tighter bounds by examining the constraints more closely, the effort involved in doing so would at least partially offset any saving in computation time that might result. We now plot $f_1(x_1) = 2\sqrt{x_1}$, as shown in Fig. 8-2(a), and form the approximating function $\hat{f}_1(x_1)$

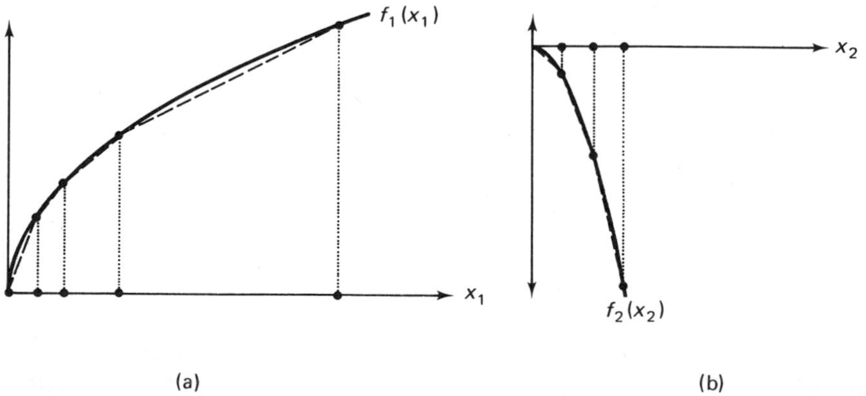

FIGURE 8-2

by connecting the five points listed in Table 8-1; the function $\hat{f}_2(x_2)$ is formed in a similar way. We make no effort to justify the particular linear approximations we have constructed. Ideally, it would be desirable to work with an actual diagram of each $f_j(x_j)$, as we have done here, and to use shorter line segments to approximate the curve more closely in the region where the optimal value of x_j is most likely to lie. In practice, such careful treatment would be much too time-consuming, so the functions $\hat{f}_j(x_j)$ are usually generated by computer, with segment lengths often being in accordance with the second derivative of f_j (the greater the magnitude of d^2f_j/dx_j^2, the faster the slope of the curve is changing, and the shorter and more numerous the linear segments must be in order to achieve a good approximation).

TABLE 8-1

DATA FOR SEPARABLE PROGRAMMING EXAMPLE

Breakpoints of $\hat{f}_1(x_1)$		*Breakpoints of $\hat{f}_2(x_2)$*	
$x_{10} = 0$	$f_{10} = 0$	$x_{20} = 0$	$f_{20} = 0$
$x_{11} = .5$	$f_{11} = 1.414$	$x_{21} = 1.0$	$f_{21} = 1.0$
$x_{12} = 1.0$	$f_{12} = 2.0$	$x_{22} = 2.0$	$f_{22} = -4.0$
$x_{13} = 2.0$	$f_{13} = 2.828$	$x_{23} = 3.0$	$f_{23} = -9.0$
$x_{14} = 6.0$	$f_{14} = 4.899$		

Plugging the data of Table 8-1 into our general formulation (8-13), we arrive at the following approximating problem:

$$\begin{aligned}
\text{Max } z = {} & 1.414\lambda_{11} + 2.0\lambda_{12} + 2.828\lambda_{13} + 4.899\lambda_{14} - 1.0\lambda_{21} - 4.0\lambda_{22} \\
& - 9.0\lambda_{23} \\
\text{subject to } & .5\lambda_{11} + \lambda_{12} + 2\lambda_{13} + 6\lambda_{14} + 2\lambda_{21} + 4\lambda_{22} + 6\lambda_{23} \leqq 6, \\
& .5\lambda_{11} + \lambda_{12} + 2\lambda_{13} + 6\lambda_{14} - \lambda_{21} - 2\lambda_{22} - 3\lambda_{23} \leqq 0, \\
& \lambda_{10} + \lambda_{11} + \lambda_{12} + \lambda_{13} + \lambda_{14} = 1, \\
& \lambda_{20} + \lambda_{21} + \lambda_{22} + \lambda_{23} = 1, \\
& \text{and } \lambda_{jk} \geqq 0 \quad \text{for all } j \text{ and } k,
\end{aligned}$$

where conditions 1 and 2 must also hold for $j = 1$ and $j = 2$. Note that $r_1 = 4$ and $r_2 = 3$ in this example and that the formulation has grown from its original two variables and two constraints to nine variables and four constraints. In general, if a separable programming problem initially has m constraints and n variables, and if $\hat{f}_j(x_j)$ comprises r_j linear segments, $j = 1, \ldots, n$, then the approximating problem has $m + n$ constraints and $\sum_{j=1}^{n} (r_j + 1)$ variables.

8.3 CONVEX SEPARABLE PROGRAMMING

In this section we shall assume that all the original functions $f_j(x_j)$ are *concave*, so that the separable program (8-5) is a convex program, in the sense of Section 6.11: Any local maximum of the objective function within the feasible region is an optimal solution to the problem. Remarkably, it then follows, as we shall see, that the simplex method can be used with no modification whatever to solve the approximating problem—that is, restrictions 1 and 2 will turn out to be satisfied by the optimal solution even though they are not explicitly enforced. To arrive at this conclusion we must prove, first, that the approximating problem is itself a convex program, and, second, that for any given value of

$$x_j = \sum_{k=0}^{r_j} \lambda_{jk} x_{jk}$$

the values of the variables λ_{jk} that maximize the approximated function

$$\hat{f}_j(x_j) = \sum_{k=0}^{r_j} \lambda_{jk} f_{jk}$$

"voluntarily" satisfy restrictions 1 and 2.

For purposes of clarity the two results we need will be set forth in lemma form. We begin with the following:

Lemma 8.1: If each of the functions $f_j(x_j)$ *in the separable program* (8-5) *is concave, then the approximating problem* (8-13), *or its equivalent* (8-14), *is a convex program.*

To prove this lemma we must show that $\hat{f}_j(x_j)$ is a concave function whenever $f_j(x_j)$ is concave. This is easily established by using a geometric argument. Given a concave function f_j, construct any piecewise linear approximation $\hat{f}_j$, as shown in Fig. 8-3, and select any two points $x_j = a$ and $x_j = b$. Then draw a straight line S connecting the two points $A = (a, \hat{f}_j(a))$ and $B = (b, \hat{f}_j(b))$; we wish to show that S lies entirely on or below

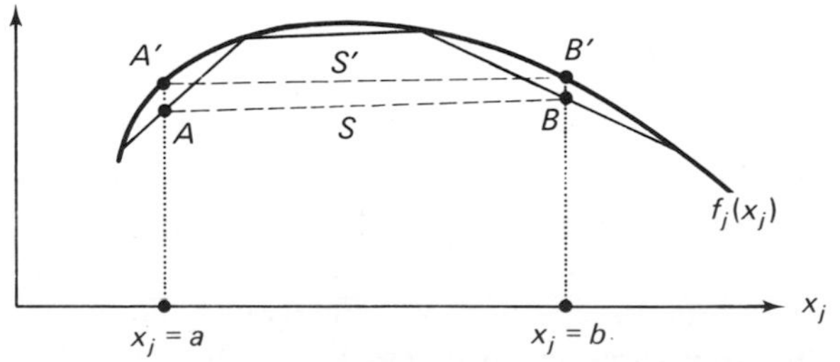

FIGURE 8-3

$\hat{f}_j$, which is equivalent to proving that $\hat{f}_j$ is concave. To expedite the proof we construct a second straight line S' between the two points $A' = (a, f_j(a))$ and $B' = (b, f_j(b))$ on the curve f_j; because f_j is a concave function, S' must lie entirely on or below it. Similarly, each of the line segments that constitute $\hat{f}_j$ must lie on or below f_j, so in particular A and B must be below (or not above) A' and B', respectively. It follows that the line S must lie entirely on or below S' and therefore on or below the curve f_j as well.

Now suppose that somewhere in the interval between $x_j = a$ and $x_j = b$ the line S is *above* the piecewise linear function $\hat{f}_j$; it must then have *first* risen above $\hat{f}_j$ (moving from left to right) at some specific point $x_j = c$, where $c \geqq a$. Subsequently, S must fall back to or through $\hat{f}_j$, say, at the point $x_j = d$, where $d \leqq b$. But c and d cannot be in the same linear segment of $\hat{f}_j$, because two straight lines cannot intersect at two different points. Therefore S must lie above one or more breakpoints of $\hat{f}_j$ between c and d, implying that S must lie above the curve f_j at those points, which is a contradiction. We conclude that S lies entirely on or below $\hat{f}_j$, so that $\hat{f}_j$ must be a concave function. It follows that the objective function of the approximating problem is concave and, because the constraints are linear, that the approximating problem is a convex program. Q.E.D.

The importance of Lemma 8.1 is theoretical: It guarantees that the approximating problem (8-13) can have no local maxima other than its global maximum. That is, in forming (8-13), we did not introduce any extraneous local maxima: The optimal solution to (8-13), once we have found it, will be an approximation to the optimal solution of the original separable program (8-5). In general, the greater the numbers of breakpoints used in constructing the piecewise linear functions $\hat{f}_j$, the more closely those functions will approximate the curves f_j, and the closer the optimal solution of the approximating problem is likely to be to the true optimum. Of course, the price of reducing the expected error in the solution is the introduction of additional variables λ_{jk} in (8-13), which will tend to increase overall computation time; this trade-off between accuracy and speed is characteristic of a great many nonlinear programming algorithms.

The other result we need in this section is that the logical restrictions 1 and 2 on the values of the variables λ_{jk} are superfluous when the approximating problem (8-13) is a convex program. The lemma below states this result, in essence, for a single concave function of one variable; the extension to any number of variables is then straightforward. To simplify notation for the lemma and its proof, we omit the subscript j from the function $\hat{f}_j$ and the variable x_j.

Lemma 8.2: Suppose a piecewise linear function $\hat{f}(x)$ *with breakpoints at* $x = x_0, x_1, \ldots, x_r$ *has been constructed in the usual manner to approximate a concave func-*

tion f(x), *so that any value* $x = X$ *satisfying* $x_0 \leqq X \leqq x_r$ *can be expressed in at least one way as a convex combination of the breakpoints,*

(8-15) $$X = \sum_{k=0}^{r} \lambda_k x_k, \quad \textit{where } \sum_{k=0}^{r} \lambda_k = 1 \quad \textit{and} \quad \lambda_k \geqq 0, \qquad \textit{all } k.$$

Then, of all the convex combinations that satisfy (8-15) *for some given* X, *the one that maximizes*

$$\hat{f}(X) = \sum_{k=0}^{r} \lambda_k f(x_k)$$

includes at most two positive components λ_k*; moreover, if two of the* λ_k *are positive, they must be adjacent.*

The first step in the proof is to demonstrate that, from any given representation of X as a convex combination of *three* of the breakpoints, it is possible to derive a different convex combination of only *two* of them that yields a greater value of $\hat{f}$. It will then follow, by extension, that a convex combination of *any* number of breakpoints can be reduced by similar means to a superior combination of two. The final step in the proof will be to show that the two breakpoints involved in the optimal convex combination must be adjacent.

Accordingly, consider the following representation of X as a convex combination of three breakpoints:

(8-16) $$X = \lambda_\alpha x_\alpha + \lambda_\beta x_\beta + \lambda_\gamma x_\gamma,$$
$$\text{where } \lambda_\alpha + \lambda_\beta + \lambda_\gamma = 1 \text{ and } \lambda_\alpha, \lambda_\beta, \lambda_\gamma > 0.$$

Assuming the three breakpoints have been ordered so that $x_\alpha < x_\beta < x_\gamma$, as shown in Fig. 8-4, we can express x_β as a unique convex combination of the other two:

(8-17) $$x_\beta = \phi x_\alpha + (1 - \phi) x_\gamma, \qquad \text{where } 0 < \phi < 1.$$

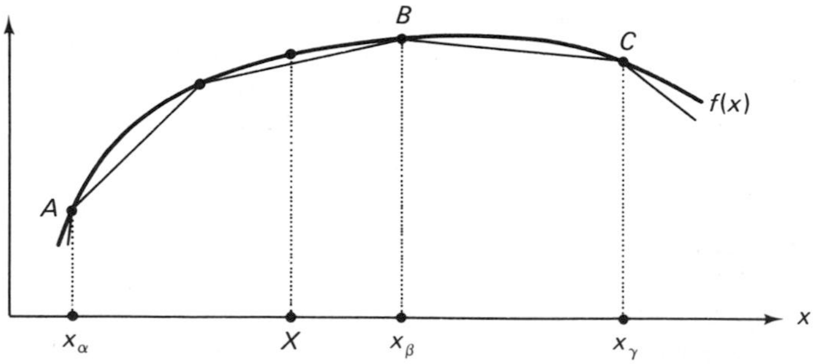

FIGURE 8-4

In addition, because $f(x)$ is a concave function, we can write

$$f(x_\beta) = f(\phi x_\alpha + (1 - \phi)x_\gamma) > \phi f(x_\alpha) + (1 - \phi)f(x_\gamma). \tag{8-18}$$

Note that the strict inequality is used in (8-18) because the point $B = (x_\beta, f(x_\beta))$—see Fig. 8-4—cannot lie exactly on the (imaginary) straight line from A to C: If it did, $f(x)$ would have to be linear between A and C, and there would then have been no need to identify x_β as a breakpoint of $\hat{f}(x)$.

Let us now assume that the given point X satisfies $x_\alpha < X \leqq x_\beta$; the proof would be analogous if X were between x_β and x_γ. We can then express X uniquely as a convex combination of x_α and x_β:

$$\begin{aligned} X &= \lambda_\alpha x_\alpha + \lambda_\beta x_\beta + (1 - \lambda_\alpha - \lambda_\beta)x_\gamma \\ &= \lambda_\alpha x_\alpha + \lambda_\beta x_\beta + (1 - \lambda_\alpha - \lambda_\beta)\left[\frac{1}{1-\phi}x_\beta - \frac{\phi}{1-\phi}x_\alpha\right], \end{aligned}$$

or

$$X = \frac{\lambda_\alpha - \phi + \phi\lambda_\beta}{1-\phi}x_\alpha + \frac{1 - \lambda_\alpha - \phi\lambda_\beta}{1-\phi}x_\beta, \tag{8-19}$$

where the first equality follows from (8-16) and the second from (8-17). The reader can easily verify that the sum of the two coefficients above is 1, so (8-19) must indeed be the desired convex combination (there is only one way to express X as a linear combination of x_α and x_β with coefficients that add to 1). It is then only necessary to observe that

$$\begin{aligned} \lambda_\alpha f(x_\alpha) + \lambda_\beta f(x_\beta) + \lambda_\gamma f(x_\gamma) &= \lambda_\alpha f(x_\alpha) + \lambda_\beta f(x_\beta) + (1 - \lambda_\alpha - \lambda_\beta)f(x_\gamma) \\ &< \lambda_\alpha f(x_\alpha) + \lambda_\beta f(x_\beta) + (1 - \lambda_\alpha - \lambda_\beta)\left[\frac{1}{1-\phi}f(x_\beta) - \frac{\phi}{1-\phi}f(x_\alpha)\right] \\ &= \frac{\lambda_\alpha - \phi + \phi\lambda_\beta}{1-\phi}f(x_\alpha) + \frac{1 - \lambda_\alpha - \phi\lambda_\beta}{1-\phi}f(x_\beta). \end{aligned} \tag{8-20}$$

The strict inequality follows from (8-18) and from the fact that $\lambda_\gamma = 1 - \lambda_\alpha - \lambda_\beta$ is positive by assumption. We have thus shown that any representation of X as a convex combination of three breakpoints (each having a positive coefficient λ_k) can be reduced to a different convex combination of only two of those points that yields a greater value of

$$\hat{f}(X) = \sum_{k=0}^{r} \lambda_k f(x_k).$$

But the same technique we used for reducing from three to two breakpoints while increasing the value of $\hat{f}$ could also be applied to reduce from four to three, or from five to four, and so on, and it follows that no representation of X as a convex combination $\sum_{k=0}^{r} \lambda_k x_k$ can possibly maximize f if more than two of the coefficients λ_k are positive.

We now come to the final step in the proof, which is to show that any representation of X as a convex combination of two breakpoints x_α and x_γ still cannot maximize $\hat{f}$ if x_α and x_γ are not adjacent. Suppose that

$$X = \lambda_\alpha x_\alpha + \lambda_\gamma x_\gamma \qquad \text{where } \lambda_\alpha + \lambda_\gamma = 1 \text{ and } \lambda_\alpha, \lambda_\gamma > 0,$$

and suppose further that there is another breakpoint x_β between x_α and x_γ:

$$x_\alpha < x_\beta < x_\gamma.$$

Then we can write

$$X = \lambda_\alpha x_\alpha + \lambda_\beta x_\beta + \lambda_\gamma x_\gamma, \tag{8-21}$$
$$\text{where } \lambda_\alpha + \lambda_\beta + \lambda_\gamma = 1,\ \lambda_\beta = 0, \text{ and } \lambda_\alpha, \lambda_\gamma > 0.$$

We now assert that the previous proof, beginning at (8-16), can be applied to (8-21), leading to a new convex combination of the form (8-19) whose coefficients satisfy the inequality (8-20). The reader can see that this assertion must be valid, because nowhere in the previous proof did we use the fact that λ_β was positive. Therefore, given any representation of X as a convex combination of two nonadjacent breakpoints, we can derive a different convex combination of two breakpoints that are closer together than the original pair and that yield an increased value of $\hat{f}$. By repeating this step as many times as is necessary, we eventually obtain a representation of X as a convex combination of two adjacent breakpoints or—if X happens to coincide with some breakpoint x_β—as a "convex combination" of x_β alone:

$$X = 1 \cdot x_\beta.$$

This completes the proof.

Suppose now that the simplex method is being used to solve the linear program obtained by deleting the logical restrictions 1 and 2 from the approximating problem (8-13), and suppose further that some current basic feasible solution $\boldsymbol{\lambda}_0$ includes several positive λ_{jk} (or two positive but nonadjacent λ_{jk}) for some x_j. We know from Lemma 8.2 that there must then exist another convex combination $\lambda_{j0}, \lambda_{j1}, \ldots, \lambda_{jr_j}$ for which the value of the objective term

$$\sum_{k=0}^{r_j} \lambda_{jk} f_j(x_{jk}) = \sum_{k=0}^{r_j} \lambda_{jk} f_{jk}$$

is greater, while the value of

$$\sum_{k=0}^{r_j} \lambda_{jk} x_{jk} \tag{8-22}$$

is the same. But (8-22) represents the net impact of the original variable x_j on the constraints: If its value is left unchanged, the constraints

$$\sum_{j=1}^{n} \sum_{k=0}^{r_j} a_{ijk} \lambda_{jk} = \sum_{j=1}^{n} a_{ij} \left(\sum_{k=0}^{r_j} \lambda_{jk} x_{jk} \right) = b_i$$

will continue to be satisfied. Thus the solution derived from $\boldsymbol{\lambda}_0$ by changing the values of $\lambda_{j0}, \lambda_{j1}, \ldots, \lambda_{jr_j}$ as indicated above must be feasible and must have a greater objective value than $\boldsymbol{\lambda}_0$, which implies that $\boldsymbol{\lambda}_0$ cannot be optimal.[2] We conclude, then, that when the approximating problem (8-13) is solved via the simplex method *without* the logical restrictions 1 and 2, the resulting optimal solution $\boldsymbol{\lambda}^*$ will turn out to satisfy those restrictions anyway. It follows that $\boldsymbol{\lambda}^*$ must be the optimal solution to (8-13) itself and therefore that the values

$$x_j^* = \sum_{k=0}^{r_j} \lambda_{jk}^* x_{jk}, \qquad j = 1, \ldots, n, \tag{8-23}$$

where the λ_{jk}^* are the components of $\boldsymbol{\lambda}^*$, constitute the optimal solution to the equivalent problem

$$\begin{cases} \text{Max } z = \sum_{j=1}^{n} \hat{f}_j(x_j) \\ \text{subject to } \mathbf{Ax} = \mathbf{b} \\ \quad \text{and } L_j \leqq x_j \leqq M_j, \qquad j = 1, \ldots, n. \end{cases} \tag{8-14}$$

Thus, as we wished to show, $\mathbf{x}^* = (x_1^*, \ldots, x_n^*)$ can be taken as an approximation to the optimal solution of the original separable program (8-5).

Note that $\mathbf{x}^*$ will *not*, in general, coincide exactly with the optimum of the original problem (8-5), essentially because the peaks of the piecewise linear approximations $\hat{f}_j$, $j = 1, \ldots, n$, do not necessarily coincide with the peaks of the original functions f_j. It would seem intuitively obvious, however, that as the numbers of breakpoints x_{jk} in the functions $\hat{f}_j$ are

[2]It is true that if the original function $f_j(x_j)$ were linear, the value of the objective term $\sum_{k=0}^{r_j} \lambda_{jk} f_{jk}$ for the newly constructed set of values would be the *same* as for the solution $\boldsymbol{\lambda}_0$, rather than greater. In such a case $\boldsymbol{\lambda}_0$ might be found to be an optimal solution, even though it did not satisfy the logical restrictions 1 and 2 (there would, of course, be an alternative optimum that did satisfy those restrictions). This ambiguity should never arise in a sensibly formulated problem, however, because when $f_j(x_j)$ is linear there is no need in the first place for a piecewise linear approximation to it.

increased, so that the $\hat{f}_j$ come to approximate the f_j more and more closely, the solution $\mathbf{x}^*$, as computed via (8-23), will approach as a limit the true optimal solution of the original problem. This statement in fact is true, provided that the density of breakpoints is increased in some regular manner throughout the full range of each variable, or at least in the vicinity of its optimal value (see Exercise 8-8).

8.4 SEPARABLE PROGRAMMING WITH RESTRICTED BASIS ENTRY

In this section we relax our assumption of a concave objective function and consider the general separable program

$$\text{(8-5)} \qquad \begin{cases} \text{Max } z = \sum_{j=1}^{n} f_j(x_j) \\ \text{subject to } \mathbf{Ax} = \mathbf{b} \text{ and } \mathbf{x} \geqq \mathbf{0}, \end{cases}$$

where the functions f_j, so long as they are continuous, may be of any algebraic form whatever. Constructing the piecewise linear approximations $\hat{f}_j$ as before, we arrive again at the approximating problem

$$\text{(8-13)} \qquad \begin{cases} \text{Max } z = \sum_{j=1}^{n} \hat{f}_j(x_j) \equiv \sum_{j=1}^{n} \sum_{k=0}^{r_j} f_{jk}\lambda_{jk} \\ \text{subject to } \sum_{j=1}^{n} \sum_{k=0}^{r_j} a_{ijk}\lambda_{jk} = b_i, \qquad i = 1, \ldots, m, \\ \qquad \sum_{k=0}^{r_j} \lambda_{jk} = 1, \qquad j = 1, \ldots, n, \\ \text{and} \qquad \lambda_{jk} \geqq 0 \qquad \text{for all } j \text{ and } k, \\ \textit{and to the logical restrictions 1 and 2 of Section 8.2.} \end{cases}$$

Now, however, the functions $\hat{f}_j$ need not all be concave; it follows that the lemmas of Section 8.3 do not apply, so we can no longer rely on solving the problem (8-13) by ordinary simplex pivoting without regard for the logical restrictions included in the formulation. That is, if we apply the simplex method to (8-13), ignoring the restrictions as before, we may obtain an "optimal" solution that does not satisfy them.

Accordingly, to solve the approximating problem (8-13) we shall use a modified version of the simplex method in which restrictions 1 and 2 are enforced at every stage of the computations. This is accomplished by rejecting as a candidate for basis entry any variable λ_{jk} for which it is true that either another *nonadjacent* variable $\lambda_{j\alpha}$ or two other variables $\lambda_{j\alpha}$ and $\lambda_{j\beta}$ are already in the basis (unless, as may sometimes happen, the offending variable will be driven to zero by the pivot, allowing restrictions 1 and 2 to remain in force as desired). Thus at each iteration the variable λ_{jk} chosen to enter the basis will be the one that, among all those not rejected by the conditions just stated, has the most favorable—that is, in a maximization problem, the most negative—reduced cost $z_{jk} - c_{jk}$. If none of the nonbasic variables eligible for basis entry has a negative reduced cost, the computations are terminated and the current basic solution $\boldsymbol{\lambda}^*$ is taken as an approximation to a constrained local maximum of the original separable program. In general, there may, of course, be several local optima in the feasible region.

This procedure for solving the approximating problem (8-13) is another illustration of the use of the simplex method with *restricted basis entry*; the same type of modification was encountered in Wolfe's quadratic programming algorithm, presented in Section 7.3. As was true in the case of Wolfe's algorithm, the only step required to justify the method is a proof that the solution obtained by it does in fact possess the desired optimality properties. The question of whether convergence to such a solution will be finite does not arise as a *practical* matter, because no simplex pivoting method has ever been observed to cycle endlessly, except in carefully contrived artificial problems.

It is not difficult to see that the solution $\boldsymbol{\lambda}^*$ obtained by the simplex method with the modifications described above must be a locally optimal solution to the approximating problem (8-13)—that is, locally optimal with respect to perturbations permitted by the logical restrictions 1 and 2. Observe that, by virtue of the basis entry restrictions, the solution $\boldsymbol{\lambda}^*$ has the property that for every j, $j = 1, \ldots, n$, either

1. Only one variable λ_{jk} is positive (necessarily equal to 1), or
2. λ_{jk} and $\lambda_{j,k+1}$ are positive for some k.

In case 1 a local perturbation in the value of λ_{jk} would require pivoting either $\lambda_{j,k-1}$ or $\lambda_{j,k+1}$ into the basis (unless $k = 0$ or $k = r_j$, in which case only one of these variables is defined). However, neither of these pivots was barred by the basis entry restrictions, so the fact that the method terminated at $\boldsymbol{\lambda}^*$ indicates that neither of them would have improved the value of the objective function. Similarly, in case 2 the variables λ_{jk} and $\lambda_{j,k+1}$ would both remain positive, and therefore in the basis, for a sufficiently small perturbation; hence such a perturbation could only result from pivoting some other variable $\lambda_{\alpha\beta}$, where $\alpha \neq j$, into the basis in accordance with the entry restrictions. But, again, such pivots were not prohibited and must therefore have been unfavorable. We conclude that $\boldsymbol{\lambda}^*$ is a locally optimal solution with respect to permitted perturbations in the variables λ_{jk}, which implies that the set of values

$$x_j^* = \sum_{k=0}^{r_j} \lambda_{jk}^* x_{jk}, \qquad j = 1, \ldots, n, \tag{8-23}$$

constitutes a locally optimal solution to the equivalent problem (8-14) and an approximation to a local optimum of the original separable program (8-5).

Insofar as the practical merits of separable programming are concerned, the relative efficiency of the technique—relative, that is, to some of the competing methods we shall discuss below—depends on the degree of linearity of the problem. The crucial advantage of the separable programming approach, whether it is applied to convex or to nonconvex problems, is that it employs the simplex method. It therefore benefits both from the unique computational efficiency of that method and from the wide general availability of simplex computer codes. Many operations research departments and computer facilities maintain simplex codes or access to them but have no other type of mathematical programming capability: In dealing with nonlinear optimization problems, such users may have no practical alternative to separable programming, even if it requires modifying a simplex code to provide for restricted basis entry.

On the other hand, the separable programming approach has several significant disadvantages:

1. The number of constraints and, less importantly, the number of variables are much greater in the approximating problem (8-13) than in the original problem[3] (8-5);

2. A significant amount of computational effort, including choice of breakpoints and calculation of ordinate values, must be expended initially in order to construct the piecewise linear approximations $\hat{f}_j(x_j)$;

3. An initial basic feasible solution to the approximating problem must be found via the phase I simplex method, itself a time-consuming procedure; and

4. The solution eventually obtained will be approximate rather than exact and may be a relatively poor approximation unless the intervals between the breakpoints are quite small.

Notice that all these disadvantages except the third are roughly proportional, in their unfavorable effects, to the degree of nonlinearity in the original problem. This follows from the fact that when $f_j(x_j)$ is linear it is not necessary to construct the piecewise linear approximation $\hat{f}_j$, nor to introduce the variables λ_{jk}, $k = 0, 1, \ldots, r_j$—the variable x_j is simply retained in the formulation instead. As a result, it is fair to say that the separable programming approach is *competitive* (i.e., approximately comparable or superior in efficiency to other competing nonlinear programming methods) only for

[3]This disadvantage of the separable programming approach is mitigated considerably by the fact that the additional constraints $\sum_k \lambda_{jk} = 1$ in the approximating problem can be handled via the so-called "generalized upper bounding" technique in such a way that they do not increase the size of the simplex basis matrix. A discussion of this technique can be found in Chapter 6 of Lasdon [29].

problems having few nonlinearities. When most or all of the $f_j(x_j)$ are nonlinear, separable programming should really be used only if other, more efficient computer codes are unavailable. It should be noted that these observations hold for both convex and nonconvex separable programs.

We conclude our discussion of separable programming with one or two remarks of a historical nature. The idea of using piecewise linear approximations to enable the simplex method to be applied to nonlinear programs seems to have originated with Charnes and Lemke [7] in 1954. They proposed a computational procedure for solving convex programs with separable objective functions and linear constraints that was roughly similar to the one presented in Section 8.3. Shortly thereafter, an alternative approach using a different set of variables to represent the piecewise linear functions $\hat{f}_j$ was devised by Dantzig [9]; his method will be developed in the exercises. Then, several years later, Miller [35] extended the separable programming technique to problems with nonlinear constraints: We shall discuss this generalized version in Chapter 9.

8.5 THE CONVEX SIMPLEX METHOD

Having completed our discussion of separable programming for linearly constrained problems, we now direct our attention to a somewhat different family of algorithms which, like the separable programming technique, are simplex-based but which also embody in a more natural and visible form the feasible-directions strategy described in Section 8.1. The reader will recall that the feasible-directions approach consists of choosing at each iteration a direction of displacement from the current feasible point and then proceeding in that direction until either the objective value stops improving or a boundary of the feasible region is encountered. In the methods to which we now turn, this strategy is implemented within the familiar simplex computational framework: A set of basic variables is identified, with only one member permitted to change during any single iteration, and a computational tableau containing the usual simplex columns $\mathbf{y}_j = \mathbf{B}^{-1}\mathbf{a}_j$ is maintained and updated in a pivot operation whenever a new variable is substituted for another in the basis. The method of Beale, discussed in Section 7.6, is one example of a simplex-based feasible-directions algorithm, although, of course, it can be used only when the objective function is quadratic. In the next few sections we shall present two more general members of this family, the convex simplex method of Zangwill [58] and the reduced gradient method of Wolfe [54], each of which can be applied to any mathematical program of the following form:

$$
\text{(8-24)} \qquad \begin{cases} \text{Max } z = f(\mathbf{x}) \\ \text{subject to } \mathbf{A}\mathbf{x} = \mathbf{b} \\ \quad\text{and} \quad \mathbf{x} \geqq \mathbf{0}, \end{cases}
$$

where $\mathbf{A}$ is an m-by-n matrix and where the objective function $f(\mathbf{x})$ is continuous and has continuous first partial derivatives. We remark that in our development of these methods we shall temporarily postpone any detailed consideration of the important theoretical questions of degeneracy and convergence, preferring to present more broadly based discussions of them later on in sections of their own.

The convex simplex method, as originally published by Zangwill [58] in 1967, was designed for solving convex programming problems—hence its name. However, it can in fact be used to seek local optima of any linearly constrained mathematical program. In this section we shall develop the method for the general maximization problem (8-24), where the objective function $f(\mathbf{x})$ is not required to be concave. Let us begin by assuming that the phase I simplex method of linear programming has been applied to (8-24), producing a basic feasible solution $\mathbf{x_B} = (x_{\mathbf{B}1}, \ldots, x_{\mathbf{B}m})$ with associated basis matrix $\mathbf{B}$, and eliminating any redundant constraints. Thus the problem to be solved by the convex simplex method must have feasible solutions, although it may have an unbounded optimum.

Having obtained a basic solution $\mathbf{x_B}$, we partition the variables $\mathbf{x} = (x_1, \ldots, x_n)$ into $\mathbf{x} \equiv (\mathbf{x_B}, \mathbf{x_R})$, where $\mathbf{x_R}$ is the $(n-m)$-component vector of nonbasic variables. Similarly, the constraint matrix $\mathbf{A}$ is partitioned into $[\mathbf{B} \quad \mathbf{R}]$, where $\mathbf{B}$ is the basis associated with $\mathbf{x_B}$ and $\mathbf{R}$ is the m-by-$(n-m)$ matrix of nonbasic columns; the ordering of the columns of $\mathbf{R}$ corresponds to the ordering of the nonbasic variables in $\mathbf{x_R}$. The constraints $\mathbf{Ax} = \mathbf{b}$ then become

$$\text{(8-25)} \qquad \mathbf{Ax} = \mathbf{Bx_B} + \mathbf{Rx_R} = \mathbf{Bx_B} + \sum_{j \in J} x_j \mathbf{a}_j = \mathbf{b},$$

where $\mathbf{a}_j$ is the jth column of $\mathbf{A}$ and J is the set of subscripts of the nonbasic variables—that is, j is in J if and only if x_j is in $\mathbf{x_R}$. Solving (8-25) for $\mathbf{x_B}$, we have

$$\text{(8-26)} \qquad \mathbf{x_B} = \mathbf{b}_0 - \sum_{j \in J} x_j \mathbf{y}_j,$$

where $\mathbf{b}_0 = \mathbf{B}^{-1}\mathbf{b}$ is the set of values of the basic variables and $\mathbf{y}_j = \mathbf{B}^{-1}\mathbf{a}_j$ is the familiar vector of the simplex method. The ith component of (8-26) is then

$$\text{(8-27)} \qquad x_{\mathbf{B}i} = b_{0i} - \sum_{j \in J} x_j y_{ij}, \qquad i = 1, \ldots, m.$$

Let us now imagine substituting the expressions (8-27) for the basic variables into the objective function $f(\mathbf{x})$, converting it into a function of the nonbasic variables only:

$$z = f(\mathbf{x_B}, \mathbf{x_R}) = \hat{f}(\mathbf{x_R}).$$

The partial derivative of z with respect to the nonbasic variable x_j, j in J, can then be calculated via the chain rule:

$$\frac{\delta z}{\delta x_j} \equiv \frac{\delta \hat{f}}{\delta x_j} = \frac{\delta f}{\delta x_j} + \sum_{i=1}^{m} \frac{\delta f}{\delta x_{\mathbf{B}i}} \cdot \frac{\delta x_{\mathbf{B}i}}{\delta x_j}$$

$$= \frac{\delta f}{\delta x_j} + \sum_{i=1}^{m} \frac{\delta f}{\delta x_{\mathbf{B}i}} \cdot (-y_{ij})$$

or

$$\text{(8-28)} \qquad \frac{\delta z}{\delta x_j} = \frac{\delta f}{\delta x_j} - \nabla \mathbf{f}(\mathbf{x})_{\mathbf{B}} \cdot \mathbf{y}_j, \qquad j \in J,$$

where the m-component row vector $\nabla \mathbf{f}(\mathbf{x})_{\mathbf{B}}$ is formed from those components of the gradient $\nabla \mathbf{f}(\mathbf{x})$ that are associated with the basic variables:

$$\text{(8-29)} \qquad \nabla \mathbf{f}(\mathbf{x})_{\mathbf{B}} \equiv \left[\frac{\delta f}{\delta x_{\mathbf{B}1}}, \ldots, \frac{\delta f}{\delta x_{\mathbf{B}m}}\right].$$

It is not difficult to see that for a linear objective function $z = f(\mathbf{x}) = c_1 x_1 + \cdots + c_n x_n$ the expression (8-28) would reduce to (the negative of) the reduced cost of linear programming:

$$\frac{\delta z}{\delta x_j} = c_j - \mathbf{c_B} \cdot \mathbf{y}_j = c_j - z_j,$$

where $\mathbf{c_B}$ is the basic cost vector.

Given the set of partial derivatives (8-28), the strategy of the convex simplex method at the first iteration is to seek an improved value of z by increasing some nonbasic variable x_k for which $\delta z/\delta x_k$ is positive; all other nonbasic variables are held at zero. As x_k increases, the values of the basic variables are adjusted so as to keep the constraints satisfied, exactly as in the ordinary or "linear" simplex method. Thus the direction of displacement in the space of the original variables $\mathbf{x}$ has a component 1 in the x_k-direction, a component 0 in the direction of every other nonbasic variable, and a component $-y_{ik}$ in the direction of each basic variable $x_{\mathbf{B}i}$. In the linear simplex method the reduced cost or partial derivative $\delta z/\delta x_k$ is constant, so the objective value will continue to increase as long as x_k does; displacement stops, therefore, only when a boundary of the feasible region is encountered, that is, when the value of one of the basic variables falls to zero. In the convex simplex method, however, the partial derivative changes as x_k increases and may at some point fall to zero; when this happens it is no longer desirable to increase x_k, even though it may still be feasible to do so. Accordingly, displacement ends, as in the method of Beale, either when

1. The value of some basic variable $x_{\mathbf{B}r}$ falls to zero, or
2. The partial derivative $\delta z/\delta x_k$ vanishes.

The point at which one of these two events first occurs is taken as the starting point of the next iteration. Thus the convex simplex method, like the method of Beale, is a member of the feasible-directions family.

If the first iteration of the convex simplex method leads to case 1, an ordinary simplex pivot is executed in which x_k replaces $x_{\mathbf{B}r}$ in the basis; the partial derivatives (8-28) are then recalculated and the next iteration begins. In case 2, however, none of the basic variables is driven to zero. The basis is therefore left unchanged, and it follows that at the next iteration there will be a nonbasic variable x_k with a positive value. This condition, in which more than m variables have positive values,[4] is quite likely to arise at some stage in solving almost any problem via the convex simplex method; in particular, it may well prevail at the optimum, which does not necessarily lie at an extreme point when the objective is nonlinear.

Having seen how positive nonbasic variables can be introduced, let us now describe in detail a single general iteration of the convex simplex method. Let $\mathbf{x_B}$ and $\mathbf{B}^{-1}$ be the current basic solution and associated basis inverse, and let J be the set of subscripts of the nonbasic variables, where each x_j, j in J, may have a positive or zero value. The initial step in the iteration is to calculate the partial derivatives

$$\text{(8-28)} \qquad \frac{\delta z}{\delta x_j} = \frac{\delta f}{\delta x_j} - \nabla \mathbf{f}(\mathbf{x})_{\mathbf{B}} \cdot \mathbf{y}_j, \qquad j \in J.$$

These values are then examined to see if any favorable perturbations of the nonbasic variables are possible. In general, it would be desirable to increase any nonbasic variable x_j for which $\delta z/\delta x_j$ is positive or to decrease any positive-valued x_j for which $\delta z/\delta x_j$ is negative. Accordingly, to determine the variable that will be adjusted, we compute

$$\text{(8-30)} \qquad \frac{\delta z}{\delta x_p} = \max_{j \in J} \left(\frac{\delta z}{\delta x_j}\right)$$

and

$$\text{(8-31)} \qquad \frac{\delta z}{\delta x_q} \cdot x_q = \min_{j \in J} \left(\frac{\delta z}{\delta x_j} \cdot x_j\right).$$

Note that we cannot wish to decrease a positive-valued nonbasic variable

[4]The reader may recall that the same condition can arise in solving linear programs with upper bound constraints of the form $x_j \leqq M_j$. Certain modifications of the simplex pivoting procedure allow upper bounds to be treated implicitly, instead of being represented explicitly as constraints; in using this approach, the variable that leaves the basis at each iteration is the one whose value is first driven either to zero or to its upper bound. Thus in the later stages of applying the "upper-bounded" simplex method a substantial number of the nonbasic variables may be at their maximum levels $x_j = M_j$.

x_j unless its partial derivative is negative; thus we are interested only in negative values of the product $(\delta z/\delta x_j)\cdot x_j$. The variable to be adjusted is then chosen as follows:

$$\text{if } \frac{\delta z}{\delta x_p} > 0 \quad \text{and} \quad \frac{\delta z}{\delta x_q}\cdot x_q \geqq 0, \qquad \text{increase } x_p;$$

$$\text{if } \frac{\delta z}{\delta x_p} \leqq 0 \quad \text{and} \quad \frac{\delta z}{\delta x_q}\cdot x_q < 0, \qquad \text{decrease } x_q;$$

$$\text{if } \frac{\delta z}{\delta x_p} > 0 \quad \text{and} \quad \frac{\delta z}{\delta x_q}\cdot x_q < 0, \qquad \text{either increase } x_p \text{ or decrease } x_q;$$

but

$$\text{if } \frac{\delta z}{\delta x_p} \leqq 0 \quad \text{and} \quad \frac{\delta z}{\delta x_q}\cdot x_q \geqq 0,$$

the current solution $\mathbf{x}_0$ satisfies the Kuhn-Tucker conditions and the method terminates. In the third case Zangwill suggests increasing x_p if

$$\left|\frac{\delta z}{\delta x_p}\right| \geqq \left|\frac{\delta z}{\delta x_q}\cdot x_q\right|$$

and decreasing x_q otherwise; this seems to be a reasonable rule.[5]

To prove that a Kuhn-Tucker point is indeed obtained in the fourth case, we observe that the termination criterion implies

$$\text{(8-32)} \qquad \frac{\delta z(\mathbf{x}_0)}{\delta x_j} \leqq 0 \quad \text{and} \quad \frac{\delta z(\mathbf{x}_0)}{\delta x_j}\cdot x_{0j} \geqq 0 \text{ for all } j \text{ in } J,$$

where $\mathbf{x}_0 = (x_{01}, \ldots, x_{0n})$ is the final solution obtained and $\delta z(\mathbf{x}_0)/\delta x_j$ denotes the partial derivative $\delta z/\delta x_j$ evaluated at $\mathbf{x}_0$. However, because $\delta z(\mathbf{x}_0)/\delta x_j$ is nonpositive and x_{0j} is nonnegative, their product cannot be positive, so the conditions (8-32) reduce to

$$\text{(8-33)} \qquad \frac{\delta z(\mathbf{x}_0)}{\delta x_j} \leqq 0 \quad \text{and} \quad \frac{\delta z(\mathbf{x}_0)}{\delta x_j}\cdot x_{0j} = 0, \qquad j \in J.$$

[5]It may be of interest to the reader that the criterion (8-31) for selecting a positive-valued nonbasic variable to be decreased was chosen by Zangwill, in preference to the simpler and more natural rule

$$\min_{j\in J}\left\{\frac{\delta z}{\delta x_j}, x_j > 0\right\},$$

not for computational reasons but in order to facilitate his theoretical convergence proof for the convex simplex algorithm.

We now recall from (8-28) that at the solution point $\mathbf{x}_0$

$$\text{(8-34)} \qquad \frac{\delta z}{\delta x_j} = \frac{\delta f}{\delta x_j} - \nabla \mathbf{f}(\mathbf{x}_0)_{\mathbf{B}} \cdot \mathbf{y}_j = \frac{\delta f}{\delta x_j} - \nabla \mathbf{f}(\mathbf{x}_0)_{\mathbf{B}} \cdot \mathbf{B}^{-1}\mathbf{a}_j$$
$$\equiv \frac{\delta f}{\delta x_j} - \boldsymbol{\lambda}_0 \cdot \mathbf{a}_j = \frac{\delta f}{\delta x_j} - \sum_{i=1}^{m} \lambda_{0i} a_{ij}, \qquad j \in J,$$

where $\mathbf{B}^{-1}$ and $\nabla \mathbf{f}(\mathbf{x}_0)_{\mathbf{B}}$ are the basis inverse and "basis gradient" associated with $\mathbf{x}_0$ and where we have defined a new m-component row vector $\boldsymbol{\lambda}_0 \equiv \nabla \mathbf{f}(\mathbf{x}_0)_{\mathbf{B}} \cdot \mathbf{B}^{-1}$ with components λ_{0i}. Moreover, if x_j is, say, the rth *basic* variable $x_{\mathbf{B}r}$, then $\mathbf{y}_j$ must be the unit vector $\mathbf{e}_r$, and the equalities

$$\text{(8-35)} \qquad \frac{\delta f}{\delta x_j} - \sum_{i=1}^{m} \lambda_{0i} a_{ij} \equiv \frac{\delta f}{\delta x_j} - \nabla \mathbf{f}(\mathbf{x}_0)_{\mathbf{B}} \cdot \mathbf{y}_j = \frac{\delta f}{\delta x_j} - \frac{\delta f}{\delta x_{\mathbf{B}r}} = 0, \qquad j \notin J,$$

must also hold at $\mathbf{x}_0$. It then follows from (8-33), (8-34), and (8-35) that the solution $\mathbf{x}_0$ and the set of values $\boldsymbol{\lambda}_0 = \nabla \mathbf{f}(\mathbf{x}_0)_{\mathbf{B}} \cdot \mathbf{B}^{-1}$ jointly satisfy the following relations:

$$\text{(8-36)} \qquad \begin{cases} \dfrac{\delta f}{\delta x_j} - \displaystyle\sum_{i=1}^{m} \lambda_i a_{ij} \leqq 0, & j = 1, \ldots, n. \\ x_j \left\{ \dfrac{\delta f}{\delta x_j} - \displaystyle\sum_{i=1}^{m} \lambda_i a_{ij} \right\} = 0, & j = 1, \ldots, n. \end{cases}$$

But (8-36) constitutes the Kuhn-Tucker conditions for the original problem

$$\text{Max } z = f(\mathbf{x})$$
$$\text{subject to } a_{i1}x_1 + \cdots + a_{in}x_n = b_i, \qquad i = 1, \ldots, m,$$
$$\text{and} \qquad x_j \geqq 0, \qquad j = 1, \ldots, n,$$

so $\mathbf{x}_0$ is in fact a Kuhn-Tucker point, as was to be shown. The student will recall that if $f(\mathbf{x})$ is a concave function, then we can conclude from Theorem 6.6 that $\mathbf{x}_0$ is an optimal solution.

Returning to the computational details of the convex simplex method, once the nonbasic variable to be adjusted has been chosen, the next step is to determine by how much the value of that variable will be changed. There are three cases to consider.

Case A. *x_k is to be increased and at least one component of $\mathbf{y}_k = \mathbf{B}^{-1}\mathbf{a}_k$ is positive* (note that the value of x_k may be positive or zero). In this case as x_k is increased the values of one or more basic variables fall toward zero; the first to reach zero will be $x_{\mathbf{B}r}$, where

$$\text{(8-37)} \qquad \frac{x_{\mathbf{B}r}}{y_{rk}} = \min_i \left(\frac{x_{\mathbf{B}i}}{y_{ik}}, \; y_{ik} > 0 \right).$$

This, of course, is the familiar exit criterion of the simplex method, and $\Delta = x_{\mathbf{B}r}/y_{rk}$ is the increase in the value of x_k that will just cause the value of the rth basic variable to reach zero. Thus to maintain feasibility x_k cannot be increased by more than Δ; it may, however, be desirable to increase it by some lesser amount if the value of the objective function stops improving. In fact, the desired increase in x_k, and therefore the new solution point $\hat{\mathbf{x}}$, can be obtained by solving the following optimization problem:

$$\text{(8-38)} \qquad \begin{cases} \text{Max } f(\hat{\mathbf{x}}) \equiv f(\mathbf{x} + \theta \mathbf{s}) \\ \text{subject to } 0 \leqq \theta \leqq \Delta, \end{cases}$$

where $\mathbf{x}$ is the current solution point, $\hat{\mathbf{x}}$ is the new point, θ is a scalar parameter (the "step length") whose value is to be determined, and $\mathbf{s}$ is the direction of displacement from $\mathbf{x}$ to $\hat{\mathbf{x}}$, with components as follows:

$$\text{(8-39)} \qquad \begin{cases} s_k = 1, & \\ s_j = 0 & \text{if } j \in J, j \neq k, \\ s_j = -y_{ik} & \text{if } j \notin J \text{ and } x_j = x_{\mathbf{B}i}. \end{cases}$$

Note that J is the set of subscripts of the nonbasic variables x_j in the current solution.

The problem (8-38) is an ordinary one-dimensional direct search problem and can be solved in various ways, using the methods of Chapters 4 and 5. For example, one might choose a small fixed increment ϵ and compute the objective value at the points $\mathbf{x}$, $\mathbf{x} + \epsilon\mathbf{s}$, $\mathbf{x} + 2\epsilon\mathbf{s}, \ldots,$ until a decrease is observed at $\mathbf{x} + t\epsilon\mathbf{s}$; a golden section search could then be used to find the optimal value θ_{opt} within the interval $(t - 2)\epsilon < \theta < t\epsilon$. Alternatively, if $f(\mathbf{x} + \theta\mathbf{s})$ happens to be linear in the $m + 1$ variables whose values are changing (i.e., x_k and the basic variables), then the objective value must continue improving at a constant rate as long as x_k is increased, and it follows immediately that $\theta_{\text{opt}} = \Delta$. Thus in the linear case no significant computational effort is required to solve the problem (8-38). This possibility of linearity is, incidentally, an important consideration: Many problems arising in the practice of operations research are linear in all but a few variables, and for such problems it is useful to have available an algorithm such as the convex simplex method (or the reduced gradient method, to be discussed two sections hence) that works with only a subset of the variables at each iteration.

At any rate, if it is found that $\theta_{\text{opt}} = \Delta$, then the basic variable $x_{\mathbf{B}r}$, as determined by (8-37), is driven to zero, and an ordinary simplex pivot is executed, with x_k replacing $x_{\mathbf{B}r}$ in the basis. The computations are based, as usual, on the familiar $\mathbf{\Phi}$-vector (2-30), although, of course, the reduced costs $z_j - c_j$ of the linear simplex method are not present. In addition, it is convenient at each stage of the convex simplex method to calculate the new values of all variables directly from the relation $\hat{\mathbf{x}} = \mathbf{x} + \theta_{\text{opt}}\mathbf{s}$, instead of trying to use the transform vector $\boldsymbol{\Phi}$; the latter approach, unless appropriately modified, will fail whenever the initial value of x_k is positive. Note that if the old basis is degenerate, it may happen that $\theta_{\text{opt}} = \Delta = 0$, in which case the pivot will be of the "zero-for-zero" variety and the new solution point will be the same as the old (although the new value of the rth basic variable will be that of x_k, which is not necessarily zero—thus degeneracy may actually be *eliminated* by the pivot). Note also that when $\theta_{\text{opt}} = \Delta > 0$ degeneracy may arise if a tie occurs in choosing the variable to leave the basis.

On the other hand, if $\theta_{\text{opt}} < \Delta$, no pivot is required. The new values of x_k and the basic variables are simply substituted for the old, and the iteration is complete.

Case B. *x_k is to be increased and no component of $\mathbf{y}_k$ is positive.* This case differs from the one preceding in that, no matter how much x_k is increased, none of the basic variables can ever be driven to zero. The new point $\hat{\mathbf{x}}$ is therefore determined by solving the problem

$$\begin{cases} \text{Max } f(\hat{\mathbf{x}}) \equiv f(\mathbf{x} + \theta\mathbf{s}) \\ \text{subject to } \theta \geqq 0, \end{cases} \tag{8-40}$$

where the components of $\mathbf{s}$ are as given in (8-39). If a finite θ_{opt} is obtained, the values of x_k and of the basic variables are adjusted and we proceed to the next iteration, without executing a pivot. But if no finite value of θ is optimal, the convex simplex method terminates and the solution is unbounded. Note that in this case the optimal value of the objective function is not necessarily infinite; for example, the maximum value of $f(x_1, x_2) = -e^{-x_1} - e^{-x_2}$ is 0, attained as x_1 and x_2 both approach infinity.

Case C. *x_k is to be decreased* (from an initially positive value). In this case, as in case A, x_k can only be changed by a finite amount: Even if no basic variable decreases in value as x_k decreases, x_k itself can fall no further than zero. The maximum permitted decrease in x_k is computed as follows:

$$\Delta' = \frac{x_{\mathbf{B}r}}{y_{rk}} = \max_i \left(\frac{x_{\mathbf{B}i}}{y_{ik}}, \; y_{ik} < 0\right) \tag{8-41}$$

and

$$\Delta = \min(-\Delta', x_k). \tag{8-42}$$

Here Δ' must be nonpositive; if $y_{ik} \geqq 0$ for all i, we take $\Delta' = -\infty$. Thus $-\Delta'$ is the amount of decrease in the value of x_k that is just sufficient to drive the value of a basic variable to zero; if $\Delta = -\Delta' < x_k$, this will happen before x_k itself reaches zero.

When Δ has been computed, the desired amount of decrease in x_k and therefore the new solution point $\hat{\mathbf{x}}$ are obtained by solving the following problem:

$$\text{(8-43)} \qquad \begin{cases} \text{Max } f(\hat{\mathbf{x}}) \equiv f(\mathbf{x} - \theta \mathbf{s}) \\ \text{subject to } 0 \leqq \theta \leqq \Delta, \end{cases}$$

where again the components of $\mathbf{s}$ are as given in (8-39). If $\theta_{\text{opt}} < \Delta$—that is, if the objective value stops improving before any variable has fallen to zero—no pivot is required: It is only necessary to compute and record the new values of the variables,

$$\text{(8-44)} \qquad \hat{\mathbf{x}} = \mathbf{x} - \theta_{\text{opt}}\mathbf{s},$$

before proceeding to the next iteration. Similarly, if $\theta_{\text{opt}} = \Delta = x_k$, again no pivot is necessary, because it is the nonbasic variable x_k that has been driven to zero (of course, if $\theta_{\text{opt}} = x_k = -\Delta'$, one of the basic variables will reach zero just as x_k does, and degeneracy will be introduced). However, if $\theta_{\text{opt}} = \Delta = -\Delta' < x_k$, the basic variable $x_{\mathbf{B}r}$ is driven to zero while x_k is still positive; in this case a simplex pivot must be performed, with x_k replacing $x_{\mathbf{B}r}$ in the basis. We remind the reader that the new values of the variables should be calculated directly from (8-44), rather than from the usual transform formula involving the vector $\mathbf{\Phi}$. Note that if the old basis is degenerate, with $x_{\mathbf{B}i} = 0$ and $y_{ik} < 0$ for some i, then $x_{\mathbf{B}r} = x_{\mathbf{B}i}$, $\Delta' = 0$, $\Delta = 0$, and $\theta_{\text{opt}} = 0 < x_k$, so the new solution point $\hat{\mathbf{x}}$ is the same as the old; however, because $x_k > 0$ replaces $x_{\mathbf{B}r} = 0$ in the basis, the value of the rth basic variable is increased by the pivot and degeneracy may in fact be eliminated.

8.6 AN EXAMPLE

We have seen that the convex simplex method, like the linear simplex method, is built around a set of m basic variables whose associated columns in the matrix $\mathbf{A}$ are linearly independent. At each iteration a single nonbasic variable x_k is identified for which an increase or a decrease in value would be feasible (i.e., if $x_k = 0$, it cannot be decreased) and would initially cause the objective value $z = f(\mathbf{x})$ to improve; if no such x_k can be found, it was shown that the current solution must be a Kuhn-Tucker point. As the value of x_k changes, with the other nonbasic variables being held fixed, the value of each basic variable $x_{\mathbf{B}i}$ increases or decreases at a constant rate y_{ik}, until either some variable is driven to zero or the objective function stops improving. Thus the objective value z must be improved by a nonzero amount at each iteration, except when the basis is degenerate, in which case z may remain unchanged. If a basic variable $x_{\mathbf{B}r}$ is driven to zero, or if displacement away from the current solution point is prevented by the fact that $x_{\mathbf{B}r} = 0$ initially, a simplex pivot is performed in which x_k replaces $x_{\mathbf{B}r}$ in the basis; otherwise, the set of basic variables remains the same for the next iteration. We have not yet examined the convergence behavior of the convex simplex method or even defined what we mean by algorithmic convergence in a constrained optimization context; a general discussion of this issue, covering all feasible-directions methods, will be presented in Section 8.14.

Let us now illustrate the convex simplex method by applying it to the following convex program:

$$\text{Max } z = f(x_1, x_2) = 10 \log(x_1 + 1) - 2x_1 + x_1(x_2 + 1)^2$$
$$\text{subject to } x_1 + x_2 \leqq 12,$$
$$-x_1 + x_2 \leqq 6,$$
$$\text{and} \quad x_1, x_2 \geqq 0,$$

where the "log" in the objective is the natural logarithm. With the addition of slack variables, the constraints become

$$x_1 + x_2 + x_3 \qquad = 12,$$
$$-x_1 + x_2 \qquad + x_4 = 6,$$

and

$$x_1, x_2, x_3, x_4 \geqq 0.$$

The first partial derivatives of the objective function, for future reference, are

$$\frac{\delta f}{\delta x_1} = \frac{10}{x_1 + 1} - 2 + (x_2 + 1)^2,$$
$$\frac{\delta f}{\delta x_2} = 2x_1(x_2 + 1),$$

and

$$\frac{\delta f}{\delta x_3} = \frac{\delta f}{\delta x_4} = 0.$$

If we take the origin as our starting point, the initial basic solution is $\mathbf{x_B} = (x_3, x_4) = (12, 6)$, with the nonbasic variables x_1 and x_2 both having values of zero. The initial basis is the identity matrix $\mathbf{I}_2$, and the first convex simplex tableau is constructed in the usual manner, omitting the reduced costs of the linear simplex method:

	x_1	x_2	x_3	x_4
$x_{\mathbf{B}1} = x_3 = 12$	1	1	1	0
$x_{\mathbf{B}2} = x_4 = \ \ 6$	−1	1	0	1

We begin the first iteration by computing the partial derivatives

$$\frac{\delta z}{\delta x_j} = \frac{\delta f}{\delta x_j} - \nabla \mathbf{f}(\mathbf{x})_{\mathbf{B}} \cdot \mathbf{y}_j, \qquad j \in J. \tag{8-28}$$

Inasmuch as $\nabla \mathbf{f}(\mathbf{x})_{\mathbf{B}} = [\delta f/\delta x_3, \delta f/\delta x_4] = [0, 0]$, these values are

$$\frac{\delta z}{\delta x_1} = \frac{10}{x_1 + 1} - 2 + (x_2 + 1)^2 = 9$$

and

$$\frac{\delta z}{\delta x_2} = 2x_1(x_2 + 1) = 0.$$

Referring to the criteria (8-30) and (8-31), we find that the only desirable adjustment in a nonbasic variable is an increase in x_1. Because $\mathbf{y}_1 = (1, -1)$ has a positive component, this adjustment falls within the domain of case A. The maximum amount by which the value of $x_k = x_1$ can be increased is found by computing

$$\Delta = \min_i \left(\frac{x_{\mathbf{B}i}}{y_{ik}}, y_{ik} > 0\right) = \frac{x_{\mathbf{B}1}}{y_{11}} = \frac{12}{1} = 12.$$

That is, to maintain feasibility, we cannot permit the value of x_1 to increase by more than 12 units—any further increase would result in a negative value of the basic variable $x_{\mathbf{B}1} = x_3$. The desired increase in x_1 is therefore determined by solving the problem

$$\begin{cases} \text{Max } f(\hat{\mathbf{x}}) = f(\mathbf{x} + \theta\mathbf{s}) \\ \text{subject to } 0 \leqq \theta \leqq 12, \end{cases} \tag{8-45}$$

where $\mathbf{x} = (0, 0, 12, 6)$ and $\mathbf{s} = (1, 0, -y_{11}, -y_{21}) = (1, 0, -1, 1)$. The objective of this problem is to maximize

$$f(\mathbf{x} + \theta\mathbf{s}) = f(\theta, 0) = 10 \log(\theta + 1) - \theta \tag{8-46}$$

[in using the notation $f(\theta, 0)$ we omitted the third and fourth arguments, $12 - \theta$ and $6 + \theta$, because x_3 and x_4 do not appear in the objective function of the original problem]. The derivative of (8-46) is

$$\frac{df}{d\theta} = \frac{10}{\theta + 1} - 1,$$

which vanishes at $\theta = 9$; this value is within the permitted range $0 \leqq \theta \leqq 12$

and is therefore the optimal solution to (8-45). The new values of the variables are

$$\hat{\mathbf{x}} = \mathbf{x} + \theta_{\text{opt}}\mathbf{s} = \begin{bmatrix} 0 \\ 0 \\ 12 \\ 6 \end{bmatrix} + 9 \cdot \begin{bmatrix} 1 \\ 0 \\ -1 \\ 1 \end{bmatrix} = \begin{bmatrix} 9 \\ 0 \\ 3 \\ 15 \end{bmatrix}.$$

Because $\theta_{\text{opt}} = 9$ is less than $\Delta = 12$, no pivot is required; however, the values of the basic variables have changed, and the tableau is now as follows:

	x_1	x_2	x_3	x_4
$x_{\mathbf{B}1} = x_3 = 3$	1	1	1	0
$x_{\mathbf{B}2} = x_4 = 15$	-1	1	0	1

To begin the second iteration we evaluate the partial derivatives $\delta z/\delta x_j$, $j \in J$, at the current point $\mathbf{x} = (9, 0, 3, 15)$:

$$\frac{\delta z}{\delta x_1} = \frac{10}{x_1 + 1} - 2 + (x_2 + 1)^2 = 0$$

and

$$\frac{\delta z}{\delta x_2} = 2x_1(x_2 + 1) = 18.$$

Of course, having optimized in the x_1-direction in the previous iteration, we knew in advance that we would find $\delta z/\delta x_1 = 0$. Because the partial derivative $\delta z/\delta x_2$ is positive, optimality has not yet been achieved, and the variable x_2 is chosen to be increased. The vector $\mathbf{y}_2$ has positive components, placing us again within the domain of case A. The maximum amount by which x_2 can be increased is

$$\Delta = \min_i\left(\frac{x_{\mathbf{B}i}}{y_{i2}}, y_{i2} > 0\right) = \min\left(\frac{3}{1}, \frac{15}{1} = 3 = \right)\frac{x_{\mathbf{B}1}}{y_{12}},$$

so we must solve the problem

(8-47) $$\text{Max } f(\mathbf{x} + \theta\mathbf{s}) \text{ subject to } 0 \leqq \theta \leqq 3,$$

where $\mathbf{x} = (9, 0, 3, 15)$ and $\mathbf{s} = (0, 1, -1, -1)$. Here the objective is

$$f(\mathbf{x} + \theta\mathbf{s}) = f(9, \theta) = 10 \log 10 - 18 + 9(\theta + 1)^2,$$

and the optimal solution to (8-47) is obviously $\theta_{\text{opt}} = 3$. The new solution point is $\mathbf{x} + \theta_{\text{opt}}\mathbf{s} = (9, 3, 0, 12)$, and the new tableau is obtained by pivoting x_2 into the basis to replace $x_{\mathbf{B}1} = x_3$, using the computational vector $\mathbf{\Phi} = (1/y_{12} - 1, -y_{22}/y_{12}) = (0, -1)$:

	x_1	x_2	x_3	x_4
$x_{\mathbf{B}1} = x_2 = 3$	1	1	1	0
$x_{\mathbf{B}2} = x_4 = 12$	−2	0	−1	1

The new values of the basic variables were entered directly, although in this case they could have been obtained via the $\mathbf{\Phi}$-transformation.

For the third iteration the "basis gradient" is

$$\nabla \mathbf{f}(\mathbf{x})_{\mathbf{B}} = \left[\frac{\delta f}{\delta x_2}, \frac{\delta f}{\delta x_4}\right] = [2x_1(x_2 + 1), 0] = [72, 0],$$

and the partial derivatives with respect to the nonbasic variables are

$$\frac{\delta z}{\delta x_1} = \frac{10}{x_1 + 1} - 2 + (x_2 + 1)^2 - [72, 0]\cdot(1, -2) = -57$$

and

$$\frac{\delta z}{\delta x_3} = 0 - [72, 0]\cdot(1, -1) = -72.$$

Inasmuch as the positive-valued nonbasic variable x_1 has a negative partial derivative, the current solution is still not optimal and x_1 is chosen to be decreased (case C). Using (8-41) and (8-42), we compute

$$\Delta' = \max_i\left(\frac{x_{\mathbf{B}i}}{y_{i1}}, y_{i1} < 0\right) = \frac{x_{\mathbf{B}2}}{y_{21}} = -6$$

and

$$\Delta = \min(-\Delta', x_1) = \min(6, 9) = 6.$$

Thus the desired amount of decrease in x_1 is obtained by solving the following problem:

$$\text{Max } f(\mathbf{x} - \theta\mathbf{s})$$
$$\text{subject to } 0 \leqq \theta \leqq 6,$$

where $\mathbf{x} = (9, 3, 0, 12)$ and $\mathbf{s} = (1, -1, 0, 2)$ and where

$$f(\mathbf{x} - \theta\mathbf{s}) = f(9 - \theta, 3 + \theta) = 10 \log(10 - \theta) + 126 + 58\theta + \theta^2 - \theta^3.$$

Following the search strategy suggested in the previous section, we evaluate this function at each of a sequence of increasing values of θ, incrementing by 1.0 at each step:

θ	$f(9 - \theta, 3 + \theta)$
.0	149.03
1.0	205.97
2.0	258.79
3.0	301.46
4.0	327.92
5.0	332.09
6.0	307.86

The first decrease in the function f is observed at the step from $\theta = 5.0$ to $\theta = 6.0$; hence, if we assume unimodality, the optimal value θ_{opt} must lie between 4.0 and 6.0. Proceeding now via golden section search, we evaluate the function (or place "experiments") at those points that are $r = .618$ of the way from either end of the interval [4.0, 6.0] to the other:

$$\theta = 4.764 \qquad f = 333.44,$$

$$\theta = 5.236 \qquad f = 329.17.$$

The maximum value found so far is at $\theta = 4.764$, so we next evaluate the function at the point that is symmetrically located with respect to $\theta = 4.764$ within the remaining interval [4.0, 5.236]:

$$\theta = 4.472 \qquad f = 333.03.$$

Continuing in this manner, we eventually obtain $\theta_{\text{opt}} = 4.671$, with $f(9 - \theta_{\text{opt}}, 3 + \theta_{\text{opt}}) = 333.55$; the new solution point is $\hat{\mathbf{x}} = \mathbf{x} - \theta_{\text{opt}}\mathbf{s} = (4.329, 7.671, 0.0, 2.658)$. Because θ_{opt} is less than $\Delta = 6$, no pivot is required; to update the tableau, it is only necessary to adjust the values of the basic variables x_2 and x_4.

For the next iteration the basis gradient is

$$\nabla \mathbf{f}(\mathbf{x})_{\mathbf{B}} = \left[\frac{\delta f}{\delta x_2}, \frac{\delta f}{\delta x_4}\right] = [2x_1(x_2 + 1), 0] = [75.07, 0],$$

and the partial derivatives of the nonbasic variables are

$$\frac{\delta z}{\delta x_1} = \frac{10}{x_1 + 1} - 2 + (x_2 + 1)^2 - [75.07, 0]\cdot(1, -2) = 0.0$$

and

$$\frac{\delta z}{\delta x_3} = 0 - [75.07, 0]\cdot(1, -1) = -15.07.$$

Observe that neither of the $\delta z/\delta x_j$ is positive and that neither of the products $(\delta z/\delta x_j)\cdot x_j$ is negative; it follows that the current solution is at least a Kuhn-Tucker point of the original problem. In fact, as can be seen from the indifference curves sketched in Fig. 8-5, the solution obtained by the convex simplex method is the global optimum. The heavy arrows in the figure trace the path taken by the algorithm from starting point to optimum.

It should be evident from this example that while the convex simplex method is to some extent based on the linear simplex method, it is rather more complex in its logic and in its calculations. Although a convex simplex computer code could borrow the tableau format and pivoting computations from a linear simplex code, it would still need several additional features, including a function statement or subprogram for each of the partial derivatives $\delta f/\delta x_j$; various logical tests for identifying a Kuhn-Tucker point, choosing a nonbasic variable to be adjusted, and computing the maximum permitted adjustment Δ; and, most critically, a computational subroutine for determining the value θ_{opt} that optimizes $f(\mathbf{x} \pm \theta\mathbf{s})$ for any $\mathbf{x}$ and $\mathbf{s}$. Notice, moreover, that the actual coding of the objective and its partial derivatives must be changed with each new problem (unless the program is to be restricted solely to solving problems whose objective functions are of a certain fixed algebraic form and differ only in the values of constant parameters); thus, to have general applicability, a convex simplex code must be able to accept user-

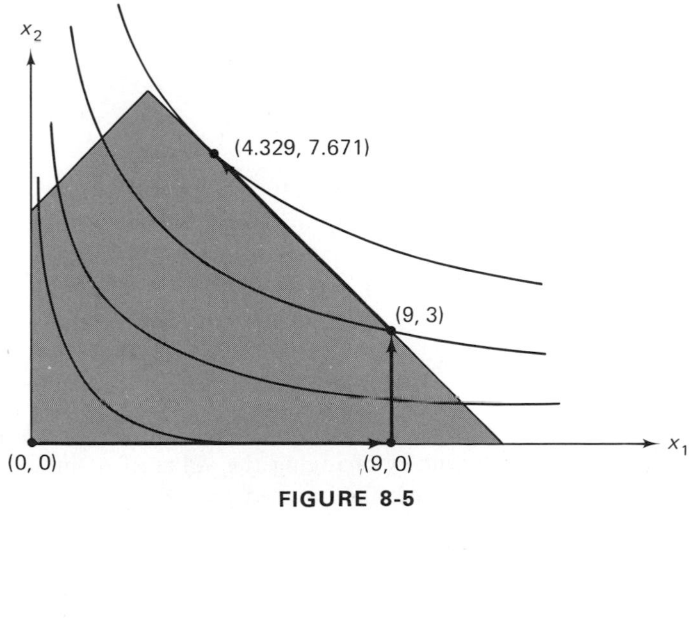

FIGURE 8-5

supplied subroutines and/or function statements. For these reasons, and also because of the lower overall level of commercial demand for nonlinear programming methods of all types, flexible convex simplex computer codes are not yet generally available. Nevertheless, the method itself appears to be quite efficient, especially for convex objective functions.

8.7 THE REDUCED GRADIENT METHOD

Like the convex simplex method, the reduced gradient method of Wolfe [54] is applicable to any problem of the form

$$\text{(8-24)} \qquad \begin{cases} \text{Max } z = f(\mathbf{x}) \\ \text{subject to } \mathbf{Ax} = \mathbf{b} \\ \qquad \text{and} \quad \mathbf{x} \geqq \mathbf{0}, \end{cases}$$

where $\mathbf{A}$ is an m-by-n matrix and where $f(\mathbf{x})$ is continuous and has continuous first partial derivatives. The two methods were developed at about the same time—the mid-1960s—and are, moreover, quite similar in their operation, differing only in the direction of displacement chosen at each iteration. The reduced gradient method begins with a basic feasible solution obtained via the usual phase I simplex approach. At any subsequent stage, then, there will be a set of basic variables $\mathbf{x_B} = (x_{\mathbf{B}1}, \ldots, x_{\mathbf{B}m})$, with associated basis matrix $\mathbf{B}$, and a group of $n - m$ nonbasic variables $x_j, j \in J$; the values of the latter may be positive or zero. Following our development of the convex simplex method, we can partition the constraints $\mathbf{Ax} = \mathbf{b}$ into basic and nonbasic portions,

$$\mathbf{Ax} = \mathbf{Bx_B} + \sum_{j \in J} x_j \mathbf{a}_j = \mathbf{b},$$

leading to

$$\text{(8-26)} \qquad \mathbf{x_B} = \mathbf{b}_0 - \sum_{j \in J} x_j \mathbf{y}_j$$

and eventually to the familiar expression for the partial derivatives of the nonbasic variables:

$$\text{(8-28)} \qquad \frac{\delta z}{\delta x_j} = \frac{\delta f}{\delta x_j} - \nabla \mathbf{f}(\mathbf{x})_{\mathbf{B}} \cdot \mathbf{y}_j, \qquad j \in J,$$

where again J denotes the set of subscripts of the nonbasic variables x_j in the current solution.

However, where the strategy of the convex simplex method was to move toward a better solution point by changing the value of a single nonbasic variable, holding the others fixed, the reduced gradient strategy is to allow all nonbasic variables whose values can change favorably to do so simultaneously, at rates that are proportional in magnitude to the partial derivatives $\delta z/\delta x_j$. Specifically, the direction of displacement $\mathbf{u}$ from the current point in nonbasic space has the following components:

$$u_j = \begin{cases} 0 & \text{if } \dfrac{\delta z}{\delta x_j} \leqq 0 \quad \text{and} \quad x_j = 0, \\ \dfrac{\delta z}{\delta x_j} & \text{otherwise.} \end{cases} \tag{8-48}$$

The $(n - m)$-component vector $\mathbf{u}$ defined by (8-48) is called the *reduced gradient*; note that when all x_j are positive $\mathbf{u}$ coincides with the gradient vector and points in the direction of most rapid improvement in nonbasic space. As the nonbasic variables x_j change in accordance with (8-48), their combined effects on the values of the basic variables can be computed easily from (8-26): An increase of u_j units in the value of every nonbasic variable x_j would cause an increase of

$$v_i = -\sum_{j \in J} u_j y_{ij} \tag{8-49}$$

units in the ith basic variable $x_{\mathbf{B}i}$. Thus the direction of displacement in the space of the variables $x_1, \ldots, x_n$ has components (8-48) in the directions of the nonbasic variables and components (8-49) in the directions of the basic variables. For convenience, we use the vector $\mathbf{s} = (s_1, \ldots, s_n)$ to denote this direction, where

$$\begin{cases} s_j = u_j & \text{if } j \in J, \\ s_j = v_i & \text{if } j \notin J \text{ and } x_j = x_{\mathbf{B}i}. \end{cases} \tag{8-50}$$

As in all feasible-directions methods, displacement from the current point in the selected direction $\mathbf{s}$ continues until either the objective stops improving or a boundary of the feasible region is encountered (i.e., a variable is driven to zero). The maximum amount of displacement that can take place before the latter event occurs is determined via the following computations:

$$\begin{cases} \Delta_1 = \max\limits_{j \in J}\left(\dfrac{x_j}{u_j}, u_j < 0\right), \\ \Delta_2 = \max\limits_{1 \leq i \leq m}\left(\dfrac{x_{\mathbf{B}i}}{v_i}, v_i < 0\right), \\ \Delta = \min(-\Delta_1, -\Delta_2). \end{cases} \tag{8-51}$$

Here $-\Delta_1$ is the (nonnegative) step length in the direction $\mathbf{s}$ that is just sufficient to drive the value of a nonbasic variable to zero; if none of the u_j

are negative, implying that the displacement under consideration causes none of the nonbasic variables to decrease in value, $\mathbf{\Delta}_1$ is set at $-\infty$. Similar remarks apply to $\mathbf{\Delta}_2$ with respect to the basic variables; thus it is possible that $\mathbf{\Delta} = \infty$.

Having established that in our displacement from the current point $\mathbf{x}$ we can go no further than $\mathbf{x} + \mathbf{\Delta s}$ without leaving the feasible region, we now wish to determine the feasible displacement that yields the best value of the objective function. This is accomplished by solving the problem

$$\text{(8-52)} \qquad \begin{cases} \text{Max } f(\mathbf{x} + \theta\mathbf{s}) \\ \text{subject to } 0 \leqq \theta \leqq \mathbf{\Delta}. \end{cases}$$

This is a one-dimensional direct search problem exactly like those that must be solved in the course of the convex simplex method, except that the number of variables involved—that is, the number of nonzero components of $\mathbf{s}$—is not limited to $m + 1$ and may be as great as n. The completion of the iteration depends, in a manner that by now should be familiar, on the value of θ that optimizes (8-52).

1. If $\mathbf{\Delta} = \infty$ and no finite value of θ is optimal, the solution to the original problem (8-24) is unbounded and the reduced gradient method terminates.
2. If θ_{opt} is finite and $\theta_{\text{opt}} = \mathbf{\Delta} = -\mathbf{\Delta}_2$, the displacement from $\mathbf{x}$ to $\mathbf{x} + \theta_{\text{opt}}\mathbf{s}$ drives at least one basic variable, call it $x_{\mathbf{B}r}$, to zero (if two or more are driven to zero simultaneously, degeneracy is introduced). A simplex pivot is then executed, with $x_{\mathbf{B}r}$ being replaced in the basis by some nonbasic variable x_j for which the element y_{rj} is nonzero. There must be at least one such x_j, because $v_r \neq 0$; if there are more than one, let the variable with the greatest value be chosen to enter the basis. As was true in the case of convex simplex pivots, the new values of all the variables should in general be computed from the relation $\hat{\mathbf{x}} = \mathbf{x} + \theta_{\text{opt}}\mathbf{s}$, because the $\mathbf{\Phi}$-transformation of ordinary linear programming fails when the initial values of any of the nonbasic variables are positive. Note that if $\theta_{\text{opt}} = \mathbf{\Delta} = -\mathbf{\Delta}_2 = 0$, which may happen if the current basis is degenerate, the new point will be the same as the old; however, in the ensuing pivot the zero-valued basic variable $x_{\mathbf{B}r}$ might well be replaced by some positive x_j, thus reducing and possibly eliminating degeneracy.
3. If θ_{opt} is finite and $\theta_{\text{opt}} < -\mathbf{\Delta}_2$, then none of the basic variables are driven to zero (and if $\theta_{\text{opt}} < -\mathbf{\Delta}_1$ as well, neither are any of the *non*basic variables). In this case no pivot is necessary; we simply adjust the values of the variables and proceed to the next iteration.

The reduced gradient method continues in this manner until at some iteration it is found that the reduced gradient vector $\mathbf{u}$ vanishes. The method then terminates, and it is easily shown that the final solution $\mathbf{x}_0$ must be a

Kuhn-Tucker point—the proof is an almost exact duplicate of the termination proof given for the convex simplex method. From (8-48), the condition $\mathbf{u} = \mathbf{0}$ that holds at $\mathbf{x}_0$ is equivalent to

$$\frac{\delta z(\mathbf{x}_0)}{\delta x_j} \leqq 0 \quad \text{and} \quad \frac{\delta z(\mathbf{x}_0)}{\delta x_j} \cdot x_{0j} = 0, \qquad j \in J. \tag{8-53}$$

If we define an m-component row vector $\boldsymbol{\lambda}_0 \equiv \nabla \mathbf{f}(\mathbf{x}_0)_{\mathbf{B}} \cdot \mathbf{B}^{-1}$, with components λ_{0i}, we can write

$$\begin{aligned} \frac{\delta z}{\delta x_j} &= \frac{\delta f}{\delta x_j} - \nabla \mathbf{f}(\mathbf{x}_0)_{\mathbf{B}} \cdot \mathbf{y}_j = \frac{\delta f}{\delta x_j} - \nabla \mathbf{f}(\mathbf{x}_0)_{\mathbf{B}} \cdot \mathbf{B}^{-1}\mathbf{a}_j \\ &= \frac{\delta f}{\delta x_j} - \boldsymbol{\lambda}_0 \cdot \mathbf{a}_j = \frac{\delta f}{\delta x_j} - \sum_{i=1}^{m} \lambda_{0i} a_{ij}, \qquad j \in J, \end{aligned} \tag{8-54}$$

where all partial derivatives are understood as being evaluated at $\mathbf{x}_0$. Moreover, if x_j is the rth basic variable $x_{\mathbf{B}r}$, then $\mathbf{y}_j$ is the unit vector $\mathbf{e}_r$, and the following equalities must also hold at $\mathbf{x}_0$:

$$\frac{\delta f}{\delta x_j} - \sum_{i=1}^{m} \lambda_{0i} a_{ij} = \frac{\delta f}{\delta x_j} - \nabla \mathbf{f}(\mathbf{x}_0)_{\mathbf{B}} \cdot \mathbf{y}_j = \frac{\delta f}{\delta x_j} - \frac{\delta f}{\delta x_{\mathbf{B}r}} = 0, \qquad j \notin J. \tag{8-55}$$

If then follows from (8-53), (8-54), and (8-55) that $\mathbf{x}_0$ and $\boldsymbol{\lambda}_0$ jointly satisfy the relations

$$\frac{\delta f}{\delta x_j} - \sum_{i=1}^{m} \lambda_i a_{ij} \leqq 0, \qquad j = 1, \ldots, n,$$

and

$$x_j \left\{ \frac{\delta f}{\delta x_j} - \sum_{i=1}^{m} \lambda_i a_{ij} \right\} = 0, \qquad j = 1, \ldots, n,$$

which constitute the Kuhn-Tucker conditions for the original problem. Q.E.D. We remark, as before, that if the objective function being maximized is concave, then the solution $\mathbf{x}_0$ is a global optimum.

The reduced gradient method differs from the convex simplex method in that it permits the value of every nonbasic variable, rather than just one, to change at each iteration. Because of this complete freedom of search in nonbasic space, the reduced gradient method selects better directions of displacement than does the convex simplex method and can thus be expected to require fewer iterations for convergence. On the other hand, more computational effort is needed at each iteration of the reduced gradient method, both in calculating the components of the direction of displacement and in determining the maximum feasible distance Δ. This difference in effort per iteration is not overwhelming, however, particularly inasmuch as the one-dimensional search for θ_{opt}, which is relatively time-consuming, is exactly the same in the two methods. Thus it is not surprising that the reduced gradient method has in fact proved somewhat more efficient overall than the convex simplex method. Moreover, its relative advantage tends to be greater when very few of the original inequality constraints (i.e., inequalities before the addition of slack variables required to convert to standard form) are binding at the optimal solution: In such cases the problem-solving process is closer to that of unconstrained optimization, a mode in which the selection of efficient direction vectors is all-important.

Regarding the nature of the solution points obtained, although we have been able to prove only that the solutions generated by the convex simplex and reduced gradient methods are Kuhn-Tucker points, in practice these methods almost always yield local optima. It was mentioned in Chapter 4 that, in seeking unconstrained local maxima of some given objective function via a gradient-following method, there is no real danger of terminating at a local minimum or saddle point unless the method begins at one, or at least begins at a starting point that is locally minimal with respect to perturbations of some variable x_p (i.e., at the starting point $\delta f/\delta x_p = 0$ but $\delta^2 f/\delta x_p^2$ is positive). Moreover, such "false terminations" are always razor's edge situations and can readily be identified by the simple means of evaluating the objective function for small perturbations of each variable. The same remarks, suitably modified, apply also to the two simplex-based algorithms we have studied in the past three sections: In general, they do yield local optima.

As a counterexample, however, consider what happens when the reduced gradient method is applied to the following problem:

$$\begin{aligned} \text{Max } z = f(x_1, x_2) &= (x_1 - 1)^2 + x_2 \\ \text{subject to } x_1 - x_2 + x_3 \qquad &= 0, \\ x_2 \qquad + x_4 &= 1, \\ \text{and} \quad x_1, x_2, x_3, x_4 &\geqq 0, \end{aligned}$$

where x_3 and x_4 are slack variables. Let the starting solution be $x_1 = x_2 = 1$, $x_3 = x_4 = 0$, and suppose the two basic variables are $x_{\mathbf{B}1} = x_1 = 1$ and $x_{\mathbf{B}2} = x_2 = 1$. Then the basis inverse and **y**-columns are

$$\mathbf{B}^{-1} = \begin{bmatrix} 1 & -1 \\ 0 & 1 \end{bmatrix}^{-1} = \begin{bmatrix} 1 & 1 \\ 0 & 1 \end{bmatrix}, \quad \mathbf{y}_3 = \mathbf{B}^{-1}\mathbf{a}_3 = \begin{bmatrix} 1 \\ 0 \end{bmatrix}, \quad \text{and} \quad \mathbf{y}_4 = \mathbf{B}^{-1}\mathbf{a}_4 = \begin{bmatrix} 1 \\ 1 \end{bmatrix}.$$

The partial derivatives of the objective function are

$$\frac{\delta f}{\delta x_1} = 2(x_1 - 1), \quad \frac{\delta f}{\delta x_2} = 1, \quad \text{and} \quad \frac{\delta f}{\delta x_3} = \frac{\delta f}{\delta x_4} = 0,$$

so the current basis gradient is

$$\nabla \mathbf{f}(\mathbf{x})_{\mathbf{B}} = \left[\frac{\delta f}{\delta x_1}, \frac{\delta f}{\delta x_2}\right] = [0, 1],$$

and the partial derivatives with respect to the nonbasic variables are as follows:

$$\frac{\delta z}{\delta x_3} = \frac{\delta f}{\delta x_3} - \nabla \mathbf{f}(\mathbf{x})_{\mathbf{B}} \cdot \mathbf{y}_3 = 0 - [0, 1] \cdot (1, 0) = 0$$

and

$$\frac{\delta z}{\delta x_4} = \frac{\delta f}{\delta x_4} - \nabla \mathbf{f}(\mathbf{x})_{\mathbf{B}} \cdot \mathbf{y}_4 = -1.$$

From (8-48) we have $u_3 = u_4 = 0$, so the reduced gradient vanishes and the current solution is a Kuhn-Tucker point. But notice that it is not a local maximum, because any small decrease in the value of x_1 is feasible and increases the value of z.

8.8 DEGENERACY IN LINEARLY CONSTRAINED PROBLEMS

Degeneracy in linear and mathematical programming is most conveniently defined in algebraic, rather than geometric, terms. In linear programming texts it is customary to explain what a basic solution is and then to define a basic solution as being *degenerate* if one or more of the basic variables has a value of zero. This definition can in fact be applied throughout the entire general class of linearly constrained mathematical programs. We have, accordingly, been using it in this text, even in discussing the convex simplex and reduced gradient methods, for which it is not necessarily true that the value of every *nonbasic* variable is zero.

In addition to its algebraic meaning, degeneracy also has an instructive and useful geometric interpretation. It will be recalled that any linear constraint in n variables

$$g_i(\mathbf{x}) = a_{i1}x_1 + \cdots + a_{in}x_n\{\leqq, =, \geqq\}b_i \tag{8-56}$$

defines a hyperplane in Euclidean n-space E^n—that is, a straight line in E^2, a plane in E^3, and so on. Any point $\mathbf{x}$ in E^n that satisfies $g_i(\mathbf{x}) = b_i$ lies in the hyperplane, and at such a point the constraint is said to be *binding*, whether it is given in equality or inequality form (thus an equality constraint is automatically binding at any point that satisfies it). It will also be recalled, from Chapter 4, that the gradient vector

$$\nabla \mathbf{g}_i^T \equiv \left[\frac{\delta g_i}{\delta x_1}, \ldots, \frac{\delta g_i}{\delta x_n}\right] = [a_{i1}, \ldots, a_{in}], \tag{8-57}$$

which is sometimes referred to as the gradient of the constraint (8-56) or the gradient of its associated hyperplane $g_i(\mathbf{x}) = b_i$, is perpendicular to that hyperplane.[6] Using these background ideas, we can state a simple geometric definition of degeneracy: A set of linear constraints of the form (8-56), including equations and/or inequalities, is *degenerate* at a point $\mathbf{x}_0$, or the point $\mathbf{x}_0$ is degenerate, *if the gradients of the constraints that are binding at* $\mathbf{x}_0$ *are linearly dependent*. Notice that the concept of degeneracy, as defined here in terms of the gradients of binding constraints, can be directly extended to include nonlinear constraint sets as well. We shall make use of this more general notion in Section 9.4; for the present, however, we restrict the discussion to linear constraint sets only.

In most cases of degeneracy the linear dependence of the constraint gradients follows *a fortiori* from the intersection of $n + 1$ constraint hyperplanes at a single point in E^n. Two examples in E^2 are diagrammed in Fig. 8-6; the constraints that are binding at the degenerate points shown are as follows:

$$\text{(a)}\quad \begin{aligned} x_1 + x_2 &\leqq 2, \\ x_2 &\leqq 1, \\ 4x_1 + x_2 &\leqq 5; \end{aligned} \qquad \text{(b)}\quad \begin{aligned} x_1 + x_2 &\geqq 1, \\ x_1 - x_2 &\leqq 1, \\ x_2 &\geqq 0. \end{aligned}$$

In view of the definition just stated, we know immediately that the point of

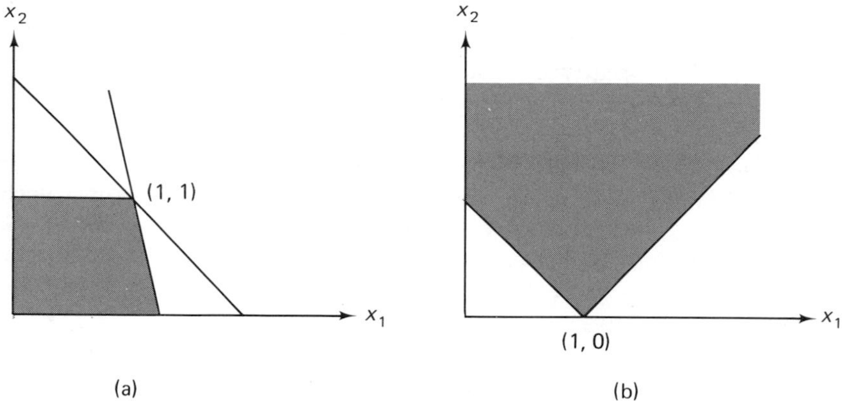

FIGURE 8-6

[6]For a simple and direct geometric proof of this proposition, let H denote the hyperplane of points $\mathbf{x}$ that satisfy $\boldsymbol{\alpha}_i\mathbf{x} = b_i$, where $\boldsymbol{\alpha}_i$ is the row vector $[a_{i1}, \ldots, a_{in}]$. Any vector lying in H can be represented as $\mathbf{x}_1 - \mathbf{x}_2$, where the points $\mathbf{x}_1$ and $\mathbf{x}_2$ are in H. But

$$\boldsymbol{\alpha}_i \cdot (\mathbf{x}_1 - \mathbf{x}_2) = \boldsymbol{\alpha}_i\mathbf{x}_1 - \boldsymbol{\alpha}_i\mathbf{x}_2 = b_i - b_i = 0,$$

so $\boldsymbol{\alpha}_i$ is perpendicular to any vector lying in the hyperplane H and is therefore perpendicular to H itself.

intersection in each of these cases must be degenerate, because three gradient vectors cannot be linearly independent in E^2. Degeneracy can also occur in E^n at points where n or fewer hyperplanes meet. One trivial example would arise if some inequality $g_i(\mathbf{x}) \leqq b_i$ were inadvertently included twice in the set of constraints; then any point $\mathbf{x}_0$ satisfying $g_i(\mathbf{x}_0) = b_i$ would automatically be a degenerate point.

To see how the algebraic and geometric definitions of degeneracy are related, consider the constraint set of a linear programming problem in standard form:

$$\mathbf{Ax} = \mathbf{b}, \qquad \mathbf{x} \geqq \mathbf{0}, \tag{8-58}$$

where $\mathbf{A}$ is an m-by-n matrix. For any given basic feasible solution generated by the linear simplex method, there are $n - m$ nonbasic variables with values of zero, implying that $n - m$ of the constraints $x_j \geqq 0$ are satisfied as equalities. These, added to the m constraints $\mathbf{Ax} = \mathbf{b}$, make a total of $(n - m) + m = n$ binding constraints. However, if the basic solution is degenerate in the algebraic or simplex sense, then at least one *basic* variable also has a value of zero, so that one additional nonnegativity restriction is satisfied as an equality. Thus, overall, at least $n + 1$ constraints are binding, and their gradients in E^n must be linearly dependent, which implies degeneracy in the geometric sense.

Both the algebraic and geometric interpretations of degeneracy have important computational implications in the development and evaluation of solution methods for linearly constrained mathematical programs. For the family of simplex-based algorithms we have studied thus far (including the simplex method itself, the algorithms of Wolfe and Beale, separable programming, and the convex simplex and reduced gradient methods) it is the algebraic aspect of degeneracy—the potential appearance of a basic variable with a value of zero—that has been of interest. These methods operate via basis pivoting, and the presence of a zero-valued basic variable at any stage introduces at least the theoretical possibility of cycling through an endlessly repeating sequence of degenerate bases, each a different representation of the same nonoptimal solution point. For practical purposes, as we have seen, it is best simply to ignore this possibility and to rely on the fact that infinite cycling has never actually been observed, except in two or three artificially contrived problems. Thus, simplex-based pivoting algorithms are usually equipped only with elementary tie-breaking rules for dealing with degeneracy. When, in using one of these methods, a sequence of degenerate pivots is encountered, it is simply allowed to play itself out, in accordance with the automatic selection criteria, until eventually it is terminated by a real pivot to a different solution point. As an alternative to accepting this remote danger of an infinite cycling disaster, certain lexicographic rules or perturbation techniques that are guaranteed to prevent the recurrence of any basic solution (see Chapter 6 of [21] for details) could be grafted onto any of the simplex-based algorithms. However, because they require additional programming sophistication as well as various amounts of extra computer time and storage, these anticycling precautions are not ordinarily used.

Although the geometric interpretation of degeneracy, in terms of the gradients of the binding constraints, was not explicitly of interest in our study of simplex-based pivoting methods, there are other solution algorithms for linearly constrained problems in which it plays an important role. Included in this latter group are the gradient projection method of Rosen [37], which will be discussed in detail below, and the variable reduction method of McCormick [34]. These algorithms employ the familiar feasible-directions strategy but actually threaten to break down when a degenerate point is reached, because in each case the computation of the next direction vector requires that the gradients of the currently binding constraints be linearly independent. Therefore, to provide theoretical justification for a method of this type, it is necessary either to exclude from consideration all problems in which any degeneracy whatever is present, or to attach a perturbation technique or some other computational device for escaping from a degenerate point. The gradient projection method, as we shall see, takes the latter approach. It is worth noting that, in this and similar methods, the occurrence of a degenerate solution does not introduce the possibility of any sort of infinite cycling: As soon as the new direction vector has been computed, the degenerate point is left behind.

8.9 METHODS OF FEASIBLE DIRECTIONS

Each of the algorithms for solving linearly constrained mathematical programming problems that we have studied in the past two chapters has been based to a significant extent on the simplex method. In particular, with the exception of Lemke's algorithm, they have all taken the simplicial ("simplex-like") approach of using the constraints to define the algebraic dependence of a subset of the variables, called the basic variables, on the remaining or nonbasic variables, which are then treated as being independent. Computationally, this idea is implemented by solving the constraints for the basic variables, thereby expressing them as functions of the nonbasic variables, and then in effect working in nonbasic space rather than in the space of the original problem. Thus in simplex-based methods the constraints are not held explicitly in view but are instead expressed implicitly in the changes in value that are forced upon the dependent basic variables in response to changes made freely in the independent variables. One implication of this arrangement is that it is impossible to adjust the values of *all* the variables simultaneously at freely chosen rates; in other words, it is impossible to move through the feasible region in some predetermined desirable direction, such as the gradient direction.

We now turn our attention to a family of computational algorithms for linearly constrained problems for which those restrictions on movement through the feasible region no longer hold. The methods are classical in origin and depend mainly on the strategy of moving at each iteration in either the gradient or some gradient-related direction, until either the objective value stops improving or a boundary of the feasible region is encountered. In the early 1960s they became known as *feasible-directions methods*, although that term, introduced by Zoutendijk [61], was subsequently applied to the convex simplex, reduced gradient, and other simplex-based algorithms that were appearing in the middle and late 1960s. In fact, the (linear) simplex method itself embodies the feasible-directions strategy, although, of course, a simplex pivot can only be terminated when a constraint hyperplane is encountered, never because the objective value stops improving. It seems reasonable, then, to divide feasible-directions algorithms for linearly constrained problems into two families, those that are based on the simplex method and those that are not; it is the latter group that we shall be considering in the next few sections.

The problem we propose to solve can be represented in the following general form:

$$\text{(8-59)} \qquad \begin{cases} \text{Max } z = f(\mathbf{x}), & \mathbf{x} = (x_1, \ldots, x_n), \\ \text{subject to } \boldsymbol{\alpha}_i\mathbf{x} \equiv \sum_{j=1}^{n} a_{ij}x_j \begin{cases} \leqq b_i, & i \text{ in } I, \\ = b_i, & i \text{ in } E, \end{cases} \end{cases}$$

where $f(\mathbf{x})$ is a scalar-valued function with continuous first partial derivatives; each $\boldsymbol{\alpha}_i \equiv [a_{i1}, \ldots, a_{in}]$ is a row vector of constant coefficients; each b_i is a real number; and I and E are, respectively, the sets of subscripts of the inequality and equality constraints. Note that the formulation does not explicitly require nonnegativity of the variables; any nonnegativity restrictions that may be present are included among the inequality constraints, in the form $-x_j \leqq 0$. It is convenient to let m be the total number of constraints, so that every i, for $i = 1, \ldots, m$, is in either I or E. We remark that the solution methods to be presented below will deal directly with the constraints as they are given in (8-59)—that is, no slack variables will be added to convert the inequalities into standard form.

Before beginning our discussion of non-simplex-based feasible-directions algorithms, we pause to state a crucial definition on which they will all be based. Given a point $\mathbf{x}_k$ in a feasible region R in E^n, a *feasible direction* of displacement from $\mathbf{x}_k$ is a direction in which it is possible to proceed for at least some small positive distance through E^n without leaving R. In mathematical terms, $\mathbf{s}_k$ is a feasible direction from $\mathbf{x}_k$ if there exists some positive number δ such that all points of the form $\mathbf{x}_k + \theta \mathbf{s}_k$, where $0 \leqq \theta \leqq \delta$, lie in R. It should be noted that this definition does not depend in any way on the linearity of the constraints defining R. In fact, as we shall see in Chapter 9, feasible-directions algorithms can also be used to solve nonlinearly constrained problems.

Returning now to the linearly constrained problem (8-59), assume that an initial feasible solution $\mathbf{x}_0$ has been obtained; the problem of identifying an initial feasible solution when one is not readily available will be discussed at the end of this section. Each iteration of a feasible-directions algorithm then consists of the following four steps:

1. Determine which of the constraints are binding at the current feasible point $\mathbf{x}_k$, noting that every equality constraint is automatically binding.
2. Determine an *improving* feasible direction of displacement $\mathbf{s}_k$ from the point $\mathbf{x}_k$, that is, a feasible direction in which the gradient $\nabla \mathbf{f}(\mathbf{x}_k)$ has a positive component. In the search for $\mathbf{s}_k$ only the binding constraints need be considered (if no constraints are binding at $\mathbf{x}_k$, then *any* $\mathbf{s}$ in E^n constitutes a feasible direction, though not necessarily an improving feasible direction).
3. Determine the maximum permitted "step length" Δ in the $\mathbf{s}_k$-direction, that is, the maximum value of θ for which $\mathbf{x}_k + \theta \mathbf{s}_k$ is still a feasible solution to (8-59).
4. Determine the feasible step length θ_{opt} that yields the greatest increase in the objective value; then set $\mathbf{x}_{k+1} = \mathbf{x}_k + \theta_{\text{opt}} \mathbf{s}_k$ and return to step 1.

The original reference work on feasible-directions methods is that of Zoutendijk [61], who derived a number of general theoretical results about them and their convergence properties and suggested various means of increasing their efficiency. His later work [63] includes studies of the computational behavior of these and other nonlinear programming algorithms.

Step 1 of the general feasible-directions method requires little in the way of discussion: A simple evaluation of the dot product $\boldsymbol{\alpha}_i \mathbf{x}_k$ will determine whether or not the ith constraint is binding at $\mathbf{x}_k$. Step 3 is also quite straightforward computationally, owing to the linearity both of the constraint surfaces and of the path to be followed from $\mathbf{x}_k$ to $\mathbf{x}_{k+1}$. Observe that if the ith constraint is binding at $\mathbf{x}_k$ and if the path of displacement is either embedded in the hyperplane $\boldsymbol{\alpha}_i \mathbf{x} = b_i$ or directed away from it into the feasible region, then that path can be extended infinitely far without ever violating the ith constraint. Therefore, given that $\mathbf{s}_k$ is an immediately feasible direction from $\mathbf{x}_k$, in determining the maximum permitted displacement Δ we need only consider those inequality constraints that are *not binding* at $\mathbf{x}_k$. For each of them it must be true that

$$\boldsymbol{\alpha}_i(\mathbf{x}_k + \theta \mathbf{s}_k) = \boldsymbol{\alpha}_i \mathbf{x}_k + \theta \boldsymbol{\alpha}_i \mathbf{s}_k \leqq b_i,$$

so the parameter θ cannot be allowed to exceed

$$\Delta = \min_i \left(\frac{b_i - \boldsymbol{\alpha}_i \mathbf{x}_k}{\boldsymbol{\alpha}_i \mathbf{s}_k}, \boldsymbol{\alpha}_i \mathbf{s}_k > 0 \right), \tag{8-60}$$

where the index i ranges over all inequality constraints that are not binding at $\mathbf{x}_k$. If all the constraints are binding at $\mathbf{x}_k$, or if $\boldsymbol{\alpha}_i \mathbf{s}_k \leqq 0$ for all those that are not, then $\Delta = \infty$.

Step 4 in the outline above requires obtaining the optimal solution to the familiar one-dimensional search problem

$$\begin{cases} \text{Max } f(\mathbf{x}_k + \theta \mathbf{s}_k) \\ \text{subject to } 0 \leqq \theta \leqq \Delta. \end{cases} \tag{8-61}$$

This problem was discussed in connection with the convex simplex method; it can be solved in any of several ways, using the methods of Chapters 4 and 5. As a general rule, it is probably most efficient to settle for a fairly rough approximation to the optimal value θ_{opt}, except in the later stages, when the distances between successive points in the sequence $\{\mathbf{x}_k\}$ begin to grow small and greater accuracy is needed. Note that if $\Delta = \infty$ and no finite value of θ is optimal in (8-61), then the original problem (8-59) has an unbounded solution.

We come, finally, to step 2 of the general feasible-directions method, the problem of determining a direction of displacement $\mathbf{s}_k$ from the present solution point $\mathbf{x}_k$. Ideally, what we would like to find is the best feasible direction, that is, the feasible direction in which the component of the gradient $\nabla \mathbf{f}(\mathbf{x}_k)$ is largest and the immediate rate of improvement therefore greatest. This direction can be obtained from the following considerations. Let B_k be the set of subscripts i of the inequality constraints $\boldsymbol{\alpha}_i \mathbf{x} \leqq b_i$ that are *binding* at $\mathbf{x}_k$; thus B_k is a subset of I. If $\mathbf{s}$ is to be a feasible direction from $\mathbf{x}_k$, then at least some nearby points of the form $\mathbf{x}_k + \theta \mathbf{s}$, where $\theta > 0$, must satisfy all constraints that are binding at $\mathbf{x}_k$:

$$\boldsymbol{\alpha}_i(\mathbf{x}_k + \theta \mathbf{s}) \equiv \boldsymbol{\alpha}_i \mathbf{x}_k + \theta \boldsymbol{\alpha}_i \mathbf{s} \begin{cases} \leqq b_i & \text{for all } i \text{ in } B_k, \\ = b_i & \text{for all } i \text{ in } E. \end{cases} \tag{8-62}$$

Inasmuch as $\boldsymbol{\alpha}_i \mathbf{x}_k = b_i$ for each of these constraints, (8-62) reduces to

$$\boldsymbol{\alpha}_i \mathbf{s} \begin{cases} \leqq 0, & i \text{ in } B_k, \\ = 0, & i \text{ in } E. \end{cases}$$

Note that we are ignoring the nonbinding constraints here: If $\boldsymbol{\alpha}_i \mathbf{x}_k < b_i$ for some i in I, it must be possible to proceed in *any* direction from $\mathbf{x}_k$ for at least some small positive distance without violating the ith constraint. Thus the nonbinding constraints impose no restrictions at all on the choice of $\mathbf{s}_k$, and it follows that the feasible direction offering the greatest immediate rate of improvement can be determined, in theory, by solving the nonlinear program

$$\text{(8-63)} \qquad \begin{cases} \text{Max } \dfrac{1}{|\mathbf{s}|} \nabla \mathbf{f}^T(\mathbf{x}_k) \cdot \mathbf{s} & \\ \text{subject to } \boldsymbol{\alpha}_i \mathbf{s} \leqq 0, & i \text{ in } B_k, \\ \text{and } \boldsymbol{\alpha}_i \mathbf{s} = 0, & i \text{ in } E. \end{cases}$$

Here $|\mathbf{s}|$ is the magnitude of the vector $\mathbf{s}$,

$$|\mathbf{s}| \equiv (\mathbf{s}^T\mathbf{s})^{1/2} \equiv \left(\sum_{j=1}^{n} s_j^2\right)^{1/2},$$

so the objective function is the component of the gradient $\nabla \mathbf{f}(\mathbf{x}_k)$ in the direction $\mathbf{s}$.

The problem (8-63) has two undesirable characteristics: Its objective function is an extremely complicated function of the unknowns s_j, and its optimal solution will not be unique (i.e., if $\hat{\mathbf{s}}$ is an optimal solution, so is any positive multiple of $\hat{\mathbf{s}}$, including the unit-length vector $|\hat{\mathbf{s}}|^{-1}\hat{\mathbf{s}}$). Both of these defects can be remedied by the addition of a normalization constraint $\mathbf{s}^T\mathbf{s} = 1$, which will restrict attention to vectors of unit length only and thereby exclude from consideration all but one of the optimal solutions to (8-63). This new constraint implies $|\mathbf{s}| = 1$, so the objective function can be simplified and the problem (8-63) becomes

$$\text{(8-64)} \qquad \begin{cases} \text{Max } \nabla \mathbf{f}^T(\mathbf{x}_k) \cdot \mathbf{s} & \\ \text{subject to } \boldsymbol{\alpha}_i \mathbf{s} \leqq 0, & i \text{ in } B_k, \\ \boldsymbol{\alpha}_i \mathbf{s} = 0, & i \text{ in } E, \\ \text{and } \mathbf{s}^T\mathbf{s} = 1. & \end{cases}$$

Notice that this problem would be a linear program if it were not for the nonlinear constraint $\mathbf{s}^T\mathbf{s} = 1$.

Ideally, then, we would like to use the optimal solution to the problem (8-64) as the next direction vector $\mathbf{s}_k$. As a practical matter, however, that single nonlinear constraint makes (8-64) rather difficult to solve, except when there are no binding constraints at $\mathbf{x}_k$—in that case B_k and E are empty and we know immediately that the optimal solution must be $\mathbf{s} = |\nabla \mathbf{f}(\mathbf{x}_k)|^{-1} \nabla \mathbf{f}(\mathbf{x}_k)$, a vector of unit length in the gradient direction. But apart from that special case the difficulty of solving (8-64) should make us willing to settle for any reasonably good feasible direction, perhaps even for any feasible direction whatever in which the gradient $\nabla \mathbf{f}(\mathbf{x}_k)$ has a positive component. There are in fact several different ways of modifying (8-64) so as to obtain a feasible direction that can be used at the next iteration. Some are more complicated and require more calculation than others, but they tend also to yield better directions $\mathbf{s}_k$ and therefore to arrive at the optimal solution of the original problem (8-59) in fewer iterations. Each of these different approaches for obtaining the direction vectors $\mathbf{s}_k$ characterizes a different member of the family of feasible-directions methods; a number of them will be discussed in some detail below. Incidentally, if the issue of degeneracy, in the geometric sense described in Section 8.8, arises at all in considering a feasible-directions method, it arises in connection with obtaining the direction vectors $\mathbf{s}_k$. In many of these methods degeneracy is, to all intents and purposes, irrelevant; one in which it is *not*—the gradient projection method—will, however, be given major coverage later on.

In general, whatever the procedure for obtaining the direction vectors $\mathbf{s}_k$, the application of a feasible-directions method to a given problem may lead to any one of three possible outcomes:

1. In solving (8-61) at some stage it may be found that $\Delta = \infty$ and no finite θ_{opt} exists.
2. In solving (8-64) at some stage it may be found that there is no feasible direction $\mathbf{s}$ such that $\nabla \mathbf{f}^T(\mathbf{x}_k) \cdot \mathbf{s}$ is positive.
3. The sequence of points $\{\mathbf{x}_k\}$ generated by the algorithm may converge toward a limit point $\mathbf{x}^*$ without outcome 1 or 2 ever occurring at any stage.

Either of the first two outcomes causes immediate and automatic termination of the algorithm. In case 1 it can be concluded that the original problem (8-59) has an unbounded optimal solution of the form $\mathbf{x}_k + \theta \mathbf{s}_k$, where $\theta \longrightarrow \infty$; note that the optimal value of $f(\mathbf{x})$, obtained by allowing θ to increase without bound, may be either finite or infinite. In case 2 it will normally be true that no displacement from the terminal point $\mathbf{x}^* = \mathbf{x}_k$ in any feasible direction can cause an increase in $f(\mathbf{x})$, which would imply that $\mathbf{x}^*$ must be a constrained local maximum. This conclusion is subject to the usual exceptions when $f(\mathbf{x})$ is not a concave function: $\mathbf{x}^*$ may then be only a constrained saddle point of some type. However, it is not particularly difficult to prove that, *if a feasible-directions method terminates in case* 2 *and* $f(\mathbf{x})$ *is a concave function, then the solution point* $\mathbf{x}^* = \mathbf{x}_k$ *is a global maximum for the original problem* (8-59). Let $\hat{\mathbf{x}}$ be any point in the feasible region, and define $\hat{\mathbf{s}} = \hat{\mathbf{x}} - \mathbf{x}^*$. The feasible region is a convex set, so $\hat{\mathbf{s}}$ lies in it and is a feasible direction from $\mathbf{x}^*$, and it follows that the normalized vector

$$\hat{\mathbf{s}}_N \equiv \left(\frac{1}{|\hat{\mathbf{s}}|}\right) \cdot \hat{\mathbf{s}},$$

where $|\hat{\mathbf{s}}| = (\hat{\mathbf{s}}^T\hat{\mathbf{s}})^{1/2}$ is the length of $\hat{\mathbf{s}}$, must satisfy all the constraints of the problem (8-64). But if the gradient $\nabla \mathbf{f}(\mathbf{x}^*)$ had a positive component in the direction $\hat{\mathbf{s}}_N$, the algorithm would not have terminated; hence

$$\nabla \mathbf{f}^T(\mathbf{x}^*) \cdot \hat{\mathbf{s}}_N \leqq 0,$$

which implies

$$\nabla \mathbf{f}^T(\mathbf{x}^*) \cdot \hat{\mathbf{s}} \leqq 0.$$

Now, because $f(\mathbf{x})$ is concave, Theorem 3.9 allows us to write

$$f(\mathbf{x}^* + \hat{\mathbf{s}}) \leqq f(\mathbf{x}^*) + \nabla \mathbf{f}^T(\mathbf{x}^*) \cdot \hat{\mathbf{s}},$$

where $\mathbf{x}^* + \hat{\mathbf{s}}$ has been substituted for $\mathbf{x}$ in Eq. (3-13). Therefore, combining the last two inequalities, we have

$$f(\mathbf{x}^* + \hat{\mathbf{s}}) \equiv f(\hat{\mathbf{x}}) \leqq f(\mathbf{x}^*),$$

and $\mathbf{x}^*$ must be a global maximum. Q.E.D.

Turning now to case 3, if the sequence $\{\mathbf{x}_k\}$ is never terminated by the occurrence of case 1 or 2, then the algorithm will theoretically require an infinite number of iterations to converge to the limit point $\mathbf{x}^*$. This type of behavior is to be expected whenever the optimum does not lie at an extreme point *and* the objective function is sufficiently complicated to prevent the exact value of θ_{opt} from being determined when the problem (8-61) is solved at each iteration. Under those circumstances the sequence $\{\mathbf{x}_k\}$ may converge toward $\mathbf{x}^*$ without ever reaching it, and in practice it is necessary to impose some sort of computational stopping rule in order to ensure finite termination. Normally the algorithm is interrupted, with the current solution $\mathbf{x}_k$ being taken as an approximation to $\mathbf{x}^*$, when either the distance $|\mathbf{x}_k - \mathbf{x}_{k-1}|$ or the optimal value of $\nabla \mathbf{f}^T(\mathbf{x}_k) \cdot \mathbf{s}$—which must be positive—is less than some prespecified small number. We cannot, however, conclude that $\mathbf{x}^*$ is a constrained local maximum or constrained saddle point, *even when* $f(\mathbf{x})$ *is concave*. This uncertainty about the nature of $\mathbf{x}^*$ is due to a form of pathological convergence behavior known as *zigzagging* or *jamming*, which will be covered in some detail in our general discussion of convergence in Section 8.14.

We have not yet considered the question of how to obtain a starting feasible solution to the original problem (8-59). Frequently there is one readily available, either a very simple and obvious solution, such as $\mathbf{x} = \mathbf{0}$, or—for problems arising in operations research contexts—a solution that is known to be feasible because it represents current operating policy. Alternatively, simple algebraic transformations can be used as necessary to convert the constraints to standard form,

$$\mathbf{Ax} = \mathbf{b}, \qquad \mathbf{x} \geqq \mathbf{0},$$

and a basic feasible solution can then be obtained via the phase I simplex procedure. Once the solution has been found, the problem can be reduced to its original size by dropping the slack variables and reversing any other transformations.

The phase I approach, however, is relatively time-consuming, particularly if the original problem does not include nonnegativity restrictions on the variables. Moreover, it cancels one of the major computational advantages of using a non-simplex-based feasible-directions method, which is that the initial feasible solution does not have to be basic. Consider, then, the following approach, which can be applied to any linearly constrained problem of the form

$$\text{(8-59)} \qquad \begin{cases} \text{Max } z = f(\mathbf{x}) \\ \text{subject to } \boldsymbol{\alpha}_i\mathbf{x} \equiv \sum_{j=1}^{n} a_{ij}x_j \begin{cases} \leqq b_i, & i \in I, \\ = b_i, & i \in E. \end{cases} \end{cases}$$

Let any equality constraint in which b_i is negative be multiplied through by -1, so that $b_i \geqq 0$ for all i in E. Next, we partition the set I into two mutually exclusive, collectively exhaustive subsets I_S and I_U such that i is in I_S if $b_i \geqq 0$ and i is in I_U if $b_i < 0$; thus I_S and I_U are the sets of inequality constraints that are satisfied and unsatisfied, respectively, at the point $\mathbf{x} = \mathbf{0}$. We now modify (8-59) by introducing the artificial variables u_0 and u_i, $i \in E$, to produce the following problem:

$$\text{(8-65)} \qquad \begin{cases} \text{Max } z' = -u_0 - \sum_{i \in E} u_i \\ \text{subject to } \boldsymbol{\alpha}_i\mathbf{x} \leqq b_i, & i \in I_S, \\ \boldsymbol{\alpha}_i\mathbf{x} - u_0 \leqq b_i, & i \in I_U, \\ \boldsymbol{\alpha}_i\mathbf{x} + u_i = b_i, & i \in E, \\ u_0 \geqq 0, \\ \text{and } \; u_i \geqq 0, & i \in E. \end{cases}$$

Observe that the set of values

$$\text{(8-66)} \qquad \mathbf{x} = \mathbf{0}, \quad u_0 = \max_{i \in I_U}(-b_i), \quad \text{and} \quad u_i = b_i, \qquad i \in E,$$

constitutes a starting feasible solution to the problem (8-65), which can thus be solved by whatever feasible-directions method we plan to use for

(8-59). In the exercises the student is asked to prove that (1) if the optimal value of z' is negative, then the original problem (8-59) has no feasible solution, and (2) if the optimal value of z' is 0, then the **x**-component of the optimal solution is a feasible solution to (8-59).

8.10 DETERMINATION OF THE DIRECTION VECTOR

We have said that a number of different methods are available for determining the direction vector $\mathbf{s}_k$ to be used at each stage of a feasible-directions method. The best feasible direction—that is, the one in which the component of the gradient $\nabla\mathbf{f}(\mathbf{x}_k)$ is greatest—can be obtained by solving the problem

$$\text{(8-64)} \quad \begin{cases} \text{Max } \nabla\mathbf{f}^T(\mathbf{x}_k) \cdot \mathbf{s} \\ \text{subject to } \boldsymbol{\alpha}_i\mathbf{s} \leqq 0, & i \in B_k, \\ \boldsymbol{\alpha}_i\mathbf{s} = 0, & i \in E, \\ \text{and } \mathbf{s}^T\mathbf{s} = 1, \end{cases}$$

where it will be recalled that B_k is the set of subscripts of the inequality constraints that are binding at $\mathbf{x}_k$. It is also possible to simplify (8-64) in various ways, so that improving but suboptimal feasible directions can be obtained with less computational labor at each iteration. In general, the simpler the method by which $\mathbf{s}_k$ is determined at each stage, the greater the number of iterations likely to be required overall to solve the original problem (8-59).

Let us suppose first that the best feasible direction, and therefore the optimal solution to the nonlinearly constrained problem (8-64), is desired at each iteration. The most efficient way of obtaining it is to transform (8-64) into a quadratic program, which can then be solved by any of the methods of Chapter 7. Consider, therefore, the following problem:

$$\text{(8-67)} \quad \begin{cases} \text{Min } \mathbf{t}^T\mathbf{t} \\ \text{subject to } \boldsymbol{\alpha}_i\mathbf{t} \leqq 0, & i \in B_k, \\ \boldsymbol{\alpha}_i\mathbf{t} = 0, & i \in E, \\ \text{and } \nabla\mathbf{f}^T(\mathbf{x}_k) \cdot \mathbf{t} = 1, \end{cases}$$

where $\mathbf{t}$ is an n-component column vector of unknowns. It is easily seen that if $\hat{\mathbf{t}}$ is a feasible solution to (8-67), then $\hat{\mathbf{s}} = \hat{\mathbf{t}}/|\hat{\mathbf{t}}|$ is a feasible solution to (8-64); similarly, if $\hat{\mathbf{s}}$ is a feasible solution to (8-64) with a *positive* objective value $\hat{\psi} = \nabla\mathbf{f}^T(\mathbf{x}_k) \cdot \hat{\mathbf{s}} > 0$, then $\hat{\mathbf{t}} = \hat{\mathbf{s}}/\hat{\psi}$ must be a feasible solution to (8-67). This implies that if (8-67) has no feasible solution the algorithm should be terminated: There can be no feasible direction $\mathbf{s}$ in which the gradient $\nabla\mathbf{f}(\mathbf{x}_k)$ has a positive component, so $\mathbf{x}_k$ must be a constrained local maximum for the original problem [or possibly a constrained saddle point if $f(\mathbf{x})$ is not concave]. But suppose instead that when (8-67) is solved it is found that $\mathbf{t}^*$ is an optimal feasible solution. We shall now show that $\mathbf{s}^* = \mathbf{t}^*/|\mathbf{t}^*|$, which is known to be a feasible solution to (8-64), is also optimal for that problem. Accordingly, assume there exists some $\hat{\mathbf{s}}$ that satisfies the constraints of (8-64) and has a greater objective value than $\mathbf{s}^*$:

$$\nabla \mathbf{f}^T(\mathbf{x}_k) \cdot \hat{\mathbf{s}} = \hat{\psi} > \nabla \mathbf{f}^T(\mathbf{x}_k) \cdot \mathbf{s}^* = \psi^*. \tag{8-68}$$

Because $\mathbf{t}^*$ is a feasible solution to (8-67),

$$\psi^* = \nabla \mathbf{f}^T(\mathbf{x}_k) \cdot \mathbf{s}^* = \nabla \mathbf{f}^T(\mathbf{x}_k) \cdot \frac{\mathbf{t}^*}{|\mathbf{t}^*|} = \frac{1}{|\mathbf{t}^*|} > 0, \tag{8-69}$$

so from (8-68) $\hat{\psi}$ is positive and $\hat{\mathbf{t}} = \hat{\mathbf{s}}/\hat{\psi}$ must be a feasible solution to (8-67). Then

$$\hat{\mathbf{t}}^T\hat{\mathbf{t}} = \frac{\hat{\mathbf{s}}^T\hat{\mathbf{s}}}{\hat{\psi}^2} = \frac{1}{\hat{\psi}^2} < \frac{1}{\psi^{*2}} = |\mathbf{t}^*|^2 = \mathbf{t}^{*T}\mathbf{t}^*, \tag{8-70}$$

where the second equality of (8-70) follows from the assumption that $\hat{\mathbf{s}}$ is a feasible solution to (8-64), the inequality follows from (8-68), and the third equality follows from (8-69). This contradicts the hypothesis that $\mathbf{t}^*$ is the optimal solution to problem (8-67); hence the assumption is false, and $\mathbf{s}^* = \mathbf{t}^*/|\mathbf{t}^*|$ is optimal for (8-64), as was to be shown. Thus the optimal solution $\mathbf{t}^*$ to (8-67), whether or not we bother to normalize it, yields the best feasible direction from the current point $\mathbf{x}_k$.

The quadratic program (8-67), which calls for the minimization of a convex function, can be solved by any of the methods discussed in Chapter 7: The first local optimum or Kuhn-Tucker point obtained will necessarily be the optimal solution. Unfortunately, the application of a quadratic programming algorithm to (8-67)—after the initial transformation of the constraints into whatever format is required—would be a rather laborious procedure, particularly inasmuch as no initial feasible solution is conveniently available. The task of obtaining an improving feasible direction at each iteration becomes substantially less difficult when a linear program is substituted for the problem (8-64). Consider the following problem:

$$\begin{cases} \text{Max } \nabla \mathbf{f}^T(\mathbf{x}_k) \cdot \mathbf{s} & \\ \text{subject to } \boldsymbol{\alpha}_i\mathbf{s} \leqq 0, & i \in B_k, \\ \boldsymbol{\alpha}_i\mathbf{s} = 0, & i \in E, \\ \text{and } -1 \leqq s_j \leqq 1, & j = 1, \ldots, n. \end{cases} \tag{8-71}$$

Because a vector of length 1 can have no single component of magnitude greater than 1, it is obvious that any direction $\mathbf{s}$ satisfying the constraints of (8-64) also satisfies those of (8-71). Hence, if at some iteration the optimal solution to (8-71) has a nonpositive objective value, the same must be true of (8-64) and the algorithm can be terminated. By the same token, if an improving feasible direction does exist [i.e., if (8-64) has at least one feasible solution with a positive objective value], then the optimal solution to (8-71) will also be an improving direction, although in general not the best direction. Thus more iterations overall are likely to be required than when (8-64) or (8-67) is solved rigorously at each stage.

A different sort of linear program can be derived if, in determining the next direction vector $\mathbf{s}_k$, we consider not only the component of the gradient, but also the maximum distance that can be traveled, in any given feasible direction. That is, we are proposing to measure the "desirability" of any direction as the product of the immediate rate of improvement multiplied by the maximum feasible displacement in that direction. This can be accomplished by solving the following problem at each iteration:

$$\text{(8-72)} \quad \begin{cases} \text{Max } \nabla \mathbf{f}^T(\mathbf{x}_k) \cdot (\mathbf{x} - \mathbf{x}_k) \\ \text{subject to } \boldsymbol{\alpha}_i \mathbf{x} \leqq b_i, \quad i \in I, \\ \text{and } \boldsymbol{\alpha}_i \mathbf{x} = b_i, \quad i \in E. \end{cases}$$

The constraints of (8-72) are those of the original problem (8-59). Because the feasible region is a convex set, if $\mathbf{x}$ is any feasible point, the vector $\mathbf{x} - \mathbf{x}_k$ defines a feasible direction of displacement from the current point $\mathbf{x}_k$. The component of the gradient in that direction would be

$$\frac{1}{|\mathbf{x} - \mathbf{x}_k|} \nabla \mathbf{f}^T(\mathbf{x}_k) \cdot (\mathbf{x} - \mathbf{x}_k),$$

so the objective function of (8-72) is the product of that component multiplied by the distance from $\mathbf{x}_k$ to $\mathbf{x}$. It should therefore be clear that, for any given *improving* feasible direction, the objective function will "prefer" the point $\mathbf{x}$ that lies as far in that direction from the current point $\mathbf{x}_k$ as the boundaries of the feasible region will permit.[7] This implies, of course, that the direction chosen by (8-72) may be relatively poor, in terms of the rate of increase of $f(\mathbf{x})$, if, by way of compensation, it is possible to travel a substantial distance in that direction without leaving the feasible region.

[7]The problem (8-72) can also be viewed as a linear approximation to the original problem (8-59). The feasible regions are the same, and the objective function of (8-72) is equivalent to a first-order Taylor expansion—recall Eq. (2-15)—of the original objective function $f(\mathbf{x})$ about the point $\mathbf{x}_k$, except that the constant term $f(\mathbf{x}_k)$ has been dropped. Thus the value of the objective function $\nabla \mathbf{f}^T(\mathbf{x}_k) \cdot (\mathbf{x} - \mathbf{x}_k)$ at any point $\mathbf{x}$ is a first-order Taylor approximation to the increase in $f(\mathbf{x})$ that would result from moving from $\mathbf{x}_k$ to $\mathbf{x}$ at the next iteration.

Observe that, because $\mathbf{x}_k$ is a feasible solution to (8-72), the optimal value of the objective function must be at least 0. It should not be difficult for the student to prove that, at any given iteration in solving (8-72),

1. If the optimal objective value is found to be 0, then there is no feasible direction of displacement from $\mathbf{x}_k$ in which the component of the gradient is positive;
2. If the optimal objective value is positive and finite, then the optimal solution $\bar{\mathbf{x}}$ defines an improving feasible direction $\mathbf{s}_k = \bar{\mathbf{x}} - \mathbf{x}_k$; and
3. If the optimal objective value is infinite, then the extreme ray along which the solution becomes unbounded (as discovered by the simplex method) itself constitutes an improving feasible direction from $\mathbf{x}_k$.

Either of the linear programs (8-71) or (8-72) would be significantly easier to solve at each iteration than would the quadratic program (8-67). Note that in either case an initial feasible solution is readily available: $\mathbf{s} = \mathbf{0}$ is a feasible solution to (8-71), and $\mathbf{x} = \mathbf{x}_k$ is feasible for (8-72). The linear program (8-71) has more constraints than (8-72), but a number of them are of the upper-bound variety and therefore do not increase the size of the basis when an "upper-bounded" simplex algorithm is used. In fact, the size of the basis would normally be smaller for (8-71) than for (8-72), because B_k is a subset of I.

Apart from these considerations, however, it is extremely laborious and, for most problems, clearly inefficient to use a feasible-directions method that requires the solution of a full-blown optimization problem, even a linear program, at every iteration. Accordingly, let us now turn our attention to another, more useful method in which an improving feasible direction is computed directly at each iteration without any mathematical program, linear or otherwise, having to be solved. In keeping with our earlier remarks, we remind the reader that the feasible direction so obtained may, in general, be quite different from that in which the component of the gradient $\nabla \mathbf{f}(\mathbf{x}_k)$ is greatest.

8.11 THE GRADIENT PROJECTION METHOD

The most widely used of the non-simplex-based feasible-directions algorithms, and the last that will be considered in this chapter, is the gradient projection method of Rosen [37], which first appeared in published form in 1960. This method is applicable to any linearly constrained mathematical program, that is, to any problem of the form

$$
(8\text{-}59) \qquad \begin{cases} \text{Max } z = f(\mathbf{x}), & \mathbf{x} = (x_1, \ldots, x_n), \\ \text{subject to } \boldsymbol{\alpha}_i \mathbf{x} \equiv \sum_{j=1}^{n} a_{ij} x_j \begin{cases} \leq b_i, & i \in I, \\ = b_i, & i \in E, \end{cases} \end{cases}
$$

where the inequality constraints include any nonnegativity restrictions that may be present. We begin our development by supposing that, having obtained an initial feasible solution $\mathbf{x}_0$ to (8-59), we have subsequently executed k iterations of the gradient projection method, arriving at the current feasible point $\mathbf{x}_k$. Let h, where $h \geqq 0$, be the number of constraints that are binding at $\mathbf{x}_k$, and let the row vectors $\boldsymbol{\alpha}_i = [a_{i1}, \ldots, a_{in}]$ associated with those constraints be collected into an h-by-n matrix $\mathbf{M}_k$. For the present, *we assume that the feasible region of* (8-59) *includes no points that are degenerate* in the geometric sense of Section 8.8. This implies that the gradients $\boldsymbol{\alpha}_i$ of all the constraints that are binding at any given feasible point, such as $\mathbf{x}_k$, are linearly independent; thus in particular *the h rows of* $\mathbf{M}_k$ *are linearly independent*. Note that, because these rows are n-component vectors, there can be no more than n of them. In a later section we shall examine the implications of relaxing this crucial nondegeneracy assumption.

If $h = 0$—that is, if there are no binding constraints at $\mathbf{x}_k$—the direction of displacement $\mathbf{s}_k$ at the next iteration will be the gradient $\nabla \mathbf{f}(\mathbf{x}_k)$. But if $h > 0$, the direction vector $\mathbf{s}_k$ must be obtained from the following considerations. Let T be the subspace of E^n spanned by the rows of $\mathbf{M}_k$. Thus T contains every point $\mathbf{x}$ in E^n that can be expressed as a linear combination of the rows of $\mathbf{M}_k$; in particular, T contains the origin of E^n. Let $\mathbf{t}_k$ denote the projection of the gradient vector $\nabla \mathbf{f}(\mathbf{x}_k)$ upon the subspace T; the gradient can then be written as

$$\nabla \mathbf{f}(\mathbf{x}_k) = \mathbf{t}_k + \mathbf{v}_k, \tag{8-73}$$

where $\mathbf{v}_k$ is perpendicular to T and therefore to $\mathbf{t}_k$. Both $\mathbf{t}_k$ and $\mathbf{v}_k$ are n-component column vectors. Because the projection $\mathbf{t}_k$ is unique, the vector $\mathbf{v}_k$ is uniquely defined by (8-73); the situation is diagrammed schematically in Fig. 8-7(a). Note that if there are n binding constraints at $\mathbf{x}_k$, then the space T spanned by the n linearly independent rows of $\mathbf{M}_k$ is E^n itself, so under these circumstances $\nabla \mathbf{f}(\mathbf{x}_k)$ lies in T, $\mathbf{t}_k = \nabla \mathbf{f}(\mathbf{x}_k)$, and $\mathbf{v}_k = \mathbf{0}$.

Because the vector $\mathbf{t}_k$ lies in T, it must be a linear combination of the rows of $\mathbf{M}_k$:

$$\mathbf{t}_k = \sum_{i=1}^{h} \lambda_i \boldsymbol{\mu}_i^T \equiv \mathbf{M}_k^T \boldsymbol{\lambda}, \tag{8-74}$$

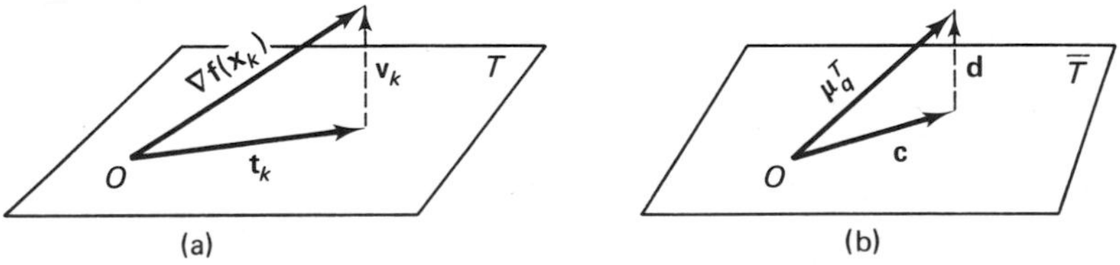

FIGURE 8-7

where $\boldsymbol{\lambda} = (\lambda_1, \ldots, \lambda_h)$ is a column vector of scalar coefficients and $\boldsymbol{\mu}_i$ denotes the ith row of $\mathbf{M}_k$. In general, $\boldsymbol{\mu}_i \neq \boldsymbol{\alpha}_i$—that is, the ith row of $\mathbf{M}_k$ is not necessarily the vector $\boldsymbol{\alpha}_i$ associated with the ith constraint. Note that, because the vectors $\boldsymbol{\mu}_i$ are linearly independent, the coefficients λ_i are uniquely defined by (8-74). Substituting (8-74) into (8-73) and premultiplying by $\mathbf{M}_k$ leads to

$$\mathbf{M}_k \nabla \mathbf{f}(\mathbf{x}_k) = \mathbf{M}_k \mathbf{M}_k^T \boldsymbol{\lambda} + \mathbf{M}_k \mathbf{v}_k = \mathbf{M}_k \mathbf{M}_k^T \boldsymbol{\lambda}, \tag{8-75}$$

where $\mathbf{M}_k \mathbf{v}_k = \mathbf{0}$ because $\mathbf{v}_k$, being perpendicular to T, is perpendicular to every row of $\mathbf{M}_k$. Using the fact that the rows of $\mathbf{M}_k$ are linearly independent, it can be proved by means of an expansion of determinants (see, for example, Chapter 3 of [20]) that the inverse $(\mathbf{M}_k \mathbf{M}_k^T)^{-1}$ exists. Therefore (8-75) yields

$$\boldsymbol{\lambda} = (\mathbf{M}_k \mathbf{M}_k^T)^{-1} \mathbf{M}_k \nabla \mathbf{f}(\mathbf{x}_k),$$

and, combining this with (8-73) and (8-74), we obtain

$$\mathbf{v}_k = \nabla \mathbf{f}(\mathbf{x}_k) - \mathbf{t}_k = \nabla \mathbf{f}(\mathbf{x}_k) - \mathbf{M}_k^T \boldsymbol{\lambda}$$

and finally

$$\mathbf{v}_k = \mathbf{P} \nabla \mathbf{f}(\mathbf{x}_k), \qquad \text{where } \mathbf{P} \equiv \mathbf{I} - \mathbf{M}_k^T (\mathbf{M}_k \mathbf{M}_k^T)^{-1} \mathbf{M}_k. \tag{8-76}$$

The matrix $\mathbf{P}$ is known as the *projection matrix*; it projects the gradient onto the $(n - h)$-dimensional subspace that constitutes the orthogonal complement of T (the orthogonal complement of a subspace T of E^n is the set of all vectors in E^n that are perpendicular to every vector in T).

If the vector $\mathbf{v}_k$ differs from $\mathbf{0}$, it can be used as the direction of displacement from the current point $\mathbf{x}_k$. We have noted that $\mathbf{v}_k$ is perpendicular to every row of $\mathbf{M}_k$; therefore $\boldsymbol{\alpha}_i \mathbf{v}_k = 0$ for every binding constraint and $\mathbf{v}_k$ must be a feasible direction—upon being normalized it would satisfy the constraints of the "direction-finding" problem (8-64). Moreover, $\mathbf{v}_k$ is an improving direction because, from (8-73),

$$\nabla \mathbf{f}^T(\mathbf{x}_k) \cdot \mathbf{v}_k = (\mathbf{t}_k + \mathbf{v}_k)^T \mathbf{v}_k = \mathbf{v}_k^T \mathbf{v}_k > 0,$$

where we have used the fact that $\mathbf{v}_k$ is perpendicular to $\mathbf{t}_k$. Thus if $\mathbf{v}_k \neq \mathbf{0}$, we have found an acceptable direction of displacement from $\mathbf{x}_k$ for the next iteration; moreover, we were able to do so via the direct computation (8-76), that is, without having to solve an optimization problem. Note, finally, that when the direction $\mathbf{v}_k$ is used, every constraint $\boldsymbol{\alpha}_i \mathbf{x} \leqq b_i$ or $\boldsymbol{\alpha}_i \mathbf{x} = b_i$ that is binding at $\mathbf{x}_k$ will continue to be binding as new points of the form $\mathbf{x}_k + \theta \mathbf{v}_k$ are generated, because

$$\boldsymbol{\alpha}_i(\mathbf{x}_k + \theta \mathbf{v}_k) = \boldsymbol{\alpha}_i \mathbf{x}_k = b_i.$$

But what if $\mathbf{v}_k = \mathbf{0}$? It then follows that the gradient lies in T and is a linear combination of the rows of $\mathbf{M}_k$:

$$\nabla \mathbf{f}(\mathbf{x}_k) = \mathbf{t}_k + \mathbf{v}_k = \mathbf{t}_k = \sum_{i=1}^{h} \lambda_i \boldsymbol{\mu}_i^T = \mathbf{M}_k^T \boldsymbol{\lambda}, \tag{8-77}$$

using (8-73) and (8-74). At this point we must distinguish between two mutually exclusive possibilities.

Case A. $\lambda_i \geqq 0$ *for all i corresponding to inequality constraints.* In this case consider *any* normalized feasible direction **s** originating at the current point $\mathbf{x}_k$. The direction **s** must satisfy the constraints of (8-64), so

$$\begin{cases} \boldsymbol{\alpha}_i \mathbf{s} \leqq 0, & i \in B_k, \\ \boldsymbol{\alpha}_i \mathbf{s} = 0, & i \in E, \end{cases} \tag{8-78}$$

where B_k and E are, respectively, the sets of subscripts of the binding inequality and equality constraints. The component of the gradient in the direction **s** is, from (8-77),

$$\nabla \mathbf{f}^T(\mathbf{x}_k) \cdot \mathbf{s} = \sum_{i=1}^{h} \lambda_i \boldsymbol{\mu}_i \mathbf{s}. \tag{8-79}$$

We now observe that the terms $\lambda_i \boldsymbol{\mu}_i \mathbf{s}$ in (8-79) that correspond to (binding) inequality constraints must be nonpositive, because $\lambda_i \geqq 0$ by hypothesis and $\boldsymbol{\mu}_i \mathbf{s} \leqq 0$ from (8-78). Similarly, (8-78) ensures that all terms corresponding to equality constraints must vanish, and it follows that

$$\nabla \mathbf{f}^T(\mathbf{x}_k) \cdot \mathbf{s} \leqq 0$$

for every feasible direction **s**. Therefore the current point $\mathbf{x}_k$ is a constrained local maximum—or possibly, if $f(\mathbf{x})$ is not a concave function, a constrained saddle point of some type—and the algorithm terminates. In Exercise 8-38 the student will be asked to verify that $\mathbf{x}_k$ satisfies the Kuhn-Tucker conditions for the original problem (8-59).

Case B. $\lambda_i < 0$ *for at least one i corresponding to an inequality constraint.* In this case we have not yet found an optimal solution, so an improving direction of displacement from $\mathbf{x}_k$ must be sought. However, because the gradient $\nabla \mathbf{f}(\mathbf{x}_k)$ lies in the subspace T, its projection $\mathbf{v}_k$ on the orthogonal complement of T is, as we have seen, of length 0 and thus does not define a usable direction. It is therefore impossible to move away from $\mathbf{x}_k$ in an improving feasible direction in such a way that all binding constraints remain binding. In most cases this will mean that the current solution $\mathbf{x}_k$ lies at an extreme point in E^n where $h = n$, that is, where n constraints with linearly independent gradients are binding. Occasionally, this situation may also arise at a non-extreme point where $h < n$; for example, if a single constraint $\boldsymbol{\alpha}_i\mathbf{x} \leqq b_i$ is binding at $\mathbf{x}_k$ and the gradient $\nabla\mathbf{f}(\mathbf{x}_k)$ is perpendicular to the hyperplane but directed into the feasible region, then $\nabla\mathbf{f}(\mathbf{x}_k) = -\boldsymbol{\alpha}_i^T$, with i corresponding to an inequality constraint, and we are in the domain of case B.

At any rate, because all the constraints that are binding at $\mathbf{x}_k$ cannot remain binding, we must select one or more of them to be relaxed. We shall now show that an improving feasible direction of displacement can be obtained by relaxing any single inequality constraint for which λ_i is negative, while requiring that all other binding constraints remain so. This assertion is obviously true when only a single inequality constraint $\boldsymbol{\alpha}_i\mathbf{x} \leq b_i$ is binding at $\mathbf{x}_k$; inasmuch as $\nabla\mathbf{f}(\mathbf{x}_k)$ is then a negative multiple of $\boldsymbol{\alpha}_i^T$ (otherwise we would not be in case B), the gradient itself points into the interior of the feasible region and thus constitutes an improving feasible direction that can be followed when the binding constraint is relaxed. In all other situations at least two constraints will be binding at $\mathbf{x}_k$, so that $\mathbf{M}_k$ will have at least two rows. Suppose $\lambda_q < 0$, where q corresponds to an inequality constraint, and let $\bar{T}$ be the subspace of E^n generated by all the rows of $\mathbf{M}_k$ other than $\boldsymbol{\mu}_q$—thus $\bar{T}$ contains all points $\mathbf{x}$ in E^n that can be expressed as linear combinations of the rows $\boldsymbol{\mu}_1, \ldots, \boldsymbol{\mu}_{q-1}, \boldsymbol{\mu}_{q+1}, \ldots, \boldsymbol{\mu}_h$. Let the n-component column vectors $\mathbf{c}$ and $\mathbf{d}$ denote, respectively, the projections of $\boldsymbol{\mu}_q$ upon the subspace $\bar{T}$ and upon its orthogonal complement, so that

$$\boldsymbol{\mu}_q^T = \mathbf{c} + \mathbf{d}, \tag{8-80}$$

where $\mathbf{d}$ is perpendicular to $\bar{T}$ and to $\mathbf{c}$. The positions of these vectors in space, diagrammed in Fig. 8-7(b), are precisely analogous to the positions of $\nabla\mathbf{f}(\mathbf{x}_k)$, $\mathbf{t}_k$, and $\mathbf{v}_k$ in Fig. 8-7(a), the difference being that $\bar{T}$ is a subspace of dimension $h - 1$ rather than h. Note that, inasmuch as the rows of $\mathbf{M}_k$ are linearly independent, $\boldsymbol{\mu}_q$ cannot lie in $\bar{T}$, so $\mathbf{d}$ must be different from $\mathbf{0}$.

Because $\mathbf{c}$ lies in $\bar{T}$, there must exist a unique set of scalar coefficients π_i, $i = 1, \ldots, h$, $i \neq q$, such that

$$\mathbf{c} = \sum_{\substack{i=1 \\ i \neq q}}^{h} \pi_i \boldsymbol{\mu}_i^T.$$

Thus (8-80) becomes

$$\boldsymbol{\mu}_q^T = \sum_{i \neq q} \pi_i \boldsymbol{\mu}_i^T + \mathbf{d},$$

and this, when substituted into (8-77), yields

$$\nabla\mathbf{f}(\mathbf{x}_k) = \lambda_q \boldsymbol{\mu}_q^T + \sum_{i \neq q} \lambda_i \boldsymbol{\mu}_i^T = \lambda_q \mathbf{d} + \sum_{i \neq q} (\lambda_i + \lambda_q \pi_i) \boldsymbol{\mu}_i^T = \lambda_q \mathbf{d} + \sum_{i \neq q} \sigma_i \boldsymbol{\mu}_i^T, \tag{8-81}$$

where $\sigma_i = \lambda_i + \lambda_q \pi_i$, $i = 1, \ldots, h$, $i \neq q$. We now define $\bar{\mathbf{M}}_k$ to be the $(h - 1)$-by-n matrix formed by deleting the row $\boldsymbol{\mu}_q$ from $\mathbf{M}_k$; Eq. (8-81) can then be represented more compactly as

$$\nabla \mathbf{f}(\mathbf{x}_k) = \lambda_q \mathbf{d} + \bar{\mathbf{M}}_k^T \boldsymbol{\sigma}, \tag{8-82}$$

where $\boldsymbol{\sigma}$ is the $(h - 1)$-component column vector whose components are the coefficients σ_i. Paralleling the development of Eq. (8-74) through (8-76), we next premultiply (8-82) by $\bar{\mathbf{M}}_k$; because every row of $\bar{\mathbf{M}}_k$ is perpendicular to $\mathbf{d}$, the product $\bar{\mathbf{M}}_k \mathbf{d}$ vanishes and we are left with

$$\bar{\mathbf{M}}_k \nabla \mathbf{f}(\mathbf{x}_k) = \bar{\mathbf{M}}_k \bar{\mathbf{M}}_k^T \boldsymbol{\sigma}.$$

As before, the inverse $(\bar{\mathbf{M}}_k \bar{\mathbf{M}}_k^T)^{-1}$ is known to exist, so

$$\boldsymbol{\sigma} = (\bar{\mathbf{M}}_k \bar{\mathbf{M}}_k^T)^{-1} \bar{\mathbf{M}}_k \nabla \mathbf{f}(\mathbf{x}_k),$$

and, substituting this into (8-82), we obtain

$$\lambda_q \mathbf{d} = \nabla \mathbf{f}(\mathbf{x}_k) - \bar{\mathbf{M}}_k^T \boldsymbol{\sigma} = \bar{\mathbf{P}} \nabla \mathbf{f}(\mathbf{x}_k), \tag{8-83}$$

where

$$\bar{\mathbf{P}} \equiv \mathbf{I} - \bar{\mathbf{M}}_k^T (\bar{\mathbf{M}}_k \bar{\mathbf{M}}_k^T)^{-1} \bar{\mathbf{M}}_k.$$

The vector $\lambda_q \mathbf{d}$ can then be taken as the direction of displacement from the current point $\mathbf{x}_k$.

Note first that, because $\lambda_q < 0$ and $\mathbf{d} \neq \mathbf{0}$, the direction vector $\mathbf{w}_k = \lambda_q \mathbf{d}$ must be different from $\mathbf{0}$. Next, observe that the inequality constraint corresponding to $\boldsymbol{\mu}_q$ will continue to be satisfied at all points $\mathbf{x}_k + \theta \mathbf{w}_k$ generated by displacement in the $\mathbf{w}_k$-direction: From (8-80), using the fact that $\mathbf{d}$ is perpendicular to $\mathbf{c}$, we have

$$\boldsymbol{\mu}_q \mathbf{w}_k = (\mathbf{c}^T + \mathbf{d}^T)(\lambda_q \mathbf{d}) = \lambda_q \mathbf{d}^T \mathbf{d} < 0,$$

and it follows that the corresponding constraint $\boldsymbol{\mu}_q \mathbf{x} = \boldsymbol{\alpha}_i \mathbf{x} \leqq b_i$, which was binding at $\mathbf{x}_k$, must remain satisfied at $\mathbf{x}_k + \theta \mathbf{w}_k$ for any $\theta > 0$. But, inasmuch as $\mathbf{d}$ is perpendicular to each of the rows of $\bar{\mathbf{M}}_k$, every other constraint $\boldsymbol{\alpha}_i \mathbf{x} \leqq b_i$ or $\boldsymbol{\alpha}_i \mathbf{x} = b_i$ that was binding at $\mathbf{x}_k$ must remain binding as the new points $\mathbf{x}_k + \theta \mathbf{w}_k$ are generated, because

$$\boldsymbol{\alpha}_i(\mathbf{x}_k + \theta \mathbf{w}_k) = \boldsymbol{\alpha}_i \mathbf{x}_k + \theta \lambda_q \boldsymbol{\alpha}_i \mathbf{d} = b_i + 0 = b_i.$$

We can therefore conclude that $\mathbf{w}_k = \lambda_q \mathbf{d}$ is a feasible direction. Finally, using (8-81),

$$\nabla \mathbf{f}^T(\mathbf{x}_k) \cdot \mathbf{w}_k = (\lambda_q \mathbf{d}^T + \sum_{i \neq q} \sigma_i \boldsymbol{\mu}_i) \cdot (\lambda_q \mathbf{d}) = \lambda_q^2 \mathbf{d}^T \mathbf{d} > 0,$$

so $\mathbf{w}_k = \lambda_q \mathbf{d}$ must be an *improving* feasible direction as well.

We have shown that in Case B, when $\mathbf{v}_k = \mathbf{0}$ and some λ_q corresponding to a binding inequality constraint is negative in (8-77), the vector $\lambda_q \mathbf{d}$ computed via (8-83) constitutes an improving feasible direction from $\mathbf{x}_k$. The gradient projection method then continues exactly as any other (non-simplex-based) feasible-directions algorithm. The steps that must be executed in a single iteration of the gradient projection method, under the assumption that the feasible region of the given problem (8-59) is *nondegenerate*, may be summarized as follows:

1. Identify all constraints that are binding at the current point $\mathbf{x}_k$ and form the matrix $\mathbf{M}_k$ whose rows are the rows $\boldsymbol{\alpha}_i$ of the binding constraints.

2. If $\nabla \mathbf{f}(\mathbf{x}_k) = \mathbf{0}$, the current solution $\mathbf{x}_k$ is an unconstrained local maximum or saddle point and the algorithm terminates. If not, but if no constraints are binding at $\mathbf{x}_k$, set $\mathbf{s}_k = \nabla \mathbf{f}(\mathbf{x}_k)$ and go to step 3. Otherwise compute

$$\boldsymbol{\lambda} = (\mathbf{M}_k \mathbf{M}_k^T)^{-1} \mathbf{M}_k \nabla \mathbf{f}(\mathbf{x}_k) \quad \text{and} \quad \mathbf{v}_k = \nabla \mathbf{f}(\mathbf{x}_k) - \mathbf{M}_k^T \boldsymbol{\lambda}.$$

If $\mathbf{v}_k \neq \mathbf{0}$, $\mathbf{v}_k$ is an improving feasible direction for which all constraints that are binding at $\mathbf{x}_k$ remain binding; set $\mathbf{s}_k = \mathbf{v}_k$ and go to step 3. But if $\mathbf{v}_k = \mathbf{0}$, examine the vector $\boldsymbol{\lambda}$:

2A. If $\lambda_i \geqq 0$ for all i corresponding to inequality constraints, $\mathbf{x}_k$ is a constrained local maximum or saddle point and the algorithm terminates.

2B. If not, form the matrix $\bar{\mathbf{M}}_k$ by removing from $\mathbf{M}_k$ the row $\boldsymbol{\mu}_q$ that corresponds to the inequality constraint having the most negative λ_i, and compute

$$\mathbf{w}_k \equiv \lambda_q \mathbf{d} = (\mathbf{I} - \bar{\mathbf{M}}_k^T (\bar{\mathbf{M}}_k \bar{\mathbf{M}}_k^T)^{-1} \bar{\mathbf{M}}_k) \nabla \mathbf{f}(\mathbf{x}_k).$$

If this computation is impossible because $\mathbf{M}_k$ had only one row, set $\mathbf{w}_k = \nabla \mathbf{f}(\mathbf{x}_k)$. In either case $\mathbf{w}_k$ is an improving feasible direction in which the inequality constraint corresponding to $\boldsymbol{\mu}_q$ is relaxed while all other binding constraints remain binding. Set $\mathbf{s}_k = \mathbf{w}_k$.

3. Compute in the usual manner the maximum permitted step length Δ in the feasible direction $\mathbf{s}_k = \nabla \mathbf{f}(\mathbf{x}_k)$, $\mathbf{s}_k = \mathbf{v}_k$, or $\mathbf{s}_k = \mathbf{w}_k$ that was chosen in step 2:

$$\Delta = \min_i \left(\frac{b_i - \boldsymbol{\alpha}_i \mathbf{x}_k}{\boldsymbol{\alpha}_i \mathbf{s}_k}, \ \boldsymbol{\alpha}_i \mathbf{s}_k > 0 \right), \tag{8-60}$$

where the index i ranges over all inequality constraints that are *not binding* at $\mathbf{x}_k$. If $\boldsymbol{\alpha}_i \mathbf{s}_k \leqq 0$ for all nonbinding inequality constraints, set $\Delta = \infty$.

4. Determine the feasible step length θ_{opt} that yields the greatest increase in the objective function $f(\mathbf{x})$ by solving the following one-dimensional optimization problem:

$$\begin{cases} \text{Max } f(\mathbf{x}_k + \theta \mathbf{s}_k) \\ \text{subject to } 0 \leqq \theta \leqq \Delta. \end{cases} \tag{8-61}$$

If $\Delta = \infty$ and no finite value of θ is optimal, the original problem (8-59) has an unbounded solution. Otherwise, set $\mathbf{x}_{k+1} = \mathbf{x}_k + \theta_{\text{opt}}\mathbf{s}_k$ and return to step 1.

8.12 GEOMETRY, COMPUTATIONS, AND AN EXAMPLE

The manner in which the gradient projection method operates can be clarified with the aid of some simple geometric illustrations. Let the shaded area in Fig. 8-8 represent the feasible region of a mathematical programming problem with two variables and five linear inequality constraints, two of which are nonnegativity restrictions; for purposes of easy reference the constraints are given the labels 1 through 5. The objective function under consideration will be to minimize the distance—or, rather, to maximize the negative of the distance—from the point Q shown in the diagram.

Consider first the feasible point $\mathbf{x}_0$, where the nonnegativity restriction 2 is the only binding constraint. The gradient $\nabla \mathbf{f}(\mathbf{x}_0)$ points into the interior of the feasible region, as shown, but is clearly not a linear combination of the gradient $\boldsymbol{\alpha}_2$ of the binding constraint at $\mathbf{x}_0$ [that is, $\nabla \mathbf{f}(\mathbf{x}_0)$ is not a constant multiple of either $\boldsymbol{\alpha}_2$ or $-\boldsymbol{\alpha}_2$]. Therefore $\nabla \mathbf{f}(\mathbf{x}_0)$ does not lie in the subspace T—refer to Fig. 8-7(a)—and it follows that $\mathbf{v}_0 \neq \mathbf{0}$ constitutes an improving feasible direction of displacement from $\mathbf{x}_0$. Recalling that $\mathbf{v}_0$ must be a direction in which the binding constraint 2 remains binding, we can conclude immediately from the orientation of $\nabla \mathbf{f}(\mathbf{x}_0)$ that $\mathbf{v}_0$ points up the x_2-axis. Let us verify this conclusion by performing the actual computations. The binding constraint 2 is

$$-x_1 \leqq 0,$$

so $\boldsymbol{\alpha}_2 = [-1 \quad 0]$ and the matrix $\mathbf{M}_0$ has only the single row $\boldsymbol{\mu}_1 = \boldsymbol{\alpha}_2 = [-1 \quad 0]$. We then find that

$$(\mathbf{M}_0\mathbf{M}_0^T)^{-1} = ([-1 \quad 0]\cdot(-1, 0))^{-1} = 1,$$

$$\mathbf{P} = \mathbf{I} - \mathbf{M}_0^T(\mathbf{M}_0\mathbf{M}_0^T)^{-1}\mathbf{M}_0 = \begin{bmatrix} 1 & 0 \\ 0 & 1 \end{bmatrix} - \begin{bmatrix} -1 \\ 0 \end{bmatrix}[-1 \quad 0] = \begin{bmatrix} 1 & 0 \\ 0 & 1 \end{bmatrix} - \begin{bmatrix} 1 & 0 \\ 0 & 0 \end{bmatrix}$$
$$= \begin{bmatrix} 0 & 0 \\ 0 & 1 \end{bmatrix},$$

and

$$\mathbf{v}_0 = \mathbf{P}\nabla\mathbf{f}(\mathbf{x}_0) = \begin{bmatrix} 0 & 0 \\ 0 & 1 \end{bmatrix} \begin{bmatrix} \delta f/\delta x_1 \\ \delta f/\delta x_2 \end{bmatrix} = \begin{bmatrix} 0 \\ \delta f/\delta x_2 \end{bmatrix}.$$

Because the gradient is oriented upward, $\delta f/\delta x_2$ must be positive and $\mathbf{v}_0$ points upward as well.

If we were now to proceed with an actual iteration of the gradient projection method, we would find that displacement from $\mathbf{x}_0$ in the $\mathbf{v}_0$-direction continues as far as the point $\mathbf{x}_1$, where, as shown in Fig. 8-8, the hyperplane of constraint 3 is encountered. At this point there are two non-parallel binding constraints 2 and 3 whose associated rows $\boldsymbol{\alpha}_2$ and $\boldsymbol{\alpha}_3$ form a basis for E^2; thus the gradient $\nabla\mathbf{f}(\mathbf{x}_1)$ lies in the subspace T, and $\mathbf{v}_1 = \mathbf{0}$. However, it is clear from Fig. 8-8 that, in expressing $\nabla\mathbf{f}(\mathbf{x}_1)$ as a linear combination of $\boldsymbol{\mu}_1 = \boldsymbol{\alpha}_2$ and $\boldsymbol{\mu}_2 = \boldsymbol{\alpha}_3$, the scalar coefficient associated with $\boldsymbol{\mu}_1 = \boldsymbol{\alpha}_2$ must be negative. Hence we are in the domain of case B: $\mathbf{x}_1$ is not a local maximum, and an improving feasible direction can be obtained by relaxing constraint 2 while forcing constraint 3 to remain binding. Displacement then proceeds along the hyperplane 3 until the point $\mathbf{x}_2$ is reached. Here again the rows $\boldsymbol{\alpha}_3$ and $\boldsymbol{\alpha}_4$ associated with the binding constraints form a basis for E^2, so $\nabla\mathbf{f}(\mathbf{x}_2)$ lies in T and $\mathbf{v}_2 = \mathbf{0}$. This time, however, the scalar coefficients λ_i in (8-77) are all nonnegative (case A), and $\mathbf{x}_2$ must be a constrained local maximum.

As a final illustration of how the gradient projection method operates, consider the point $\mathbf{x}_3$ in Fig. 8-8. Because the rows $\boldsymbol{\alpha}_1$ and $\boldsymbol{\alpha}_5$ associated with the binding constraints form a basis for E^2, the gradient $\nabla\mathbf{f}(\mathbf{x}_3)$ can be

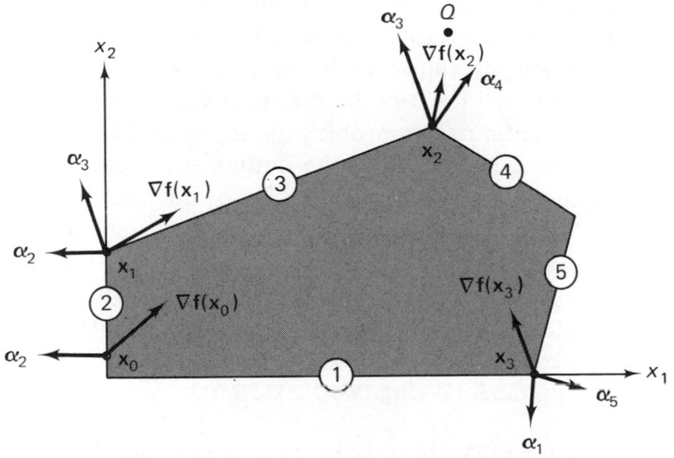

FIGURE 8-8

expressed as a linear combination of $\boldsymbol{\mu}_1 = \boldsymbol{\alpha}_1$ and $\boldsymbol{\mu}_2 = \boldsymbol{\alpha}_5$. Moreover, it can be seen from the diagram that *both* of the scalar coefficients λ_i in this linear combination are negative; hence an improving feasible direction can be obtained by relaxing *either* of the two binding constraints. In practice, the coefficient corresponding to $\boldsymbol{\mu}_1 = \boldsymbol{\alpha}_1$ would be found to be greater in magnitude (more negative), and it would accordingly be preferable to relax constraint 1.

The case just discussed, involving $\mathbf{x}_3$, illustrates the principal disadvantage of the gradient projection method: The direction of displacement that it chooses will often be markedly inferior to the best feasible direction. This shortcoming is due to the fact that no more than one binding constraint can be relaxed in any single iteration of the gradient projection method; thus, while the best feasible direction of displacement from $\mathbf{x}_3$ in Fig. 8-8 was the gradient $\nabla \mathbf{f}(\mathbf{x}_3)$, which pointed into the interior of the feasible region, the best available direction was along the hyperplane of constraint 5. The degree of inferiority of the direction chosen by the gradient projection method becomes potentially much more significant as the numbers of variables and binding constraints increase. Note, for example, that if h inequality constraints are binding at the current or initial solution $\mathbf{x}_k$, then at least h iterations must be executed before it becomes possible to enter the interior of the feasible region, whereas a "best-feasible-direction" algorithm, which solved the problem (8-64) exactly at every iteration, could enter the interior directly from $\mathbf{x}_k$ if it were desirable to do so. Even the linear approximation procedures discussed in Section 8.10 would be expected to yield significantly better directions of displacement than the gradient projection method, especially during the early stages, when it is more likely to be desirable to cut across the interior of the feasible region.

In return for this handicap, however, the gradient projection method enjoys the significant advantage of being able to calculate an improving feasible direction at each iteration by means of algebraic formulas, without having to solve an optimization problem of any kind. The required computations are summarized in step 2 of the outline at the end of Section 8.11. It should be noted that these computations are by no means trivial, inasmuch as the inverses $(\mathbf{M}_k\mathbf{M}_k^T)^{-1}$ and possibly $(\bar{\mathbf{M}}_k\bar{\mathbf{M}}_k^T)^{-1}$ must somehow be obtained. For problems involving few variables, and therefore (assuming nondegeneracy) few binding constraints at any feasible point, it is probably easiest simply to compute the inverses directly. For larger problems, transformation formulas based on the partitioned form of the matrix inverse are available for performing the following two tasks:

1. Given the inverse $(\mathbf{M}\mathbf{M}^T)^{-1}$, where $\mathbf{M}$ is an h-by-n matrix, $h < n$, compute $(\mathbf{M}_+\mathbf{M}_+^T)^{-1}$, where $\mathbf{M}_+$ is formed by adding to $\mathbf{M}$ a single row that is linearly independent of the rows of $\mathbf{M}$.

2. Given $(\mathbf{M}\mathbf{M}^T)^{-1}$, where $\mathbf{M}$ is h-by-n, $h \leqq n$, compute $(\mathbf{M}_-\mathbf{M}_-^T)^{-1}$, where $\mathbf{M}_-$ is formed by deleting any single row of $\mathbf{M}$.

The reader should be able to verify that, under the assumption of nondegeneracy, these two transformations are sufficient for the computation of all the inverses that are required in obtaining the direction of displacement at any stage of the gradient projection method. That is, $\bar{\mathbf{M}}_k$ is formed from $\mathbf{M}_k$, when necessary, by deleting a row; $\mathbf{M}_{k+1}$ may be the same as $\mathbf{M}_k$ or $\bar{\mathbf{M}}_k$ or may require the addition of a row; and so on (what would be the procedure when displacement in some direction is stopped because *two* new constraint hyperplanes are simultaneously encountered?). Exercise 8-43 asks for an algebraic derivation of the two transform formulas.

In view of the arguments presented in the last two paragraphs, it would probably be most efficient overall to modify the gradient projection method *during the early stages only* so as to allow displacement to proceed in the gradient direction whenever that direction is feasible. In using this variation, the procedure would be to compute the gradient vector at the current point and then test to see whether it constitutes a feasible direction; if so, it would be used as the direction of displacement, and the inverses $(\mathbf{M}_k\mathbf{M}_k^T)^{-1}$ and $(\bar{\mathbf{M}}_k\bar{\mathbf{M}}_k^T)^{-1}$ would simply not be calculated at all. This would mean that whenever the gradient direction is *not* feasible the inverse $(\mathbf{M}_k\mathbf{M}_k^T)^{-1}$ would have to be completely recalculated—that is, the transformation formulas could not be used. In the long run this should be an acceptable price to pay: Relatively few constraints are likely to be binding during the early stages, so the h-by-h matrix $\mathbf{M}_k\mathbf{M}_k^T$ can be expected to be manageably small. In the later stages, however, the probability of the gradient direction being feasible is greatly diminished (except when *none* of the constraints are binding at the optimum), and the "pure" gradient projection method, as described in Section 8.11, should be used.

We shall now illustrate the gradient projection method by applying it to the following example problem:

$$\begin{aligned}
\text{Max } f(x_1, x_2) &= -(x_1 - 3)^2 - (x_2 - 7)^2 \\
\text{subject to } ① \quad x_1 - 2x_2 &\leqq 0, \\
② \quad x_1 + 2x_2 &\leqq 12, \\
③ \; -x_1 + 6x_2 &\leqq 24, \\
④ \; -x_1 \qquad &\leqq 0, \\
\text{and } ⑤ \qquad -x_2 &\leqq -1.
\end{aligned}$$

The feasible region is diagrammed in Fig. 8-9, with the bounding hyperplanes labeled from 1 to 5 to correspond with the constraints listed above.

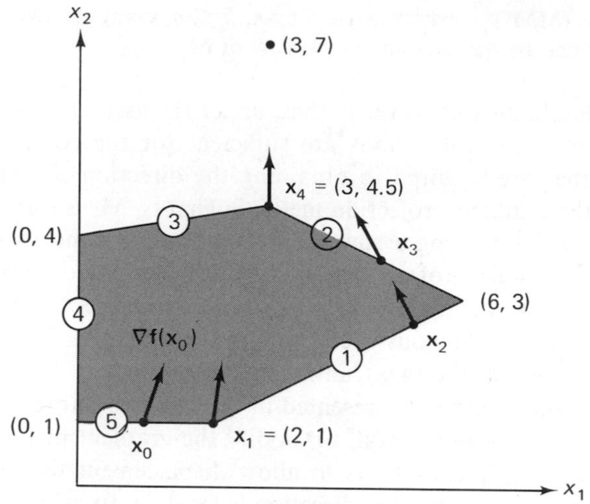

FIGURE 8-9

The objective is to minimize the distance from the point ($x_1 = 3$, $x_2 = 7$,) and its gradient is

$$\nabla \mathbf{f}(\mathbf{x}) = (6 - 2x_1, 14 - 2x_2).$$

We take as our starting solution the point $\mathbf{x}_0 = (1, 1)$, where the value of the gradient is $\nabla \mathbf{f}(\mathbf{x}_0) = (4, 12)$ and where only a single constraint,

$$\boldsymbol{\alpha}_5 \mathbf{x} = -x_2 \leqq -1,$$

is binding. Although the gradient constitutes a feasible direction, we shall use the "pure" gradient projection approach of requiring the binding constraint $-x_2 \leqq -1$ to remain binding, if possible. The direction of displacement from $\mathbf{x}_0$ is then computed as follows:

$$\boldsymbol{\mu}_1 = \boldsymbol{\alpha}_5 = [0 \quad -1],$$

$$\mathbf{M}_0 = [0 \quad -1],$$

$$(\mathbf{M}_0 \mathbf{M}_0^T)^{-1} = 1,$$

$$\boldsymbol{\lambda} = (\mathbf{M}_0 \mathbf{M}_0^T)^{-1} \mathbf{M}_0 \nabla \mathbf{f}(\mathbf{x}_0) = [0 \quad -1] \begin{bmatrix} 4 \\ 12 \end{bmatrix} = -12,$$

and

$$\mathbf{v}_0 = \nabla \mathbf{f}(\mathbf{x}_0) - \mathbf{M}_0^T \boldsymbol{\lambda} = \begin{bmatrix} 4 \\ 12 \end{bmatrix} - \begin{bmatrix} 0 \\ -1 \end{bmatrix} (-12) = \begin{bmatrix} 4 \\ 0 \end{bmatrix}.$$

Because $\mathbf{v}_0 \neq \mathbf{0}$, we take $\mathbf{s}_0 = \mathbf{v}_0 = (4, 0)$ as the direction of displacement; as would be expected from the orientation of the gradient $\nabla\mathbf{f}(\mathbf{x}_0)$ in Fig. 8-9, this involves moving along the hyperplane 5 in the direction of increasing x_1. Next, we compute the maximum permitted step length Δ in the direction $\mathbf{s}_0$:

$$\Delta = \min_i \left(\frac{b_i - \boldsymbol{\alpha}_i \mathbf{x}_0}{\boldsymbol{\alpha}_i \mathbf{s}_0}, \ \boldsymbol{\alpha}_i \mathbf{s}_0 > 0 \right)$$

$$= \min \left(\frac{b_1 - \boldsymbol{\alpha}_1 \mathbf{x}_0}{\boldsymbol{\alpha}_1 \mathbf{s}_0}, \frac{b_2 - \boldsymbol{\alpha}_2 \mathbf{x}_0}{\boldsymbol{\alpha}_2 \mathbf{s}_0} \right) = \frac{b_1 - \boldsymbol{\alpha}_1 \mathbf{x}_0}{\boldsymbol{\alpha}_1 \mathbf{s}_0} = \frac{0 - (-1)}{4} = \frac{1}{4}.$$

The new point $\mathbf{x}_1$ can therefore be obtained by solving the following problem:

$$\text{Max } f(\mathbf{x}_0 + \theta \mathbf{s}_0) = f(1 + 4\theta, 1) = 16\theta - 16\theta^2 - 40$$
$$\text{subject to } 0 \leqq \theta \leqq \tfrac{1}{4}.$$

Because the derivative $df/d\theta = 16 - 32\theta$ vanishes at $\theta = \frac{1}{2}$, it is evident that $\theta_{\text{opt}} = \frac{1}{4}$, so that $\mathbf{x}_1 = (2, 1)$; the value of the objective function thus increases from $f(1, 1) = -40$ to $f(2, 1) = -37$. Note that for a more complicated objective function, some sort of sequential search procedure would have had to be used to determine θ_{opt}.

The gradient of the objective at the new point $\mathbf{x}_1 = (2, 1)$ is $\nabla\mathbf{f}(\mathbf{x}_1) = (2, 12)$, which is directed into the interior of the feasible region, as indicated by the arrow attached to $\mathbf{x}_1$ in Fig. 8-9 (each arrow attached to a point $\mathbf{x}_k$ in the diagram indicates the direction of the gradient $\nabla\mathbf{f}$ at $\mathbf{x}_k$). The two constraints 1 and 5 are binding at $\mathbf{x}_1$, and we compute

$$\boldsymbol{\mu}_1 = \boldsymbol{\alpha}_1 = [1 \quad -2], \qquad \boldsymbol{\mu}_2 = \boldsymbol{\alpha}_5 = [0 \quad -1],$$

$$\mathbf{M}_1 = \begin{bmatrix} 1 & -2 \\ 0 & -1 \end{bmatrix},$$

$$(\mathbf{M}_1\mathbf{M}_1^T)^{-1} = \begin{bmatrix} 5 & 2 \\ 2 & 1 \end{bmatrix}^{-1} = \begin{bmatrix} 1 & -2 \\ -2 & 5 \end{bmatrix},$$

$$\boldsymbol{\lambda} = (\mathbf{M}_1\mathbf{M}_1^T)^{-1}\mathbf{M}_1\nabla\mathbf{f}(\mathbf{x}_1) = \begin{bmatrix} 1 & -2 \\ -2 & 5 \end{bmatrix}\begin{bmatrix} 1 & -2 \\ 0 & -1 \end{bmatrix}\begin{bmatrix} 2 \\ 12 \end{bmatrix} = \begin{bmatrix} 2 \\ -16 \end{bmatrix},$$

and

$$\mathbf{v}_1 = \nabla\mathbf{f}(\mathbf{x}_1) - \mathbf{M}_1^T\boldsymbol{\lambda} = \begin{bmatrix} 2 \\ 12 \end{bmatrix} - \begin{bmatrix} 1 & 0 \\ -2 & -1 \end{bmatrix}\begin{bmatrix} 2 \\ -16 \end{bmatrix} = \begin{bmatrix} 0 \\ 0 \end{bmatrix}.$$

Discovering that $\mathbf{v}_1 = \mathbf{0}$, we examine the components of $\boldsymbol{\lambda}$: Because $\lambda_2 = -16$, the current solution is not a local maximum, and an improving feasible direction can be obtained by relaxing the inequality constraint 5, which corresponds to $\boldsymbol{\mu}_2$ and λ_2. The computations then proceed as follows:

$$\bar{\mathbf{M}}_1 = \boldsymbol{\alpha}_1 = [1 \quad -2],$$

$$(\bar{\mathbf{M}}_1\bar{\mathbf{M}}_1^T)^{-1} = .2,$$

$$\bar{\mathbf{P}} = \mathbf{I} - \bar{\mathbf{M}}_1^T(\bar{\mathbf{M}}_1\bar{\mathbf{M}}_1^T)^{-1}\bar{\mathbf{M}}_1 = \begin{bmatrix} 1 & 0 \\ 0 & 1 \end{bmatrix} - (.2)\begin{bmatrix} 1 \\ -2 \end{bmatrix}\cdot[1 \quad -2]$$

$$= \begin{bmatrix} 1 & 0 \\ 0 & 1 \end{bmatrix} - \begin{bmatrix} .2 & -.4 \\ -.4 & .8 \end{bmatrix} = \begin{bmatrix} .8 & .4 \\ .4 & .2 \end{bmatrix},$$

and

$$\lambda_2\mathbf{d} = \bar{\mathbf{P}}\nabla\mathbf{f}(\mathbf{x}_1) = \begin{bmatrix} 6.4 \\ 3.2 \end{bmatrix}.$$

The direction of displacement $\mathbf{s}_1 = \lambda_2\mathbf{d} = (6.4, 3.2)$ is, as expected, along the hyperplane of constraint 1. The maximum permitted step length in this direction is found to be

$$\Delta = \frac{b_2 - \boldsymbol{\alpha}_2\mathbf{x}_1}{\boldsymbol{\alpha}_2\mathbf{s}_1} = \frac{12 - 4}{12.8} = 0.625,$$

and the new point $\mathbf{x}_2$ is then obtained by solving the following problem:

$$\text{Max } f(\mathbf{x}_1 + \theta\mathbf{s}_1) = f(2 + 6.4\theta, 1 + 3.2\theta) = 51.2\theta - 51.2\theta^2 - 37$$
$$\text{subject to } 0 \leqq \theta \leqq .625.$$

The derivative $df/d\theta$ vanishes at $\theta_{\text{opt}} = .5$, and we find that $\mathbf{x}_2 = (5.2, 2.6)$, with $f(\mathbf{x}_2) = -24.2$.

The gradient of the objective function at the new point is $\nabla\mathbf{f}(\mathbf{x}_2) = (-4.4, 8.8)$; this is perpendicular to the hyperplane of constraint 1, as expected (otherwise the objective value would not have stopped increasing at $\mathbf{x}_2$). The only binding constraint is now constraint 1, and, proceeding with the third iteration, we compute

$$\boldsymbol{\mu}_1 = \boldsymbol{\alpha}_1 = [1 \quad -2],$$

$$\mathbf{M}_2 = [1 \quad -2],$$

$$(\mathbf{M}_2\mathbf{M}_2^T)^{-1} = .2,$$

$$\boldsymbol{\lambda} = (0.2)[1 \quad -2]\cdot(-4.4, 8.8) = -4.4,$$

and

$$\mathbf{v}_2 = \begin{bmatrix} -4.4 \\ 8.8 \end{bmatrix} - \begin{bmatrix} 1 \\ -2 \end{bmatrix}(-4.4) = \begin{bmatrix} 0 \\ 0 \end{bmatrix}.$$

Testing for optimality, we find that, because λ_1 is negative, an improving feasible direction can be obtained by relaxing constraint 1, which will leave no binding constraints at all. Note that this is an instance of the exceptional case in which a binding constraint must be relaxed even though the current solution is not an extreme point. Inasmuch as $\nabla \mathbf{f}(\mathbf{x}_2)$ is a negative multiple of $\boldsymbol{\alpha}_1^T$, we know that $\nabla \mathbf{f}(\mathbf{x}_2)$ points into the interior of the feasible region and constitutes an improving feasible direction. Hence we can immediately set $\mathbf{s}_2 = \nabla \mathbf{f}(\mathbf{x}_2)$. After determining the maximum permitted step length in the $\mathbf{s}_2$-direction and solving the familiar one-dimensional optimization problem (8-61), we find that displacement proceeds until the hyperplane of constraint 2 is encountered; the new point is $\mathbf{x}_3 = (14/3, 11/3)$, with $f(\mathbf{x}_3) = -125/9$.

The gradient at the new point is $\nabla \mathbf{f}(\mathbf{x}_3) = (-10/3, 20/3)$, which points in the same direction as $\nabla \mathbf{f}(\mathbf{x}_2)$ (why?). The computations for the fourth iteration are as follows:

$$\boldsymbol{\mu}_1 = \boldsymbol{\alpha}_2 = [1 \quad 2],$$

$$\mathbf{M}_3 = [1 \quad 2],$$

$$\boldsymbol{\lambda} = (.2)[1 \quad 2]\cdot(-10/3, 20/3) = 2,$$

and

$$\mathbf{v}_3 = \begin{bmatrix} -10/3 \\ 20/3 \end{bmatrix} - \begin{bmatrix} 1 \\ 2 \end{bmatrix}(2) = \begin{bmatrix} -16/3 \\ 8/3 \end{bmatrix}.$$

The direction of displacement $\mathbf{s}_3 = \mathbf{v}_3$ is along the hyperplane of constraint 2, and, completing the iteration, we find that displacement proceeds until the extreme point $(x_1 = 3, x_2 = 4.5)$ is reached; thus the new point is $\mathbf{x}_4 = (3, 4.5)$, with $f(\mathbf{x}_4) = -6.25$ and $\nabla \mathbf{f}(\mathbf{x}_4) = (0, 5)$. Moving on to the fifth iteration, we compute

$$\boldsymbol{\mu}_1 = \boldsymbol{\alpha}_2 = [1 \quad 2], \qquad \boldsymbol{\mu}_2 = \boldsymbol{\alpha}_3 = [-1 \quad 6],$$

$$\mathbf{M}_4 = \begin{bmatrix} 1 & 2 \\ -1 & 6 \end{bmatrix},$$

$$(\mathbf{M}_4\mathbf{M}_4^T)^{-1} = \begin{bmatrix} 5 & 11 \\ 11 & 37 \end{bmatrix}^{-1} = \frac{1}{64}\begin{bmatrix} 37 & -11 \\ -11 & 5 \end{bmatrix},$$

$$\boldsymbol{\lambda} = \frac{1}{64}\begin{bmatrix} 37 & -11 \\ -11 & 5 \end{bmatrix}\begin{bmatrix} 1 & 2 \\ -1 & 6 \end{bmatrix}\begin{bmatrix} 0 \\ 5 \end{bmatrix} = \begin{bmatrix} 5/8 \\ 5/8 \end{bmatrix},$$

and

$$\mathbf{v}_4 = \begin{bmatrix} 0 \\ 5 \end{bmatrix} - \begin{bmatrix} 1 & -1 \\ 2 & 6 \end{bmatrix}\begin{bmatrix} 5/8 \\ 5/8 \end{bmatrix} = \begin{bmatrix} 0 \\ 0 \end{bmatrix}.$$

Examining $\boldsymbol{\lambda}$, we discover that $\lambda_i \geqq 0$ for all i corresponding to inequality constraints. Therefore case A obtains: The current point $\mathbf{x}_4 = (3, 4.5)$ is a (global) optimal solution—it cannot be any other type of Kuhn-Tucker point because the objective function is concave—and the gradient projection algorithm terminates.

8.13 DEGENERACY IN THE GRADIENT PROJECTION METHOD

Let us now discard the nondegeneracy assumption made at the beginning of Section 8.11 and admit the possibility that some degenerate points may be included in the feasible region of the following problem:

$$(8\text{-}59)\qquad \begin{cases} \text{Max } z = f(\mathbf{x}), & \mathbf{x} = (x_1, \ldots, x_n), \\ \text{subject to } \boldsymbol{\alpha}_i\mathbf{x} \equiv \sum_{j=1}^{n} a_{ij}x_j \begin{cases} \leqq b_i, & i \in I, \\ = b_i, & i \in E. \end{cases} \end{cases}$$

That is, we shall no longer assume that the gradients of all the constraints that are binding at any given feasible point must necessarily be linearly independent. It then follows that the gradient projection method must be modified in certain respects in order to take into account the possibility of degeneracy.

We begin by assuming that we know or have obtained an initial feasible solution $\mathbf{x}_0$ to the given problem (8-59) and that we have formed the matrix $\mathbf{M}_0$ whose rows are the rows $\boldsymbol{\alpha}_i$ of those constraints that are binding at $\mathbf{x}_0$. Next, we examine sequentially the rows of $\mathbf{M}_0$ and delete each one that is found to be a linear combination of those already examined. For purposes of convenience, let us continue to use the same notation as before. Thus the h-by-n matrix $\mathbf{M}_0$ is now composed of the linearly independent rows $\boldsymbol{\alpha}_i$ associated with h constraints of (8-59) that are binding at $\mathbf{x}_0$; if any other constraint is binding at $\mathbf{x}_0$, its row $\boldsymbol{\alpha}_i$ must be a linear combination of the rows of $\mathbf{M}_0$.

Having seen how a matrix $\mathbf{M}_0$ with linearly independent rows can be produced initially, let us now suppose that at some later stage of the gradient projection method we have arrived at the feasible point $\mathbf{x}_k$ and have formed the h-by-n matrix $\mathbf{M}_k$ with properties analogous to those of $\mathbf{M}_0$ above (i.e., for any constraint $\boldsymbol{\alpha}_i\mathbf{x} \leqq b_i$ or $\boldsymbol{\alpha}_i\mathbf{x} = b_i$ that is binding at $\mathbf{x}_k$, the vector $\boldsymbol{\alpha}_i$ must be either a row of $\mathbf{M}_k$ or a linear combination of the rows of $\mathbf{M}_k$). Proceeding as before, let T be the subspace spanned by the rows of $\mathbf{M}_k$—or, equivalently, by the rows $\boldsymbol{\alpha}_i$ of all the binding constraints at $\mathbf{x}_k$—and compute $\boldsymbol{\lambda}$ and $\mathbf{v}_k$. If $\mathbf{v}_k$ differs from $\mathbf{0}$, it can again be used as the direction of displacement from $\mathbf{x}_k$: Inasmuch as $\mathbf{v}_k$ is perpendicular to every row of $\mathbf{M}_k$, it is perpendicular to the row $\boldsymbol{\alpha}_i$ of every binding constraint, and all binding constraints will therefore remain so during any displacement in the

$\mathbf{v}_k$-direction. On the other hand, if $\mathbf{v}_k = \mathbf{0}$, the gradient $\nabla\mathbf{f}(\mathbf{x}_k)$ lies in T and can, as before, be represented as a linear combination of the rows of $\mathbf{M}_k$:

$$\nabla\mathbf{f}(\mathbf{x}_k) = \sum_{i=1}^{h} \lambda_i \boldsymbol{\mu}_i^T = \mathbf{M}_k^T\boldsymbol{\lambda}. \tag{8-77}$$

Then if $\lambda_i \geqq 0$ for all i corresponding to inequality constraints, the proof of case A in Section 8.11 can be applied verbatim, as the reader can easily verify, and we can conclude that $\mathbf{x}_k$ must be a constrained local maximum or saddle point.

We are left with the possibility that $\mathbf{v}_k = \mathbf{0}$ but that some λ_q corresponding to an inequality constraint is negative. Again proceeding as before, we form the $(h - 1)$-by-n matrix $\bar{\mathbf{M}}_k$ by dropping the row $\boldsymbol{\mu}_q$ from $\mathbf{M}_k$ and compute the direction vector $\mathbf{w}_k = \lambda_q\mathbf{d}$ via (8-83). From our discussion of case B, we know that $\mathbf{w}_k$ is an improving feasible direction and that displacement in the $\mathbf{w}_k$-direction will violate none of the constraints whose rows $\boldsymbol{\alpha}_i$ were included in $\mathbf{M}_k$. But consider a binding constraint $\boldsymbol{\alpha}_r\mathbf{x}\{\leq, =\}b_r$ for which $\boldsymbol{\alpha}_r$ was *not* included among the rows of $\mathbf{M}_k$—will that constraint remain satisfied? We know that there exist scalars $\eta_1, \ldots, \eta_h$ such that

$$\boldsymbol{\alpha}_r = \sum_{i=1}^{h} \eta_i \boldsymbol{\mu}_i \tag{8-84}$$

(if not, $\boldsymbol{\alpha}_r$ would have been included in $\mathbf{M}_k$), and we can then write

$$\boldsymbol{\alpha}_r\mathbf{w}_k = \lambda_q \sum_{i=1}^{h} \eta_i\boldsymbol{\mu}_i\mathbf{d} = \lambda_q\eta_q\boldsymbol{\mu}_q\mathbf{d} = \lambda_q\eta_q(\mathbf{c}^T + \mathbf{d}^T)\mathbf{d} = \lambda_q\eta_q\mathbf{d}^T\mathbf{d}, \tag{8-85}$$

where the first equality follows from (8-84), the second from the fact that $\mathbf{d}$ is perpendicular to every row of $\bar{\mathbf{M}}_k$, the third from (8-80), and the fourth from the fact that $\mathbf{d}$ is perpendicular to $\mathbf{c}$. Because λ_q is negative and $\mathbf{d}^T\mathbf{d}$ is positive, the constraint $\boldsymbol{\alpha}_r\mathbf{x}\{\leqq, =\}b_r$ will remain satisfied as displacement proceeds in the $\mathbf{w}_k$-direction if and only if one of the following two conditions holds:

1. The constraint is an inequality and $\eta_q \geqq 0$, or
2. The constraint is an equality and $\eta_q = 0$.

The student can verify these statements by evaluating the constraint at any point of the form $\mathbf{x}_k + \theta\mathbf{w}_k$, where $\theta > 0$.

In the event that each of the binding constraints not included in $\mathbf{M}_k$ (if there are any) passes the test just described, the vector $\mathbf{w}_k = \lambda_q\mathbf{d}$ is an improving feasible direction and we can proceed exactly as in the nondegenerate case. However, if one or more constraints fail the test, we must proceed via perturbation, as follows. For every constraint $\boldsymbol{\alpha}_i\mathbf{x}\{\leq, =\}b_i$ of the original problem (8-59) that is binding at $\mathbf{x}_k$ but is not included among the rows of $\mathbf{M}_k$, imagine increasing the right-hand side from b_i to $b_i + \epsilon^i$, where ϵ is a very small positive number. After such a perturbation, none of those excluded constraints would any longer be binding at $\mathbf{x}_k$. Moreover, if ϵ were chosen sufficiently small, it would then be theoretically possible to "short-step" around the corner in a sequence of iterations and thus eventually to break away from the degenerate point $\mathbf{x}_k$. This process is illustrated in Fig. 8-10(a) and (b) for a two-variable problem in which four constraints meet at the current point $\mathbf{x}_k$. Suppose the constraints included in $\mathbf{M}_k$ are 1 and 2; then the perturbation of 3 and 4 might lead to the situation shown in (b), where the path around the corner requires a total of four iterations instead of two and includes two short steps along hyperplanes 2 and 3. Note that each of the points $\mathbf{x}_k, \mathbf{x}_{k+1}, \ldots$ produced in this sequence of iterations is nondegenerate.[8]

As a practical matter, it is not actually necessary to perform all the computations implied by the perturbation approach described above. The perturbation scheme itself must be used in order to determine which of the more-than-h constraints that are binding at $\mathbf{x}_k$ will be included in each of the h-by-n matrices $\mathbf{M}_{k+1}, \mathbf{M}_{k+2}, \ldots$ that are generated during the short-step iterations. However, there is no need to determine the various points at which the perturbed constraints intersect, nor to compute a new gradient vector at each. Instead, all computations can be based on the point $\mathbf{x}_k$, with the distances covered in the short-step iterations being reduced to zero. In effect, $\bar{\mathbf{M}}_k$ is formed and the direction vector $\mathbf{w}_k$ is used at each of these iterations, but at the same time a maximum permitted displacement of length

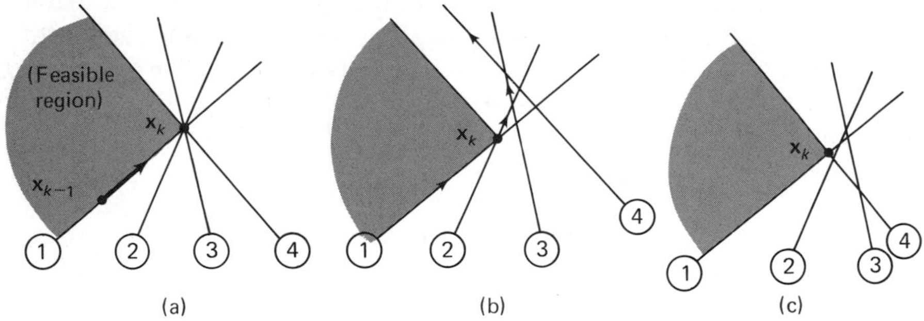

FIGURE 8-10

[8]For the perturbations to be such that, say, the hyperplanes 2, 3, and 4 would *still* meet at a single (degenerate) point, the perturbations would have to be in a certain precise ratio, depending on the relative orientations of the three hyperplanes. To avoid this complication, it is only necessary to select a value of ϵ small enough to be less than any such ratio.

0 is imposed by the presence of one or more constraint hyperplanes that were not included in $\mathbf{M}_k$. Then, after a "step" of length 0 has been taken in the $\mathbf{w}_k$-direction, one of the "newly encountered" constraints is added to $\bar{\mathbf{M}}_k$ to form $\mathbf{M}_{k+1}$ for the next iteration.

The behavior of the gradient projection method during the short-step iterations is quite similar to that of the linear simplex method during a sequence of degenerate or "zero-for-zero" pivots: The solution point and objective value remain the same for two or more consecutive iterations, while auxiliary computational changes take place that eventually make it possible to break away from the degenerate solution point. It is important to observe that the perturbation scheme of the gradient projection method provides a unique path of short steps and thereby determines a unique sequence of matrices $\mathbf{M}_k$, $\mathbf{M}_{k+1}$, $\mathbf{M}_{k+2}$, . . . that must eventually lead to an escape from the degenerate point—there is no possibility of infinite cycling (just as the linear simplex method can never become trapped in an unending cycle of pivots if this type of perturbation technique is used). Note that the number of iterations required by the gradient projection method to escape from the degenerate point $\mathbf{x}_k$ depends unpredictably on the spatial arrangement of the various hyperplanes and on which of them happen to be included in $\mathbf{M}_k$. Referring to Fig. 8-10(c), if $\mathbf{M}_k$ included constraints 1 and 4 instead of 1 and 2, no short steps at all would be required to break away from $\mathbf{x}_k$, and two iterations would be saved.

One further computational point remains to be discussed. Whenever an iteration of the gradient projection method is terminated because a new constraint hyperplane is encountered, it is necessary to determine whether the row $\boldsymbol{\alpha}_r$ of the new constraint is linearly independent of the rows of $\mathbf{M}_k$ (if $\mathbf{v}_k$ was being used as the direction of displacement) or $\bar{\mathbf{M}}_k$ (if $\mathbf{w}_k$ was being used). If $\boldsymbol{\alpha}_r$ is linearly independent, it will be added to $\mathbf{M}_k$ or $\bar{\mathbf{M}}_k$ to form $\mathbf{M}_{k+1}$; on the other hand, if it is linearly dependent, $\mathbf{M}_{k+1}$ will be the same as $\mathbf{M}_k$ or $\bar{\mathbf{M}}_k$. This determination can be made by computing $\mathbf{P}\boldsymbol{\alpha}_r^T$ (or $\bar{\mathbf{P}}\boldsymbol{\alpha}_r^T$—henceforth we shall omit the references to $\bar{\mathbf{M}}_k$, $\bar{T}$, $\bar{\mathbf{P}}$, and $\mathbf{w}_k$), where $\mathbf{P}$ is the projection matrix that was used in the computation of $\mathbf{v}_k$. Recall that premultiplication by $\mathbf{P}$ should project the vector $\boldsymbol{\alpha}_r^T$ onto the $(n - h)$-dimensional subspace that constitutes the orthogonal complement of T, where T is the subspace spanned by the rows of $\mathbf{M}_k$. Thus if $\mathbf{P}\boldsymbol{\alpha}_r^T \neq \mathbf{0}$, $\boldsymbol{\alpha}_r$ has a nonvanishing projection on the orthogonal complement and so cannot lie in T; it then follows that $\boldsymbol{\alpha}_r$ is linearly independent of the rows of $\mathbf{M}_k$ and should be added to $\mathbf{M}_k$ to form $\mathbf{M}_{k+1}$. But if $\mathbf{P}\boldsymbol{\alpha}_r^T = \mathbf{0}$, $\boldsymbol{\alpha}_r$ lies in T, is expressible as a linear combination of the rows of $\mathbf{M}_k$, and should therefore not be added to $\mathbf{M}_k$. As a final note, it may occasionally happen that an iteration is terminated when two or more constraint hyperplanes are *simultaneously* encountered; in this event, the new rows $\boldsymbol{\alpha}_i$ are tested, and possibly added to $\mathbf{M}_k$, one at a time.

The coverage given to the gradient projection method in the past three sections is somewhat out of proportion to its actual importance in either the theory or the practice of mathematical programming. The gradient projection method is, after all, but one of several competing algorithms for solving linearly constrained problems. However, although it is one of the most difficult methods from the theoretical point of view, it is also one of the most efficient computationally and therefore one of the most widely used. Its great advantage is its ability to compute directly, without solving an optimization problem of any kind, an improving feasible direction of displacement at each iteration. This advantage more than compensates for the limitation imposed upon the choice of direction (i.e., at most one binding constraint can be relaxed), particularly in the later stages, when freedom of maneuver in the vicinity of the optimum is greatly restricted anyway by the presence of binding constraint hyperplanes.

A further reason for covering the gradient projection method in such great detail was to provide the student with a thorough introduction to the general properties and behavior of non-simplex-based feasible-directions methods. The treatment of degeneracy should be particularly instructive, inasmuch as some version of the perturbation approach described above is employed in a great many types of mathematical programming algorithms.

8.14 CONVERGENCE OF FEASIBLE-DIRECTIONS ALGORITHMS

The final section of this chapter will deal with the general topic of algorithmic convergence as it applies to feasible-directions methods, both simplex- and non-simplex-based. Before proceeding further, the reader may wish to review the discussion in Section 4.6 on the convergence of iterative algorithms for *unconstrained* problems; the definition we are about to state for the case of *constrained* problems is almost exactly the same, except that the phrase "Kuhn-Tucker point," or "constrained stationary point," is substituted for the phrase "stationary point" wherever the latter appears. In particular, suppose that some iterative computational algorithm is applied to a constrained mathematical program, starting at the point $\mathbf{x}_0$ and generating a sequence of solution points $\{\mathbf{x}_i\} \equiv \mathbf{x}_0, \mathbf{x}_1, \mathbf{x}_2, \ldots$. Then that algorithm is said to *converge*, or to be *theoretically convergent*, if for any given problem and starting point the following three conditions are satisfied:

1. If the sequence $\{\mathbf{x}_i\}$ terminates at a solution point $\mathbf{x}_k$ after a finite number of iterations, then either $\mathbf{x}_k$ is a Kuhn-Tucker point or evidence that no bounded optimal solution exists is provided;

2. If the sequence $\{\mathbf{x}_i\}$ converges to a limit point $\mathbf{x}^*$ after an infinite number of iterations, then $\mathbf{x}^*$ is a Kuhn-Tucker point; and

3. If the sequence $\{\mathbf{x}_i\}$ fails to converge, then no bounded optimal solution exists.

In conditions 1 and 3 the conclusion that "no bounded optimal solution exists" is meant to imply that either the optimal solution is unbounded or no feasible solution whatever exists. The reader is reminded that mathematical programming algorithms are usually supplied with computational stopping rules that cause termination at the point $\mathbf{x}_k$ when $\mathbf{x}_k$ is found to be sufficiently close to $\mathbf{x}_{k-1}$; in such a case $\mathbf{x}_k$ is taken as an approximation to the limit point of the sequence $\{\mathbf{x}_i\}$.

The definition of algorithmic convergence just stated applies to solution methods for *all* types of mathematical programming problems, regardless of the nature of their constraints—in fact, for unconstrained problems, as noted above, it reduces essentially to the definition given in Chapter 4. In our treatment of algorithmic convergence in this section, however, we shall restrict our attention to algorithms for linearly constrained problems and, in particular, to feasible-directions methods, including both the simplex-based and the non-simplex-based types. Thus each of the solution points in the sequence $\{\mathbf{x}_i\}$ will be assumed here to be a feasible point, and the possibility that a given problem may have no feasible solution at all will be ignored.

Let us begin with a brief discussion of part 1 of the convergence definition. As a general rule, the conclusion of part 1 is relatively easy to prove for any reasonable mathematical programming algorithm; we have, in fact, proved it for each of the feasible-directions methods presented in the past two chapters. Of course, when the objective function is not concave there is no guarantee that the Kuhn-Tucker point obtained by the algorithm is a constrained local maximum (we continue, as before, to frame our discussion in terms of maximization problems). If, for example, a constrained saddle point of some type happened to be chosen as the starting solution, most feasible-directions methods would terminate immediately. Apart from this rather unlucky possibility, extreme points that are also saddle points can sometimes act as "traps" for feasible-directions methods, as the following problem will illustrate:

$$\begin{aligned} \text{Max } f(\mathbf{x}) &= x_1 + (x_2 - 9)^2 \\ \text{subject to } x_1 + 9x_2 &\leqq 90, \\ x_1 \qquad &\leqq 9, \\ \text{and} \qquad x_1, x_2 &\geqq 0. \end{aligned}$$

Referring to Fig. 8-11, the extreme point $\mathbf{x}^* = (9, 9)$ is evidently not a local maximum, inasmuch as any small downward displacement would be feasible and would increase the value of the objective function. However, at $\mathbf{x}^*$ the gradient $\nabla\mathbf{f}(\mathbf{x}) = (1, 2x_2 - 18)$ has the value $(1, 0)$, which evidently has no positive component in any feasible direction; thus, upon arriving at $\mathbf{x}^*$,

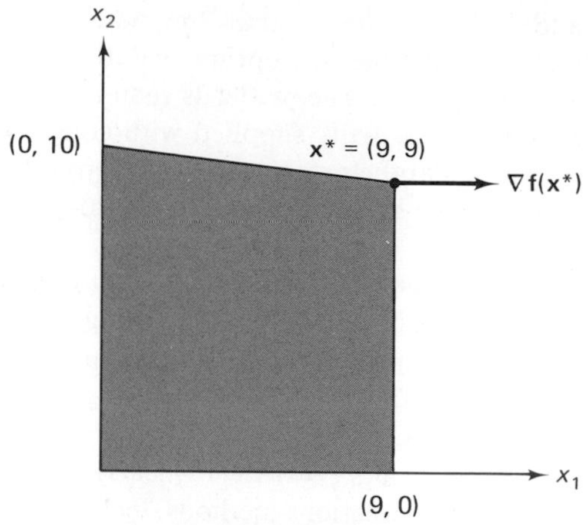

FIGURE 8-11

almost any feasible-directions method would terminate there. Moreover, at any feasible point $\mathbf{x}$ lying *above* the line $x_2 = 9$, the gradient is directed upward and to the right, and it can be seen that, given *any* starting point on or above that line, most feasible-directions methods would be unable to escape the trap of eventually terminating at $\mathbf{x}^*$. The reader may wish to verify that this is indeed true for (1) the gradient projection method and (2) any method that obtains the best feasible direction at each iteration by solving the problem (8-64). It should be added, incidentally, that the better computer codes use perturbation tests to be sure that any solution point obtained is indeed a local optimum of the type desired; if it is not, an improving feasible direction of displacement is identified, and the optimization process is resumed.

Although the discussion has thus far been restricted to finite solutions, part 1 of the convergence definition also covers most instances in which an unbounded optimal solution is discovered. Typically, at some stage it is found to be both possible and desirable to proceed an unlimited distance from the current point $\mathbf{x}_k$ in some feasible direction $\mathbf{s}_k$; thus "evidence" of an unbounded solution "is provided" and the method terminates, after having generated only a finite number of points $\mathbf{x}_i$. On occasion, however, unboundedness may manifest itself in a different way with the generation of an unending sequence of feasible points $\{\mathbf{x}_i\}$ whose magnitudes $|\mathbf{x}_i|$ eventually become infinitely large. As long as each point leads to another, the algorithm cannot terminate itself automatically, and, because $\{\mathbf{x}_i\}$ does not converge to a finite limit point, the situation falls within the domain of case 3. In practice, nonconvergence of $\{\mathbf{x}_i\}$ is usually diagnosed, or asserted to have occurred, when $|\mathbf{x}_i|$ exceeds either the tolerance of the computer or some limiting value prespecified by the analyst.

In addition to this second type of unboundedness, case 3 also covers certain pathological situations involving oscillation of the sequence $\{\mathbf{x}_i\}$. For practical purposes, however, the circumstances included under case 3 are not especially interesting, particularly inasmuch as well-formulated problems in engineering and operations research generally have bounded optimal solutions. Accordingly, we shall give no further consideration to part 3, beyond remarking that in practical situations it is very seldom encountered.

This leaves part 2 of the convergence definition still to be discussed. Although we have noted that the proof of part 1 is relatively straightforward for almost every mathematical programming algorithm, it turns out that part 2 is usually far more difficult to establish—that is, when it can be proved at all. It is an interesting fact that some of the best-known and most useful algorithms in mathematical programming have never been proved to converge, in the sense of the definition stated above, and it is in the attempt to establish part 2 that the crucial theoretical difficulties have normally arisen. For some solution methods clever counterexamples have been devised to demonstrate *failure* of convergence, while in other cases the question of convergence actually remains unresolved.

The number of algorithms in the "unresolved" category has been reduced significantly as a result of the very important work of Zangwill in the middle and late 1960s (see [59] and [60]): During that period he identified certain general properties that all solution algorithms have in common and showed how they could frequently be used to prove or disprove convergence. Among the algorithms for which he was able to provide convergence proofs are his convex simplex method, the Frank and Wolfe algorithm [16], and a well-known method developed by Fiacco and McCormick [15] for solving nonlinearly constrained problems, which will be discussed in Chapter 9. On the other hand, neither the reduced gradient nor the gradient projection method satisfies the convergence conditions stated by Zangwill; nevertheless, while the former is known to be nonconvergent, the latter, with a suitable degeneracy device attached, does in fact appear to converge.

Useful as all this research has been, however, it is still somewhat beyond the scope of an introductory text in nonlinear programming. Accordingly, we shall confine our efforts to presenting a rather broad description of convergence behavior—and misbehavior—and not attempt to develop rigorous convergence proofs for specific algorithms. Students who are interested in a thorough and sophisticated treatment of convergence theory should consult the texts of Zangwill [60] and Luenberger [31]; the latter also contains extensive material on *rates* of convergence, which obviously become important in those practical situations in which a choice must be made from among competing solution algorithms.

Returning to our discussion of the convergence definition stated at the beginning of this section, the difficulty in establishing part 2 for most mathematical programming algorithms appears to be related to the phenomenon of *zigzagging*, or *jamming*. These terms refer to a special type of convergence behavior in which the sequence of points $\{\mathbf{x}_i\}$ describes a zigzag path whose segments (the vectors $\mathbf{x}_i - \mathbf{x}_{i-1}$) become shorter and shorter, until the entire sequence converges to a point $\mathbf{x}^*$ that *may or may not* be a Kuhn-Tucker point. We have already encountered this phenomenon in our study of unconstrained optimization—refer to the optimal gradient example in Section 4.9, illustrated in Fig. 4-2. In most cases, however, the zigzag path is formed by being reflected back and forth between two or more constraint hyperplanes that come together to define an extreme point, edge, or the like, with the path normally converging toward the junction of those hyperplanes. As a simple illustration, consider the following example problem, due to Zoutendijk [63]:

$$\begin{aligned} \text{Max } f(\mathbf{x}) &= -x_1^2 - 2x_2^2 \\ \text{subject to } -x_1 + 4x_2 &\leqq 0 \\ \text{and } -x_1 - 4x_2 &\leqq 0. \end{aligned}$$

The feasible region is illustrated in Fig. 8-12. Observe that at any feasible point on either of the two hyperplanes, excepting the origin, the gradient $\nabla \mathbf{f}(\mathbf{x}) = (-2x_1, -4x_2)$ points into the interior of the feasible region; accordingly, a "best-feasible-direction" method [i.e., one that obtains the best feasible direction of displacement by solving the problem (8-64) at each stage] would select the gradient direction at every iteration. If we choose as a starting point $\mathbf{x}_0 = (4, 1)$, the direction of displacement at the first iteration is $\mathbf{s}_0 = \nabla \mathbf{f}(\mathbf{x}_0) = (-8, -4)$. The value of the objective function along the

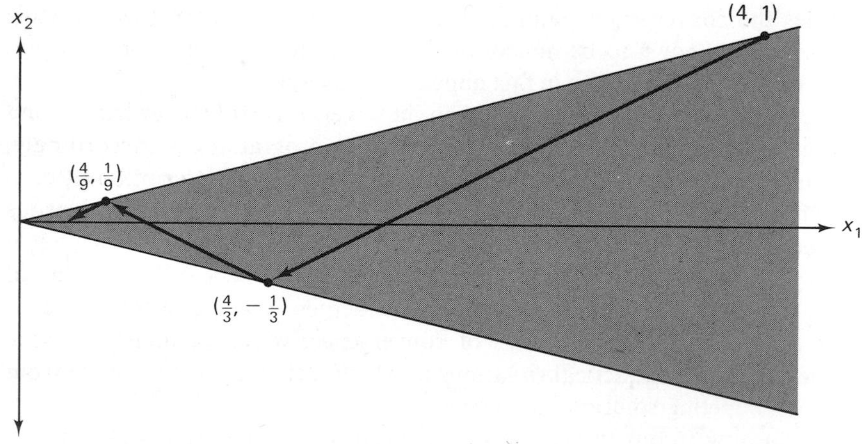

FIGURE 8-12

gradient path is given by

$$f(\mathbf{x}_0 + \theta \mathbf{s}_0) = f(4 - 8\theta, 1 - 4\theta) = -18 + 80\theta - 96\theta^2,$$

which is maximized at $\theta = \frac{5}{12}$. But the maximum feasible length of displacement, computed from the general formula (8-60), is

$$\Delta = \frac{b_2 - \boldsymbol{\alpha}_2 \mathbf{x}_0}{\boldsymbol{\alpha}_2 \mathbf{s}_0} = \frac{0 - (-8)}{8 + 16} = \frac{1}{3}.$$

Thus displacement is stopped when the lower constraint hyperplane is encountered: $\theta_{\text{opt}} = \Delta = \frac{1}{3}$ and $\mathbf{x}_1 = (\frac{4}{3}, -\frac{1}{3})$. For the second iteration the direction of displacement is $\mathbf{s}_1 = \nabla \mathbf{f}(\mathbf{x}_1) = (-\frac{8}{3}, \frac{4}{3})$, and we eventually find that $\theta_{\text{opt}} = \Delta = \frac{1}{3}$ and $\mathbf{x}_2 = (\frac{4}{9}, \frac{1}{9})$. It is not difficult to show that, in general,

$$\mathbf{x}_i = \left(\frac{4}{3^i}, \frac{1}{(-3)^i}\right),$$

so the sequence $\{\mathbf{x}_i\}$ approaches (0, 0), which is in fact the desired maximum. The first few segments of the zigzag path generated by this best-feasible-direction approach are diagrammed in Fig. 8-12. Note that if the gradient projection method had been used, the initial direction of displacement from $\mathbf{x}_0$ would have been along the constraint hyperplane $-x_1 + 4x_2 = 0$ (why?) and the optimum would have been reached in a single iteration.

This example was not too dismaying because at least the sequence $\{\mathbf{x}_i\}$ did converge, albeit very slowly, to a Kuhn-Tucker point. Unfortunately, however, it is also possible for a useful and otherwise perfectly unobjectionable algorithm to converge via a zigzag path to a solution that is *not* a Kuhn-Tucker point. One very clever example devised by Wolfe [55] is presented in [63]:

$$\text{Max } f(\mathbf{x}) = -\tfrac{4}{3}(x_1^2 - x_1x_2 + x_2^2)^{3/4} - x_3$$
$$\text{subject to } x_1, x_2, x_3 \geqq 0.$$

The quadratic form $x_1^2 - x_1x_2 + x_2^2$ is positive definite, so the origin (0, 0, 0) evidently constitutes the optimal feasible solution. Let us now attempt to solve this problem via the best-feasible-direction approach, taking as our starting point $\mathbf{x}_0 = (\frac{1}{4}, 0, 1)$—actually, any starting point of the form $\mathbf{x}_0 = (p, 0, q)$, where $p \leqq \sqrt{2}/4$ and $q > (1 + \sqrt{2}/2)\sqrt{p}$, would serve equally well. At the starting point the gradient vector

$$\nabla \mathbf{f}(\mathbf{x}) = \left(\frac{-2x_1 + x_2}{(x_1^2 - x_1x_2 + x_2^2)^{1/4}}, \frac{x_1 - 2x_2}{(x_1^2 - x_1x_2 + x_2^2)^{1/4}}, -1\right)$$

has the value $\nabla \mathbf{f}(\mathbf{x}_0) = (-1, \frac{1}{2}, -1)$; inasmuch as this is a feasible direction, we choose $\mathbf{s}_0 = \nabla\mathbf{f}(\mathbf{x}_0)$. The maximum permissible step length is $\Delta = \frac{1}{4}$, and, assuming it is desirable to proceed as far as possible in the $\mathbf{s}_0$-direction, the new point is $\mathbf{x}_1 = \mathbf{x}_0 + \Delta\mathbf{s}_0 = (0, \frac{1}{8}, \frac{3}{4})$, with $\nabla\mathbf{f}(\mathbf{x}_1) = (\sqrt{2}/4, -\sqrt{2}/2, -1)$. Note that the gradient at $\mathbf{x}_1$ has a positive component in the $\mathbf{s}_0$-direction,

$$\nabla\mathbf{f}^T(\mathbf{x}_1) \cdot \mathbf{s}_0 = 1 - \frac{\sqrt{2}}{2} > 0,$$

so $\mathbf{s}_0$ still constitutes an improving direction at $\mathbf{x}_1$ and our assumption was justified.

For the next iteration the gradient direction is again feasible and permits a maximum step length of $\Delta = \sqrt{2}/8$. Making the same assumption as before, we obtain $\mathbf{x}_2 = (\frac{1}{16}, 0, (6 - \sqrt{2})/8)$, with $\nabla\mathbf{f}(\mathbf{x}_2) = (-\frac{1}{2}, \frac{1}{4}, -1)$; because $\nabla\mathbf{f}^T(\mathbf{x}_2) \cdot \mathbf{s}_1 = 1 - \sqrt{2}/4 > 0$, we see again that our assumption was valid. Continuing in this manner, we find that at each iteration the gradient direction is feasible and the maximum permitted step is taken; the third point is $\mathbf{x}_3 = (0, \frac{1}{32}, (5 - \sqrt{2})/8)$, and as i becomes infinitely large the general point $\mathbf{x}_i$ approaches, in the limit,

$$\mathbf{x}^* = \left(0, 0, 1 - \frac{1}{4}\left(1 + \frac{1}{\sqrt{2}} + \frac{1}{2} + \frac{1}{2\sqrt{2}} + \cdots\right)\right) = \left(0, 0, \frac{2 - \sqrt{2}}{4}\right)$$
$$\cong (0, 0, .146).$$

But note that $\delta f/\delta x_3$ has the value -1 everywhere in the feasible region; thus $\mathbf{x}^*$ cannot be a Kuhn-Tucker point, or constrained stationary point, because $x_3 = 0$ must hold at all such points. A more rigorous proof that $\mathbf{x}^*$ is not a constrained stationary point is developed in Exercise 8-45.

As this example suggests, the sequence $\{\mathbf{x}_i\}$ can converge by zigzagging to a non-Kuhn-Tucker point $\mathbf{x}^*$ only if $\mathbf{x}^*$ is a point on the boundary of the feasible region at which two or more constraint hyperplanes come together. When zigzagging begins, after a finite number of iterations, every new point $\mathbf{x}_i$ generated by the algorithm is determined when one or another of those constraint hyperplanes is encountered, rather than when the objective function stops improving. Thus in the later stages the directional derivative $df(\mathbf{x}_i + \theta\mathbf{s}_i)/d\theta$ never reaches a value of zero. Regarding this last observation, the reader may recall that the proof of convergence of the optimal gradient method for unconstrained problems, given in Section 4.7, depended on the vanishing of $df/d\theta$ at each iteration (i.e., on the fact that successive gradients would be perpendicular) in the vicinity of the limit point $\mathbf{x}^*$. It is because this convenient circumstance, the vanishing of $df/d\theta$, cannot be relied on in the case of constrained optimization that the proof of algorithmic convergence breaks down for a great many feasible-directions methods.

To repair this deficiency and guarantee convergence, certain computational modifications are sometimes appended to computer-coded versions of the non-simplex-based feasible-directions algorithms discussed in Sections 8.9 and 8.10. The principal motivation in doing so is not actually to prevent convergence to a non-Kuhn-Tucker point—this is a rather rare event and is not likely to occur in any naturally arising engineering problem—but rather to accelerate convergence in those cases in which the sequence $\{\mathbf{x}_i\}$ approaches a legitimate Kuhn-Tucker point, but via a very tedious zigzag path. Zoutendijk was the first to investigate ways and means of preventing zigzagging, or jamming.[9] In his 1960 text [61] he proposed a scheme that required remembering at each stage which of the constraint hyperplanes had been encountered in the previous few iterations. In using that approach, when a given hyperplane is encountered for the *second* time, indicating the possibility of zigzagging, its associated inequality constraint is then held binding during the ensuing iterations, even though the gradient vector might at various stages point away from the hyperplane and into the interior of the feasible region. This maneuver effectively prevents zigzagging by ensuring that the path generated by the sequence $\{\mathbf{x}_i\}$ is held in the hyperplane instead of being allowed to reflect or rebound from it. The constraint in question should in fact remain binding until it is no longer possible to find an improving feasible direction without relaxing it.

In a later work [63] Zoutendijk described a slightly different procedure using the same tactic: As soon as any hyperplane is encountered for the *first* time, its associated inequality constraint is held binding until it becomes necessary to relax it in order to obtain an improving feasible direction. In either of these two approaches, particularly the latter, there is a tendency to build up an "inventory" of binding inequality constraints, until eventually no further improvement is possible—perhaps because the current solution is completely bound at some extreme point. The binding inequality constraints are then relaxed in sequence, "eldest first" (i.e., starting with the one that has been binding for the greatest number of iterations), until an improving feasible direction of displacement becomes available. If no such direction can be found after every inequality constraint has been relaxed, the algorithm terminates.

Another antizigzagging technique, called *bending*, has been suggested by McCormick [33] for application to problems of the form

$$\text{Max } f(\mathbf{x}) \text{ subject to } \mathbf{x} \geqq \mathbf{0}. \qquad (8\text{-}86)$$

[9]The terms "zigzagging" and "jamming" were introduced into the literature by Zoutendijk [61] and Zangwill [60], respectively. It might seem desirable to use the former term to denote *any* case in which the sequence $\{\mathbf{x}_i\}$ converges to a limit point via a zigzag path and the latter to refer only to those cases in which the limit point is not a Kuhn-Tucker point. In this usage jamming would become a subtype of zigzagging. To avoid confusion in this text, however, we shall continue to use the term "zigzagging" exclusively, further specifying the nature of the limit point only when it is relevant to the discussion.

When, during the $(k+1)$th iteration, displacement from $\mathbf{x}_k$ in the $\mathbf{s}_k$-direction is stopped at some boundary point because the hyperplane $x_j = 0$ has been encountered, optimization continues from that boundary point in a new direction $\mathbf{s}'_k$ that differs from $\mathbf{s}_k$ only in that its jth component is zero. Thus the vector $\mathbf{s}'_k$, which is produced by "bending" $\mathbf{s}_k$, lies embedded in the hyperplane $x_j = 0$. Moreover, if another boundary is encountered while the objective value is still improving, another bending operation is performed, and, if necessary, yet another, so that in general during the $(k+1)$th iteration displacement proceeds along a sequence of directional vectors $\mathbf{s}_k, \mathbf{s}'_k, \mathbf{s}''_k, \ldots$, each of which differs from its predecessor in only a single component. When no further improvement is possible—perhaps because the direction vector has been reduced to $\mathbf{0}$—the iteration is complete: The current solution point is taken to be $\mathbf{x}_{k+1}$, and only then is the gradient $\nabla \mathbf{f}(\mathbf{x}_{k+1})$ recomputed.

It is not difficult to see that any of these techniques of Zoutendijk or McCormick would suffice to prevent the type of zigzagging behavior that was observed in the two example problems considered earlier in this section. The crucial aspect of all three techniques is that the path described by the sequence $\{\mathbf{x}_i\}$ is absorbed into any newly encountered hyperplane instead of being reflected away from it. Because zigzagging to a non-Kuhn-Tucker point is thereby prevented, it becomes possible to prove convergence for various non-simplex-based feasible-directions algorithms with "bending" or "absorbing" devices attached. These devices thus play a role similar to that of the lexicographic and perturbation techniques that are sometimes attached to simplex-based algorithms, or to such algorithms as the gradient projection method, in order to eliminate any chance of infinite cycling.

It is interesting to reflect that, of all the non-simplex-based feasible-directions algorithms we have considered in this chapter, the only one whose normal mode of operation—apart from any peripheral devices—seems to preclude the possibility of zigzagging between hyperplanes is the gradient projection method.[10] Another algorithm in the same category is the variable reduction method, developed by McCormick [34]; this very complicated algorithm for linearly constrained mathematical programs is based essentially on a generalization of the bending technique (i.e., bending occurs when *any* hyperplane, not necessarily of the form $x_j = 0$, is encountered). Note, however, that although the variable reduction method is immune to zigzagging difficulties, it must still, like the gradient projection method, be equipped with a special perturbation device in order to be able to cope successfully with degeneracy.

[10] We use the phrase "seems to preclude" because although zigzagging has never been observed to occur in applications of the gradient projection method, no proof of its impossibility has yet been advanced.

EXERCISES

Section 8.4

8-1. Verify that the separable programming problem presented at the end of Section 8.2 is a convex program, and obtain the optimal solution $\mathbf{x}^*$ by differential calculus, using the fact that the gradient of the objective function at $\mathbf{x}^*$ must be perpendicular to the constraint hyperplane on which $\mathbf{x}^*$ lies. Then solve the given approximating problem via the simplex method, and compare the solution to the true optimum $\mathbf{x}^*$. What relationship, if any, exists between the basic feasible solutions generated by the simplex method and the extreme points of the feasible region of the original problem?

Exercises 8-2 and 8-3: Solve each of the following problems via separable programming, starting with the solution that corresponds to the origin $(x_1, x_2) = (0, 0)$. In constructing the functions $\hat{f}_j(x_j)$, use breakpoints x_{jk} that are 1 unit apart.

8-2. Min $z = (x_1 - 3)^2 + (x_2 - 3)^2$
subject to $x_1 + x_2 \leqq 4$
and $x_1, x_2 \geqq 0$.

8-3. Max $z = x_1^2 + x_2^2$
subject to $2x_1 - x_2 \leqq 2,$
$-x_1 + 2x_2 \leqq 2,$
and $x_1, x_2 \geqq 0$.

8-4. (a) Can the objective

$$\text{Min } z = \sum_i \sum_{j>i} |x_i - x_j|,$$

where $|x_i - x_j|$ denotes the absolute value of the difference $x_i - x_j$, be expressed as a separable objective function without introducing nonlinear constraints?
(b) Repeat part (a) for the objective function

$$\text{Max } z = \max\{x_1, x_2, \ldots, x_n\}.$$

8-5. Suppose that each of the constraints $\sum_{k=0}^{r_j} \lambda_{jk} = 1$ in the approximating problem (8-13) were solved for one variable, say, λ_{j0}, in terms of the others and that the resulting expressions $1 - \lambda_{j1} - \lambda_{j2} - \cdots - \lambda_{jr_j}$ were then substituted for the variables λ_{j0} wherever the latter occur in the formulation of (8-13). Would this be a legitimate way of reducing the total number of constraints in the approximating problem? Why or why not?

8-6. If the function $f_j(x_j)$ in the separable program (8-5) is concave and has a continuous first derivative, prove that its piecewise linear approximating function $\hat{f}_j(x_j)$ is also concave by using the law of the mean. This in effect constitutes an alternative proof of Lemma 8.1. The law of the mean may be stated as follows: Let the function of one variable $f(x)$ be continuous in the closed interval $[a, b]$, and let its first derivative $f'(x)$ exist at every point in the open interval (a, b). Then there exists a point $x = c$ satisfying $a < c < b$ such that $f(b) - f(a) = (b - a)f'(c)$.

8-7. In solving the approximating problem of a linearly constrained convex separable program via the simplex method, with no restrictions on basis entry, does each of the basic feasible solutions $\boldsymbol{\lambda}_B$ necessarily satisfy the logical restrictions 1 and 2 of Section 8.2? That is, does every $\boldsymbol{\lambda}_B$ correctly represent, for each j, both a feasible value $\bar{x}_j$ of the jth variable and the associated value of the approximating function $\hat{f}_j(\bar{x}_j)$? Is your answer the same when restricted basis entry is being used?

8-8. Given a convex separable program of the form (8-5) with a bounded optimal solution, let an approximating problem A_0 be constructed in the usual manner. Now consider the sequence of approximating problems $A_0, A_1, A_2, \ldots$, in which each A_i, $i \geqq 1$, is created from A_{i-1} by adding enough breakpoints to exactly bisect each of the intervals $[x_{jk}, x_{j,k+1}]$ along every x_j-axis of A_{i-1}; thus A_i has twice as many intervals and about twice as many breakpoints as A_{i-1}. Prove that as i approaches infinity the value of $\mathbf{x}$ implied by the optimal solution $\boldsymbol{\lambda}_i^*$ to the problem A_i approaches the true optimal solution of the original separable program.

8-9. In this exercise we shall develop an alternative form of separable programming (see Chapter 4 of [22]) that uses as problem data the differences between successive breakpoints rather than the points themselves. Assume that the break points $x_0, x_1, \ldots, x_r$ have been selected for the variable x (for the moment we can simplify by omitting the subscript j). Then in general any point x between x_0 and x_r can be represented as a sum of differences,

$$x = x_0 + \sum_{k=1}^{r} \delta_k(x_k - x_{k-1}), \tag{8-87}$$

where if, say, $x_q \leqq x \leqq x_{q+1}$, the variables δ_k have the following values:

$$\begin{cases} \delta_1 = \delta_2 = \cdots = \delta_q = 1, \\ \delta_{q+1} = \dfrac{x - x_q}{x_{q+1} - x_q}, \\ \delta_{q+2} = \delta_{q+3} = \cdots = \delta_r = 0. \end{cases} \tag{8-88}$$

Thus the value of x is accumulated by summing x_0, $(x_1 - x_0)$, $(x_2 - x_1)$, $\ldots$, $(x_q - x_{q-1})$ and finally a fraction δ_{q+1} of the difference $(x_{q+1} - x_q)$. Now, given some nonlinear function $f(x)$, this set of breakpoints can be used to construct a piecewise linear approximation to $f(x)$ in the manner described in the text, that is, by computing the ordinates $f_k = f(x_k)$ and then joining the points (x_0, f_0), $(x_1, f_1), \ldots, (x_r, f_r)$ with r successive straight-line segments. Let $\hat{f}(x)$ denote that approximating function.

(a) Given the representation of the general point x via (8-87) and (8-88), how might the value of the approximating function $\hat{f}(x)$ be represented in terms of the variables δ_k?

(b) In view of (8-87) and (8-88), suggest a set of logical and/or algebraic restrictions on the variables $\delta_1, \ldots, \delta_r$ such that any value of x between x_0 and x_r can be represented by one and only one feasible set of values of those variables, and such that any feasible set of values for $\delta_1, \ldots, \delta_r$ represents a unique value of x satisfying $x_0 \leqq x \leqq x_r$.

(c) Given a separable programming problem (8-5) for which piecewise linear approximations $\hat{f}_j(x_j)$ have been constructed, substitute expressions of the form (8-87) for the variables x_j and expressions of the type determined in part (a) for the approximating functions $\hat{f}_j(x_j)$ to create the *δ-form of the approximating problem.* Be sure to include all necessary restrictions on the variables δ_{jk}. It will be convenient to use the symbols Δx_{jk} and Δf_{jk} to denote the differences $x_{jk} - x_{j,k-1}$ and $f_{jk} - f_{j,k-1}$.
(d) Sketch out a proof that, when the original separable programming problem is a convex program, the δ-form of the approximating problem can be solved by the simplex method with no additional modifications or restrictions attached.
(e) Describe a simple set of logical restrictions that can be added to a simplex code to enable it to solve the δ-form of the approximating problem for *any* separable program.

8-10. It is well known that the simplex method can be modified so as to enable it to handle upper-bound restrictions of the form $x_j \leqq M_j$ much more efficiently than ordinary constraints involving two or more variables. In fact, the number of rows and columns in the basis matrix at each iteration is simply equal to the number of non-upper-bound constraints in the problem being solved; thus the size of the basis is not affected by the upper-bound restrictions (just as it is not affected by the lower-bound restrictions $x_j \geqq 0$). Given a separable program with n variables and m linear constraints, suppose that for each x_j, $j = 1, \ldots, n$, the approximating function $\hat{f}_j(x_j)$ is constructed from $r_j + 1$ breakpoints in the usual manner. How many variables, ordinary constraints, and upper-bound restrictions will the δ-form of the approximating problem have? In view of these considerations, what are the relative advantages and disadvantages of the λ-form (presented in the text) and the δ-form of separable programming?

8-11. Solve the problem of Exercise 8-2 via the δ-form of separable programming, using the same set of breakpoints and the same starting solution.

8-12. Solve the problem of Exercise 8-3 via the δ-form of separable programming, using the same breakpoints and starting solution.

Section 8.6

8-13. Use the convex simplex method to solve the example problem of Section 7.7:

$$\text{Max } z = 18x_1 - x_1^2 + x_1x_2 - x_2^2$$
$$\text{subject to} \quad x_1 + x_2 \leqq 12,$$
$$-x_1 + x_2 \leqq 6,$$
$$\text{and } x_1, x_2 \geqq 0,$$

taking the origin as the starting point. Then compare and contrast the method of Beale and the convex simplex method with respect to solving quadratic programming problems. Note that the numbers of iterations required by the two methods for this particular example are not necessarily a valid general indicator of their relative efficiencies.

8-14. Consider the parameterized convex program

$$\begin{aligned} \text{Max } z &= Kx_1 - x_1^2 + x_1x_2 - x_2^2 \\ \text{subject to } \quad x_1 + x_2 &\leqq 12, \\ -x_1 + x_2 &\leqq 6, \\ \text{and } x_1, x_2 &\geqq 0, \end{aligned}$$

where K is a scalar parameter. This problem with $K = 18$ constituted the example of Section 7.7. Solve it via the convex simplex method, using the origin as the starting point, for each of the following values of K:

(a) $K = 20$.
(b) $K = 24$.
(c) $K = 26$.

8-15. Solve the following by the convex simplex method, using the origin as the starting point:

$$\begin{aligned} \text{Max } z &= 2x_1 - x_1^2 + x_2^2 \\ \text{subject to } \quad x_1 - 2x_2 &\leqq 0, \\ -x_1 + x_2 &\leqq 2, \\ \text{and } x_1, x_2 &\geqq 0. \end{aligned}$$

8-16. Solve the following via the convex simplex method, starting at the origin:

$$\begin{aligned} \text{Max } z = 10 \log_e(x_1 + 1) + 4x_2 &+ (x_3 + 1)^2 \\ \text{subject to } 4x_1 + 2x_2 + x_3 &\leqq 28, \\ x_1 + x_2 + x_3 &\leqq 12, \\ x_2 + 3x_3 &\leqq 24, \\ \text{and } x_1, x_2, x_3 &\geqq 0. \end{aligned}$$

8-17. Determine a general transformation formula for updating the values of the basic variables in a convex simplex pivot when
(a) The nonbasic variable $x_k \geqq 0$ is chosen to be increased.
(b) The nonbasic variable $x_k > 0$ is chosen to be decreased.

8-18. In solving the one-dimensional optimization problem

$$\text{(8-38)} \qquad \begin{cases} \text{Max } f(\mathbf{x} + \theta \mathbf{s}) \\ \text{subject to } 0 \leqq \theta \leqq \Delta, \end{cases}$$

one might wish to use a golden section search over the entire interval $[0, \Delta]$. However, it is quite possible that $f(\mathbf{x} + \theta\mathbf{s})$ is not a unimodal function of θ within that interval. Show that under those circumstances a golden section search using N sequential experiments will still approach some local maximum of $f(\mathbf{x} + \theta\mathbf{s})$ as N approaches infinity.

8-19. Suppose the objective function $f(\mathbf{x})$ is quadratic in the $m + 1$ variables $\mathbf{x}_\mathbf{B}^+ \equiv (\mathbf{x}_\mathbf{B}, x_k)$ whose values are to change during some convex simplex pivot. That is, suppose that if the nonbasic variables other than x_k are held constant, the objective function $f(\mathbf{x})$ reduces to

$$f^+(\mathbf{x}_\mathbf{B}^+) = \mathbf{c}^T\mathbf{x}_\mathbf{B}^+ + (\mathbf{x}_\mathbf{B}^+)^T\mathbf{D}\mathbf{x}_\mathbf{B}^+,$$

where $\mathbf{c}$ and $\mathbf{D}$ are a column vector and square matrix of appropriate dimensions. Outline an efficient procedure for obtaining the optimal solution θ_{opt} to the problem

$$\text{(8-38)} \qquad \begin{cases} \text{Max } f(\mathbf{x} + \theta\mathbf{s}) \\ \text{subject to } 0 \leqq \theta \leqq \Delta. \end{cases}$$

8-20. Describe in some detail how the convex simplex method can be modified along the lines of the transportation algorithm of linear programming so as to form an algorithm for solving the *convex transportation problem*

$$\begin{aligned} &\text{Min } z = f(\mathbf{x}) \\ \text{subject to } &\sum_{j=1}^{n} x_{ij} = a_i, \qquad i = 1, \ldots, m, \\ &\sum_{i=1}^{m} x_{ij} = b_j, \qquad j = 1, \ldots, n, \\ &\text{and } x_{ij} \geqq 0 \qquad \text{for all } i \text{ and } j, \end{aligned}$$

where $\sum_i a_i = \sum_j b_j$ and where $f(\mathbf{x})$ is any convex function of the variables x_{ij}, $i = 1, \ldots, m$, $j = 1, \ldots, n$. Use the resulting "convex transportation algorithm" to solve the following problem (devised by Zangwill):

$$\begin{aligned} \text{Min } f(\mathbf{x}) = x_{11} + 2x_{12} + x_{13}^2 + x_{21}^2 + 3x_{22} + 2x_{23}^2 + e^{x_{11}x_{21}} \\ \text{subject to } x_{11} + x_{12} + x_{13} \qquad\qquad\qquad &= 3, \\ x_{21} + x_{22} + x_{23} &= 2, \\ x_{11} \qquad\qquad + x_{21} \qquad\qquad &= 1, \\ x_{12} \qquad\qquad + x_{22} \qquad &= 2, \\ x_{13} \qquad\qquad + x_{23} &= 2, \\ \text{and } x_{ij} &\geqq 0 \qquad \text{for all } i \text{ and } j. \end{aligned}$$

Take as the starting solution $x_{11} = 1$, $x_{12} = 2$, $x_{23} = 2$, all others $= 0$.

Section 8.7

8-21. Solve the following problem via (a) the convex simplex method and (b) the reduced gradient method, starting at the origin in each case:

$$\begin{aligned} \text{Max } z = 20x_1 + 10x_2 - x_1^2 + x_1x_2 - x_2^2 \\ \text{subject to } \quad x_1 + x_2 &\leqq 12, \\ -x_1 + x_2 &\leqq 6, \\ \text{and } x_1, x_2 &\geqq 0. \end{aligned}$$

Generalizing from this example, comment on the relative advantages and disadvantages of each.

8-22. Given the starting solution $x_1 = 9$, $x_2 = 0$, $x_3 = 7$, $x_4 = 13$, perform one complete reduced gradient pivot for the problem

$$\begin{aligned} \text{Max } z &= 22x_1 + 16x_2 - x_1^2 - x_1x_2 - x_2^2 \\ \text{subject to } & x_1 + x_2 + x_3 = 16, \\ & -x_1 + x_2 + x_4 = 4, \\ \text{and } & x_1, x_2, x_3, x_4 \geqq 0 \end{aligned}$$

using each of the following pairs of variables as the initial basis: (a) x_3 and x_4; (b) x_1 and x_4; (c) x_2 and x_4.

8-23. Solve the problem of Exercise 8-16 via the reduced gradient method, using the origin as the starting point.

8-24. Describe in detail how the reduced gradient method, including all its logical tests and computations, must be modified in order to make it directly applicable to *minimization* problems.

8-25. What changes would have to be made to an ordinary linear simplex computer code in order to convert it into a code for the reduced gradient method? Does this appear to be a practical way of developing a reduced gradient code?

Section 8.8

8-26. Let the feasible region R be the set of all points $\mathbf{x}$ in E^2 that satisfy the following constraints:

$$x_1 + x_2 \leqq 2, \quad x_1 - x_2 \leqq 0, \quad x_2 \leqq 1, \quad \text{and} \quad x_1, x_2 \geqq 0.$$

Identify all points in R that are degenerate in the geometric sense. Then add a slack variable to each of the first three constraints and find all basic feasible solutions that are degenerate in the algebraic or simplex sense. Discuss how the two sets of degenerate points are related.

8-27. Given a set of linear constraints $\mathbf{Ax} \leqq \mathbf{b}$, where $\mathbf{A}$ is m-by-n and $m \geqq n$, suppose that no set of $n + 1$ or fewer of the row vectors $\boldsymbol{\alpha}_i^+ \equiv [a_{i1}, \ldots, a_{in}, b_i]$ is linearly dependent. Prove that no more than n constraints can be binding at any feasible point. Does this constitute a proof that no feasible point is degenerate?

8-28. Suppose that three linear inequality constraints are binding at some given point in a feasible region in E^2, as illustrated in the diagrams of Fig. 8-6. Is it necessarily true that (at least) one of the binding constraints is totally superfluous, in the sense that the feasible region would not be enlarged if it were deleted? Prove or give a counterexample. Answer the same question if four linear inequality constraints are binding at some feasible point in E^3.

Section 8.10

8-29. With regard to the problem (8-65) constructed at the end of Section 8.9, prove that
(a) A bounded optimal solution exists.

(b) If the optimal value of the objective z' is negative, then the original problem (8-59) has no feasible solution.
(c) If the optimal value of z' is 0, then the **x**-component of the optimal solution constitutes a feasible solution to (8-59).

8-30. Verify that the Kuhn-Tucker conditions for the problem (8-59) state that if $\mathbf{x}^*$ is a locally optimal solution to (8-59), then there must exist a set of real values λ_i^*, $i \in I$ or $i \in E$, such that

$$\nabla \mathbf{f}^T(\mathbf{x}^*) = \sum_{\text{all } i} \lambda_i^* \boldsymbol{\alpha}_i,$$

$$\left.\begin{aligned} \lambda_i^*(\boldsymbol{\alpha}_i\mathbf{x}^* - b_i) &= 0 \\ \lambda_i^* &\geqq 0 \end{aligned}\right\} \quad \text{for all } i \text{ in } I,$$

where the summation in the first equation is taken over all i in I and E. Prove that these conditions are satisfied at the point $\mathbf{x}_k$ generated by a feasible-directions algorithm if the optimal value of $\nabla \mathbf{f}^T(\mathbf{x}_k) \cdot \mathbf{s}$ in problem (8-64) is nonpositive. This essentially requires proving that the gradient $\nabla \mathbf{f}(\mathbf{x}_k)$ must then be expressible as a linear combination of the row vectors $\boldsymbol{\alpha}_i$ of the binding constraints, both inequalities and equalities, in which every scalar coefficient λ_i^* corresponding to an inequality is nonnegative. [*Hint:* Letting β denote the total number of constraints that are binding at $\mathbf{x}_k$, show that this statement is true for $\beta = 0$ and $\beta = 1$, and proceed via induction.]

8-31. In what sense can it be said that the reduced gradient method chooses the best feasible direction in nonbasic space? Show that the direction in the original space E^n that is defined by this reduced gradient direction is not necessarily the best feasible direction in E^n. [*Hint:* This can be done, in the case where all nonbasic variables are slacks and $u_j = \delta z/\delta x_j$ for all j, by substituting (8-28) into (8-49) and then observing that the resulting expressions for the components v_i are not in the same ratio as the $\delta f/\delta x_{\mathbf{B}i}$.] To illustrate your answer, consider the following problem:

$$\begin{aligned} \text{Max } f(x_1, x_2) &= (x_1 - 4)^2 + (x_2 - 4)^4 \\ \text{subject to} \quad x_1 + x_2 + x_3 &= 6, \\ -x_1 + 3x_2 + x_4 &= 6, \\ \text{and } x_1, x_2, x_3, x_4 &\geqq 0, \end{aligned}$$

where x_3 and x_4 are slack variables, so that the original space of the problem is E^2. Let the current basic solution be $x_{\mathbf{B}1} = x_1 = 3$, $x_{\mathbf{B}2} = x_2 = 3$, with $x_3 = x_4 = 0$, and verify that the current simplex tableau is

	x_1	x_2	x_3	x_4
$x_{\mathbf{B}1} = x_1 = 3$	1	0	.75	−.25
$x_{\mathbf{B}2} = x_2 = 3$	0	1	.25	.25

What is now the best feasible direction of displacement in the space of x_1 and x_2? What would be the direction in that space selected by the reduced gradient method at the next iteration?

8-32. Prove that if $\mathbf{s}^*$ is the optimal solution to (8-64) and its objective value $\psi^* = \nabla \mathbf{f}^T(\mathbf{x}_k) \cdot \mathbf{s}^*$ is positive, then $\mathbf{t}^* = \mathbf{s}^*/\psi^*$ must be the optimal solution to (8-67). In what sense, if any, can it be said that the mathematical programs (8-64) and (8-67) are equivalent?

8-33. Prove that if at any given iteration in solving the linear approximation problem (8-72) the optimal objective value is found to be
(a) Zero, then there is no improving feasible direction from the current point $\mathbf{x}_k$.
(b) Positive and finite, then $\bar{\mathbf{x}} - \mathbf{x}_k$ is an improving feasible direction, where $\bar{\mathbf{x}}$ is the optimal solution to (8-72).
(c) Infinite, then the extreme ray along which the solution becomes unbounded constitutes an improving feasible direction.

8-34. Given the mathematical programming problem

$$\begin{aligned} \text{Max } z &= -(x_1 - 4)^2 - x_2^2 \\ \text{subject to} \quad x_1 + 2x_2 &\leqq 9, \\ -x_1 + x_2 &\leqq 3, \\ \text{and } x_1, x_2 &\geqq 0, \end{aligned}$$

and the starting solution $x_1 = 1$, $x_2 = 4$, perform one complete feasible-directions iteration, using each of the following three methods of obtaining an improving feasible direction:
(a) Solution of the linear program (8-71).
(b) Solution of the linear program (8-72).
(c) Solution of the quadratic program (8-67).
Each of these direction-finding subproblems can be solved graphically. Compare the directions obtained via the three methods.

8-35. Repeat Exercise 8-34, using the starting solution $x_1 = 0$, $x_2 = 3$.

8-36. Repeat Exercise 8-34, using the starting solution $x_1 = 3$, $x_2 = 2$.

8-37. Using the starting solution $x_1 = 4$, $x_2 = 1$, solve the following problem via a feasible-directions method in which the best feasible direction is used at each iteration:

$$\begin{aligned} \text{Min } z &= x_1^2 + 2x_2^2 \\ \text{subject to } -x_1 + 4x_2 &\leqq 0 \\ \text{and } -x_1 - 4x_2 &\leqq 0. \end{aligned}$$

Draw a diagram of the feasible region, showing the first few points generated by the algorithm. Comment in general terms on the path described by the sequence $\mathbf{x}_0, \mathbf{x}_1, \mathbf{x}_2, \ldots$ and on the efficiency of the method in obtaining the solution. Can you suggest some means of accelerating the solution process?

Section 8.13

8-38. Write the Kuhn-Tucker conditions for the general linearly constrained mathematical program

$$(8\text{-}59)\qquad \begin{cases} \text{Max } z = f(\mathbf{x}), & \mathbf{x} = (x_1, \ldots, x_n), \\ \text{subject to } \boldsymbol{\alpha}_i\mathbf{x} \equiv \sum_{j=1}^{n} a_{ij}x_j \begin{cases} \leqq b_i, & i \in I, \\ = b_i, & i \in E. \end{cases} \end{cases}$$

Suppose now that the gradient projection method has terminated at a solution $\mathbf{x}_k$ at which $\mathbf{v}_k = \mathbf{P}\nabla\mathbf{f}(\mathbf{x}_k) = \mathbf{0}$ and $\lambda_i \geqq 0$ for all i corresponding to inequality constraints (case A). Show that there exists a set of generalized Lagrange multiplier values which, together with $\mathbf{x}_k$, satisfy the Kuhn-Tucker conditions for the problem (8-59). [*Hint:* A subset of the multiplier values was derived, though not identified, in the course of our development of the gradient projection method.]

8-39. Use the gradient projection method to solve the example problem of Section 7.7 and Exercise 8-13, taking the origin as the starting point. Compare the method of Beale and the gradient projection method with respect to solving quadratic programming problems, being careful not to generalize unjustifiably from the particular example under consideration.

8-40. In solving the mathematical programming problem

$$\begin{aligned} \text{Max } z = 2(x_1 - 1)^2 - x_2^2 + 16x_3 \\ \text{subject to } x_1 + x_2 \qquad\quad \leqq 8, \\ x_1 \qquad + 3x_3 \leqq 9, \\ x_2 - 2x_3 \leqq 0, \\ \text{and } x_1, x_2, x_3 \geqq 0, \end{aligned}$$

what direction of displacement from each of the following points would the gradient projection method select?

(a) $\mathbf{x} = (3, 4, 2)$ (c) $\mathbf{x} = (0, 0, 3)$

(b) $\mathbf{x} = (6, 2, 1)$ (d) $\mathbf{x} = (1, 4, 2)$

8-41. Solve the problem of Exercise 8-16 via the gradient projection method, using the origin as the starting point.

8-42. Solve the following problem via the gradient projection method, starting at the origin:

$$\begin{aligned} \text{Max } z = 2x_1 + x_2 + x_1^3x_2 \\ \text{subject to } 2x_1 + x_2 \leqq 8, \\ x_1 - 2x_2 \leqq 4, \\ -x_1 + x_2 \leqq 0, \\ \text{and } x_1, x_2 \geqq 0. \end{aligned}$$

8-43. (a) Given a nonsingular square matrix $\mathbf{X}$ partitioned as follows,

$$\mathbf{X} = \begin{bmatrix} \mathbf{A} & \mathbf{B} \\ \mathbf{C} & \mathbf{D} \end{bmatrix},$$

where $\mathbf{D}$ is square and nonsingular, verify that the inverse of $\mathbf{X}$ is given by the following partitioned matrix:

$$\mathbf{X}^{-1} = \begin{bmatrix} \mathbf{E} & \mathbf{F} \\ \mathbf{G} & \mathbf{H} \end{bmatrix},$$

where

$$\mathbf{E} = (\mathbf{A} - \mathbf{B}\mathbf{D}^{-1}\mathbf{C})^{-1},$$
$$\mathbf{F} = -\mathbf{E}\mathbf{B}\mathbf{D}^{-1},$$
$$\mathbf{G} = -\mathbf{D}^{-1}\mathbf{C}\mathbf{E},$$

and

$$\mathbf{H} = \mathbf{D}^{-1} - \mathbf{D}^{-1}\mathbf{C}\mathbf{F}.$$

(b) Given an inverse $(\mathbf{M}\mathbf{M}^T)^{-1}$ of the type generated by the gradient projection method—where $\mathbf{M}$ is an h-by-n matrix, with $h < n$—use the result of part (a) to write a partitioned formula for the inverse $(\mathbf{M}_+\mathbf{M}_+^T)^{-1}$, where $\mathbf{M}_+$ is formed by adding to $\mathbf{M}$ a new row that is linearly independent of the rows of $\mathbf{M}$.
(c) Repeat part (b) to obtain a formula for the inverse $(\mathbf{M}_-\mathbf{M}_-^T)^{-1}$, where $\mathbf{M}_-$ is formed by deleting the hth row of $\mathbf{M}$.
(d) Show how the result of part (c) can be used when $\mathbf{M}_-$ is formed by deleting *any* single row of $\mathbf{M}$.

8-44. Suppose the gradient projection method is to be applied to a problem about which it is known that, for any given point in the feasible region, the gradients of either all or all but one of the binding constraints are linearly independent. That is, at the beginning of any iteration of the method, there can be at most one binding constraint whose row vector $\boldsymbol{\alpha}_i$ is not included in $\mathbf{M}_k$. Devise a procedure *not* involving explicit or implicit perturbation of the constraints that is capable of determining an improving feasible direction of displacement from any degenerate point $\mathbf{x}_k$.

Section 8.14

8-45. Referring to Wolfe's example of convergence to a non-Kuhn-Tucker point that was presented in Section 8.14, show that the function

$$f(x_1, x_2, x_3) = -\tfrac{4}{3}(x_1^2 - x_1x_2 + x_2^2)^{3/4} - x_3$$

is concave in the first quadrant. Then prove that the limit point

$$\mathbf{x}^* = (0, 0, (2 - \sqrt{2})/4) \cong (0, 0, .146)$$

obtained via the best-feasible-direction approach cannot satisfy the Kuhn-Tucker conditions for the problem

$$\text{Max } f(x_1, x_2, x_3)$$
$$\text{subject to } x_1, x_2, x_3 \geqq 0.$$

8-46. Solve Wolfe's convergence example in Section 8.14 via the gradient projection method, using the same starting point $\mathbf{x}_0 = (\frac{1}{4}, 0, 1)$. Is zigzagging observed? Comment on why or why not.

8-47. Inasmuch as an antizigzagging device inevitably increases the computation time required per iteration by a feasible-directions algorithm, it would be desirable, in the course of solving any given problem, to refrain from using such a device until after zigzagging has actually begun. To adopt a strategy of this sort would, of course, introduce a recognition problem. Zoutendijk suggested that a "mindless" computer might recognize the onset of zigzagging by remembering which of the constraint hyperplanes had been encountered in previous iterations; zigzagging could then be diagnosed, rightly or wrongly, when a given constraint hyperplane is encountered for the second time. Discuss the relative merits of withholding the diagnosis of zigzagging until a hyperplane has been reentered for, say, the third time (or the fourth, etc.). Suggest other logical tests, based on such factors as how the successive direction vectors $\mathbf{s}_k$ are oriented, or how the points $\mathbf{x}_k$ are determined (i.e., whether by the vanishing of the directional derivative $df/d\theta$ or by the encountering of a new hyperplane), that a computer might use to recognize zigzagging. Discuss in general the pros and cons of using *any* type of diagnostic test to determine whether or not to bypass the antizigzagging computations.

8-48. Another type of antizigzagging strategy involves taking into consideration, in choosing a direction of displacement from the current point $\mathbf{x}_k$, not only the constraints that are binding at $\mathbf{x}_k$ but also those that are "nearly binding." Can you suggest a direction-finding procedure that is based on this general approach? How would your procedure have to be modified in the later stages, that is, in the vicinity of the optimum?

General

8-49. In solving linear programming problems, does the convex simplex method reduce to the ordinary (linear) simplex method? If not, what are the differences? Answer the same questions for the reduced gradient method.

8-50. (a) Describe briefly the steps required at each iteration in solving a linear program via a feasible-directions method of the general type discussed in Sections 8.9 and 8.10.
(b) How, if at all, would the linearity of the objective function affect each of the direction-finding problems (8-67), (8-71), and (8-72)? How well suited does each of these problems seem to be for use in solving linear programs? Consider in your answer to what extent the optimal solution of the direction-finding problem at one iteration can be used to help solve it at the next.
(c) What are the overall advantages and disadvantages of the feasible-directions approach vis-à-vis the linear simplex method?

8-51. Prove that a best-feasible-direction method, in which the optimal direction of displacement is obtained at each iteration by solving (8-64), can solve any linear program in a finite number of iterations. [*Hint:* A crucial step in the proof is to establish that the optimal value of the objective function of (8-64) decreases monotonically from one iteration to the next. To derive this result, begin by considering the reduced problem that would be obtained at any iteration by deleting from (8-64) the constraints that were not binding at the optimal solution $\mathbf{s}_k$. These are precisely the constraints that will not be included in the direction-finding problem to be solved at the next iteration.]

8-52. (a) Construct a stepwise outline of a single iteration of the gradient projection method as it would be applied in solving a nondegenerate linear program. Insofar as possible, your outline should parallel that which appears at the end of Section 8.11.
(b) Does the gradient projection method in fact reduce to the linear simplex method in solving the standard-form linear program

$$\text{Max } z = \mathbf{c}^T\mathbf{x}$$

$$\text{subject to } \mathbf{A}\mathbf{x} = \mathbf{b} \text{ and } \mathbf{x} \geqq \mathbf{0}?$$

If not, what are the differences?

8-53. If a linear constraint $g_i(\mathbf{x}) \equiv \boldsymbol{\alpha}_i\mathbf{x} \leqq b_i$ is binding at a feasible point $\mathbf{x}_k$, we know that it is possible to move any distance we want in a straight-line direction perpendicular to the gradient $\nabla g_i(\mathbf{x}_k) = \boldsymbol{\alpha}_i^T$—that is, along the hyperplane $g_i(\mathbf{x}) = b_i$—without ever violating the ith constraint. However, if a *nonlinear* constraint $g_i(\mathbf{x}) \leqq b_i$, where $g_i(\mathbf{x})$ is a strictly convex function, is binding at $\mathbf{x}_k$, it is then impossible to move any distance whatever in the tangent direction, which is perpendicular to the gradient $\nabla g_i(\mathbf{x}_k)$, without leaving the feasible region. Thus the tangent direction *cannot* be a feasible direction of displacement from $\mathbf{x}_k$. What are the implications of this conclusion with regard to solving nonlinearly constrained problems via feasible-directions algorithms? Given a current point $\mathbf{x}_k$ on the boundary of a nonlinear feasible region, what considerations will be important in choosing an efficient or desirable direction of displacement? In answering this question, explain the trade-off that will frequently exist between (a) the rate of improvement of the objective function and (b) the maximum permitted displacement that are offered by various feasible directions.

9

Algorithms for Nonlinearly Constrained Problems

9.1 NONLINEARITY IN THE CONSTRAINTS

In this chapter we turn our attention to mathematical programming problems of the general form

$$\text{(9-1)} \qquad \text{Max } z = f(\mathbf{x}), \qquad \mathbf{x} = (x_1, \ldots, x_n),$$

$$\text{(9-2)} \qquad \text{subject to } g_i(\mathbf{x}) \begin{pmatrix} \leqq \\ = \\ \geqq \end{pmatrix} b_i, \qquad i = 1, \ldots, m,$$

where f and the g_i are scalar-valued functions having continuous first partial derivatives and the b_i are real numbers. The constraints (9-2) include any nonnegativity restrictions that may be present.

Most of the algorithms presented in Chapter 8 for solving linearly constrained problems were of the feasible-directions type. Unfortunately, the feasible-directions strategy is not nearly so well suited to problems having nonlinear constraints. There are two general computational reasons why this is so:

1. Whereas searching along a hyperplane corresponding to a binding linear constraint involves simple straight-line movement, searching along a curved boundary hypersurface requires a complicated parameterization of variables that is likely to be computationally prohibitive.

2. Whereas, given a current point $\mathbf{x}_k$ and direction $\mathbf{s}_k$, a simple long division suffices to determine the maximum displacement that is just possible before a constraint hyperplane $\boldsymbol{\alpha}_i\mathbf{x} \leqq b_i$ is encountered, the same computation in the case of a nonlinear constraint hypersurface would require the solution, by some iterative procedure, of a nonlinear equation $g_i(\mathbf{x}_k + \theta\mathbf{s}_k) = b_i$.

In view of these considerations—both of which were discussed in Section 7.1—nonlinearly constrained problems are not usually solved by feasible-directions methods; other strategies and approaches, developed for the most part in the mid- and late 1960s, are used instead. In this chapter we shall consider a number of these general nonlinear programming methods. In every case it will, of course, be true that the general method is applicable to linearly constrained problems, including, for that matter, linear programs. However, for such applications the general methods are, on the whole, far less efficient than those of Chapter 8, which strongly exploit the linearity of the constraints. Thus, although it is quite reasonable and sometimes fully competitive to solve a quadratic program via the convex simplex or some other feasible-directions method, it can never be desirable to use an algorithm designed for the general nonlinear program (9-1) and (9-2) in solving a linearly constrained problem.

We begin with two relatively straightforward procedures, both based on the simplex method, that are of somewhat limited applicability in nonlinear programming. The separable programming approach, insofar as its technical requirements are concerned, can be used whenever the functions $f(\mathbf{x})$ and $g_i(\mathbf{x})$ are all completely separable, or can be made so; however, it seems to be relatively efficient, and thus competitive, only when the feasible region of the separable program is a convex set. Kelley's cutting-plane algorithm may be of even more limited utility: It can theoretically be employed in solving any convex program but is in fact competitive only for those problems that are almost entirely linear and include only a few nonlinear constraint terms. After discussing these two methods we shall proceed to others that can be applied efficiently to somewhat broader classes of problems.

9.2 SEPARABLE PROGRAMMING

In this section we extend the separable programming approach described in Sections 8.2 through 8.4 in order to make it applicable to nonlinearly constrained problems as well. Credit for this extension is due to Miller [35]. The procedure involved is essentially the same as for linearly constrained problems, except that piecewise linear approximations must now be constructed for the constraint functions as well as for the objective. Thus every function appearing in the problem statement must be *separable*, in the sense defined in Section 8.2, either immediately or after some type of algebraic substitution or transformation.

Although the methods of this section can be applied to any fully separable mathematical programming problem, significant difficulties arise when the problem is not also a *convex* program. It is probably fair to say that whereas the separable programming approach is at least competitive and probably superior for solving any convex separable program, it should be used on a nonconvex problem only when alternative computer codes are not available. We shall therefore devote the bulk of our discussion to the *convex separable program*, whose general formulation is as follows:

$$
(9\text{-}3)\quad \begin{cases} \text{Max } z = \sum_{j=1}^{n} f_j(x_j) \\ \text{subject to} \quad \sum_{j=1}^{n} g_{ij}(x_j) \leqq b_i, & i = 1, \ldots, u, \\ \sum_{j=1}^{n} g_{ij}(x_j) \equiv \sum_{j=1}^{n} a_{ij}x_j = b_i, & i = u + 1, \ldots, v, \\ \text{and} \quad \sum_{j=1}^{n} g_{ij}(x_j) \geqq b_i, & i = v + 1, \ldots, m, \end{cases}
$$

where the functions f_j are concave; the functions g_{ij}, $i = 1, \ldots, u$, all j, are convex; the functions g_{ij}, $i = v + 1, \ldots, m$, all j, are concave; and the a_{ij} and b_i are real numbers. As can be readily verified from Theorem 3.6, the feasible region of (9-3) is a convex set, and the problem itself is a convex program. Note that it is not necessary to formulate the problem in terms of nonnegative variables.

Although nonlinear programming problems usually do not arise in separable form, as given above, various algebraic transformations are available for obtaining separability. For example, it was shown in Section 8.2 that the term x_1x_2 could be replaced by the separable expression $y_1^2 - y_2^2$ provided that the linear equations

$$y_1 = \tfrac{1}{2}(x_1 + x_2) \quad \text{and} \quad y_2 = \tfrac{1}{2}(x_1 - x_2)$$

were added to the set of constraints. More generally, any polynomial term—or, for that matter, any product of individually separable functions—can be separated by means of logarithmic transforms as long as the arguments of all logarithms involved are restricted to positive values. Thus the term $(x_1 - 2)^2(\sin x_2)^3(x_3 + \log x_3)^{-4}$ can be replaced by the single new variable y if the following equation is added to the constraint set:

$$\log y = 2\log(x_1 - 2) + 3\log \sin x_2 - 4\log(x_3 + \log x_3).$$

Note that the new constraint equation has the desired separability property. Other types of terms and functions for which complete separability can be achieved are covered in the exercises. It should be pointed out, however, that most of these transformations of variables cause logarithmic or other nonlinear equations to be introduced into the constraint set, thereby ensuring that the feasible region can no longer be convex. Inasmuch as this condition makes the separable programming approach relatively undesirable, for reasons to be explained below, the use of transformations for achieving separability is a topic of rather more theoretical than practical interest.

To solve the problem (9-3), we begin by forming an approximating problem; the procedure for doing so is essentially the same as that used in approximating the linearly constrained problem in Section 8.2. For each j, $j = 1, \ldots, n$, let L_j and M_j be a lower and upper bound on the value of x_j, and choose a set of $r_j + 1$ breakpoints x_{jk}, $k = 0, 1, \ldots, r_j$, satisfying

$$x_{j0} = L_j < x_{j1} < x_{j2} < \cdots < x_{jr_j} = M_j.$$

For each of these values compute the ordinates

$$f_{jk} = f_j(x_{jk}) \tag{9-4}$$

and

$$g_{ijk} = \begin{cases} g_{ij}(x_{jk}), & i = 1, \ldots, u \\ a_{ij}x_{jk}, & i = u + 1, \ldots, v, \\ g_{ij}(x_{jk}), & i = v + 1, \ldots, m. \end{cases} \tag{9-5}$$

The ordinates f_{jk} and g_{ijk} define piecewise linear functions $\hat{f}_j(x_j)$ and $\hat{g}_{ij}(x_j)$ that can be taken as approximations to the original functions f_j and g_{ij}. Let a new set of variables $\lambda_{j0}, \lambda_{j1}, \ldots, \lambda_{jr_j}$ be defined for each $x_j, j = 1, \ldots, n$. Then, making substitutions of the form (8-9) through (8-11) for each variable x_j, including a substitution

$$\hat{g}_{ij}(x_j) = \sum_{k=0}^{r_j} \lambda_{jk} g_{ijk}$$

for each of the constraint functions $\hat{g}_{ij}$, $i = 1, \ldots, m$, we arrive at the following *approximating problem* in the variables λ_{jk}:

$$\begin{cases} \text{Max } z = \sum_{j=1}^{n} \hat{f}_j(x_j) \equiv \sum_{j=1}^{n} \sum_{k=0}^{r_j} f_{jk}\lambda_{jk} \\ \text{subject to } \sum_{j=1}^{n} \hat{g}_{ij}(x_j) \equiv \sum_{j=1}^{n} \sum_{k=0}^{r_j} g_{ijk}\lambda_{jk} \begin{Bmatrix} \leqq \\ = \\ \geqq \end{Bmatrix} b_i, & i = 1, \ldots, m, \\ \sum_{k=0}^{r_j} \lambda_{jk} = 1, & j = 1, \ldots, n, \\ \text{and} \quad \lambda_{jk} \geqq 0 & \text{for all } j \text{ and } k, \end{cases} \tag{9-6}$$

and subject to the following restrictions for each j, $j = 1, \ldots, n$:

1. *At most two of the λ_{jk} can be positive, and*
2. *If two are positive, they must be adjacent.*

The values of the variables x_j associated with any particular solution $\boldsymbol{\lambda}$ are given by

$$x_j = \sum_{k=0}^{r_j} \lambda_{jk} x_{jk}, \qquad j = 1, \ldots, n. \tag{9-7}$$

Note that the approximating problem (9-6) is identical to the problem

$$\begin{cases} \text{Max } z = \sum_{j=1}^{n} \hat{f}_j(x_j) \\ \text{subject to } \sum_{j=1}^{n} \hat{g}_{ij}(x_j) \begin{Bmatrix} \leqq \\ = \\ \geqq \end{Bmatrix} b_i, \qquad i = 1, \ldots, m, \\ \text{and } L_j \leqq x_j \leqq M_j, \qquad j = 1, \ldots, n, \end{cases} \tag{9-8}$$

which is itself an approximation to the original problem (9-3), assuming that the lower and upper bounds L_j and M_j do not encroach on the feasible region.

If it were not for the logical restrictions 1 and 2, the problem (9-6) would be a linear program in the variables λ_{jk}. In fact, it can be shown that when the original problem is a convex separable program, as we are assuming here, the approximating problem (9-6) can be solved directly via the simplex method, with no computational or logical modifications whatsoever; that is, restrictions 1 and 2 will turn out to be satisfied by the optimal solution (or by an alternative optimal solution—see below) even though they are not explicitly enforced. This result is analogous to that obtained in Section 8.3 for linearly constrained convex separable programs and depends, moreover, on virtually the same line of theoretical development. From the proof of Lemma 8.1 we know that the functions $\hat{f}_j$ and $\hat{g}_{ij}$ have the same convexity/concavity characteristics as f_j and g_{ij}; hence the approximating problem (9-6) must be a convex program, and its unique optimal solution must be an approximation to the optimal solution of the original problem. Furthermore, it can be shown, by revising the proof of Lemma 8.2, either that the set of values of $\lambda_{j0}, \lambda_{j1}, \ldots, \lambda_{jr_j}$ chosen by the simplex method to represent the optimal value of any given variable x_j will "voluntarily" satisfy the logical restrictions 1 and 2, or if they do not, that an alternative optimal set of values will exist that does. The argument can be briefly outlined as follows. If the final set of values $\lambda_{jk} = \bar{\lambda}_{jk}$, $k = 0, 1, \ldots, r_j$, did *not* satisfy restrictions 1 and 2 for some j, there would then exist another set of values of those variables (obtainable as shown in the proof of Lemma 8.2) for which the objective term

$$\sum_{k=0}^{r_j} f_{jk}\lambda_{jk}$$

would be equal or greater in value, while each constraint term

$$\hat{g}_{ij}(x_j) = \sum_{k=0}^{r_j} g_{ijk}\lambda_{jk}$$

would be no greater if convex, equal if linear, and no less if concave. It follows that the alternative set of values would satisfy all constraints while yielding an objective value at least as great as that associated with the $\bar{\lambda}_{jk}$, so that these latter values could not, in fact, have been uniquely optimal. If the alternative set of values did not satisfy restrictions 1 and 2, repetition of the argument would eventually lead to another set that did.[1] It can therefore be concluded that the solution obtained by the simplex method—or an alternative version of it—must be the optimal solution to the approximating problem (9-6). This implies that *an approximation to the optimal solution of any convex separable program* (9-3) *can be obtained by solving its approximating problem via the simplex method*, with no additional modifications or attachments being needed to enforce the logical restrictions 1 and 2. A more rigorous theoretical development of this result is requested in Exercise 9-1.

Although we have thus far restricted our attention to convex separable programs, it is also possible to use separable programming techniques to seek constrained local optima of nonconvex problems. In doing so, however, we cannot exploit the theoretical results discussed in the foregoing paragraph. In particular, we can no longer rely on solving the approximating problem (9-6) by ordinary simplex pivoting, without regard for the logical restrictions 1 and 2—if those restrictions are not enforced, the "optimal" solution freely obtained by the simplex method will not necessarily satisfy them. The antidote to this theoretical difficulty is exactly the same as in the special case of

[1] Incidentally, it will very seldom happen that the set of optimal values $\bar{\lambda}_{jk}$, $k = 0, 1, \ldots, r_j$, chosen by the simplex method for some variable x_j will fail to satisfy the logical restrictions 1 and 2. In fact, this can occur only when the original function $f_j(x_j)$ is linear *and* one of the following two conditions exists:

1. Every $g_{ij}(x_j)$, $i = 1, \ldots, m$, is linear as well (although in that case no piecewise linear approximations would have been needed for f_j and the g_{ij}), or
2. None of the nonlinear constraints are binding at the optimum (so that the additional relaxation of those constraints achieved by replacing the $\bar{\lambda}_{jk}$ with an alternative set of values satisfying restrictions 1 and 2 would not allow a better solution to be obtained).

Under all other circumstances the optimal solution $\bar{\lambda}_{jk}$ generated by the simplex method will satisfy restrictions 1 and 2 immediately.

linearly constrained separable programs: The simplex method is still used, but with *restricted basis entry*, to solve the problem (9-6). The modified basis entry criteria are precisely those described in Section 8.4; in choosing the variable λ_{jk} to enter the basis at the next iteration, they reject automatically any variable whose entry would lead to a basic solution that violated the logical restrictions 1 and 2. When a stage is reached at which none of the nonbasic variables eligible for basis entry has a favorable (i.e., negative in a maximization problem) reduced cost $z_{jk} - c_{jk}$, the method terminates and the current basic solution is taken as an approximation to a constrained local optimum of the original separable program. In general, of course, there may be several local optima in the feasible region.

But, whereas the application of separable programming techniques to nonconvex problems is quite possible from a theoretical point of view, in practice it can lead to serious computational difficulties. The advantages and disadvantages of the separable programming approach to *linearly constrained* problems were outlined at the end of Section 8.4, and it was emphasized there that the separable programming approach was to be preferred only for those problems having relatively few nonlinear objective terms $f_j(x_j)$. For *nonlinearly constrained* problems, however, the overall degree of nonlinearity, while still important, appears to be less so than the question of whether or not the feasible region of the approximating problem is convex. When it is not, the difficulty of finding a starting basic feasible solution is greatly increased, because the linear program solved in phase I is itself not a convex program. This introduces the possibility that a local optimum with a negative objective value could be found in phase I, causing termination (when all artificial variables cannot be driven to zero, the interpretation is that the original problem has no feasible solution), despite the fact that the global optimum of the phase I problem might have an objective value of zero (implying the existence of a feasible solution after all). Even if feasible solutions were known to exist, it might be necessary to solve the phase I problem several times, using different starting solutions or slightly varied objective functions, simply in order to obtain an initial solution for phase II. Furthermore, to locate the various local optima of the approximating problem, several starting feasible solutions would have to be found, each possibly requiring several phase I executions. It should thus be evident that the total computational effort required to make a thorough search over a nonconvex feasible region could easily become quite prohibitive, and it follows that, as a general rule, such problems should not be solved via separable programming unless no convenient alternatives are available. Note that the difficulty just discussed did not arise in the case of linearly constrained separable programs; for that class of problems the feasible region is, of course, automatically convex, and the degree of nonlinearity thus becomes the dominant factor in determining the relative efficiency of the separable programming approach.

9.3 KELLEY'S CUTTING-PLANE METHOD

The algorithm of Kelley [25] is designed for application to any problem of the form

$$\text{(9-9)} \qquad \begin{cases} \text{Max } z = \mathbf{c}^T\mathbf{x}, & \mathbf{x} = (x_1, \ldots, x_n), \\ \text{subject to } g_i(\mathbf{x}) \leqq 0, & i = 1, \ldots, m, \end{cases}$$

where $\mathbf{c}$ is an n-component column vector of real numbers and the $g_i(\mathbf{x})$ are convex functions having continuous first partial derivatives. It follows from Theorem 3.6 that (9-9) is a convex programming problem. Although the objective function is specified as being linear, it is easily shown that any convex program

$$\text{(9-10)} \qquad \begin{cases} \text{Max } f(\mathbf{x}), & \mathbf{x} = (x_1, \ldots, x_p), \\ \text{subject to } g_i(\mathbf{x}) \leqq 0, & i = 1, \ldots, q, \end{cases}$$

where $f(\mathbf{x})$ is concave and all the $g_i(\mathbf{x})$ are convex, can be transformed into an equivalent problem that satisfies the format of (9-9). Given a convex program (9-10), consider the problem

$$\text{(9-11)} \qquad \begin{cases} \text{Max } z = x_{p+1} \\ \text{subject to } x_{p+1} - f(\mathbf{x}) \leqq 0 \\ \text{and} \qquad g_i(\mathbf{x}) \leqq 0, \qquad i = 1, \ldots, q, \end{cases}$$

where we have defined a new variable x_{p+1} and where we continue to use $\mathbf{x}$ to represent the original vector of variables $(x_1, \ldots, x_p)$. Note first that, inasmuch as $f(\mathbf{x})$ is concave, the new constraint function $x_{p+1} - f(\mathbf{x})$ is convex; thus (9-11), which calls for the maximization of a linear objective function over a convex feasible region, satisfies all the requirements of the format (9-9). To prove the equivalence of (9-10) and (9-11), suppose that $\mathbf{x} = \mathbf{x}_0$ is the optimal solution to (9-10); then the set of values

$$\text{(9-12)} \qquad \mathbf{x} = \mathbf{x}_0, \qquad x_{p+1} = f(\mathbf{x}_0)$$

constitutes a feasible solution to (9-11) having the same objective value. Moreover, if (9-11) had a superior feasible solution $\mathbf{x} = \hat{\mathbf{x}}, x_{p+1} = f(\hat{\mathbf{x}})$, with $f(\hat{\mathbf{x}}) > f(\mathbf{x}_0)$, then $\hat{\mathbf{x}}$ would be a feasible solution to (9-10) and $\mathbf{x}_0$ could not have been optimal for that problem. Therefore, the solution (9-12) must be optimal for (9-11), as was to be shown.

To facilitate the development of Kelley's algorithm, we shall represent the constraints of the problem (9-9) in a different but equivalent form. Let a new scalar-valued function $G(\mathbf{x})$ be defined as follows:

$$\text{(9-13)} \qquad G(\mathbf{x}) \equiv \max\{g_1(\mathbf{x}), \ldots, g_m(\mathbf{x})\}.$$

We then assert that the single inequality $G(\mathbf{x}) \leqq 0$ defines precisely the same feasible region as does the set of constraints $g_i(\mathbf{x}) \leqq 0, i = 1, \ldots, m$; that is, any point $\mathbf{x}$ satisfying the former must also satisfy the latter and vice versa. The reader should have no difficulty in verifying that this is so. We may therefore represent (9-9) in the equivalent form

(9-14) $$\begin{cases} \text{Max } z = \mathbf{c}^T\mathbf{x} \\ \text{subject to } G(\mathbf{x}) \leqq 0. \end{cases}$$

Note that $G(\mathbf{x})$ is a continuous, scalar-valued convex function[2] but that its first partial derivatives are not all uniquely defined at every point. Geometrically, the function $G(\mathbf{x})$ is a *piecewise convex* hypersurface, that is, a hypersurface composed of a number of *facets* each of which is itself a convex hypersurface of the same dimensionality. A one-dimensional example is displayed in Fig. 9-1(a).

We now proceed to the development of Kelley's algorithm for solving the convex program

(9-14) $$\begin{cases} \text{Max } z = \mathbf{c}^T\mathbf{x} \\ \text{subject to } G(\mathbf{x}) \leqq 0. \end{cases}$$

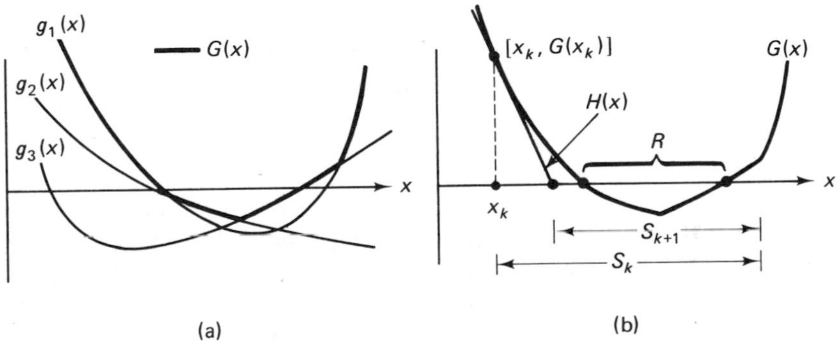

FIGURE 9-1

[2]It is easily shown that the maximum of a finite number of convex functions is itself convex. Choose any two points $\mathbf{x}_1$ and $\mathbf{x}_2$ and any value of λ satisfying $0 \leqq \lambda \leqq 1$, and suppose that at the point $\mathbf{x} = \lambda\mathbf{x}_1 + (1 - \lambda)\mathbf{x}_2$ the largest of the values $g_i(\mathbf{x})$ is $g_r(\mathbf{x})$. Then we can write

$$\begin{aligned} G(\lambda\mathbf{x}_1 + (1 - \lambda)\mathbf{x}_2) &\equiv \max\{g_1(\lambda\mathbf{x}_1 + (1 - \lambda)\mathbf{x}_2), \ldots, g_m(\lambda\mathbf{x}_1 + (1 - \lambda)\mathbf{x}_2)\} \\ &= g_r(\lambda\mathbf{x}_1 + (1 - \lambda)\mathbf{x}_2) \\ &\leqq \lambda g_r(\mathbf{x}_1) + (1 - \lambda)g_r(\mathbf{x}_2) \\ &\leqq \lambda G(\mathbf{x}_1) + (1 - \lambda)G(\mathbf{x}_2), \end{aligned}$$

where the first inequality follows from the fact that $g_r(\mathbf{x})$ is a convex function and the second from the definition of G and the nonnegativity of λ and $1 - \lambda$. Thus $G(\mathbf{x})$ satisfies the definition of convexity. Q.E.D.

The algorithm will generate a sequence of *infeasible* points $\mathbf{x}_0, \mathbf{x}_1, \ldots, \mathbf{x}_k, \mathbf{x}_{k+1}, \ldots$ that converges to the optimal solution $\mathbf{x}^*$ after either a finite or, more usually, an infinite number of iterations. In this connection it is important to note that $\mathbf{x}^*$ must be a boundary point of the feasible region of (9-14): No interior point of *any* feasible region, convex or not, can maximize or minimize a linear objective function. (This follows from the fact that if the objective hyperplane passes through any interior point $\hat{\mathbf{x}}$, it must be possible to translate that hyperplane parallel to itself through some small distance in either direction—toward greater or lesser values of z—without expelling it entirely from the feasible region. Thus any interior point can be improved upon, whether the objective is to maximize or minimize, and all local optima must be boundary points.) Observe that, in specifying the existence of $\mathbf{x}^*$, we are excluding from consideration all problems with unbounded optimal solutions.

Let R denote the feasible region of (9-14); that is, the point $\hat{\mathbf{x}}$ is in R if and only if $G(\hat{\mathbf{x}}) \leqq 0$. Note that R lies in Euclidean n-space E^n. We now choose a set of linear constraints $\mathbf{Ax} \leqq \mathbf{b}$ having the following two properties:

1. The optimal solution $\mathbf{x}^*$ satisfies every constraint in the set (that is, $\mathbf{Ax}^* \leqq \mathbf{b}$), and
2. The linear program

$$\text{Max } z = \mathbf{c}^T\mathbf{x}$$
$$\text{subject to } \mathbf{Ax} \leqq \mathbf{b}$$

has a bounded optimal solution.

Let S_0 denote the convex polyhedron in E^n defined by these constraints: $\hat{\mathbf{x}}$ is in S_0 if and only if $\mathbf{A}\hat{\mathbf{x}} \leqq \mathbf{b}$. It should be emphasized that any set of one or more constraints satisfying requirements 1 and 2 may be chosen; note, for example, that the polyhedron S_0 need not be finite in extent. In many practical problems a reasonable procedure for choosing S_0 is to determine lower and upper bounds L_j and M_j on the value of each variable x_j, defining S_0 to be the rectangular hypersolid

$$S_0 = \{\mathbf{x} \mid L_j \leqq x_j \leqq M_j, j = 1, \ldots, n\}.$$

Another possibility might be to let S_0 be the set of all points $\mathbf{x}$ satisfying the *linear* constraints $g_i(\mathbf{x}) \leqq 0$ of the original problem (9-9); a set chosen in this manner would certainly meet the first requirement, although not necessarily the second.

Given the problem (9-14) and a suitably chosen set of linear constraints $\mathbf{Ax} \leqq \mathbf{b}$, Kelley's algorithm begins by using some variety of the simplex method to solve the linear program

$$\text{(9-15)} \qquad \begin{cases} \text{Max } z = \mathbf{c}^T\mathbf{x} \\ \text{subject to } \mathbf{A}\mathbf{x} \leqq \mathbf{b}. \end{cases}$$

Let $\mathbf{x}_0$ denote the optimal solution, which is, of course, an extreme point of S_0. If $\mathbf{x}_0$ is in the original feasible region R, then it must be an optimal solution to (9-14)—this is easily proved—and the method terminates. If not, that is, if $G(\mathbf{x}_0) > 0$, consider the linear constraint

$$\text{(9-16)} \qquad H_0(\mathbf{x}) \equiv G(\mathbf{x}_0) + \nabla\mathbf{G}^T(\mathbf{x}_0) \cdot (\mathbf{x} - \mathbf{x}_0) \leqq 0,$$

where $\nabla\mathbf{G}^T(\mathbf{x}_0)$, as usual, represents the gradient of G evaluated at $\mathbf{x}_0$. For the moment we ignore the fact that the components of $\nabla\mathbf{G}$ may not all be uniquely defined at $\mathbf{x}_0$: This difficulty will be resolved later on. Because $G(\mathbf{x})$ is a convex function, we know from Theorem 3.9 that

$$H_0(\mathbf{x}) \equiv G(\mathbf{x}_0) + \nabla\mathbf{G}^T(\mathbf{x}_0) \cdot (\mathbf{x} - \mathbf{x}_0) \leqq G(\mathbf{x})$$

at any point $\mathbf{x}$ in E^n (tangent hyperplanes "underestimate" convex hypersurfaces). In particular, at any point $\hat{\mathbf{x}}$ in the original feasible region R it must be true that

$$H_0(\hat{\mathbf{x}}) \leqq G(\hat{\mathbf{x}}) \leqq 0,$$

so that the new constraint (9-16) is evidently satisfied at all points in R. However, it is obviously *not* satisfied at the solution point $\mathbf{x}_0$, inasmuch as $G(\mathbf{x}_0)$ is known to be positive. Thus the constraint (9-16) slices off a portion of the convex polyhedron S_0 that contains the optimal solution of the linear program (9-15) but not the optimum nor any feasible solutions of the original problem (9-14). For this reason the constraint hyperplane (9-16) is known as a *cutting plane*, and the various mathematical programming algorithms that are based on the addition of constraints in this manner are called *cutting-plane methods.*

For the next iteration of Kelley's algorithm, we let S_1 denote the set of all points in S_0 that satisfy the new cutting-plane constraint. It is clear that S_1 is again a convex polyhedron, being defined by the set of linear constraints

$$\text{(9-17)} \qquad \begin{cases} \mathbf{A}\mathbf{x} \leqq \mathbf{b} \\ \text{and } H_0(\mathbf{x}) \leqq 0. \end{cases}$$

If we now maximize $z = \mathbf{c}^T\mathbf{x}$ over this feasible region, we shall obtain a new optimal solution $\mathbf{x}_1 \neq \mathbf{x}_0$ such that $\mathbf{c}^T\mathbf{x}_1 \leqq \mathbf{c}^T\mathbf{x}_0$; moreover, the optimal objective value $\mathbf{c}^T\mathbf{x}_1$ for this second linear program will be strictly less than

$\mathbf{c}^T\mathbf{x}_0$ whenever, as will usually be the case, $\mathbf{x}_0$ was a uniquely optimal solution to (9-15). If $\mathbf{x}_1$ is in R, it is optimal for the original problem and the procedure terminates. If not, proceeding as before, we form a third and still smaller convex polyhedron S_2 by adding to the constraint set (9-17) another cutting plane

$$H_1(\mathbf{x}) \equiv G(\mathbf{x}_1) + \nabla \mathbf{G}^T(\mathbf{x}_1) \cdot (\mathbf{x} - \mathbf{x}_1) \leqq 0.$$

We then solve a third linear program

$$\begin{aligned} \text{Max } z &= \mathbf{c}^T\mathbf{x} \\ \text{subject to} \quad \mathbf{Ax} &\leqq \mathbf{b}, \\ H_0(\mathbf{x}) &\leqq 0, \\ \text{and } H_1(\mathbf{x}) &\leqq 0. \end{aligned}$$

Continuing in this manner, we generate a sequence of nested convex polyhedrons $\{S_k\}$, each of which contains every point of R that was contained in S_0, and a sequence of solution points $\{\mathbf{x}_k\}$ that approaches the optimal solution $\mathbf{x}^*$ of the original problem (recall that $\mathbf{x}^*$ is a boundary point of R). The last assertion requires an analytical convergence proof, given in [25], in order to eliminate the possibility that, as k becomes infinitely large, successive cutting planes remove smaller and smaller slices of territory while remaining some limiting positive distance away from the boundary of R. It should be noted that in practical applications of Kelley's algorithm termination occurs when $G(\mathbf{x}_k)$ is found to be no greater than some preselected small positive constant.

The cutting-plane algorithm of Kelley has a straightforward and instructive geometrical interpretation. The regions R, S_0, $S_1, \ldots$ and the solution points $\mathbf{x}_0, \mathbf{x}_1, \ldots$ all lie in Euclidean n-space E^n, which can be visualized as an n-dimensional hyperplane of infinite extent lying in E^{n+1}; the equation of that hyperplane in E^{n+1} would be $x_{n+1} = 0$. Suspended in E^{n+1} over (and under) this hyperplane is the constraint hypersurface $G(\mathbf{x})$, which takes the form of an irregularly shaped but convex "cereal bowl." The equation defining the hypersurface would be $x_{n+1} = G(\mathbf{x})$, where, as before, $\mathbf{x} = (x_1, \ldots, x_n)$; the value of x_{n+1} at any point $\mathbf{x}$ is simply the height of the hypersurface $G(\mathbf{x})$ above the hyperplane $x_{n+1} = 0$. For the original problem (9-14) to have feasible solutions, $G(\mathbf{x})$ must intersect this hyperplane so that the bottom of the cereal bowl lies on or below it; that is, there must be points $\mathbf{x}$ in E^n such that $x_{n+1} = G(\mathbf{x})$ is nonpositive. The portion of the hyperplane $x_{n+1} = 0$ lying within the bowl is the feasible region R of the problem (9-14): It includes all points $\mathbf{x} = (x_1, \ldots, x_n)$ satisfying $G(\mathbf{x}) \leqq 0$, and no others.

Suppose now that at the kth iteration the algorithm has generated a current point $\mathbf{x}_k$ in E^n that lies in S_k but not in R. The convex hypersurface

$x_{n+1} = G(\mathbf{x})$ must then lie *above* the hyperplane $x_{n+1} = 0$ at the point $\mathbf{x}_k$—that is, $G(\mathbf{x}_k)$ must be positive. Consider the first-order Taylor approximation to $x_{n+1} = G(\mathbf{x})$ at the point $\mathbf{x} = \mathbf{x}_k$:

$$x_{n+1} = G(\mathbf{x}_k) + \nabla \mathbf{G}^T(\mathbf{x}_k) \cdot (\mathbf{x} - \mathbf{x}_k) \equiv H_k(\mathbf{x}).$$

This function is linear in all variables and is simply the hyperplane that is tangent to the hypersurface $G(\mathbf{x})$ at the point $[\mathbf{x}_k, G(\mathbf{x}_k)]$ in E^{n+1}. Because $G(\mathbf{x})$ is convex, the tangent hyperplane $H_k(\mathbf{x})$ must lie on or below it at every point, so that $H_k(\mathbf{x}) \leqq G(\mathbf{x})$ for all $\mathbf{x}$. This implies that the tangent hyperplane either passes through or lies *below* every point $\mathbf{x}$ in the original feasible region R, inasmuch as $H_k(\mathbf{x}) \leqq G(\mathbf{x}) \leqq 0$ for any such $\mathbf{x}$. On the other hand, $H_k(\mathbf{x})$ lies *above* the point $\mathbf{x}_k$, and it follows that the two hyperplanes $x_{n+1} = H_k(\mathbf{x})$ and $x_{n+1} = 0$ must intersect in an $(n - 1)$-dimensional hyperplane or cutting plane $H_k(\mathbf{x}) = 0$ that is embedded in S_k and lies *between* the point $\mathbf{x}_k$ and the feasible region R. This hyperplane thus defines a constraint $H_k(\mathbf{x}) \leqq 0$ that is satisfied by every point in R but not by $\mathbf{x}_k$. An example for $n = 1$ is illustrated in Fig. 9-1(b); it should be clear from the diagram that Kelley's algorithm is more or less analogous to Newton's method for finding roots of nonlinear equations.

Thus far in the discussion we have used the single constraint relation

$$G(\mathbf{x}) \equiv \max\{g_1(\mathbf{x}), \ldots, g_m(\mathbf{x})\} \leqq 0$$

to represent the entire set of constraints of the original convex program (9-9). The function $G(\mathbf{x})$ is composed of a number of convex hypersurfaces or facets, each of which is a part of one of the original constraint hypersurfaces $g_i(\mathbf{x})$, as illustrated for the one-dimensional case in Fig. 9-1(a). At any point $\mathbf{x}_k$ the gradient $\nabla \mathbf{G}(\mathbf{x})$ is simply the gradient of that function $g_i(\mathbf{x})$ whose value is greatest at $\mathbf{x}_k$; that is,

$$\text{(9-18)} \qquad \nabla \mathbf{G}(\mathbf{x}_k) = \nabla \mathbf{g}_r(\mathbf{x}_k), \qquad \text{where } g_r(\mathbf{x}_k) = \max_i\{g_i(\mathbf{x}_k)\}.$$

Note that if the point $[\mathbf{x}_k, G(\mathbf{x}_k)]$ lies on an edge or boundary between two or more facets of the hypersurface $G(\mathbf{x})$, then the maximum in (9-18) will not be unique and one or more components of the gradient $\nabla \mathbf{G}(\mathbf{x}_k)$ will be undefined. However, in such a case, where $g_\alpha(\mathbf{x}_k) = g_\beta(\mathbf{x}_k) = \cdots = \max\{g_i(\mathbf{x}_k)\}$, any convex combination of the gradients $\nabla \mathbf{g}_\alpha(\mathbf{x}_k)$, $\nabla \mathbf{g}_\beta(\mathbf{x}_k), \ldots$ can be used in place of $\nabla \mathbf{G}(\mathbf{x}_k)$, and the resulting cutting-plane constraint will still be satisfied, as desired, by all points in R but not by $\mathbf{x}_k$. Exercise 9-10 asks for a formal proof of this assertion.

Kelley's algorithm has several computational aspects that should be discussed here. The linear program that is solved at the kth iteration, $k \geqq 1$, is of the form

$$(9\text{-}19)\quad \begin{cases} \text{Max } z = \mathbf{c}^T\mathbf{x} \\ \text{subject to } \mathbf{Ax} \leqq \mathbf{b} \\ \text{and } H_j(\mathbf{x}) \equiv G(\mathbf{x}_j) + \nabla \mathbf{G}^T(\mathbf{x}_j) \cdot (\mathbf{x} - \mathbf{x}_j) \leqq 0, \\ \qquad\qquad\qquad\qquad\qquad\qquad j = 0, 1, \ldots, k-1. \end{cases}$$

Inasmuch as each of these problems is obtained from its predecessor by the addition of a single constraint, the standard postoptimality techniques of linear programming (see, for example, Chapter 7 of [42]) can be exploited in solving them. The most straightforward approach is to apply the dual simplex method to (9-19) at each stage, using the previous optimum as the starting solution. Another possibility, less direct but usually more efficient, involves working with the dual of (9-19); when this approach is used, successive problems are formed by the addition of new variables rather than constraints and are solved via the primal rather than the dual simplex method [the solution to the primal problem (9-19) is, of course, obtained as a by-product when the dual is solved].

Regardless of which of the two approaches is adopted, computational efficiency can be increased by dropping some of the cutting-plane constraints after they have become redundant, that is, after they have ceased to impinge upon the feasible regions S_k. One extreme technique is to delete at each stage *all* cutting-plane constraints that are not binding on the optimal solution to the linear program (9-19). This procedure preserves the crucial convergence properties of Kelley's algorithm: The objective values $\mathbf{c}^T\mathbf{x}_k$ are still monotonically nonincreasing, and the sequence $\{\mathbf{x}_k\}$ still converges eventually to the optimal solution of the original problem. Although the number of iterations required to arrive within any given distance of the desired optimal solution will tend to increase when this technique is used, the overall computation time will in general be reduced.

As an illustration of the cutting-plane algorithm, consider the following example problem, presented by Kelley [25]:

$$\text{Min } z = x_1 - x_2$$
$$\text{subject to } G(\mathbf{x}) = 3x_1^2 - 2x_1x_2 + x_2^2 - 1 \leqq 0.$$

The feasible region R, an elliptical disk, is diagrammed in Fig. 9-2, and it can be seen there that the optimal solution is $\mathbf{x}^* = (0, 1)$, with $z_{\min} = -1$. Because $G(\mathbf{x})$ is a convex function, the algorithm of Kelley may be applied directly. Let the initial region S_0 be defined by the constraints $-2 \leqq x_1 \leqq 2$

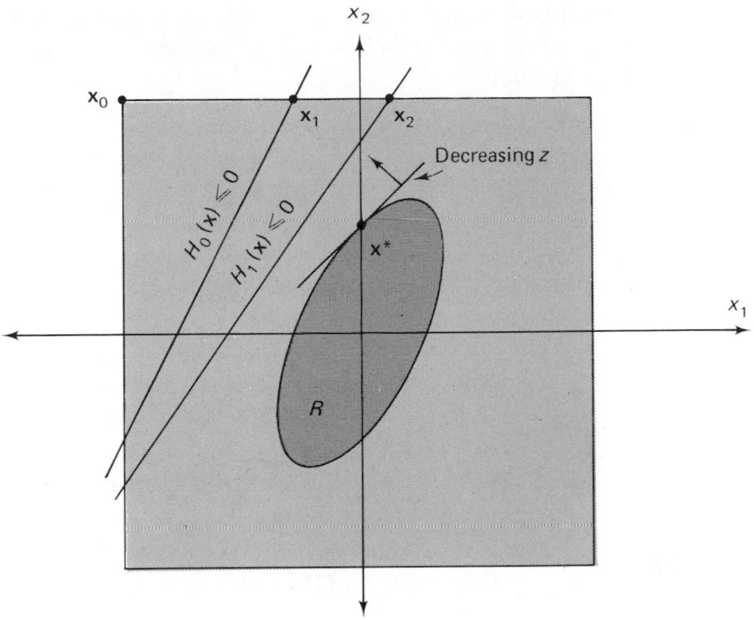

FIGURE 9-2

and $-2 \leqq x_2 \leqq 2$; note that S_0 includes all of R. We begin by solving the linear program

$$\begin{aligned} \text{Min } z &= x_1 - x_2 \\ \text{subject to } &-2 \leqq x_1 \leqq 2 \\ \text{and } &-2 \leqq x_2 \leqq 2. \end{aligned}$$

By inspection, the optimal solution is the extreme point $\mathbf{x}_0 = (-2, 2)$, where $z = -4$. Inasmuch as $G(\mathbf{x}_0) = 23 > 0$, the current point is not feasible. The value of the constraint gradient at $\mathbf{x}_0$ is

$$\nabla \mathbf{G}^T(\mathbf{x}_0) = [6x_1 - 2x_2, -2x_1 + 2x_2]_{\mathbf{x}=\mathbf{x}_0} = [-16, 8],$$

and the first cutting-plane constraint is

$$\begin{aligned} H_0(\mathbf{x}) &\equiv G(\mathbf{x}_0) + \nabla \mathbf{G}^T(\mathbf{x}_0) \cdot (\mathbf{x} - \mathbf{x}_0) \\ &= -25 - 16x_1 + 8x_2 \leqq 0 \end{aligned}$$

or

$$-16x_1 + 8x_2 \leqq 25.$$

This constraint hyperplane is shown in Fig. 9-2; the new feasible region S_1 is simply the portion of S_0 that lies below and to the right of H_0. We must now solve the linear program

$$\text{Min } z = x_1 - x_2$$
$$\text{subject to } (x_1, x_2) \in S_1,$$

where S_1 is the set of all points satisfying the constraints

$$-2 \leqq x_1 \leqq 2$$
$$-2 \leqq x_2 \leqq 2$$

and

$$-16x_1 + 8x_2 \leqq 25.$$

Again solving graphically, we find that the optimal point is $\mathbf{x}_1 = (-.5625, 2.0)$, with $z = -2.5625$. Because $G(\mathbf{x}_1) = 6.199$, the new point is still not feasible, and another iteration is required. The value of the constraint gradient is $\nabla \mathbf{G}^T(\mathbf{x}_1) = [-7.375, 5.125]$, and the next cutting-plane constraint is found to be

$$\begin{aligned} H_1(\mathbf{x}) &= G(\mathbf{x}_1) + \nabla \mathbf{G}^T(\mathbf{x}_1) \cdot (\mathbf{x} - \mathbf{x}_1) \\ &= -7.375x_1 + 5.125x_2 - 8.199 \leqq 0. \end{aligned}$$

This hyperplane is also shown in Fig. 9-2. The next linear program to be solved is then

$$\begin{aligned} \text{Min } z = x_1 - x_2 & \\ \text{subject to} \quad & -2 \leqq x_1 \leqq 2, \\ & -2 \leqq x_2 \leqq 2, \\ & -16x_1 + 8x_2 \leqq 25, \\ \text{and} \quad & -7.375x_1 + 5.125x_2 \leqq 8.199. \end{aligned}$$

Continuing as before, we develop the data shown in Table 9-1. It is evident that the optimal solution, indicated in the last row of the table, cannot be obtained in a finite number of iterations; this, in fact, is frequently the case in applications of Kelley's algorithm. Note that, although the value of z changes monotonically from one iteration to the next, the value of $G(\mathbf{x}_k)$ does not.

Quite apart from the fact that it is applicable only to convex programming problems, Kelley's algorithm has a number of other limitations and

TABLE 9-1

DATA FOR EXAMPLE

k	$\mathbf{x}_k$	z	$G(\mathbf{x}_k)$
0	(−2.000, 2.000)	−4.000	23.000
1	(− .563, 2.000)	−2.563	6.199
2	(.279, 2.000)	−1.722	2.120
3	(− .530, .838)	−1.367	1.431
4	(− .053, 1.160)	−1.213	.478
5	(.427, 1.485)	−1.058	.484
6	(.171, 1.207)	−1.036	.132
.	.	.	.
.	.	.	.
.	.	.	.
∞	(.000, 1.000)	−1.000	.000

disadvantages. Although the amount of computation time involved in solving the postoptimal linear program at each iteration is rather modest, a substantial number of iterations is usually required in order to obtain an acceptable solution, and the overall rate of convergence is somewhat slower than that of other competing algorithms. Another unattractive feature is that all the intermediate solution points, and in most cases the final solution obtained as well, are infeasible (unless the feasible region is artificially shrunk for problem-solving purposes). Furthermore, when the constraint hypersurface $x_{n+1} = G(\mathbf{x})$ makes a very small angle with the hyperplane $x_{n+1} = 0$ at the optimal point $\mathbf{x}_\infty = \mathbf{x}^*$, it is possible for termination to occur at some point $\mathbf{x}_k$ that is quite far from the feasible region R, and from $\mathbf{x}^*$ in particular, simply because $G(\mathbf{x}_k)$, the height of the hypersurface above $\mathbf{x}_k$, is small enough to satisfy the stopping rule. The expected difference between the solution eventually obtained and the true optimum can, of course, be reduced by using a stricter termination criterion, but only at the price of increasing the overall number of iterations needed.

Because of all these various disadvantages, Kelley's algorithm is no longer really competitive today, except for solving convex programs that are almost entirely linear and include only a few nonlinear constraints. Its chief advantage when it first appeared in 1960 was that it could be computer-coded fairly easily by making the appropriate alterations to already-existing simplex codes. The development and coding of other, more efficient nonlinear programming methods in the ensuing decade have to a significant extent neutralized this advantage. Even for those convex programs that possess only a few nonlinearities—the class of problems for which Kelley's algorithm is best suited—it is more efficient to use the separable programming approach whenever it is possible to do so, that is, whenever the constraint functions are completely separable. Thus Kelley's method has really come to be of quite limited practical utility. However, in justification of our including a discussion of it in this chapter, we should point out that the cutting-plane approach, which it embodies, is of considerable theoretical and practical importance in mathematical programming. Several other cutting-plane methods exist that can be applied either to nonlinear programs (see Chapter 14 of Zangwill's text [60]) or to *integer programming problems* (i.e., linear programs in which some or all of the variables are additionally restricted to integer values only). Although the other cutting-plane methods for nonlinear programs are, in general, neither more efficient nor more useful than that of Kelley, those that are applicable to integer programs are still, at this writing, of some practical value and remain in wide use. A discussion of them can be found in any textbook on integer programming.

9.4 ZOUTENDIJK'S FEASIBLE-DIRECTIONS METHOD

In this section and the next we shall see how the familiar feasible-directions strategy can be used to solve the nonlinear programming problem

$$\text{(9-20)} \qquad \begin{cases} \text{Max } z = f(\mathbf{x}), & \mathbf{x} = (x_1, \ldots, x_n), \\ \text{subject to } g_i(\mathbf{x}) \leqq b_i, & i \in I, \\ \text{and } \quad \boldsymbol{\alpha}_i \mathbf{x} = b_i, & i \in E, \end{cases}$$

where $f(\mathbf{x})$ and the $g_i(\mathbf{x})$ are scalar-valued functions with continuous first partial derivatives, the $\boldsymbol{\alpha}_i$ are row vectors of constant coefficients, the b_i are any real numbers, and I and E are, respectively, the sets of subscripts of the inequality and equality constraints. Each constraint function $g_i(\mathbf{x})$ may be linear or nonlinear, but notice that no nonlinear equations are permitted in the constraint set. The reason for this prohibition is that a nonlinear constraint equation would narrowly circumscribe or totally eliminate any possibility of straight-line movement through the feasible region: If the point $\hat{\mathbf{x}}$ satisfies the nonlinear constraint $g_i(\mathbf{x}) = b_i$, then $\hat{\mathbf{s}}$ can be a feasible direction of displacement from $\hat{\mathbf{x}}$ only if the constraint hypersurface happens to have no curvature in the $\hat{\mathbf{s}}$-direction. (Recall in this connection that $\hat{\mathbf{s}}$ is a feasible direction of displacement from the point $\hat{\mathbf{x}}$ in the feasible region R if there exists some positive scalar δ such that the point $\hat{\mathbf{x}} + \theta\hat{\mathbf{s}}$ lies in R for every value of θ satisfying $0 \leqq \theta \leqq \delta$.) Notice also with regard to the problem (9-20) that the constraint functions $g_i(\mathbf{x})$ and the objective $f(\mathbf{x})$ are not required to have any particular convexity/concavity properties; that is, we are not restricting our attention exclusively to convex programs.

The methods we now propose to consider for solving the nonlinear program (9-20) employ the same general *feasible-directions* strategy that was discussed in Section 8.9. This strategy and its applications to both linearly constrained and nonlinearly constrained problems were first studied in detail by Zoutendijk [61], who established a number of important and useful results. Given a current feasible solution $\mathbf{x}_k$ to the problem (9-20), the $(k + 1)$th iteration of any feasible-directions method proceeds according to the following general outline:

1. Determine an improving feasible direction of displacement $\mathbf{s}_k$ from the current point $\mathbf{x}_k$, that is, a direction that yields an immediate increase in the value of the objective function without violating any of the constraints; if no such direction can be found, the method terminates.
2. Given the improving feasible direction $\mathbf{s}_k$, find θ_{opt}, the value of θ that maximizes or approximately maximizes $f(\mathbf{x}_k + \theta\mathbf{s}_k)$ subject to the requirement that $\mathbf{x}_k + \theta\mathbf{s}_k$ be feasible; then the new point is $\mathbf{x}_{k+1} = \mathbf{x}_k + \theta_{\text{opt}}\mathbf{s}_k$.

Different computational procedures for obtaining the direction $\mathbf{s}_k$ characterize different feasible-directions algorithms. Algorithms in this class are also known in general as "large-step" methods because they permit large distances (i.e., from $\mathbf{x}_k$ to some $\mathbf{x}_{k+1}$ on the other side of the feasible region) to be covered in single straight-line moves, without any pauses along the way for recomputation of direction. By contrast, there also exists a family of "small-step" methods in which tight upper limits are imposed on step length at each iteration; these methods are more complicated and less efficient than those using large steps and will not be considered in this text.

In using feasible-directions algorithms to solve nonlinear programs of the form (9-20), significant difficulties are encountered that did not arise in dealing with linearly constrained problems, difficulties both in the selection of a suitable direction of displacement and in the determination of the maximum allowable step length $\Delta = \theta_{\max}$. Taking the latter first, recall that it was quite easy to determine the maximum displacement from $\mathbf{x}_k$ in the $\mathbf{s}_k$-direction that was just permitted by the nonbinding linear constraint $\boldsymbol{\alpha}_i\mathbf{x} \leqq b_i$: Simply set

$$\boldsymbol{\alpha}_i(\mathbf{x}_k + \theta\mathbf{s}_k) \leqq b_i$$

and compute

$$\theta \leqq \frac{b_i - \boldsymbol{\alpha}_i\mathbf{x}_k}{\boldsymbol{\alpha}_i\mathbf{s}_k}.$$

The minimum of these quotients over all nonbinding constraints $\boldsymbol{\alpha}_i\mathbf{x} \leqq b_i$ for which $\boldsymbol{\alpha}_i\mathbf{s}_k > 0$ was then the maximum permitted value of θ. But consider now how much more difficult it would be to make the same maximum-step-length determination when there are several nonlinear constraints $g_i(\mathbf{x}) \leqq b_i$ to contend with. In general, for each of those constraints an inequality of the form $g_i(\mathbf{x}_k + \theta\mathbf{s}_k) \leqq b_i$ would have to be solved by means of some time-consuming iterative procedure such as Newton's method, to yield an upper limit on θ, with the smallest of those upper limits then being taken as the maximum permitted value of θ. The computational effort involved in this approach would normally be prohibitive. Instead, it is more efficient overall to solve the one-dimensional search problem

$$\text{(9-21)} \qquad \begin{cases} \text{Max}\, f(\hat{\mathbf{x}}) = f(\mathbf{x}_k + \theta \mathbf{s}_k) \\ \text{subject to } \theta \geqq 0 \text{ and } \hat{\mathbf{x}} \text{ feasible} \end{cases}$$

by incrementing θ in successive small amounts, testing for improvement in $f(\hat{\mathbf{x}})$ *and* for continued satisfaction of all constraints at each step, and stopping finally when one of these tests fails. This approach to solving (9-21) requires significantly more computation time than was needed when the constraint set was entirely linear. Note that in testing for violation of constraints we cannot afford to ignore those that are binding at $\mathbf{x}_k$, as we could in the case of linear constraints: Even though $\mathbf{s}_k$ is directed initially away from the binding hypersurface $g_i(\mathbf{x}) \leqq b_i$, that surface could still curve around to meet the ray $\mathbf{x}_k + \theta \mathbf{s}_k$ at some positive value of θ.

Apart from the increased computational difficulty associated with solving the search problem (9-21), extension of the feasible-directions approach to the general nonlinear program also introduces significant theoretical complications into the problem of determining an improving feasible direction at each iteration. Recall that in the linearly constrained case the direction-finding problem was of the following form:

$$\text{(8-64)} \qquad \begin{cases} \text{Max } \nabla \mathbf{f}^T(\mathbf{x}_k) \cdot \mathbf{s} & \\ \text{subject to } \boldsymbol{\alpha}_i \mathbf{s} \leqq 0, & i \in B_k, \\ \boldsymbol{\alpha}_i \mathbf{s} = 0, & i \in E, \\ \text{and } \mathbf{s}^T \mathbf{s} = 1, & \end{cases}$$

where B_k is the set of subscripts of those inequality constraints $\boldsymbol{\alpha}_i \mathbf{x} \leqq b_i$ that are binding at the current point $\mathbf{x}_k$; E is the set of subscripts of the equality constraints, which are automatically binding at all feasible points; and $\nabla \mathbf{f}^T(\mathbf{x}_k) \cdot \mathbf{s}$ is the initial rate of increase of the objective function $f(\mathbf{x})$ in the (variable) direction $\mathbf{s}$. Notice in (8-64) that in order for a direction $\mathbf{s}$ to be feasible with respect to some binding inequality constraint $\boldsymbol{\alpha}_i \mathbf{x} \leqq b_i$, $i \in B_k$, it is only necessary that $\boldsymbol{\alpha}_i \mathbf{s}$ be nonpositive. In particular, $\boldsymbol{\alpha}_i \mathbf{s} = 0$ is permitted; that is, movement in a direction perpendicular to the constraint gradient $\boldsymbol{\alpha}_i$ (i.e., movement along the constraint hyperplane) must always be feasible, at least with respect to the ith constraint.

This is no longer true in the case of nonlinear constraints. In Fig. 9-3, for example, it is impossible to move any positive distance whatever from the boundary point $\mathbf{x}_k$ in the tangent direction $\mathbf{t}$, where $\nabla \mathbf{g}_i^T(\mathbf{x}_k) \cdot \mathbf{t} = 0$, without leaving the feasible region. More generally, when the constraint $g_i(\mathbf{x}) \leqq b_i$, where $g_i(\mathbf{x})$ is any strictly convex function, is binding at the current point $\mathbf{x}_k$,

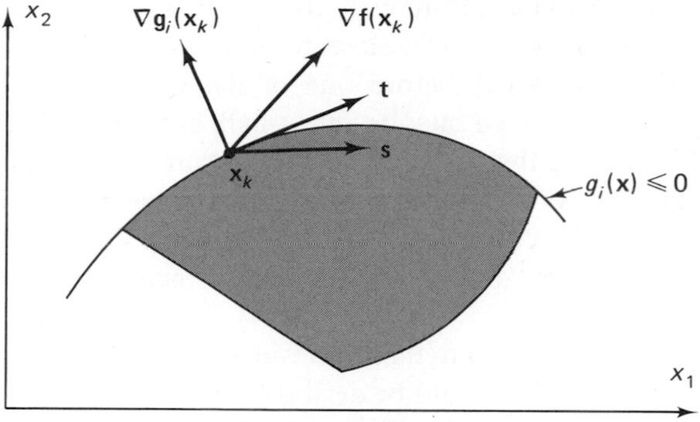

FIGURE 9-3

movement in any tangent direction—that is, any direction perpendicular to the gradient $\nabla \mathbf{g}_i(\mathbf{x}_k)$—is infeasible. In fact, only those directions $\mathbf{s}$ for which $\nabla \mathbf{g}_i^T(\mathbf{x}_k) \cdot \mathbf{s}$ is *negative* are immediately feasible (with respect to the ith constraint); this follows because when $\nabla \mathbf{g}_i^T(\mathbf{x}_k) \cdot \mathbf{s} < 0$, movement from $\mathbf{x}_k$ in the $\mathbf{s}$-direction initially causes the value of $g_i(\mathbf{x})$ to decrease. We conclude that, from a current point $\mathbf{x}_k$ in the feasible region of the problem (9-20), any vector $\mathbf{s}$ that satisfies the following set of conditions constitutes an immediately feasible direction of displacement:

$$\text{(9-22)} \qquad \begin{cases} \nabla \mathbf{g}_i^T(\mathbf{x}_k) \cdot \mathbf{s} < 0, & i \in B_k, \\ \boldsymbol{\alpha}_i \mathbf{s} = 0, & i \in E, \end{cases}$$

where B_k is the set of subscripts of the inequality constraints of (9-20) that are binding at $\mathbf{x}_k$. Note that when the feasible region is not strictly convex in the vicinity of $\mathbf{x}_k$, there may be feasible directions $\mathbf{s}$ that violate the conditions (9-22); for example, displacement in the tangent direction from the point (0, 0) along the x_1-axis in Fig. 6-6 is immediately feasible. Thus the conditions (9-22) are sufficient for defining feasible directions, but not necessary.

The presence of strict inequalities makes the conditions (9-22) rather difficult to work with. In general, if we formulate a direction-finding problem analogous to (8-64),

$$\text{(9-23)} \qquad \begin{cases} \text{Max } \nabla \mathbf{f}^T(\mathbf{x}_k) \cdot \mathbf{s} & \\ \text{subject to } \nabla \mathbf{g}_i^T(\mathbf{x}_k) \cdot \mathbf{s} < 0, & i \in B_k, \\ \boldsymbol{\alpha}_i \mathbf{s} = 0, & i \in E, \\ \text{and} \quad \mathbf{s}^T\mathbf{s} = 1, & \end{cases}$$

we may find that no specifiable optimal solution exists, just as no largest negative number exists. A more onerous difficulty is that for some of the directions $\mathbf{s}$ that satisfy (9-22)—namely, those for which one or more of the dot products $\nabla \mathbf{g}_i^T(\mathbf{x}_k) \cdot \mathbf{s}$ are very close to zero—the maximum amount of displacement that can occur before one of the binding constraint hypersurfaces is again encountered may be very small, even infinitesimal. This can be illustrated by letting the direction vector $\mathbf{s}$ approach the tangent $\mathbf{t}$ in the diagram of Fig. 9-3: As the angle between $\mathbf{s}$ and $\mathbf{t}$ decreases, the direction $\mathbf{s}$ "improves" [i.e., the component of the gradient $\nabla \mathbf{f}(\mathbf{x}_k)$ in the $\mathbf{s}$-direction increases] but the maximum feasible displacement in the $\mathbf{s}$-direction approaches zero.

It should be evident from these considerations that, for purposes of computational efficiency, it would be desirable to modify (9-23) in such a way as to take into account the trade-off between the rate of improvement offered by a feasible direction $\mathbf{s}$ and the maximum amount of displacement it permits. Consider, therefore, the following direction-finding problem:

$$
(9\text{-}24)\quad \begin{cases} \text{Max } \sigma & \\ \text{subject to } \nabla \mathbf{g}_i^T(\mathbf{x}_k) \cdot \mathbf{s} + \sigma \leqq 0, & i \in B_k, \\ -\nabla \mathbf{f}^T(\mathbf{x}_k) \cdot \mathbf{s} + \sigma \leqq 0, & \\ \boldsymbol{\alpha}_i \mathbf{s} = 0, & i \in E, \\ s_j \leqq 1, & j = 1, \ldots, n, \\ \text{and} \quad -s_j \leqq 1, & j = 1, \ldots, n, \end{cases}
$$

where $\mathbf{x}_k$ is the current solution point; $s_1, \ldots, s_n$ are the components of the unknown vector $\mathbf{s}$; σ is a new scalar variable; and B_k and E are as defined previously. For computational convenience we have replaced the normalizing constraint $\mathbf{s}^T\mathbf{s} = 1$ by the set of bounds $-1 \leqq s_j \leqq 1, j = 1, \ldots, n$, exactly as was done in Section 8.10 in the linearly constrained case; thus (9-24) is a linear program. Observe that $(\mathbf{s} = \mathbf{0}, \sigma = 0)$ constitutes a feasible solution to (9-24). It is clear that large values of σ are associated with directions $\mathbf{s}$ that offer sizable rates of improvement in the value of the objective function while moving rapidly away, at least initially, from all binding constraint hypersurfaces. However, because $\sigma \leqq \nabla \mathbf{f}^T(\mathbf{x}_k) \cdot \mathbf{s}$ and every component of $\mathbf{s}$ is bounded, σ cannot approach infinity. Therefore the direction-finding problem (9-24) has a bounded optimal solution, which we denote $(\hat{\mathbf{s}}, \hat{\sigma})$.

Consider first the case in which $\hat{\sigma} > 0$. The constraints of (9-24) then guarantee that $\hat{\mathbf{s}} \neq \mathbf{0}$ and that $\nabla \mathbf{g}_i^T(\mathbf{x}_k) \cdot \hat{\mathbf{s}} \leqq -\hat{\sigma} < 0$ for all i in B_k; thus $\hat{\mathbf{s}}$ satisfies the conditions (9-22), which implies that $\hat{\mathbf{s}}$ is a feasible direction. Moreover, inasmuch as $\hat{\sigma} \leqq \nabla \mathbf{f}^T(\mathbf{x}_k) \cdot \hat{\mathbf{s}}$, the gradient of the original objective function $f(\mathbf{x})$ has a positive component in the $\hat{\mathbf{s}}$-direction. We therefore conclude that when $\hat{\sigma} > 0$ in the optimal solution to (9-24), the vector $\hat{\mathbf{s}}$ defines an improving feasible direction from the current point $\mathbf{x}_k$ in the space of the original problem (9-20).

But suppose, on the other hand, that $\hat{\sigma} = 0$. It can then be shown that $\mathbf{x}_k$ must be a Kuhn-Tucker point, provided that it is "nondegenerate" in the following sense: A set of linear and/or nonlinear constraints is said to be *degenerate* at the feasible point $\mathbf{x}_k$ in E^n, or $\mathbf{x}_k$ is said to be *degenerate*, if the gradients of the constraints that are binding at $\mathbf{x}_k$ are linearly dependent. This is precisely the definition of degeneracy that was stated for the special case of linear constraints in Section 8.8, and we shall continue to use the term in the same way in referring to nonlinear or general constraint sets.[3] As was true in that special case, the linear dependence of the constraint gradients in most instances of degeneracy is due simply to the intersection of $n + 1$ or more constraint hypersurfaces at a single point in E^n.

To prove that, under the nondegeneracy assumption, $\mathbf{x}_k$ must be a Kuhn-Tucker point when $\hat{\sigma} = 0$, we begin by writing the dual of the direction-finding linear program (9-24). Let the dual variables corresponding to the five constraint groups displayed in (9-24) be, respectively,

$$u_i, i \in B_k; \quad u_0; \quad u_i, i \in E; \quad v_j, j = 1, \ldots, n; \quad \text{and} \quad w_j, j = 1, \ldots, n.$$

Then the dual program is

$$\text{(9-25)} \quad \begin{cases} \text{Min } z' = \sum_{j=1}^{n} (v_j + w_j) \\ \text{subject to } \sum_{i \in B_k} u_i \nabla \mathbf{g}_i(\mathbf{x}_k) - u_0 \nabla \mathbf{f}(\mathbf{x}_k) \\ \qquad + \sum_{i \in E} u_i \boldsymbol{\alpha}_i^T + \mathbf{v} - \mathbf{w} = \mathbf{0}, \\ \qquad \sum_{i \in B_k} u_i + u_0 = 1, \\ \qquad u_i \geqq 0, \quad i \in B_k, \\ \qquad u_0 \geqq 0, \\ \text{and} \qquad v_j, w_j \geqq 0, \quad j = 1, \ldots, n. \end{cases}$$

Given that the primal problem (9-24) has an optimal feasible solution with an objective value of zero, it follows from the duality theorem 2.4 that there must exist an optimal solution to (9-25) for which $z' = 0$. Inasmuch as the v_j and w_j are nonnegative variables, the value of each of them must then be zero in that optimal solution, and we can therefore conclude that there must

[3]Other writers have preferred to reserve the terms "degeneracy" and "degenerate" to refer only to sets of linear constraints. This preference is due to the fact that the algebraic concept of degeneracy—involving basic solutions and zero-valued basic variables—is meaningful only in the context of linear constraints. What we call a "nondegenerate" point in the nonlinear context must then be given some other name, such as "regular" point.

exist a set of values u_i^*, $i \in B_k$, $i \in E$, and u_0^* such that

$$\sum_{i \in B_k} u_i^* \nabla \mathbf{g}_i(\mathbf{x}_k) + \sum_{i \in E} u_i^* \boldsymbol{\alpha}_i^T = u_0^* \nabla \mathbf{f}(\mathbf{x}_k), \tag{9-26}$$

$$\sum_{i \in B_k} u_i^* + u_0^* = 1, \tag{9-27}$$

and

$$u_0^* \geqq 0, \qquad u_i^* \geqq 0, \qquad i \in B_k. \tag{9-28}$$

Under the nondegeneracy assumption the gradients $\nabla \mathbf{g}_i(\mathbf{x}_k)$ and $\boldsymbol{\alpha}_i^T$ appearing on the left-hand side of (9-26), which are in fact the gradients of the binding constraints at $\mathbf{x}_k$, must be linearly independent. Suppose now that $u_0^* = 0$; then the linear independence of those gradient vectors would force $u_i^* = 0$ for all i in B_k and E, leading to a violation of (9-27). Hence u_0^* must be positive, and we can define a new set of values

$$\lambda_i^* = \frac{u_i^*}{u_0^*}, \qquad i \in B_k, i \in E,$$

which must satisfy the following relations:

$$\sum_{i \in B_k} \lambda_i^* \nabla \mathbf{g}_i(\mathbf{x}_k) + \sum_{i \in E} \lambda_i^* \boldsymbol{\alpha}_i^T = \nabla \mathbf{f}(\mathbf{x}_k) \tag{9-29}$$

and

$$\lambda_i^* \geqq 0, \qquad i \in B_k. \tag{9-30}$$

Moreover, for each constraint $g_i(\mathbf{x}) \leqq b_i$ that is binding at $\mathbf{x}_k$ it is trivially true that

$$\lambda_i^*[g_i(\mathbf{x}_k) - b_i] = 0, \qquad i \in B_k. \tag{9-31}$$

Let us now augment the set of values λ_i^* defined thus far by adding the values

$$\lambda_i^* = 0, \qquad i \in I \text{ but } i \notin B_k. \tag{9-32}$$

In view of (9-29) through (9-32), it is then easily verified that the set of values λ_i^*, $i \in I$, $i \in E$, must satisfy the following:

$$\begin{cases} \nabla \mathbf{f}(\mathbf{x}_k) = \sum_{i \in I} \lambda_i^* \nabla \mathbf{g}_i(\mathbf{x}_k) + \sum_{i \in E} \lambda_i^* \boldsymbol{\alpha}_i^T \\ \qquad \lambda_i^*[g_i(\mathbf{x}_k) - b_i] = 0, \qquad i \in I, \\ \text{and} \qquad\qquad \lambda_i^* \geqq 0, \qquad i \in I. \end{cases} \tag{9-33}$$

But these are precisely the Kuhn-Tucker conditions for the original problem (9-20)—Exercise 9-15 asks the student to verify that this is so—and we conclude that $\mathbf{x}_k$ is a Kuhn-Tucker point. Q.E.D.

The foregoing proof establishes the basic theoretical foundation of Zoutendijk's general feasible-directions algorithm for solving the nonlinearly constrained mathematical program (9-20). At each iteration the algorithm proceeds as follows:

1. Given the current feasible point $\mathbf{x}_k$, identify the binding inequality constraints; then formulate and solve the linear program (9-24), obtaining the optimal solution $\mathbf{s} = \hat{\mathbf{s}}$, $\sigma = \hat{\sigma}$.

2. If $\hat{\sigma} = 0$, then $\mathbf{x}_k$ is a Kuhn-Tucker point and the algorithm terminates. If $\hat{\sigma} > 0$, then $\mathbf{s}_k = \hat{\mathbf{s}} \neq \mathbf{0}$ is an improving feasible direction of displacement from $\mathbf{x}_k$. Using some iterative computational scheme, solve the one-dimensional search problem

$$\text{Max } f(\mathbf{x}_k + \theta \mathbf{s}_k)$$
$$\text{subject to } \theta \geqq 0 \text{ and } \mathbf{x}_k + \theta \mathbf{s}_k \text{ feasible.}$$

If no finite value of θ is optimal, the original problem has an unbounded solution and the algorithm terminates; otherwise set $\mathbf{x}_{k+1} = \mathbf{x}_k + \theta_{\text{opt}} \mathbf{s}_k$ and return to step 1.

In practice the method would presumably be terminated when at some iteration the optimal value $\hat{\sigma}$ is found to be less than a preselected small positive constant. It should be noted, however, that the feasible-directions algorithm is not normally implemented in the above very simple form, which is unfortunately susceptible to zigzagging. Zoutendijk suggested several different devices and modifications for curtailing zigzagging and enhancing convergence; we shall discuss one of them in detail in the next section.

To initiate the algorithm, a starting feasible solution to the problem (9-20) is required. If no starting solution is readily available, one can be obtained exactly as described at the end of Section 8.9 for the linearly constrained problem. Assume that the equality constraints $\boldsymbol{\alpha}_i \mathbf{x} = b_i$ have been formulated so that $b_i \geqq 0$, $i \in E$; this can be achieved by multiplying through by -1 wherever necessary. Next, partition I, the set of subscripts of the inequality constraints, into two subsets I_S and I_U, such that i is in I_S if $g_i(\mathbf{0}) \leqq b_i$ and i is in I_U if $g_i(\mathbf{0}) > b_i$. Finally, introduce the artificial variables u_0 and u_i, $i \in E$, and use the feasible-directions algorithm to solve the following "phase I" problem:

$$\text{(9-34)} \quad \begin{cases} \text{Max } z' = -u_0 - \sum_{i \in E} u_i \\ \text{subject to } g_i(\mathbf{x}) \leqq b_i, & i \in I_S, \\ \qquad g_i(\mathbf{x}) - u_0 \leqq b_i, & i \in I_U, \\ \qquad \boldsymbol{\alpha}_i \mathbf{x} + u_i = b_i, & i \in E, \\ \qquad u_0 \geqq 0, \\ \text{and} \qquad u_i \geqq 0, & i \in E. \end{cases}$$

Observe that the set of values

$$\mathbf{x} = \mathbf{0}, \quad u_0 = \max_{i \in I_U}(g_i(\mathbf{0}) - b_i) \quad \text{and} \quad u_i = b_i, \qquad i \in E,$$

constitutes a starting feasible solution to (9-34). Let z'_{opt} denote the optimal value of the objective of (9-34); it is then easily shown that if $z'_{\text{opt}} = 0$, the **x**-component of the optimal solution constitutes a feasible solution to the original problem (9-20). If $z'_{\text{opt}} < 0$, and if every constraint function $g_i(\mathbf{x})$, $i \in I$, is convex, it follows that the original problem has no feasible solution. In general, however, when the feasible region of (9-20) is nonconvex, so is that of (9-34); thus a local optimal solution to (9-34) at which $z'_{\text{opt}} < 0$ might be obtained even though global optima exist at which $z' = 0$ [i.e., even though (9-20) has feasible solutions].

We conclude this section by working through an iteration of the feasible-directions algorithm for the problem

$$\begin{aligned} \text{Max } f(\mathbf{x}) &= x_1 + x_2^3, \\ \text{subject to } g_1(\mathbf{x}) &\equiv x_1 - 2x_2 \leqq 0 \\ \text{and } \quad g_2(\mathbf{x}) &\equiv x_1^2 + x_2^2 \leqq 5, \end{aligned}$$

taking as our current solution $\mathbf{x}_k = (2, 1)$. The feasible region is diagrammed in Fig. 9-4. Inasmuch as both constraints are binding at $\mathbf{x}_k$, it is necessary to evaluate all three gradients there:

$$\begin{aligned} \nabla \mathbf{f}(\mathbf{x}_k) &= (1, 3x_2^2)\,|_{\mathbf{x}_k} = (1, 3), \\ \nabla \mathbf{g}_1(\mathbf{x}_k) &= (1, -2), \\ \nabla \mathbf{g}_2(\mathbf{x}_k) &= (2x_1, 2x_2)\,|_{\mathbf{x}_k} = (4, 2). \end{aligned}$$

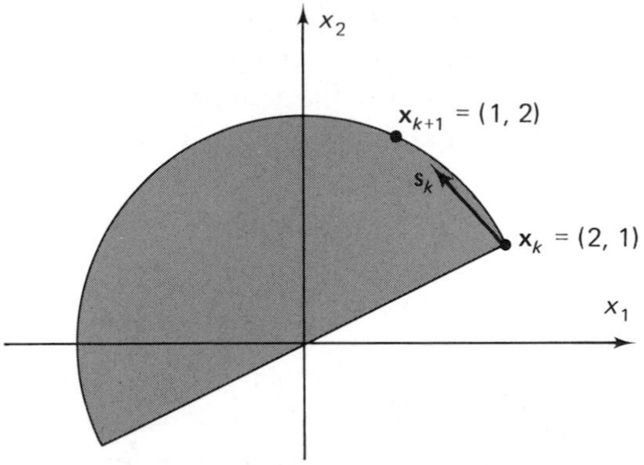

FIGURE 9-4

The linear program (9-24) is then written as follows:

$$\begin{aligned} & \text{Max } \sigma \\ \text{subject to} \quad & s_1 - 2s_2 + \sigma \leqq 0, \\ & 4s_1 + 2s_2 + \sigma \leqq 0, \\ & -s_1 - 3s_2 + \sigma \leqq 0, \\ & -1 \leqq s_1 \leqq 1, \\ \text{and} \quad & -1 \leqq s_2 \leqq 1. \end{aligned}$$

The optimal solution is found to be $s_1 = -1$, $s_2 = 1$, $\sigma = 2$, so the current point $\mathbf{x}_k$ is not optimal and an improving feasible direction is $\mathbf{s}_k = (-1, 1)$. Notice in Fig. 9-4 that there are other feasible directions offering more rapid rates of improvement than $(-1, 1)$; however, none of them permits displacement to proceed for as great a distance. To determine the next solution point, we perform a stepwise search along the ray $\mathbf{x} = \mathbf{x}_k + \theta \mathbf{s}_k$, $\theta \geqq 0$, eventually discovering that the maximum feasible value of $f(\mathbf{x}_k + \theta \mathbf{s}_k)$ occurs at $\theta_{\text{opt}} = 1$; thus $\mathbf{x}_{k+1} = \mathbf{x}_k + \theta_{\text{opt}} \mathbf{s}_k = (1, 2)$, and the iteration is complete.

9.5 ZIGZAGGING AND THE ϵ-PERTURBATION METHOD

Although the feasible-directions algorithm described in the previous section is quite well behaved in most instances, converging to optimal solutions or Kuhn-Tucker points according to whether or not the problem (9-20) is a convex program, it is also capable on occasion of zigzagging, or jamming. This phenomenon, when it occurs, is precisely the same as in the case of linearly constrained problems: The sequence of solutions $\{\mathbf{x}_i\}$ generated by the algorithm describes a zigzag path that is reflected back and forth between two or more constraint hypersurfaces or hyperplanes in segments of ever-decreasing length, with the entire path converging to a limit point $\mathbf{x}^*$ at the junction of those hypersurfaces. The difficulty is that there can be no guarantee that $\mathbf{x}^*$ satisfies the Kuhn-Tucker conditions, regardless of whether or not (9-20) is a convex program. Moreover, even when a Kuhn-Tucker point is duly obtained, the rate of convergence of the overall process can be severely retarded by zigzagging.

It is therefore important to devise some antizigzagging procedure for use in solving nonlinearly constrained problems via the feasible-directions approach. Unfortunately, the measures described in Section 8.14, all of which depend on forcing the path $\{\mathbf{x}_i\}$ generated by the algorithm to remain within one or more binding constraint hyperplanes at certain stages, obviously cannot be applied in the case of nonlinearly constrained problems, inasmuch as straight-line displacement within curved hypersurfaces is, in general, impossible. There is, however, an alternative type of antizigzagging strategy, also proposed by Zoutendijk [61], that can be used instead. Given a current feasible point $\mathbf{x}_k$, the general idea is to choose a displacement vector $\mathbf{s}_k$ that is directed away from all constraint hypersurfaces that either contain $\mathbf{x}_k$ *or lie close to it*. More specifically, this approach, known as *ϵ-perturbation*, requires a positive constant ϵ to be chosen such that the inequality constraint $g_i(\mathbf{x}) \leqq b_i$ is considered "almost binding" at $\mathbf{x}_k$ if $g_i(\mathbf{x}_k) + \epsilon \geqq b_i$. Then, a constraint of the form

$$\nabla \mathbf{g}_i^T(\mathbf{x}_k) \cdot \mathbf{s} + \sigma \leqq 0$$

is used in the direction-finding problem (9-24) for every constraint of the original problem that is either binding or almost binding. By thus forcing the selection of a vector $\mathbf{s}_k$ that leads initially away from *all* constraint hypersurfaces in the immediate vicinity of $\mathbf{x}_k$, this procedure prevents zigzagging from continuing for more than a relatively small—or at most a finite—number of iterations.

As an example, refer to Fig. 9-5, which shows a cross section of a three-dimensional feasible region containing a portion of a zigzag path generated by a feasible-directions algorithm. At each point up to and including $\mathbf{x}_{k-1}$, the nonbinding constraint hypersurface was evidently far enough away to have had no effect on the choice of direction. If, however, the inequality constraints associated with the nonlinear surface in the diagram were "almost binding" at $\mathbf{x}_k$, then at the next iteration an improving feasible direction $\mathbf{s}_k$ would have to be found that would lead away from (or at least not toward) both surfaces, that is, either upward or downward out of the plane of Fig. 9-5. Thus the zigzagging would be curtailed.

To make these ideas more specific, we shall introduce some definitions and construct a new direction-finding problem. Recall that we are concerned with solving the nonlinear programming problem

$$\text{(9-20)} \qquad \begin{cases} \text{Max } z = f(\mathbf{x}), & \mathbf{x} = (x_1, \ldots, x_n), \\ \text{subject to} \quad g_i(\mathbf{x}) \leqq b_i, & i \in I, \\ \text{and} \quad \boldsymbol{\alpha}_i \mathbf{x} = b_i, & i \in E, \end{cases}$$

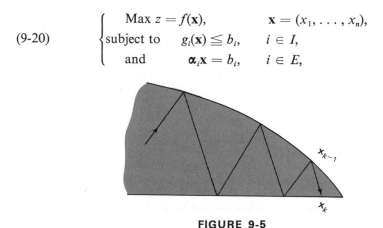

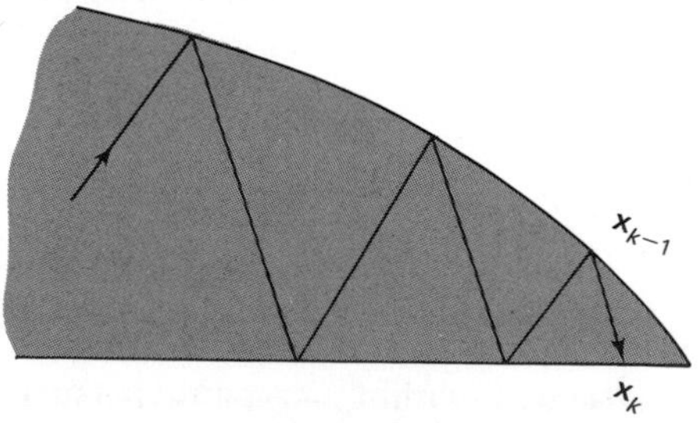

FIGURE 9-5

by means of a feasible-directions algorithm that generates a sequence of solution points $\{\mathbf{x}_i\}$. Let ϵ be any nonnegative scalar constant; then an inequality constraint $g_i(\mathbf{x}) \leqq b_i$ will be said to be *ϵ-binding* at a solution point $\mathbf{x}_k$ if $g_i(\mathbf{x}_k) + \epsilon \geqq b_i$. Such a constraint, in effect, is "within ϵ" of being binding at $\mathbf{x}_k$. Next, let $B_k(\epsilon)$ be the set of subscripts of all inequality constraints of (9-20) that are ϵ-binding at $\mathbf{x}_k$. Note that $B_k(\epsilon)$ is a subset of I for any k and ϵ and that $B_k(0)$ is simply equivalent to B_k, the set of subscripts of all constraints that are binding at $\mathbf{x}_k$. The direction-finding problem that will be used at each iteration is then of the following general form:

$$
\text{(9-35)} \qquad \begin{cases} \text{Max } \sigma & \\ \text{subject to } \nabla \mathbf{g}_i^T(\mathbf{x}_k) \cdot \mathbf{s} + \sigma \leqq 0, & i \in B_k(\epsilon) \\ -\nabla \mathbf{f}^T(\mathbf{x}_k) \cdot \mathbf{s} + \sigma \leqq 0, & \\ \boldsymbol{\alpha}_i \mathbf{s} = 0, & i \in E, \\ \text{and} \qquad -1 \leqq s_j \leqq 1, & j = 1, \ldots, n. \end{cases}
$$

Note that, as before, if $\hat{\sigma} > 0$ in the optimal solution $(\hat{\mathbf{s}}, \hat{\sigma})$, then $\hat{\mathbf{s}}$ constitutes an improving feasible direction from $\mathbf{x}_k$; moreover, in that case $\hat{\mathbf{s}}$ points initially away from all ϵ-binding constraint surfaces.

We are now ready to state the *ϵ-perturbation method*, a feasible-directions algorithm for solving the nonlinear program (9-20) that incorporates the antizigzagging modification described above. The procedure was designed by Zangwill [60], who proved convergence for it, but it is based substantially on the original work of Zoutendijk [61]. To initiate the algorithm, a starting feasible solution $\mathbf{x}_0$, an initial positive "perturbation constant" ϵ_0, and a very small positive "termination constant" δ must be specified. Then, given that the algorithm has generated the sequence of solution points $\mathbf{x}_0, \mathbf{x}_1, \ldots, \mathbf{x}_k$ and the sequence of perturbation constants $\epsilon_0, \epsilon_1, \ldots, \epsilon_k$, the $(k + 1)$th iteration proceeds as follows:

1. Using $\epsilon = \epsilon_k$, solve the direction-finding problem (9-35), obtaining the optimal solution $(\hat{\mathbf{s}}, \hat{\sigma})$. One of the following three conditions must obtain:
   a. $\hat{\sigma} \geqq \epsilon_k$, in which case set $\epsilon_{k+1} = \epsilon_k$ and proceed to step 3;
   b. $\hat{\sigma} < \epsilon_k$ and every constraint of (9-20) that is included in $B_k(\epsilon_k)$ is actually binding at $\mathbf{x}_k$, in which case proceed to step 2; or
   c. $\hat{\sigma} < \epsilon_k$ and one or more nonbinding constraints are included in $B_k(\epsilon_k)$, in which case set $\mathbf{x}_{k+1} = \mathbf{x}_k$ and $\epsilon_{k+1} = \epsilon_k/2$ and proceed to the next iteration.
2. If $\hat{\sigma} \leqq \delta$, the algorithm terminates and $\mathbf{x}_k$ is taken to be a Kuhn-Tucker point; otherwise, set $\epsilon_{k+1} = \epsilon_k/2$ and go on to step 3.
3. The vector $\mathbf{s}_k = \hat{\mathbf{s}}$ constitutes an improving feasible direction from $\mathbf{x}_k$. Use some iterative computational scheme to determine θ_{opt}, the value of θ that maximizes $f(\mathbf{x}_k + \theta \mathbf{s}_k)$ subject to $\theta \geqq 0$ and to the requirement that $\mathbf{x}_k + \theta \mathbf{s}_k$ satisfy all constraints of (9-20). If no finite value of θ is optimal, the original

problem has an unbounded solution and the algorithm terminates; otherwise, set $\mathbf{x}_{k+1} = \mathbf{x}_k + \theta_{\text{opt}}\mathbf{s}_k$ and go on to the next iteration.

This procedure differs slightly from that of Zangwill.

During the early stages of the ϵ-perturbation algorithm the optimal value $\hat{\sigma}$ obtained from solving the direction-finding problem (9-35) will in general be substantially greater than ϵ_k; thus at each iteration step 3 will be executed and a new solution point will be generated (unless unboundedness is demonstrated), with the perturbation constant ϵ_k remaining unchanged. Because ϵ_k is held at its original level during these early iterations, the directions $\mathbf{s}_k$ determined via (9-35) will avoid nearby constraint hypersurfaces and progress will be fairly rapid. If the sequence $\{\mathbf{x}_i\}$ should begin to zigzag toward a *non*-Kuhn-Tucker point, the same cycle will continue, with step 3 being executed at each iteration, until a point is reached at which *all* constraints involved in the zigzagging are binding or ϵ_k-binding. When this occurs, the next vector $\hat{\mathbf{s}}$ determined by solving (9-35) will be directed away from the danger area and the zigzagging will be disrupted, as explained in connection with Fig. 9-5.

As the sequence $\{\mathbf{x}_i\}$ begins to approach a local maximum or other Kuhn-Tucker point, the values of $\hat{\sigma}$ determined from successive direction-finding problems will decrease, until at some stage either case 1b or 1c obtains. In the former case, unless termination occurs, a new point $\mathbf{x}_{k+1}$ is generated, as in case 1a. In the latter case the solution point remains unchanged but the perturbation constant is decreased, which may, by causing a constraint to be ejected from $B_k(\epsilon_k)$, allow a larger value of $\hat{\sigma}$ to be obtained at the next iteration. Even if this does not occur, however, case 1c can persist for only a finite number of iterations: Sooner or later a stage must be reached at which ϵ_k is smaller than the positive quantity $b_i - g_i(\mathbf{x}_k)$ for every nonbinding constraint. Case 1b will then be entered and a new solution point generated. Eventually, after the sequence $\{\mathbf{x}_i\}$ has converged, perhaps via a zigzag path, to within a very small distance of a Kuhn-Tucker point, and after the perturbation constant has been reduced repeatedly until only the binding constraints are left in $B_k(\epsilon_k)$, step 2 will be entered and it will be found that $\hat{\sigma} \leqq \delta$. Termination will then occur.

Given a current solution point $\mathbf{x}_k$, it is clear that when ϵ_k is sufficiently close to zero the set $B_k(\epsilon_k)$ includes only those constraints of (9-20) that are binding at $\mathbf{x}_k$; thus the direction-finding problem (9-35) reduces to the problem (9-24) studied in Section 9.4. Under these circumstances, if $\hat{\sigma} = 0$ in the optimal solution to (9-35), we know that $\mathbf{x}_k$ must be a Kuhn-Tucker point, provided that the original set of constraints can be assumed nondegenerate at $\mathbf{x}_k$. Moreover, when $B_k(\epsilon_k)$ includes no nonbinding constraints and $\hat{\sigma}$ is *very close* to zero (i.e., $\hat{\sigma} \leqq \delta$), it follows from the continuity of the functions $f(\mathbf{x})$ and $g_i(\mathbf{x})$ and of their gradients that the current solution $\mathbf{x}_k$ must lie in the immediate vicinity of a Kuhn-Tucker point. To prove algorithmic convergence for the ϵ-perturbation method, however, it is also necessary to demonstrate that the sequence of solution points $\{\mathbf{x}_i\}$ cannot converge toward a *non*-Kuhn-Tucker point. We have argued heuristically that any zigzag path toward such a point will be interrupted when a point $\mathbf{x}_k$ is reached at which all the nearby constraints are ϵ_k-binding. However, this intuitive argument does not take into account the possibility that before the zigzagging begins the value of ϵ_k may have become so small that no solution point $\mathbf{x}_k$ can ever be reached at which the nonbinding constraint hypersurfaces in the vicinity will be ϵ_k-binding. In such an event the ϵ-perturbation algorithm would continue to cycle through case 1a and the zigzagging would never be interrupted. To eliminate this possibility it is necessary to prove that the sequence of positive values $\epsilon_0, \epsilon_1, \epsilon_2, \ldots$ cannot approach zero as a limit. The proof of this proposition is rather lengthy, requiring two or three intermediate results, and will not be given here; the interested reader is referred to Chapter 13 of Zangwill's text [60].

9.6 PRIMAL NONLINEAR PROGRAMMING ALGORITHMS: A SUMMARY

The various feasible-directions algorithms of mathematical programming, which include, for example, the linear simplex method, the reduced gradient method, and the ϵ-perturbation method just discussed, may also be referred to in general as *primal* solution algorithms. Such algorithms take what may be called the *primal* approach to solving a mathematical program: That is, they begin with a feasible point and proceed iteratively to generate a further sequence of feasible points, each having a better objective value than its predecessor, until a local optimum or Kuhn-Tucker point is reached. Feasibility is maintained at each stage by direct examination and manipulation of the constraints, while improvement in the objective value is achieved by direct utilization of the objective gradient or some closely related function of it. Thus an essential characteristic of primal methods is that they work with the given problem in its original form (except for trivial algebraic maneuvers required to convert to some standardized format). This is in contrast to the penalty and barrier methods, to be discussed in the next few sections, which operate by transforming a given nonlinearly constrained mathematical program into a sequence of completely unconstrained problems. Another group of solution algorithms that do not take the primal approach are the so-called "dual" methods, which include the dual simplex algorithm, as well as the cutting-plane methods of Kelley and others. Each of the algorithms in this group operates not on the given mathematical program but rather on some related problem that can be interpreted as its dual (in this connection see Section 9.9 and Exercise 9-14); in general, the intermediate solutions generated by dual methods are "superoptimal" but infeasible with respect to the original problem.

Primal solution algorithms, and in particular the feasible-directions methods studied in the past two chapters, have a number of practical advantages and favorable characteristics that are not unanimously shared by penalty and barrier methods and the various dual approaches. In the first place, primal methods can be applied to all types of problems, convex or nonconvex, and do not require any special structural properties such as separability. Second, the intermediate solution points generated by primal methods lie in the feasible region, so that, if the computations must be interrupted before the optimum is reached, the final solution point obtained is at least feasible and may well be near-optimal. This can be a useful advantage on occasion, although not so frequently as is commonly supposed. Third, and extremely important in the practice of operations research, the feasible-directions approach is ideally suited to linearly constrained problems and well suited to problems whose constraints are nearly linear. The reasons for this have been sufficiently explained (most recently in the early paragraphs of Section 9.4). The practical importance of feasible-directions methods is due to the fact that most optimization problems that can be *realistically* and *accurately* formulated in business, engineering, and other decision-making contexts are linear or nearly so. To be sure, the problem

$$\begin{aligned} \text{Max } z &= \sin x_1 x_2 + x_2^{-4} \log(x_3 + 1) \\ \text{subject to} \quad & (x_1 + x_2 + x_3)^3 - x_1 x_2 x_3 \leqq 0 \\ \text{and} \quad & \arctan x_1^2 + x_2 x_3 \leqq 20 \end{aligned}$$

is by definition a mathematical program, and doubtless it is comforting to reflect that algorithms for solving it exist, but it should nevertheless be realized that such problems are little more than figments of the imagination of those of us who write nonlinear programming texts. In the real world there are very few functional relationships that can accurately be modeled, say, as logarithmic or sinusoidal, although admittedly such functional models are sometimes used when, in situations of great uncertainty, they are felt to be "acceptable." Nevertheless, it is fair to say that, of all those relationships that can reasonably be represented by *any* type of analytic function, a majority are (nearly) linear and a great majority are either linear or quadratic.

Although, as we have seen, primal solution algorithms have certain very favorable properties, they also have some serious disadvantages. One is that a primal method cannot be initiated until a starting feasible solution has been obtained; this sometimes requires solving an additional phase I problem. Another, more important disadvantage is that every primal method must be supplied with computational machinery for ensuring that feasibility is maintained at each iteration. This problem of maintaining feasibility is, as has been pointed out, much more bothersome in the case of nonlinear constraints. Nevertheless, it has two unfortunate general consequences for all types of primal methods: It complicates the computer program being used, and it tends to increase the amount of computation time needed at each iteration. A further difficulty characteristic of feasible-directions algorithms is the ever-present possibility of zigzagging, which may bring with it the possibility of convergence to a non-Kuhn-Tucker point. This difficulty can be overcome by means of antizigzagging devices, such as bending or ϵ-perturbation, but the overall effect is to increase further the computational effort required for solution.

Assessing these pros and cons, we can conclude that the advantages of primal methods are decisive in the case of linearly constrained or nearly linear problems, but that their disadvantages become relatively more significant when the constraints are substantially nonlinear. In the latter case the initial feasible solution may be more difficult to find; the repeated testing required to maintain feasibility at each iteration inevitably becomes much more onerous; and, finally, the programming logic must be somewhat more complicated. Nevertheless, primal solution methods are, in general, quite useful even for nonlinearly constrained problems: Their overall convergence rates are competitive, and they have the additional useful advantage of being applicable to all types of problems. They in fact constitute by far the most important class of mathematical programming algorithms.

Before leaving the subject of primal and feasible-directions algorithms, we should remark that there exists a modified version of the gradient projection method, due to Rosen [38], which can in theory be applied to nonlinearly constrained problems. At each stage the method proceeds, as in the case of linear constraints, to compute the projection of the objective gradient $\nabla \mathbf{f}(\mathbf{x}_k)$ onto the orthogonal complement of the subspace spanned by the gradients $\nabla \mathbf{g}_i(\mathbf{x}_k)$ and $\boldsymbol{\alpha}_i^T$ of the binding constraints. This direction, denoted $(\mathbf{P}\,\nabla \mathbf{f})_k$, will be tangent to all—or, if necessary, all but one—of the binding constraint hypersurfaces, as illustrated in Fig. 9-6(a). Note, however, that the

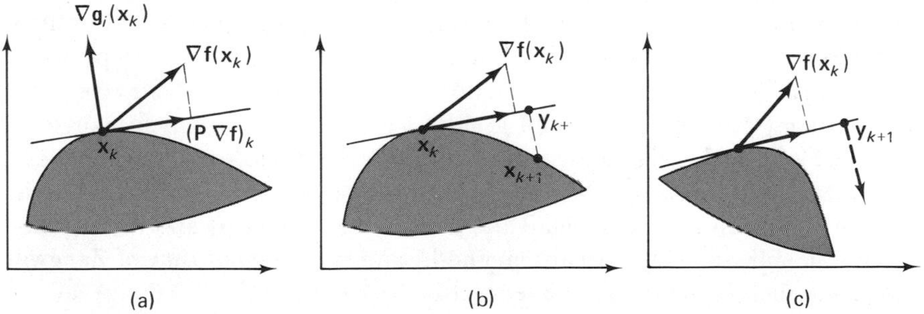

FIGURE 9-6

projected gradient need not, in general, constitute a feasible direction. When it does not, it becomes necessary, after proceeding some distance in the direction $(\mathbf{P}\,\nabla\mathbf{f})_k$ to an infeasible point $\mathbf{y}_{k+1}$, to return safely to the feasible region. This is done by moving in a direction perpendicular to $(\mathbf{P}\,\nabla\mathbf{f})_k$ until a feasible point is first encountered—that point becomes $\mathbf{x}_{k+1}$. The return to the feasible region, illustrated in Fig. 9-6(b), is accomplished via an iterative computational procedure that is unfortunately rather time-consuming. In addition to this disadvantage of having to locate two points instead of one at various iterations, the gradient projection method for nonlinear constraints also possesses several other unfavorable and complicating features. It is possible, for example, that the value of the objective function at $\mathbf{x}_{k+1}$ may prove to be less than it was at $\mathbf{x}_k$, in which case new values of $\mathbf{y}_{k+1}$ and $\mathbf{x}_{k+1}$ must be determined. Another possibility is illustrated in Fig. 9-6(c): The feasible region may curve away so sharply that the perpendicular path from $\mathbf{y}_{k+1}$ does not pass through any feasible points whatever.

Somewhat surprisingly, in view of all these difficulties, the gradient projection method has been implemented successfully and found to be competitive, or nearly so; a thorough study of its convergence properties can be found in Luenberger [31]. Nevertheless, because of the complexity of its logic and computational procedures, and in view of the availability of other, more straightforward primal solution algorithms, the gradient projection method seems to be of rather limited practical interest, insofar as solving nonlinearly constrained problems is concerned. The same remarks apply as well to a generalized version of the reduced gradient method, which is also discussed in Luenberger.

9.7 SEQUENTIAL UNCONSTRAINED OPTIMIZATION: BARRIER METHODS

In the next four sections we shall be discussing a family of nonlinear programming algorithms embodying a computational approach known as *sequential unconstrained optimization*. Given a nonlinearly constrained program satisfying some general format, the algorithms in this family proceed by generating and solving a sequence of unconstrained problems whose optimal solutions converge toward the desired solution of the original problem. Sequential unconstrained optimization techniques divide themselves into two classes, *barrier methods* and *penalty methods*, each of which includes a number of closely related algorithms that are all based on a single characteristic solution strategy. We begin with a discussion of barrier methods, which have been studied in detail and popularized by Fiacco and McCormick (see [13], [14], and [15]); penalty methods, and in particular that of Zangwill [57], will then be treated in a later section.

Barrier methods are applicable to nonlinear programs of the form

$$\text{(9-36)} \qquad \begin{cases} \text{Max } z = f(\mathbf{x}), & \mathbf{x} = (x_1, \ldots, x_n), \\ \text{subject to } g_i(\mathbf{x}) \leqq 0, & i = 1, \ldots, m, \end{cases}$$

where $f(\mathbf{x})$ and the $g_i(\mathbf{x})$ are continuous and have continuous first partial derivatives and where there exists at least one point $\hat{\mathbf{x}}$ satisfying $g_i(\hat{\mathbf{x}}) < 0$, $i = 1, \ldots, m$. That is, we are requiring that the *interior* of the feasible region (recall Section 3.2) be nonempty; this precludes the appearance of any equalities in the constraint set, either explicitly, in the form $g_i(\mathbf{x}) = 0$, or implicitly, as the pair of simultaneous inequalities $g_i(\mathbf{x}) \leqq 0$ and $g_j(\mathbf{x}) \equiv -g_i(\mathbf{x}) \leqq 0$. For convenience, we let S denote the feasible region of (9-36) and S_0 denote its interior:

$$S \equiv \{\mathbf{x} \mid g_i(\mathbf{x}) \leqq 0, i = 1, \ldots, m\}$$

and

$$S_0 \equiv \{\mathbf{x} \mid g_i(\mathbf{x}) < 0, i = 1, \ldots, m\}.$$

The *boundary* of S then includes all points $\mathbf{x}$ that lie in S but not in S_0. Note finally, with regard to (9-36), that at this stage no convexity/concavity requirements are being placed on any of the functions $f(\mathbf{x})$ and $g_i(\mathbf{x})$.

The characteristic "barrier approach" to solving the problem (9-36) involves replacing it by a sequence of unconstrained optimization problems, each of the form

$$\text{(9-37)} \qquad \text{Max } C(\mathbf{x}, r) \equiv f(\mathbf{x}) + rB(\mathbf{x}),$$

where r is a positive-valued scalar parameter and $B(\mathbf{x})$ is a *barrier function* whose value is nonpositive at every point $\mathbf{x}$ in the interior of the feasible region S and decreases toward $-\infty$ as the boundary of S is approached. Thus, when (9-37) is solved via one of the usual direct climbing methods, starting from a point in the interior S_0, the function $B(\mathbf{x})$ acts as a barrier against any movement across the boundary of S and ensures that the solution obtained will also lie in the interior. A typical barrier function, and the one most frequently used, is

$$\text{(9-38)} \qquad B(\mathbf{x}) = \sum_{i=1}^{m} \frac{1}{g_i(\mathbf{x})}, \qquad \mathbf{x} \in S_0.$$

Note that at any point $\hat{\mathbf{x}}$ in the interior S_0 each of the values $g_i(\hat{\mathbf{x}})$ is negative, so $B(\hat{\mathbf{x}})$ must be negative as well; moreover, as $\hat{\mathbf{x}}$ approaches the boundary of S, one or more of the values $g_i(\hat{\mathbf{x}})$ approaches zero from below, and $B(\hat{\mathbf{x}})$ decreases toward $-\infty$.

To specify fully the sequence of unconstrained optimization problems to be solved by a barrier algorithm, it is necessary to choose (1) a barrier function and (2) a monotonically decreasing sequence of positive numbers $\{r_k\} \equiv r_1, r_2, r_3, \ldots$ that satisfies $\lim_{k\to\infty} r_k = 0$. A standard type of sequence is one in which each term is a constant fractional multiple of the one preceding it; that is, $r_2 = \phi r_1$, $r_3 = \phi r_2 = \phi^2 r_1, \ldots, r_k = \phi r_{k-1} = \phi^{k-1} r_1, \ldots,$ where ϕ might have a value of, say, .2. Given any such sequence $\{r_k\}$ and a barrier function $B(\mathbf{x})$, the barrier approach to solving (9-36) then consists of solving each of a sequence of unconstrained optimization problems, the kth of which is of the form

$$\text{(9-39)} \qquad \text{Max } C(\mathbf{x}, r_k) \equiv f(\mathbf{x}) + r_k B(\mathbf{x}).$$

Letting $\mathbf{x}_k$ denote the optimal solution to the kth problem, it can be shown—the proof will be supplied below—that the sequence of unconstrained optima $\mathbf{x}_1, \mathbf{x}_2, \ldots$ approaches in the limit the optimal solution of the original unconstrained problem (9-36).

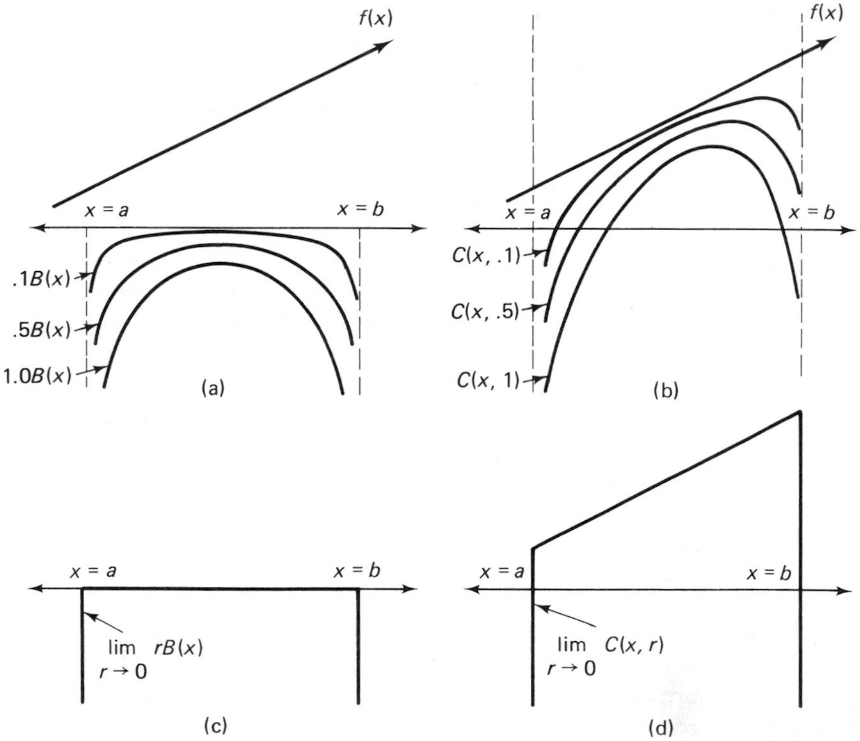

FIGURE 9-7

This somewhat surprising result is illustrated heuristically in Fig. 9-7 for the very simple one-dimensional problem

$$\text{Max } f(x) = \tfrac{1}{2}x$$
$$\text{subject to } a \leqq x \leqq b,$$

where a and b are scalar constants with $0 < a < b$. The constraints may be rewritten as

$$g_1(x) = a - x \leqq 0 \quad \text{and} \quad g_2(x) = x - b \leqq 0,$$

and the barrier function (9-38) becomes

$$B(x) = \frac{1}{a - x} + \frac{1}{x - b}, \qquad \text{defined for } a < x < b.$$

The functions $rB(x)$ for $r = 1.0$, .5, and .1 are diagrammed in Fig. 9-7(a), while the unconstrained maximization functions

$$C(x, r) \equiv f(x) + rB(x)$$

for the same three values of r are shown in Fig. 9-7(b). It is evident that as r decreases toward zero the function $rB(x)$ approaches the "ideal barrier," whose value would be zero at every interior point of the feasible region and $-\infty$ on the boundary, while the unconstrained maximum of $C(x, r)$ approaches $x = b$, the optimal solution to the original problem. The limiting functions

$$\lim_{r \to 0} rB(x) \quad \text{and} \quad \lim_{r \to 0} C(x, r)$$

are displayed in Fig. 9-7(c) and (d).

We now proceed to a more rigorous examination of the general barrier algorithm for solving the nonlinear program

$$\text{(9-36)} \qquad \begin{cases} \text{Max } z = f(\mathbf{x}), & \mathbf{x} = (x_1, \ldots, x_n), \\ \text{subject to } g_i(\mathbf{x}) \leqq 0, & i = 1, \ldots, m. \end{cases}$$

The following assumptions will be made about (9-36):

1. The functions $f(\mathbf{x})$ and $g_i(\mathbf{x})$ are continuous and have continuous first partial derivatives;
2. Either the feasible region S is bounded, or if not, $f(\mathbf{x})$ approaches $-\infty$ whenever the magnitude $|\mathbf{x}|$ is increased without bound along any path within S;

3. The interior S_0 of the feasible region is nonempty; and

4. There exists at least one (global) optimal solution $\mathbf{x}^*$ to (9-36) such that points of S_0 lie within any positive distance of $\mathbf{x}^*$.

Assumption 2 ensures that any function of the form $C(\mathbf{x}, r) \equiv f(\mathbf{x}) + rB(\mathbf{x})$ will be maximized at a finite point in S_0, thus restricting the field of search to within some bounded portion of the feasible region. Assumption 4 guarantees that $\mathbf{x}^*$ can be approached as a limit along a path lying entirely in S_0; note that $\mathbf{x}^*$ may be either a boundary point or an interior point of the feasible region.

Given a problem of the form (9-36), we now define a *barrier function* $B(\mathbf{x})$ to be any scalar-valued function of $\mathbf{x}$ that possesses the following properties:

a. $B(\mathbf{x})$ exists, is continuous, and has continuous first partial derivatives at every point $\mathbf{x}$ in the interior S_0;

b. $B(\mathbf{x}) \leqq 0$ for every $\mathbf{x}$ in S_0; and

c. $B(\mathbf{x})$ decreases toward $-\infty$ as $\mathbf{x}$ approaches the boundary of the feasible region S.

It follows from property c that when the unconstrained function

$$C(\mathbf{x}, r) \equiv f(\mathbf{x}) + rB(\mathbf{x}),$$

where r is any positive scalar parameter, is maximized by means of a direct climbing algorithm, starting from a point in the interior S_0, a finite optimal solution point is obtained that also lies in S_0. Note that, because of the way in which $B(\mathbf{x})$ is defined, the direct climbing procedure can only search over S_0; that is, the function $B(\mathbf{x})$ enforces an *implicit* constraint $\mathbf{x} \in S_0$, although for computational purposes this constraint need not be explicitly stated.

Before outlining the general barrier algorithm and proving that it converges to the optimal solution of (9-36), we must obtain an important intermediate result.

Lemma 9.1: Given a nonlinear program of the form (9-36), *with feasible region* S *and interior* S_0, *let* $B(\mathbf{x})$ *be any suitably defined barrier function for that problem. Further, let* r_1 *and* r_2 *be two scalar parameters satisfying* $r_1 > r_2 \geqq 0$, *and suppose that the function*

$$C(\mathbf{x}, r_k) \equiv f(\mathbf{x}) + r_k B(\mathbf{x})$$

takes on its maximum value over S_0 *at the point* $\mathbf{x}_k$, *for* $k = 1$ *and* $k = 2$. *It then follows that*

$$C(\mathbf{x}_1, r_1) \leqq C(\mathbf{x}_2, r_2), \tag{9-40}$$

$$B(\mathbf{x}_1) \geqq B(\mathbf{x}_2), \tag{9-41}$$

and

$$\mathrm{f}(\mathbf{x}_1) \leqq \mathrm{f}(\mathbf{x}_2). \tag{9-42}$$

The proof is straightforward. Because the function $C(\mathbf{x}, r_2)$ is maximized by $\mathbf{x}_2$, we have

$$f(\mathbf{x}_1) + r_2 B(\mathbf{x}_1) \leqq f(\mathbf{x}_2) + r_2 B(\mathbf{x}_2) \equiv C(\mathbf{x}_2, r_2). \tag{9-43}$$

This inequality will be used in obtaining each of the three conclusions of the lemma. First, using the given relations $r_1 > r_2$ and $B(\mathbf{x}_1) \leqq 0$, we can write

$$C(\mathbf{x}_1, r_1) \equiv f(\mathbf{x}_1) + r_1 B(\mathbf{x}_1) \leqq f(\mathbf{x}_1) + r_2 B(\mathbf{x}_1),$$

and this in conjunction with (9-43) yields (9-40). Next, because $\mathbf{x}_1$ maximizes $C(\mathbf{x}, r_1)$, we have

$$f(\mathbf{x}_2) + r_1 B(\mathbf{x}_2) \leq f(\mathbf{x}_1) + r_1 B(\mathbf{x}_1).$$

By adding this to the inequality (9-43) and manipulating algebraically, we arrive at

$$r_2[B(\mathbf{x}_1) - B(\mathbf{x}_2)] \leqq r_1[B(\mathbf{x}_1) - B(\mathbf{x}_2)].$$

But $r_2 < r_1$, so the difference $B(\mathbf{x}_1) - B(\mathbf{x}_2)$ cannot be negative, and (9-41) follows immediately. Finally, we can write

$$f(\mathbf{x}_1) - f(\mathbf{x}_2) \leqq r_2[B(\mathbf{x}_2) - B(\mathbf{x}_1)] \leqq 0,$$

where the first inequality follows from (9-43) and the second from (9-41) and from the nonnegativity of r_2. Thus we have obtained (9-42) as well, and the lemma is proved.

The *general barrier algorithm* for solving nonlinear programs may now be stated as follows:

1. Given a problem (9-36) satisfying the assumptions listed above, choose a barrier function $B(\mathbf{x})$ and a monotonically decreasing sequence of positive numbers $\{r_k\} \equiv r_1, r_2, \ldots$ such that $\lim_{k\to\infty} r_k = 0$.
2. *First iteration:* Use a direct climbing algorithm, starting at any point $\mathbf{x}_0$ in the interior S_0 of the feasible region of (9-36), to maximize the unconstrained function

$$C(\mathbf{x}, r_1) \equiv f(\mathbf{x}) + r_1 B(\mathbf{x}).$$

The maximum will be taken on at some finite point $\mathbf{x}_1$ in S_0.

3. Continue by repeating step 2; at the kth iteration the starting point is $\mathbf{x}_{k-1}$, the function being maximized is

$$C(\mathbf{x}, r_k) \equiv f(\mathbf{x}) + r_k B(\mathbf{x}),$$

and the maximizing point is $\mathbf{x}_k$ in S_0.

The algorithm is terminated when a solution point $\mathbf{x}_k$ is generated that lies within some preselected small distance of its predecessor $\mathbf{x}_{k-1}$. A procedure for obtaining a starting point $\mathbf{x}_0$ in S_0 when one is not readily available will be given below.

Convergence of the barrier algorithm to an optimal solution of (9-36) will now be established by proving the following theorem.

Theorem 9.1: The sequence of solution points $\{\mathbf{x}_k\} \equiv \mathbf{x}_0, \mathbf{x}_1, \mathbf{x}_2, \ldots$ *generated by the barrier algorithm as described above converges to a limit point* $\mathbf{x}^*$ *that is an optimal solution to the given problem* (9-36).

By extension of Lemma 9.1 it is easily established that $\{f(\mathbf{x}_k)\} \equiv f(\mathbf{x}_1), f(\mathbf{x}_2), \ldots$ is a monotonically nondecreasing sequence. Inasmuch as each of the $\mathbf{x}_k$ is a feasible solution to (9-36), if the sequence diverges to infinity, the original problem has an unbounded optimum, which, according to the second of our initial four assumptions about (9-36), must lie at a finite point in the feasible region S. Otherwise, $\{f(\mathbf{x}_k)\}$ must converge to a finite limiting value, and it follows that the sequence $\{\mathbf{x}_k\}$ must approach a finite limit point,[4] which we denote $\mathbf{x}^*$.

To prove that $\mathbf{x}^*$ is an optimal solution to (9-36), we proceed by contradiction. It has been assumed that there exists an optimal solution, call it $\hat{\mathbf{x}}$, such that points in the interior S_0 are arbitrarily close to $\hat{\mathbf{x}}$. Suppose $\mathbf{x}^*$ is not optimal; that is, suppose $f(\mathbf{x}^*) < f(\hat{\mathbf{x}})$. Then, because $f(\mathbf{x})$ is a continuous function, we can choose a point $\mathbf{y}$ in S_0 so close to $\hat{\mathbf{x}}$ that

$$f(\mathbf{x}^*) < f(\mathbf{y}) \leqq f(\hat{\mathbf{x}}).$$

Because $\mathbf{y}$ is in S_0, $B(\mathbf{y})$ has a finite negative value. Inasmuch as $\{r_k\}$ converges to 0, we can choose a value of k, say $k = K$, that is large enough to satisfy

$$C(\mathbf{y}, r_K) \equiv f(\mathbf{y}) + r_K B(\mathbf{y}) > f(\mathbf{x}^*).$$

[4]This conclusion is not rigorously true, nor is the theorem stated quite correctly. There exists the theoretical possibility that the sequence $\{\mathbf{x}_k\}$ may, after a finite number of terms, begin to oscillate between two or more solution points $\mathbf{x}', \mathbf{x}'', \ldots$ for which $f(\mathbf{x}') = f(\mathbf{x}'') = \ldots$. In such a case each of these points would be an optimal solution to (9-36); however, it would then be true only that a subsequence of $\{\mathbf{x}_k\}$, rather than the full sequence itself, converged to a limit point. To deal rigorously with this possibility would require the use of compact set theory, which is not a stated prerequisite for this text.

But $C(\mathbf{x}, r_K)$ is maximized by the solution point $\mathbf{x}_K$, so

$$C(\mathbf{x}_K, r_K) \equiv f(\mathbf{x}_K) + r_K B(\mathbf{x}_K) \geqq C(\mathbf{y}, r_K),$$

and because the product $r_K B(\mathbf{x}_K)$ has a nonpositive value, it follows that

$$f(\mathbf{x}_K) \geqq C(\mathbf{y}, r_K) > f(\mathbf{x}^*).$$

But we know from Lemma 9.1 that $\{f(\mathbf{x}_k)\}$ is a nondecreasing sequence; hence the limiting term cannot have a smaller value than any one of the earlier terms, and a contradiction has been obtained. The original supposition was therefore false, and $\mathbf{x}^*$ must indeed be an optimal solution to (9-36), as was to be shown.

It has thus been established that the sequence $\{\mathbf{x}_k\}$ generated by the barrier algorithm converges to the optimal solution $\mathbf{x}^*$ of the original problem (9-36), which implies that $\{f(\mathbf{x}_k)\}$ approaches $f(\mathbf{x}^*)$ as a limit. We shall now prove the further result that the sequence of nonpositive values $\{r_k B(\mathbf{x}_k)\}$ generated by the same process converges to zero as k becomes infinitely large. This result, interesting enough in its own right, will be used in Section 9.9 to help establish a useful computational property of the barrier algorithm that holds under certain frequently satisfied conditions.

Theorem 9.2: Unless unboundedness is discovered at some stage, the sequence of values $\{r_k B(\mathbf{x}_k)\}$ *generated by the barrier algorithm converges to a limiting value of zero.*

For convenience we define

$$Q \equiv \lim_{k\to\infty} r_k B(\mathbf{x}_k) \leqq 0,$$

observing that because r_k is nonnegative and $B(\mathbf{x}_k)$ nonpositive for all k, the limiting value Q must be nonpositive. Because $C(\mathbf{x}_k, r_k) \equiv f(\mathbf{x}_k) + r_k B(\mathbf{x}_k)$ is defined for all $k \geqq 1$, we can take the limit on both sides to obtain

$$\lim_{k\to\infty} C(\mathbf{x}_k, r_k) = f(\mathbf{x}^*) + Q, \tag{9-44}$$

where $\mathbf{x}^*$ is an optimal solution to the original problem (we exclude the possibility of unboundedness). From Lemma 9.1 the terms $C(\mathbf{x}_k, r_k)$ are known to form a nondecreasing sequence; inasmuch as $C(\mathbf{x}_1, r_1)$ is finite, the limiting term cannot be negatively infinite and, from (9-44), neither can Q. Thus either $Q = 0$ or Q has a finite negative value.

Assume the latter, and select a point $\mathbf{y}$ in the interior S_0 that is so close to $\mathbf{x}^*$ that

$$f(\mathbf{y}) \geqq f(\mathbf{x}^*) + \frac{Q}{2}$$

(this is the same sort of maneuver that was used in the proof of the previous theorem). Because $\mathbf{x}_k$ maximizes $C(\mathbf{x}, r_k)$, we may write

$$C(\mathbf{x}_k, r_k) \geqq C(\mathbf{y}, r_k) \equiv f(\mathbf{y}) + r_k B(\mathbf{y}) \geqq f(\mathbf{x}^*) + \frac{Q}{2} + r_k B(\mathbf{y}) \tag{9-45}$$

for any $k \geqq 1$. Now, by choosing $\mathbf{y}$ in the interior we have ensured that $B(\mathbf{y})$ has a finite nonpositive value. If $B(\mathbf{y}) = 0$, the inequality (9-45) reduces immediately to

$$C(\mathbf{x}_k, r_k) \geqq f(\mathbf{x}^*) + \frac{Q}{2} > f(\mathbf{x}^*) + Q,$$

which must hold for, say, $k = 1$. But $\{C(\mathbf{x}_k, r_k)\}$ is a nondecreasing sequence, so

$$\lim_{k \to \infty} C(\mathbf{x}_k, r_k) > f(\mathbf{x}^*) + Q,$$

which contradicts (9-44). If, on the other hand, $B(\mathbf{y}) < 0$, let a value of k, say $k = K$, be chosen that is large enough to satisfy

$$r_K < \frac{Q}{2B(\mathbf{y})}. \tag{9-46}$$

This is possible because the sequence $\{r_k\}$ converges to zero. Noting that (9-46) implies $r_K B(\mathbf{y}) > Q/2$, we can extend the inequality chain (9-45) for $k = K$ as follows:

$$C(\mathbf{x}_K, r_K) \geqq f(\mathbf{x}^*) + \frac{Q}{2} + r_K B(\mathbf{y}) > f(\mathbf{x}^*) + Q.$$

We have again obtained a contradiction of (9-44); therefore the assumption that $Q < 0$ must have been false, and we conclude that

$$Q \equiv \lim_{k \to \infty} r_k B(\mathbf{x}_k) = 0.$$

Of course, it then follows from (9-44) that

$$\lim_{k \to \infty} C(\mathbf{x}_k, r_k) = f(\mathbf{x}^*).$$

9.8 COMPUTATIONAL ASPECTS OF THE BARRIER ALGORITHM

The barrier approach to solving nonlinear programs was first proposed by Carroll [6] in 1961. He conjectured that when the scalar parameter r is allowed to decrease toward a limiting value of zero the optimal solution to the unconstrained problem

$$\text{Max } f(\mathbf{x}) + r\sum_{i=1}^{m} \frac{1}{g_i(\mathbf{x})},$$

as determined via direct climbing from a starting point in the interior of the feasible region, approaches the optimal solution of the original constrained problem (9-36). Proof of this conjecture was supplied three years later by Fiacco and McCormick [13], who then went on to study barrier methods in great detail, obtaining several interesting theoretical results and documenting a fair amount of computational experience [14, 15]. It should be pointed out that, ever since Carroll's original article appeared, by far the most important and widely used barrier function has been $\sum_i [1/g_i(\mathbf{x})]$. However, many other legitimate barrier functions exist that possess all the properties required by Lemma 9.1 and Theorem 9.1. Most prominent among them is the family of functions

$$B(\mathbf{x}) = \sum_{i=1}^{m} \frac{-1}{[-g_i(\mathbf{x})]^{\epsilon}},$$

where the parameter ϵ can be assigned any positive value; note that Carroll's original function is a member of this family (for which $\epsilon = 1$).

It is important to observe that convergence of the general barrier algorithm to an optimal solution of (9-36) can be guaranteed, via Lemma 9.1 and Theorem 9.1, only if the *global* maximum of the function $C(\mathbf{x}, r_k)$ within the interior of the feasible region is obtained at every stage. When (9-36) is a convex program, the unconstrained maximization functions $C(\mathbf{x}, r_k) \equiv f(\mathbf{x}) + r_k B(\mathbf{x})$ are concave for all commonly used barrier functions—this will be proved for the function $B(\mathbf{x}) = \sum_i [1/g_i(\mathbf{x})]$ in the next section—and thus cannot have suboptimal local maxima. The first local maximum obtained via direct climbing for any such function must therefore be the global maximum. When (9-36) is *not* convex, however, repeated applications of a direct climbing procedure from several starting points would presumably be necessary in order to identify, with a high degree of confidence, the global maximum of any function $C(\mathbf{x}, r_k)$. This would obviously become quite tedious for any problem requiring more than a very few iterations of the barrier algorithm. It is more efficient in practice to obtain a single local maximum of each function $C(\mathbf{x}, r_k)$, producing a sequence $\{\mathbf{x}_k\}$ that eventually converges to a locally optimal solution of (9-36), and then to repeat the overall procedure several times from different starting points within the feasible region. In this way the analyst can discover all or most of the local optima of $f(\mathbf{x})$. Nevertheless, as the foregoing considerations suggest, barrier algorithms can in general deal far more easily and efficiently with convex programming problems than with nonconvex programs, particularly because the former have certain additional theoretical properties that can be exploited computationally; these properties will be derived and discussed in Section 9.9.

The general barrier algorithm has other interesting and important computational aspects that should be mentioned here. One matter of primary importance is the choice of an unconstrained optimization procedure to be used in maximizing the function $C(\mathbf{x}, r_k) \equiv f(\mathbf{x}) + r_k B(\mathbf{x})$ at each stage. As r decreases toward zero, the function $rB(\mathbf{x})$, where $B(\mathbf{x})$ is any properly defined barrier function, begins to resemble the "ideal barrier" diagrammed (for a one-variable problem) in Fig. 9-7(c). Observe in particular that the first partial derivatives of $rB(\mathbf{x})$ change extremely rapidly—that is, the second partial derivatives become quite large in magnitude—as the boundary of the feasible region is approached. Thus, whenever the optimal solution of the original problem lies on the boundary, as will usually be the case, direct climbing via steepest ascent, which is based on a first-order Taylor approximation to the function being optimized, is extremely inaccurate and inefficient and yields a very slow rate of convergence. Second-order direct climbing methods, such as that of Newton, are appreciably superior in this respect and should normally be used instead. Note that, inasmuch as the starting solution $\mathbf{x}_{k-1}$ will already be in the vicinity of the point $\mathbf{x}_k$ that maximizes $C(\mathbf{x}, r_k)$, for all $k > 1$, overall convergence of the second-order climbing procedure at each stage is assured.

In addition to the tactical matter of deciding between first- and second-order methods for maximizing the functions $C(\mathbf{x}, r_k)$, there is also the strategic question of choosing the successive values of the parameters r_k. Because the barrier algorithm requires only that the sequence $\{r_k\}$ converge to zero, it might seem attractive to begin with a rather small value for r_1, in the hope that the problem can be solved in only two or three iterations. This strategy can in fact be quite successful and is not infrequently employed in its most extreme form, in which a very small value of r_1 is chosen and the resulting unconstrained maximum $\mathbf{x}_1$ is taken immediately as an approximation to the optimal solution of the given problem. One difficulty with this approach, however, is that there is no way of knowing just how small a value of r_1 is needed to achieve a desired level of accuracy in the solution; if the value chosen is not small enough, the error in the solution $\mathbf{x}_1$ may be unacceptably large. A more serious difficulty is that the search for the approximate solution $\mathbf{x}_1$ via direct climbing from a remote starting point $\mathbf{x}_0$ is likely to be extremely laborious (assuming $\mathbf{x}_1$ is close to the boundary of the feasible region). The steepness of the function $C(\mathbf{x}, r_1)$ in the vicinity of $\mathbf{x}_1$ requires that some version of Newton's method be used, except perhaps for the first few iterations, and this in turn requires frequent inversion of the Hessian matrix associated with the function C. Moreover, because the second partial derivatives $\delta^2 C/\delta x_i\, \delta x_j$ themselves change rapidly in the vicinity of $\mathbf{x}_1$, the displacement directions chosen via second-order approximation are still not particularly accurate, so that a great many iterations of Newton's method are likely to be needed. It can be concluded, in summary, that the question of whether the desired optimal solution ought to be approached gradually, via sequential optimization of a number of functions $C(\mathbf{x}, r_k)$, or abruptly, in one or two bold steps, cannot be given a single general answer; either strategy might be superior for any given problem.

We shall conclude this section by describing a procedure based on the barrier algorithm for obtaining an initial feasible solution to the nonlinear program

$$\text{Max } z = f(\mathbf{x})$$
$$\text{subject to } g_i(\mathbf{x}) \leqq 0, \qquad i = 1, \ldots, m.$$

Given any point $\mathbf{x}'$ in E^n, let S' be the set of subscripts of all constraints that are strictly satisfied at $\mathbf{x}'$:

$$S' \equiv \{i \mid g_i(\mathbf{x}') < 0\}.$$

Choose any constraint $g_u(\mathbf{x}) \leqq 0$ that is *not* strictly satisfied at $\mathbf{x}'$—that is, choose any subscript u not in S'—and begin solving the following subproblem via the barrier algorithm:

$$\text{Max } -g_u(\mathbf{x})$$
$$\text{subject to } g_i(\mathbf{x}) \leqq 0, \qquad i \in S',$$

where the point $\mathbf{x}'$ can be used as a starting feasible solution. Stop as soon as a point $\mathbf{x}''$ is reached at which the value of the objective function is positive: $\mathbf{x}''$ will then strictly satisfy the uth constraint as well as all those constraints that were included in S'. For the next step of the procedure, define another set whose elements are the subscripts of all constraints that are strictly satisfied at the new point $\mathbf{x}''$:

$$S'' \equiv \{i \mid g_i(\mathbf{x}'') < 0\}.$$

The set S'' contains, at a minimum, the subscript u and all the elements of S'; in addition, it includes the subscript of any other constraint that happens by good luck to be strictly satisfied at $\mathbf{x}''$. Now select any subscript not in S'' and repeat the procedure outlined above, continuing in this manner until a point $\mathbf{x}_0$ in the interior of the original feasible region is obtained. If at any stage the optimal solution to the subproblem is found to have a nonpositive objective value, it follows immediately that there can exist no point $\mathbf{x}$ satisfying $g_i(\mathbf{x}) < 0$, $i = 1, \ldots, m$, and the procedure is terminated. Note that termination implies either

1. That the original problem has no feasible solution, or
2. That the feasible region is nonempty but has no interior.

In the latter case, of course, the barrier algorithm cannot be applied.

9.9 CONVEX PROGRAMS, DUALITY, AND THE BARRIER ALGORITHM

In Section 9.7 we developed the general barrier algorithm for solving mathematical programming problems of the form

$$\text{(9-36)} \qquad \begin{cases} \text{Max } z = f(\mathbf{x}), & \mathbf{x} = (x_1, \ldots, x_n), \\ \text{subject to } g_i(\mathbf{x}) \leqq 0, & i = 1, \ldots, m, \end{cases}$$

where it is required that the interior S_0 of the feasible region be nonempty [i.e., there must exist at least one point $\hat{\mathbf{x}}$ satisfying $g_i(\hat{\mathbf{x}}) < 0$ for all i]. Convergence of the algorithm was proved for any problem of the form (9-36), regardless of its convexity/concavity characteristics, and for any barrier function $B(\mathbf{x})$ having the following properties:

1. $B(\mathbf{x})$ exists, is continuous, and has continuous first partial derivatives at every point $\mathbf{x}$ in S_0;
2. $B(\mathbf{x}) \leqq 0$ for all $\mathbf{x}$ in S_0; and
3. $B(\mathbf{x})$ decreases toward $-\infty$ as $\mathbf{x}$ approaches the boundary of the feasible region.

It was later observed that the barrier approach is computationally more efficient when (9-36) is a convex program than when it is not.

In this section we continue to discuss the barrier algorithm but narrow our focus in two important ways:

1. It will be assumed that the function $f(\mathbf{x})$ is concave and that the functions $g_i(\mathbf{x})$, $i = 1, \ldots, m$, are convex, so that (9-36) is a convex programming problem; and
2. Attention will be restricted throughout to the barrier function

$$\text{(9-47)} \qquad B(\mathbf{x}) \equiv \sum_{i=1}^{m} \frac{1}{g_i(\mathbf{x})}.$$

Inasmuch as (9-47) has the virtue of simplicity as well as several other convenient properties, it is not surprising that it should be the best-known and most frequently used of all barrier functions.

One such convenient property, which it shares with many other barrier functions, is that it "preserves concavity": Given the convexity/concavity characteristics assumed above, the unconstrained maximization function $C(\mathbf{x}, r) \equiv f(\mathbf{x}) + rB(\mathbf{x})$, where $B(\mathbf{x})$ is defined by (9-47), is concave over the interior S_0 of the feasible region for any positive value of r. This of course implies that none of the functions $C(\mathbf{x}, r_k)$ constructed at the various stages of the barrier algorithm can have extraneous local optima; thus each can be maximized with a single application of whatever direct climbing technique is being used. In proving the concavity of $C(\mathbf{x}, r)$, the crucial step is to show that if $g(\mathbf{x})$ is a convex function having a negative value at every point $\mathbf{x}$ throughout some convex region S_0, then the function $h(\mathbf{x}) \equiv 1/g(\mathbf{x})$ is concave over S_0. Let $\mathbf{x}_1$ and $\mathbf{x}_2$ be two points in S_0 and choose any value of λ satisfying $0 \leqq \lambda \leqq 1$. We first observe that because $\lambda - \lambda^2$ cannot have a negative value, the following inequality must hold:

$$0 \leqq (\lambda - \lambda^2)[g(\mathbf{x}_1) - g(\mathbf{x}_2)]^2.$$

Algebraic manipulation then leads to

$$g(\mathbf{x}_1)g(\mathbf{x}_2) \leqq \lambda^2 g(\mathbf{x}_1)g(\mathbf{x}_2) + (\lambda - \lambda^2)g^2(\mathbf{x}_1) + (\lambda - \lambda^2)g^2(\mathbf{x}_2) + (1 - \lambda)^2 g(\mathbf{x}_1)g(\mathbf{x}_2),$$

which is equivalent to

$$g(\mathbf{x}_1)g(\mathbf{x}_2) \leqq [\lambda g(\mathbf{x}_2) + (1 - \lambda)g(\mathbf{x}_1)][\lambda g(\mathbf{x}_1) + (1 - \lambda)g(\mathbf{x}_2)].$$

Dividing this inequality on both sides by the positive quantity $g(\mathbf{x}_1)g(\mathbf{x}_2)$ and by the negative quantity $\lambda g(\mathbf{x}_1) + (1 - \lambda)g(\mathbf{x}_2)$, we arrive at

$$\text{(9-48)} \qquad \frac{\lambda g(\mathbf{x}_2) + (1 - \lambda)g(\mathbf{x}_1)}{g(\mathbf{x}_1)g(\mathbf{x}_2)} \leqq \frac{1}{\lambda g(\mathbf{x}_1) + (1 - \lambda)g(\mathbf{x}_2)}.$$

But the left-hand side of (9-48) is equivalent to

$$\frac{\lambda}{g(\mathbf{x}_1)} + \frac{1 - \lambda}{g(\mathbf{x}_2)} \equiv \lambda h(\mathbf{x}_1) + (1 - \lambda)h(\mathbf{x}_2),$$

while the right-hand side leads to a further inequality

$$\frac{1}{\lambda g(\mathbf{x}_1) + (1 - \lambda)g(\mathbf{x}_2)} \leqq \frac{1}{g(\lambda\mathbf{x}_1 + (1 - \lambda)\mathbf{x}_2)} \equiv h(\lambda\mathbf{x}_1 + (1 - \lambda)\mathbf{x}_2),$$

where we have used the fact that $g(\mathbf{x})$ is convex. We have thus shown that

$$h(\lambda\mathbf{x}_1 + (1 - \lambda)\mathbf{x}_2) \geqq \lambda h(\mathbf{x}_1) + (1 - \lambda)h(\mathbf{x}_2),$$

which implies that $h(\mathbf{x})$ is concave over S_0. By applying the familiar convexity/concavity theorems of Chapter 3, we can then conclude successively that $B(\mathbf{x})$ is concave, $rB(\mathbf{x})$ is concave for any $r > 0$, and $C(\mathbf{x}, r) \equiv f(\mathbf{x}) + rB(\mathbf{x})$ is concave as well. Q.E.D. A more general proof of concavity for the family of barrier functions

$$B(\mathbf{x}) = \sum_{i=1}^{m} \frac{-1}{[-g_i(\mathbf{x})]^{\epsilon}}$$

is developed in Exercise 9-26.

Another important property of the barrier function (9-47), when used in solving a convex nonlinear program (9-36), is that it permits an upper bound to be placed at each iteration of the algorithm on the optimal value f^* of the objective function $f(\mathbf{x})$. Moreover, as we shall see, this upper bound decreases toward f^* with each succeeding iteration and approaches f^* in the limit as r_k approaches zero. But $f(\mathbf{x}_k)$, the objective value at the current (feasible) point, constitutes an obvious lower bound on f^* at any stage, a lower bound that increases toward f^* as k becomes infinite. Thus the difference between the two bounds shrinks toward zero as the algorithm proceeds, providing an updated measure of how close the current point $\mathbf{x}_k$ is to the optimal solution and how much further effort is likely to be required to obtain it.

To establish this upper bound on f^*, we need two or three additional theoretical results. We begin by defining a dual problem for the original convex program: Given a *primal* nonlinear program

$$\text{(9-36)} \qquad \begin{cases} \text{Max } z = f(\mathbf{x}), & \mathbf{x} = (x_1, \ldots, x_n), \\ \text{subject to } g_i(\mathbf{x}) \leqq 0, & i = 1, \ldots, m, \end{cases}$$

where $f(\mathbf{x})$ is concave and the $g_i(\mathbf{x})$, $i = 1, \ldots, m$, are convex, the *dual* problem is

$$\text{(9-49)} \qquad \begin{cases} \text{Min } z' = f(\mathbf{x}) - \sum_{i=1}^{m} u_i g_i(\mathbf{x}) \\ \text{subject to } \nabla \mathbf{f}(\mathbf{x}) = \sum_{i=1}^{m} u_i \, \nabla \mathbf{g}_i(\mathbf{x}) \\ \text{and} \qquad \mathbf{u} \geqq \mathbf{0}, \end{cases}$$

where $\mathbf{u} = (u_1, \ldots, u_m)$ are the *dual variables*. In general, (9-49) is a problem in the $n + m$ variables $\mathbf{x}$ *and* $\mathbf{u}$; however, note that if a particular value of $\mathbf{x}$ is specified, it then reduces to a linear program in the dual variables $\mathbf{u}$. The problem (9-49) was originally defined and studied by Wolfe [53] in 1961. It is in fact one of several nonlinear programming "duals" that have been defined with respect to primal problems of various types during the past 15 years. For example, another, more general dual that can be written for any primal problem, convex or not, is discussed in Chapter 2 of Zangwill's text [60].

The following two lemmas, based on the dual problem (9-49), are also due to Wolfe.

Lemma 9.2: If $\mathbf{x}^*$ *and* $(\mathbf{x}_0, \mathbf{u}_0)$ *are any feasible solutions to the primal* (9-36) *and dual* (9-49), *respectively, then the primal objective value* $f(\mathbf{x}^*)$ *cannot exceed the dual objective value* $f(\mathbf{x}_0) - \sum_i u_{0i} g_i(\mathbf{x}_0)$. *That is, any feasible value of the dual objective can be taken as an upper bound on the optimal value of the primal objective.*

The result is obtained via a chain of inequalities, as follows:

$$f(\mathbf{x}^*) - f(\mathbf{x}_0) \leqq \nabla \mathbf{f}^T(\mathbf{x}_0) \cdot (\mathbf{x}^* - \mathbf{x}_0) = \sum_{i=1}^{m} u_{0i}[\nabla \mathbf{g}_i^T(\mathbf{x}_0) \cdot (\mathbf{x}^* - \mathbf{x}_0)]$$
$$\leqq \sum_{i=1}^{m} u_{0i}[g_i(\mathbf{x}^*) - g_i(\mathbf{x}_0)] \leqq -\sum_{i=1}^{m} u_{0i} g_i(\mathbf{x}_0),$$

where the first inequality follows from the concavity of the function $f(\mathbf{x})$, using Theorem 3.9; the equality holds because $(\mathbf{x}_0, \mathbf{u}_0)$ satisfies the dual constraints; the second inequality follows from the convexity of the $g_i(\mathbf{x})$ and the nonnegativity of the u_{0i}; and the final inequality is due jointly to the primal constraints $g_i(\mathbf{x}^*) \leqq 0$ and the nonnegativity of the u_{0i}. The expressions at the ends of the chain can then be combined to yield

$$f(\mathbf{x}^*) \leqq f(\mathbf{x}_0) - \sum_{i=1}^{m} u_{0i} g_i(\mathbf{x}_0),$$

as was to be shown.

Lemma 9.3: If $\mathbf{x}^*$ *is a (finite) optimal solution to the primal problem, then there exists a set of values* $\boldsymbol{\lambda}^* = (\lambda_1^*, \ldots, \lambda_m^*)$ *such that* $(\mathbf{x}^*, \boldsymbol{\lambda}^*)$ *constitutes an optimal solution to the dual having an objective value of* $f(\mathbf{x}^*)$.

The proof simply requires an application of the Kuhn-Tucker conditions, which necessarily hold at the primal optimum $\mathbf{x}^*$: From Theorem 6.3, there must exist $\boldsymbol{\lambda}^* = (\lambda_1^*, \ldots, \lambda_m^*)$ satisfying

$$\nabla \mathbf{f}(\mathbf{x}^*) = \sum_{i=1}^{m} \lambda_i^* \nabla \mathbf{g}_i(\mathbf{x}^*),$$
$$\lambda_i^* g_i(\mathbf{x}^*) = 0, \qquad i = 1, \ldots, m,$$

and

$$\lambda_i^* \geqq 0, \qquad i = 1, \ldots, m.$$

It follows immediately from these relations that $(\mathbf{x}^*, \boldsymbol{\lambda}^*)$ is a feasible solution to the dual problem (9-49) and that its objective value is $f(\mathbf{x}^*)$. Moreover, from Lemma 9.2 there cannot exist a feasible solution to the dual having an objective value less than $f(\mathbf{x}^*)$; hence $(\mathbf{x}^*, \boldsymbol{\lambda}^*)$ optimizes the dual problem. Q.E.D.

It is interesting to observe how closely these two lemmas parallel the first duality lemma and duality theorem of linear programming (discussed in Section 2.6). In fact, the properties just established are what justifies calling (9-49) a "dual" problem; other nonlinear programming duals that have been proposed and investigated have similar or equivalent properties.

Having obtained these theoretical results, let us return to our consideration of the barrier algorithm. During the kth iteration of the algorithm the following unconstrained optimization problem is solved:

$$\text{Max } C(\mathbf{x}, r_k) \equiv f(\mathbf{x}) + r_k B(\mathbf{x}) = f(\mathbf{x}) + \sum_{i=1}^{m} \frac{r_k}{g_i(\mathbf{x})}.$$

The point $\mathbf{x}_k$ that (approximately) maximizes this function is known to be a feasible solution to the primal problem (9-36); in fact, it lies in the interior of the feasible region, so that $g_i(\mathbf{x}_k) < 0$ for all i. As has been noted, $f(\mathbf{x}_k)$ constitutes a lower bound on the optimal value f^* of the primal objective function $f(\mathbf{x})$. One way of obtaining an *upper* bound on f^* would be to substitute $\mathbf{x}_k$ into the dual problem (9-49) and solve the resulting linear program in the variables $\mathbf{u}$. If an optimal feasible solution $(\mathbf{x}_k, \mathbf{u}_k)$ could be found, its objective value

$$f(\mathbf{x}_k) - \sum_{i=1}^{m} u_{ki} g_i(\mathbf{x}_k)$$

could be taken, in view of Lemma 9.2, as an upper bound on f^*; this would in fact be the least upper bound based on, or associated with, $\mathbf{x}_k$.

Fortunately, it is also possible, after the kth iteration, to determine an upper bound on f^* without solving a linear program. Inasmuch as $\mathbf{x}_k$ is an unconstrained local maximum of $C(\mathbf{x}, r_k)$, the partial derivatives of that function must vanish at $\mathbf{x}_k$:

$$0 = \frac{\delta C(\mathbf{x}_k, r_k)}{\delta x_j} = \frac{\delta f(\mathbf{x}_k)}{\delta x_j} - \sum_{i=1}^{m} \frac{r_k}{[g_i(\mathbf{x}_k)]^2} \cdot \frac{\delta g_i(\mathbf{x}_k)}{\delta x_j}, \qquad j = 1, \ldots, n, \tag{9-50}$$

Observe now that the set of values

$$\mathbf{x} = \mathbf{x}_k, \qquad u_i = \frac{r_k}{[g_i(\mathbf{x}_k)]^2}, \qquad i = 1, \ldots, m, \tag{9-51}$$

constitutes a feasible solution to the dual problem (9-49); this follows from (9-50) and from the nonnegativity of r_k. The objective value of the dual-feasible solution (9-51) is

$$f(\mathbf{x}_k) - \sum_{i=1}^{m} \frac{r_k}{g_i(\mathbf{x}_k)},$$

which must be greater than or equal to the optimal (minimal) value of the dual objective. But Lemma 9.3 guarantees that the optimal values of the primal and dual objectives are equal, and we may therefore write

$$f(\mathbf{x}_k) \leqq f^* \leqq f(\mathbf{x}_k) - \sum_{i=1}^{m} \frac{r_k}{g_i(\mathbf{x}_k)} = f(\mathbf{x}_k) - r_k B(\mathbf{x}_k). \tag{9-52}$$

Thus the upper bound on f^* can be easily computed at any stage as the (algebraic) difference between, rather than the sum of, the two terms comprising the unconstrained maximization function $C(\mathbf{x}_k, r_k)$.

The bounds on f^* given by (9-52) hold for any k; in particular, they continue to hold as k becomes very large. But as k approaches infinity and the sequence of solution points $\{\mathbf{x}_k\}$ converges to the optimal solution $\mathbf{x}^*$ of the primal problem, we know from Theorem 9.2 that the terms $r_k B(\mathbf{x}_k) < 0$ approach a limiting value of zero. Thus, as the barrier algorithm proceeds, the upper and lower bounds on f^* both converge toward $f(\mathbf{x}^*) = f^*$ itself. At any iteration the value $-r_k B(\mathbf{x}_k)$, which is equal to the difference between the two bounds, can be taken as a measure of how close $\mathbf{x}_k$ is to the optimal solution of the original problem. This measure can in fact be used by an analyst in deciding when to terminate the algorithm, or as an automatic stopping rule in a computer code [e.g., stop when $-r_k B(\mathbf{x}_k) < .0001$].

The upper-bounding procedure just outlined can be applied in a somewhat different way when the barrier approach is being used to seek a starting feasible solution to the convex program (9-36). Recall from the end of Section 9.8 that at each step the barrier algorithm is applied to a problem of the form

$$\begin{aligned} &\text{Max } -g_u(\mathbf{x}) \\ &\text{subject to } g_i(\mathbf{x}) \leqq 0, \qquad i \in S', \end{aligned}$$

using a starting point at which all constraints included in the index set S' are strictly satisfied but the constraint $g_u(\mathbf{x}) \leqq 0$ is not. Computation proceeds until a point $\hat{\mathbf{x}}$ is obtained at which $-g_u(\hat{\mathbf{x}}) > 0$; the subscript u is then added to S', another unsatisfied constraint is chosen, and the process is repeated. If at any stage the maximum value of the objective $-g_u(\mathbf{x})$ is found to be nonpositive, it can be concluded that the original problem has no feasible solution, or at least no *interior* feasible solution. Although this overall procedure is satisfactory, the diagnosis of infeasibility can in fact be arrived at more quickly, when it is appropriate, by making use of the upper-bounding technique described above. When, in maximizing any of the objectives $-g_u(\mathbf{x})$, the upper bound is found at some stage to be negative or zero, the conclusion of infeasibility can be drawn immediately; this eliminates the need to toil on with the computations until the point that actually maximizes $-g_u(\mathbf{x})$ is finally reached.

9.10 PENALTY METHODS

The second major class of sequential unconstrained optimization procedures comprises the various *penalty methods*. These methods, like those using the barrier approach, are built around a common basic strategy in which an auxiliary function—the "penalty"—is formed from the constraints, multiplied by a scalar parameter, and added to the objective to produce a single unconstrained parametric function. This function is then optimized repeatedly for each of a sequence of parameter values, generating a sequence of unconstrained optima that converges to the solution of the original constrained problem. The various penalty methods in use today differ among themselves only in the form of the penalty function and in certain other computational details. Although computational approaches based on the penalty strategy can be traced as far back as the early 1940s, the major algorithmic results that will concern us here were first published by Zangwill [57] in 1967.

Zangwill's *method of penalties*, like the barrier algorithm, is applicable to nonlinear programs of the following from:

$$\text{(9-36)} \qquad \begin{cases} \text{Max } z = f(\mathbf{x}), & \mathbf{x} = (x_1, \ldots, x_n), \\ \text{subject to } g_i(\mathbf{x}) \leqq 0, & i = 1, \ldots, m, \end{cases}$$

where the functions $f(\mathbf{x})$ and $g_i(\mathbf{x})$ are continuous and have continuous first partial derivatives. In addition, for the penalty method to operate it is necessary that the optimal solution to (9-36) occur at a finite point having a finite objective value; the reason for this will be made clear below. Note that while this implies the existence of one or more feasible points, the feasible region is *not* required to have an interior. It follows that equality-constrained problems can be solved via the penalty method by first converting each constraint equation into a pair of inequalities satisfying the format of (9-36). Actually, the penalty method to be presented below can be modified to enable it to deal with equality constraints directly; this topic will be explored in Exercise 9-31.

As before, we let S denote the feasible region of the problem (9-36):

$$S \equiv \{\mathbf{x} \mid g_i(\mathbf{x}) \leqq 0, \quad i = 1, \ldots, m\}.$$

The characteristic *penalties approach* to solving (9-36) consists of replacing it by a sequence of unconstrained optimization problems of the form

$$\text{(9-53)} \qquad \text{Max } D(\mathbf{x}, r) \equiv f(\mathbf{x}) + \frac{1}{r}P(\mathbf{x}),$$

where r is a positive-valued scalar parameter and $P(\mathbf{x})$ is a *penalty function* whose value is zero at every point in S and negative at all other points in E^n. A typical example of a penalty function, and the one most frequently used in practice, is

$$\text{(9-54)} \qquad P(\mathbf{x}) = -\sum_{i=1}^{m} [\max\{g_i(\mathbf{x}), 0\}]^2.$$

Note that the ith constraint contributes a nonzero penalty term if and only if $g_i(\mathbf{x}) > 0$—that is, if and only if the ith constraint is unsatisfied. To obtain the optimal solution to (9-36), a sequence of unconstrained optimization problems of the form (9-53) is solved, with the parameter r being assigned successively smaller positive values $\{r_k\} = r_1, r_2, \ldots$. As r_k approaches zero, the value of the penalty $P(\mathbf{x})/r_k$ remains at zero throughout the feasible region but becomes more and more unfavorable, approaching $-\infty$, at any given infeasible point. Typically, the unconstrained maximum of $D(\mathbf{x}, r_k)$, denoted $\mathbf{x}_k$, may lie well outside the feasible region S during the early stages, when r_k is large and the penalty relatively small, but as r_k decreases toward zero the penalty term becomes more significant and the solution points $\mathbf{x}_k$ are forced closer and closer to the feasible region, actually reaching it in the limit (if not earlier).

As an example, consider the simple one-dimensional problem used to illustrate the barrier algorithm in Section 9.7:

$$\text{Max } f(x) = \tfrac{1}{2}x$$
$$\text{subject to } a \leqq x \leqq b,$$

where a and b are scalar constants. The penalty function (9-54) becomes

$$P(x) = -[\max\{a - x, 0\}]^2 - [\max\{x - b, 0\}]^2,$$

which is equivalent to

$$P(x) = \begin{cases} -(a - x)^2 & \text{if } x \leqq a, \\ 0 & \text{if } a \leqq x \leqq b, \\ -(x - b)^2 & \text{if } x \geqq b. \end{cases}$$

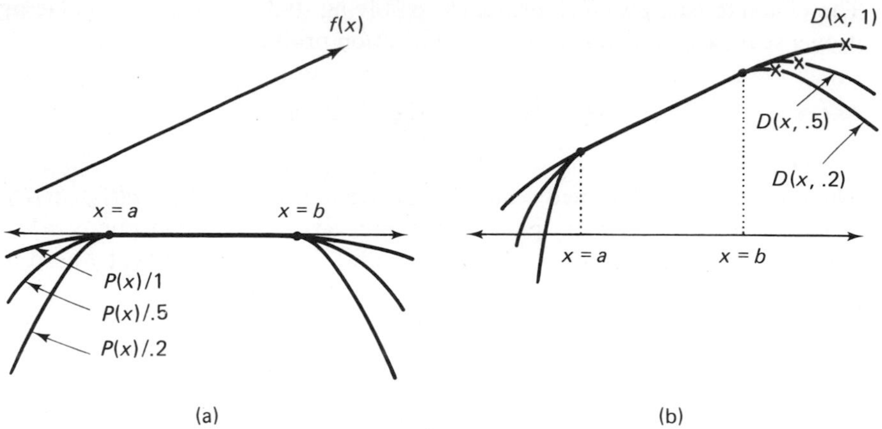

FIGURE 9-8

Several curves of the form $P(x)/r$, for $r = 1., .5$, and $.2$, are displayed in Fig. 9-8(a), while the corresponding unconstrained maximization functions

$$D(x, r) \equiv f(x) + \frac{1}{r} P(x)$$

are shown in Fig. 9-8(b). In the vicinity of its maximum the function $D(x, r)$ takes the form

$$D(x, r) = \frac{1}{2} x - \frac{(x - b)^2}{r},$$

and it is easily shown that the maximizing value is $x = b + r/4$. Observe that as r approaches zero the unconstrained maximum, indicated by a cross on each of the curves of Fig. 9-8(b), approaches $x = b$, which is the optimal solution to the original problem.

It is interesting to note that as r approaches zero the functions $P(x)/r$ and $D(x, r)$ approach the limiting barrier functions diagrammed in Fig. 9-7(c) and (d):

$$\lim_{r \to 0} \frac{1}{r} P(x) = \lim_{r \to 0} rB(x) \quad \text{and} \quad \lim_{r \to 0} D(x, r) = \lim_{r \to 0} C(x, r).$$

These limiting relationships [which actually hold only throughout the feasible region, inasmuch as $rB(x)$ does not have a negatively infinite value at every infeasible point] are characteristic of barrier and penalty methods in general and are valid for all suitably defined nonlinear programs (9-36). Figures 9-7 and 9-8 also illustrate the basic operational difference between the two approaches: Although the sequence of unconstrained maxima $\{\mathbf{x}_k\}$ converges toward the same limiting solution in both cases, the sequence generated by the barrier algorithm begins at an interior point of the feasible region and works outward toward the boundary, while that generated by the penalty method begins (normally) well outside the feasible region and works in toward it.

The penalty (9-54) is the most commonly used member of a family of penalty functions having the general form

$$P(\mathbf{x}) = -\sum_{i=1}^{m} [\max\{g_i(\mathbf{x}), 0\}]^{1+\epsilon}, \tag{9-55}$$

where the parameter ϵ can have any finite *positive* value. It is important to note that every function in this family possesses continuous first partial derivatives. This is evident in the case where (9-36) has only a single constraint:

$$P(\mathbf{x}) = \begin{cases} 0 & \text{if } g_1(\mathbf{x}) \leqq 0, \\ -[g_1(\mathbf{x})]^{1+\epsilon} & \text{if } g_1(\mathbf{x}) \geqq 0, \end{cases}$$

and

$$\frac{\delta P}{\delta x_j} = \begin{cases} 0 & \text{if } g_1(\mathbf{x}) < 0, \\ -(1+\epsilon)[g_1(\mathbf{x})]^{\epsilon}\dfrac{\delta g_1}{\delta x_j} = 0 & \text{if } g_1(\mathbf{x}) = 0, \\ -(1+\epsilon)[g_1(\mathbf{x})]^{\epsilon}\dfrac{\delta g_1}{\delta x_j} & \text{if } g_1(\mathbf{x}) > 0, \end{cases}$$

The general result for m constraints follows by extension. Continuity of the first partial derivatives is important for computational reasons: It guarantees that the unconstrained maximum of each function $D(\mathbf{x}, r_k) \equiv f(\mathbf{x}) + P(\mathbf{x})/r_k$ can be obtained via steepest-ascent and other direct climbing methods.

We now proceed to a rigorous development of the general method of penalties for solving the nonlinear program

$$\begin{cases} \text{Max } z = f(\mathbf{x}), & \mathbf{x} = (x_1, \ldots, x_n), \\ \text{subject to } g_i(\mathbf{x}) \leqq 0, & i = 1, \ldots, m. \end{cases} \tag{9-36}$$

Given a problem (9-36), let a *penalty function* $P(\mathbf{x})$ be defined to be any scalar-valued function having the following properties:

1. $P(\mathbf{x})$ exists, is continuous, and has continuous first partial derivatives for all $\mathbf{x}$ in E^n (or at least for all $\mathbf{x}$ within some positive distance of the feasible region S);
2. $P(\mathbf{x}) = 0$ for every $\mathbf{x}$ in S; and
3. $P(\mathbf{x}) < 0$ for every $\mathbf{x}$ not in S.

Based on this definition, we can now establish a result analogous to the one derived for the barrier algorithm in Section 9.7:

Lemma 9.4: Given a nonlinear program of the form (9-36), *let* $\mathrm{P}(\mathbf{x})$ *be any suitably defined penalty function. Further, let* r_1 *and* r_2 *be two scalar parameters satisfying* $\mathrm{r}_1 > \mathrm{r}_2 > 0$, *and suppose that the function*

$$\mathrm{D}(\mathbf{x}, \mathrm{r}_k) \equiv \mathrm{f}(\mathbf{x}) + \frac{1}{\mathrm{r}_k}\mathrm{P}(\mathbf{x})$$

takes on its global maximum value at the point $\mathbf{x}_k$, *for* $k = 1$ *and* $k = 2$. *It then follows that*

$$\mathrm{D}(\mathbf{x}_1, \mathrm{r}_1) \geqq \mathrm{D}(\mathbf{x}_2, \mathrm{r}_2), \tag{9-56}$$

$$\mathrm{P}(\mathbf{x}_1) \leqq \mathrm{P}(\mathbf{x}_2), \tag{9-57}$$

and

$$\mathrm{f}(\mathbf{x}_1) \geqq \mathrm{f}(\mathbf{x}_2). \tag{9-58}$$

The initial result (9-56) can be obtained directly via a chain of inequalities:

$$D(\mathbf{x}_1, r_1) \geqq D(\mathbf{x}_2, r_1) \equiv f(\mathbf{x}_2) + \frac{1}{r_1}P(\mathbf{x}_2) \geqq f(\mathbf{x}_2) + \frac{1}{r_2}P(\mathbf{x}_2) \equiv D(\mathbf{x}_2, r_2),$$

where the first inequality follows from the fact that $\mathbf{x}_1$ maximizes $D(\mathbf{x}, r_1)$ and the second from the relations $r_1 > r_2$ and $P(\mathbf{x}_2) \leqq 0$. Next, because $\mathbf{x}_1$ and $\mathbf{x}_2$ maximize $D(\mathbf{x}, r_1)$ and $D(\mathbf{x}, r_2)$, respectively, we can write

$$f(\mathbf{x}_2) + \frac{1}{r_1}P(\mathbf{x}_2) \leqq f(\mathbf{x}_1) + \frac{1}{r_1}P(\mathbf{x}_1) \tag{9-59}$$

and

$$f(\mathbf{x}_1) + \frac{1}{r_2}P(\mathbf{x}_1) \leqq f(\mathbf{x}_2) + \frac{1}{r_2}P(\mathbf{x}_2).$$

Adding these two inequalities and rearranging terms, we arrive at

$$\frac{1}{r_2}[P(\mathbf{x}_1) - P(\mathbf{x}_2)] \leqq \frac{1}{r_1}[P(\mathbf{x}_1) - P(\mathbf{x}_2)].$$

But $1/r_2 > 1/r_1$, so the difference $P(\mathbf{x}_1) - P(\mathbf{x}_2)$ cannot be positive, and (9-57) follows immediately. Finally, to establish (9-58), we can rearrange the terms of (9-59) to obtain

$$f(\mathbf{x}_2) - f(\mathbf{x}_1) \leqq \frac{1}{r_1}[P(\mathbf{x}_1) - P(\mathbf{x}_2)] \leqq 0,$$

where the second inequality follows from (9-57). The proof is complete.

We are now ready to give a precise statement of the penalty algorithm for solving nonlinear programs.

1. Given a problem (9-36) with a bounded optimal solution, choose a suitable penalty function $P(\mathbf{x})$ and a monotonically decreasing sequence of positive numbers $\{r_k\}$ satisfying $\lim_{k\to\infty} r_k = 0$.

2. *First iteration:* Use a direct climbing algorithm, starting from any point $\mathbf{x}_0$ in E^n, to maximize the unconstrained function

$$D(\mathbf{x}, r_1) \equiv f(\mathbf{x}) + \frac{1}{r_1}P(\mathbf{x}).$$

The maximum will be taken on at some finite point $\mathbf{x}_1$.

3. Continue by repeating step 2; at the kth iteration the starting point is $\mathbf{x}_{k-1}$, the function being maximized is $D(\mathbf{x}, r_k) \equiv f(\mathbf{x}) + P(\mathbf{x})/r_k$, and the maximizing point is $\mathbf{x}_k$.

The algorithm is terminated when a solution point $\mathbf{x}_k$ is generated that lies either (a) within some preselected small distance of its predecessor $\mathbf{x}_{k-1}$ or (b) within the feasible region S of the original problem. In the latter case it follows immediately that $\mathbf{x}_k$ must be an optimal solution to (9-36)—if there were some other point $\hat{\mathbf{x}}$ in S such that $f(\hat{\mathbf{x}}) > f(\mathbf{x}_k)$, then $D(\hat{\mathbf{x}}, r_k) > D(\mathbf{x}_k, r_k)$ would hold also [inasmuch as $P(\mathbf{x}) = 0$ throughout S] and $\mathbf{x}_k$ would not be the global maximum of $D(\mathbf{x}, r_k)$.

The penalty algorithm as stated above obviously requires the existence of finite maximizing points $\mathbf{x}_k$. Moreover, to prove convergence it is also necessary to rule out certain pathological cases in which the functions $D(\mathbf{x}, r_k)$ attain their maximum values at points $\mathbf{x}_k$ where either $f(\mathbf{x}_k)$ or $P(\mathbf{x}_k)$ or both are infinite in magnitude. We therefore state a further assumption or set of conditions required for applicability and convergence of the algorithm: There must exist a value $r_1 > 0$ small enough that

1. $D(\mathbf{x}, r_1)$ is maximized at a finite point $\mathbf{x}_1$ at which $f(\mathbf{x}_1)$ and $P(\mathbf{x}_1)$ are finite, and

2. Each function $D(\mathbf{x}, r_k)$ for $k > 1$ is maximized at a finite point.

Inasmuch as $\{f(\mathbf{x}_k)\}$ and $\{D(\mathbf{x}_k, r_k)\}$ are decreasing sequences and $\{P(\mathbf{x}_k)\}$ increases toward zero, none of these functions or terms can ever become infinite in value. In particular, it follows that the original problem cannot have an unbounded optimal solution, an assumption that was stated explicitly at the outset.

We now present a convergence theorem for the penalty method; the proof is essentially that of Zangwill [57].

Theorem 9.3: The sequence of solution points $\{\mathbf{x}_k\}$ *generated by the penalty algorithm converges to a limit point* $\mathbf{x}^*$ *that is a globally optimal solution to the given nonlinear program* (9-36).

In view of the supplementary assumption stated above, the sequence $\{\mathbf{x}_k\}$ must approach a finite limit point $\mathbf{x}^*$ (as in our proof of Theorem 9.1 for the barrier algorithm, we disregard the possibility of oscillation). We first prove that $\mathbf{x}^*$ satisfies the constraints of (9-36). Consider any feasible solution $\hat{\mathbf{x}}$ to (9-36): Inasmuch as the penalty function $P(\mathbf{x})$ vanishes at $\hat{\mathbf{x}}$, we may write

$$f(\hat{\mathbf{x}}) = f(\hat{\mathbf{x}}) + \frac{1}{r_k}P(\hat{\mathbf{x}}) \equiv D(\hat{\mathbf{x}}, r_k) \leqq D(\mathbf{x}_k, r_k) \tag{9-60}$$

for every $k \geqq 1$, where the inequality holds because $\mathbf{x}_k$ maximizes $D(\mathbf{x}, r_k)$. Thus the monotonically nonincreasing sequence $\{D(\mathbf{x}_k, r_k)\}$ is bounded from below and must converge to a finite limiting value D^*:

$$\lim_{k\to\infty} D(\mathbf{x}_k, r_k) = D^*.$$

Because the penalty function $P(\mathbf{x})$ is always nonpositive, the chain of relations (9-60) can be continued:

$$f(\hat{\mathbf{x}}) \leqq D(\mathbf{x}_k, r_k) \equiv f(\mathbf{x}_k) + \frac{1}{r_k}P(\mathbf{x}_k) \leqq f(\mathbf{x}_k)$$

for all k. Thus $\{f(\mathbf{x}_k)\}$ is also bounded from below and approaches a finite limit:

$$\lim_{k\to\infty} f(\mathbf{x}_k) = f^*.$$

It then follows from the definition of $D(\mathbf{x}_k, r_k)$ that

$$\lim_{k\to\infty} \frac{1}{r_k}P(\mathbf{x}_k) \equiv \lim_{k\to\infty} D(\mathbf{x}_k, r_k) - \lim_{k\to\infty} f(\mathbf{x}_k) = D^* - f^*.$$

But because the quotients $1/r_k$ become infinitely large, the sequence $\{P(\mathbf{x}_k)/r_k\}$ could not converge to a finite limit unless

$$\lim_{k\to\infty} P(\mathbf{x}_k) = P(\mathbf{x}^*) = 0.$$

Thus, inasmuch as $P(\mathbf{x}^*)$ vanishes, the limit point $\mathbf{x}^*$ must lie in the feasible region of (9-36).

To complete the proof, let $\mathbf{x}_{\text{opt}}$ denote any optimal feasible solution to (9-36). Then for any k

$$f(\mathbf{x}_{\text{opt}}) = f(\mathbf{x}_{\text{opt}}) + \frac{1}{r_k}P(\mathbf{x}_{\text{opt}}) \leqq f(\mathbf{x}_k) + \frac{1}{r_k}P(\mathbf{x}_k) \leqq f(\mathbf{x}_k), \tag{9-61}$$

where the equality holds because $P(\mathbf{x}_{\text{opt}}) = 0$, the first inequality follows from the fact that $\mathbf{x}_k$ maximizes $D(\mathbf{x}, r_k)$, and the second follows from the relations $P(\mathbf{x}_k) \leqq 0$ and $r_k > 0$. Taking the limit of (9-61) yields

$$f(\mathbf{x}_{\text{opt}}) \leqq \lim_{k\to\infty} f(\mathbf{x}_k) = f(\mathbf{x}^*).$$

But $\mathbf{x}_{\text{opt}}$ is optimal and $\mathbf{x}^*$ is feasible, so $f(\mathbf{x}_{\text{opt}}) < f(\mathbf{x}^*)$ is impossible. Thus $f(\mathbf{x}_{\text{opt}}) = f(\mathbf{x}^*)$, and $\mathbf{x}^*$ is an optimal solution to (9-36), as was to be shown.

As a corollary to this theorem, let us also prove that the penalty term $P(\mathbf{x}_k)/r_k$ converges to zero in the limit [recall from Theorem 9.2 that the same is true of the barrier term $r_k B(\mathbf{x}_k)$ as well]. Continuing to let $\mathbf{x}^*$ denote the limit point of the sequence $\{\mathbf{x}_k\}$ generated by the penalty algorithm, we can write, for every $k \geqq 1$,

$$f(\mathbf{x}^*) = f(\mathbf{x}^*) + \frac{1}{r_k}P(\mathbf{x}^*) \equiv D(\mathbf{x}^*, r_k) \leqq D(\mathbf{x}_k, r_k),$$

where the first equality holds because $P(\mathbf{x})$ vanishes at the optimal (hence feasible) solution $\mathbf{x}^*$. Taking the limit yields

$$f(\mathbf{x}^*) \leqq \lim_{k\to\infty} D(\mathbf{x}_k, r_k) = \lim_{k\to\infty} f(\mathbf{x}_k) + \lim_{k\to\infty} \frac{1}{r_k}P(\mathbf{x}_k) = f(\mathbf{x}^*) + \lim_{k\to\infty} \frac{1}{r_k}P(\mathbf{x}_k),$$

which reduces to

$$\lim_{k\to\infty} \frac{1}{r_k}P(\mathbf{x}_k) \geqq 0. \tag{9-62}$$

But the quotients $1/r_k$ are positive, while $P(\mathbf{x}_k) \leqq 0$ for all k, so the limit (9-62) cannot be positive, and we conclude that

$$\lim_{k\to\infty} \frac{1}{r_k}P(\mathbf{x}_k) = 0,$$

as was to be shown. Furthermore, it follows that

$$\lim_{k\to\infty} D(\mathbf{x}_k, r_k) = f(\mathbf{x}^*).$$

Again, an analogous result was obtained for the sequences $\{C(\mathbf{x}_k, r_k)\}$ generated by the barrier algorithm.

It was observed in Section 9.9 that the most commonly used barrier functions "preserve concavity," in the sense that the unconstrained optimization function $C(\mathbf{x}, r) \equiv f(\mathbf{x}) + rB(\mathbf{x})$ associated with a convex program is a concave function, at least over the region containing the points $\mathbf{x}_k$. Perhaps not surprisingly, the standard penalty functions also possess this desirable property, as we now proceed to show.

Lemma 9.5: Let the objective $f(\mathbf{x})$ *be concave and the constraint functions* $g_i(\mathbf{x})$, $i = 1, \ldots, m$, *be convex in the nonlinear program* (9-36). *Then the unconstrained optimization function* $D(\mathbf{x}, r_k) \equiv f(\mathbf{x}) + P(\mathbf{x})/r_k$ *is concave for any* $r_k > 0$ *whenever the penalty function is of the form*

(9-55) $$P(\mathbf{x}) = -\sum_{i=1}^{m} [\max\{g_i(\mathbf{x}), 0\}]^{1+\epsilon}, \qquad \epsilon > 0.$$

The proof requires two intermediate results:

1. The maximum of two convex functions is itself a convex function, and
2. If $h(\mathbf{x})$ is a convex function of $\mathbf{x} = (x_1, \ldots, x_n)$ and $p(y)$ is a convex and nondecreasing[5] univariate function, then $\phi(\mathbf{x}) \equiv p[h(\mathbf{x})]$ is a convex function of $\mathbf{x}$.

The former of these results was proved as Exercise 3-11(a), while the latter can be established without difficulty, as follows:

(9-63) $$\begin{aligned} \phi(\lambda\mathbf{x}_1 + (1-\lambda)\mathbf{x}_2) &\equiv p[h(\lambda\mathbf{x}_1 + (1-\lambda)\mathbf{x}_2)] \\ &\leqq p[\lambda h(\mathbf{x}_1) + (1-\lambda)h(\mathbf{x}_2)] \\ &\leqq \lambda p[h(\mathbf{x}_1)] + (1-\lambda)p[h(\mathbf{x}_2)] \\ &\equiv \lambda\phi(\mathbf{x}_1) + (1-\lambda)\phi(\mathbf{x}_2). \end{aligned}$$

The first inequality holds because h is convex—implying that $h(\lambda\mathbf{x}_1 + (1-\lambda)\mathbf{x}_2) \leqq \lambda h(\mathbf{x}_1) + (1-\lambda)h(\mathbf{x}_2)$—and p is nondecreasing; the second inequality follows from the convexity of p. Turning now to the lemma itself, we observe that, for every i, the nonnegative function $h_i(\mathbf{x}) \equiv \max\{g_i(\mathbf{x}), 0\}$ is convex because of result 1. Moreover, inasmuch as $p(y) = y^{1+\epsilon}$, where $\epsilon > 0$, is a nondecreasing convex function for nonnegative y (proof?), the function

$$\phi_i(\mathbf{x}) \equiv [h_i(\mathbf{x})]^{1+\epsilon}, \qquad \epsilon > 0,$$

is convex because of result 2. It then follows from the elementary theorems of

[5] By definition, $p(y)$ is a *nondecreasing* function over an interval I of the real line if for any two points y_1 and y_2 in I, $p(y_1) \leqq p(y_2)$ whenever $y_1 \leqq y_2$.

Section 3.4 that both

$$P(\mathbf{x}) = -\sum_{i=1}^{m} [\max\{g_i(\mathbf{x}), 0\}]^{1+\epsilon} = -\sum_{i=1}^{m} \phi_i(\mathbf{x})$$

and

$$D(\mathbf{x}, r_k) \equiv f(\mathbf{x}) + \frac{1}{r_k} P(\mathbf{x})$$

are concave functions, as was to be shown.

The importance of this "preservation-of-concavity" result is precisely the same as in the case of the barrier algorithm. When a convex program (9-36) is being solved via the penalty approach, using a penalty function of the general form (9-55), the first local maximum obtained for each of the unconstrained functions $D(\mathbf{x}, r_k)$ is the desired global maximum. Thus only a single application of whatever direct climbing technique is being used is required at each iteration. With nonconvex problems, on the other hand, the possibility of suboptimal local maxima of the functions $D(\mathbf{x}, r_k)$ arises, and the difficulties in obtaining and identifying the global maxima are in general compounded. For this reason the penalty algorithm operates much more effectively on convex programs. Of course, this is not to say that the algorithm should never be applied to nonconvex problems—the latter type are always more difficult to solve, regardless of what method is being used.

Insofar as the choice of a direct climbing technique for maximizing the functions $D(\mathbf{x}, r_k)$ is concerned, the considerations are again essentially the same as in the case of the barrier algorithm. As r_k approaches zero, the penalty function $P(\mathbf{x})/r_k$ becomes steeper and steeper in the vicinity of the maximizing points $\mathbf{x}_k$ (refer to Fig. 9-8), so that first-order or steepest-ascent methods become extremely unreliable and inefficient. Attention therefore focuses on other climbing techniques, particularly the second-order or Newton methods. It may seem that such methods cannot be used with penalty functions of the form (9-55) unless $\epsilon > 1$, because of the discontinuity of the second partial derivatives of $P(\mathbf{x})$ at the boundary of the feasible region when $\epsilon \leq 1$ (for such ϵ the second derivatives of $P(\mathbf{x})$, which include terms in $[g_i(\mathbf{x})]^{\epsilon-1}$, do not vanish—as continuity would require—at boundary points where $g_i(\mathbf{x}) = 0$). This discontinuity would imply that the Hessian matrix, whose inverse is used in the computation of the direction vector, is not well defined anywhere on the boundary. However, it is generally true that each solution point of the sequence $\{\mathbf{x}_k\}$ generated by the penalty algorithm lies outside the feasible region, where the Hessian *is* continuous; it is only the limit point of the sequence that lies on the boundary. Thus Newton's method and its variants can in fact be used in applying the penalty algorithm to most nonlinear programs.

9.11 CONCLUSION

As has been repeatedly emphasized throughout, this textbook has been oriented toward the practitioner of operations research rather than toward the theoretician. The algorithms presented have been those with solid practical utility, as opposed to those with the greatest theoretical elegance or those most recently developed. In evaluating mathematical programming algorithms with regard to practical applicability, there are a number of general criteria that can be considered; those listed here are arranged in their approximate order of importance, according to the judgment and experience of the author:

1. How simple is the algorithm to work with and to code for computer use?
2. To what degree is it subject to computational error?
3. How rapidly—based on a sample of test problems—can it be expected to obtain a solution?
4. How efficiently does the algorithm (if designed for the general nonlinear program) exploit linearity wherever it occurs in the constraints or objective?
5. Can the algorithm be used to solve very large problems?
6. Is convergence to an optimal solution guaranteed for every problem?

These considerations have been kept prominently in view throughout this textbook and were in fact the determining factors (with influence diminishing in the order given) in the author's choice of solution methods to be covered. It seems appropriate to close with a few comments on the criteria listed above and on the order in which they have been arranged.

No doubt many readers will be somewhat surprised at the top-priority selection on the list. Simplicity in most enterprises is less valued by those who design than by those who execute, and the solution of mathematical programs is no exception. In working manually with a complicated algorithm, a great deal of time can be wasted when difficulties or ambiguities are encountered or when numerous logical subcases have to be investigated. Moreover, computer-coded versions of many of the most useful algorithms are not generally available—or are available only on a job-by-job basis at an exorbitant cost—and on occasion an operations research analyst may well find it convenient to write a make-shift computer code himself, perhaps borrowing subroutines from a simplex-method code or from other sources. Very few experiences of this sort are required to instill an appreciation of simplicity. Criteria 2 and 3 are obviously of considerable significance in evaluating any mathematical solution method; their ordering is based on the judgment that the man-hours expended in verifying or improving the accuracy of a dubious solution are likely to be more costly and inconvenient than the extra computer hours consumed in using an algorithm that is not the fastest available.

Criterion 4 draws its importance from the predominance of linear (or "linearize-able") relationships in the real world and in models thereof. Problems in which all but a few of the constraints are linear are quite common in operations research, and an algorithm that provides special and more efficient treatment for linear constraints, and for nonnegativity requirements in particular, has obvious and useful advantages.

The rather low position on the priority list of an algorithm's capacity to handle very large problems may, like the top-priority choice, also appear somewhat surprising. The reason for this is not that truly large-scale problems seldom occur (although they are indeed relatively rare, their size tends to give each of them greater importance), but rather that they simply cannot be solved economically with the same methods that are used for smaller problems. The difference in size requires qualitatively different approaches, and large-scale methematical programming has recently become by itself a popular subject for advanced study. The last of the six criteria, the guarantee of convergence, is normally of very little concern to a practitioner of operations research. This is because the algebraic functions and constraints that are used in modeling most real-world problems are sufficiently smooth and well behaved to provide de facto assurance of convergence no matter what solution method is being employed. In the author's opinion, the question of whether or not a tricky example problem can be contrived to prevent convergence of an otherwise perfectly useful algorithm is best left to the theoreticians and the professional journals.

EXERCISES

Section 9.2

9-1. Let the functions $f_j(x_j)$ in the objective of the convex separable program (9-3) be *strictly* concave. Prove formally that if the associated approximating problem (9-6) is solved directly by the simplex method, *ignoring* the logical restrictions 1 and 2, then the solution obtained is in fact the optimal solution to (9-6) and yields, via the substitutions (9-7), an approximation to the optimal solution of (9-3). Can the same conclusion be derived if the functions $f_j(x_j)$ are merely concave? If so, how must the proof be modified?

9-2. Solve the following problem via separable programming, starting with the solution that corresponds to the origin $(x_1, x_2) = (0, 0)$. In constructing the functions $\hat{f}_j$ and $\hat{g}_{ij}$, use breakpoints x_{jk} that are one unit apart.

$$\begin{aligned} \text{Min } z &= (x_1 - 4)^2 + (x_2 - 3)^2 \\ \text{subject to } & x_1^2 + 4x_2^2 \leqq 16 \\ \text{and } & x_1, x_2 \geqq 0, \end{aligned}$$

9-3. Find a transformation of variables or a sequence of substitutions that achieves separability for each of the following expressions, where the values of the variables are restricted as indicated.

(a) $x_1^2x_2$, where $x_1 > 0$, $x_2 > 0$
(b) $e^{x_1^2+x_2}$, where $x_2 > 0$
(c) $x_1^{x_2}$, where $x_1 > 1$, $x_2 > 0$
(d) $x_1^2x_2$, where $x_1 \geqq 0$, $x_2 \geqq 0$
(e) $x_1^{x_2}$, where $x_1 > 0$, $x_2 \geqq 0$

9-4. Find a transformation of variables or some other maneuver that will achieve separability for the following convex program while preserving its convexity:

$$\text{Min } z = x_1 + x_2$$
$$\text{subject to } x_1x_2 \geqq 25 \text{ and } x_1 \geqq 5.$$

9-5. Using the transformation of variables (8-87) and (8-88) that was introduced in Exercise 8-9, write the δ-form of the approximating problem for the general separable program (9-3). Then outline a proof that when (9-3) is a convex program, the δ-form of the approximating problem can be solved directly via the simplex method, with no additional modifications or restrictions attached.

9-6. Solve the problem of Exercise 9-2 via the δ-form of separable programming, using the same breakpoints and starting solution.

Section 9.3

9-7. Carry the example problem of Section 9.3 through one additional complete iteration, solving the linear program graphically and checking your calculations where possible against the data in Table 9-1.

9-8. Apply Kelley's method to the following problem, performing at least two complete iterations and using simple graphical techniques to solve the linear program at each iteration:

$$\text{Max } z = x_1 - x_2$$
$$\text{subject to } x_1 + x_2 \leqq 5,$$
$$x_1x_2 \geqq 4,$$
$$\text{and } \quad x_1, x_2 \geqq 0.$$

Let S_0 be the set of all points satisfying the *linear* constraints of the problem; that is,

$$S_0 = \{(x_1, x_2) \mid x_1 + x_2 \leqq 5, x_1 \geqq 0, x_2 \geqq 0\}.$$

Can this technique for defining S_0 be used for *any* problem?

9-9. Can Kelley's cutting-plane algorithm be used to solve a problem having (a) an unbounded optimal solution? (b) An unbounded feasible region? Why or why not?

9-10. Suppose that Kelley's algorithm is being applied to the nonlinear program

$$\text{Max } z = \mathbf{c}^T\mathbf{x}$$
$$\text{subject to } g_i(\mathbf{x}) \leqq 0, \qquad i = 1, \ldots, m,$$

and define $G(\mathbf{x}) \equiv \max_i\{g_i(\mathbf{x})\}$ as usual. Prove that if at some stage $G(\mathbf{x}_k) = g_\alpha(\mathbf{x}_k) = g_\beta(\mathbf{x}_k) = \cdots$, then any convex combination of the gradients $\nabla \mathbf{g}_\alpha(\mathbf{x}_k)$, $\nabla \mathbf{g}_\beta(\mathbf{x}_k), \ldots$ can be used in place of $\nabla \mathbf{G}(\mathbf{x}_k)$ to form a legitimate cutting-plane constraint

$$H_k(\mathbf{x}) = G(\mathbf{x}_k) + \nabla \mathbf{G}^T(\mathbf{x}_k) \cdot (\mathbf{x} - \mathbf{x}_k) \leqq 0.$$

That is, prove that with any such substitution the constraint $H_k(\mathbf{x}) \leqq 0$ will be satisfied at every feasible point of the original problem but not at $\mathbf{x}_k$.

9-11. Write the dual of the linear program solved at the kth iteration, $k \geqq 1$, of Kelley's algorithm:

$$\text{Max } z = \mathbf{c}^T\mathbf{x}$$
$$\text{subject to } \mathbf{A}\mathbf{x} \leq \mathbf{b}$$
$$\text{and } H_j(\mathbf{x}) = G(\mathbf{x}_j) + \nabla \mathbf{G}^T(\mathbf{x}_j) \cdot (\mathbf{x} - \mathbf{x}_j) \leqq 0, \qquad j = 0, 1, \ldots, k-1.$$

Why would solving the dual problem at each stage tend to be more efficient then solving the primal?

9-12. Under what circumstances would the optimal solution to a convex programming problem be obtained after only a finite number of iterations of Kelley's cutting-plane method?

9-13. When Kelley's algorithm is applied to a convex program having no feasible solutions, how is this infeasibility diagnosed?

9-14. (a) Prove heuristically that any given closed convex set X in E^n can be written as the intersection of all the (infinitely many) closed half-spaces in E^n that completely contain it. That is, prove that X is the set of points in E^n satisfying the constraints

$$\boldsymbol{\alpha}_i\mathbf{x} \leqq b_i, \qquad i \in I,$$

where I includes *every* linear inequality constraint that is satisfied by all points $\mathbf{x}$ in X.
(b) Part (a) suggests that the nonlinearly constrained convex program

$$\text{Max } z = \mathbf{c}^T\mathbf{x}$$
$$\text{subject to } \mathbf{x} \in X$$

can be written as the "infinitely constrained" linear program

$$\text{Max } z = \mathbf{c}^T\mathbf{x}$$
$$\text{subject to } \boldsymbol{\alpha}_i\mathbf{x} \leqq b_i, \qquad i \in I.$$

Write the dual of this linear program and explain how Kelley's cutting-plane algorithm can be interpreted within the context of duality.

Section 9.5

9-15. Making direct use of Theorem 6.3, verify that the relations (9-33) do indeed constitute the Kuhn-Tucker conditions for the nonlinear programming problem (9-20).

9-16. Prove that if the current solution $\mathbf{x}_k$ is a Kuhn-Tucker point, then $\hat{\sigma} = 0$ in the optimal solution $(\hat{\mathbf{s}}, \hat{\sigma})$ to the next direction-finding problem (9-24).

9-17. Prove that $x_1 = 1$, $x_2 = 4$ is the optimal solution (i.e., not merely a Kuhn-Tucker point) to the following problem:

$$\text{Max } z = x_2$$
$$\text{subject to} \quad x_1^2 + 3x_2^2 \leqq 49,$$
$$(x_1 - 1)^2 - x_1x_2 + 2x_2^2 \leqq 28,$$
$$\text{and} \quad 2x_1 + 5x_2 \leqq 22.$$

9-18. Perform four or five iterations of the feasible-directions method on the problem

$$\text{Max } z = x_1$$
$$\text{subject to} \quad x_1^2 - x_2 \leqq 0$$
$$\text{and} \quad -x_1 + x_2 \leqq 0,$$

taking the origin as the starting solution and using the direction-finding problem (9-24) at each iteration. The line search problem

$$\text{(9-21)} \qquad \text{Max } f(\hat{\mathbf{x}}) = f(\mathbf{x}_k + \theta \mathbf{s}_k)$$
$$\text{subject to } \theta \geqq 0 \text{ and } \hat{\mathbf{x}} \text{ feasible}$$

can be solved by inspection at each stage, using a simple sketch of the feasible region. Identify the optimal solution. Does the phenomenon of zigzagging begin to appear? What would happen in the ensuing iterations if the procedure were continued? Contrast this development with the sequence of events that would be expected to occur during the approach toward the optimum if the direction-finding problem (9-35)—that is, the ϵ-perturbation method—were being used instead.

9-19. Suppose that we are attempting to solve the nonlinear program

$$\text{Max } z = x_1 - x_2^2$$
$$\text{subject to} \quad -x_1^2 + x_2 \leqq 0,$$
$$-x_1^2 - x_2 \leqq 0,$$
$$\text{and} \quad x_1 \geqq 0$$

by the feasible-directions method and that a current solution point $\mathbf{x}_k = (0, 0)$ has been reached. Show that when the next direction-finding problem (9-24) is solved, it will be found that the optimal value of σ is zero. Determine whether or not the Kuhn-Tucker conditions are satisfied at $\mathbf{x}_k$, and explain your findings.

9-20 Show that in formulating the direction-finding problem (9-24) or (9-35) the variable σ can be omitted from each of the constraints $\nabla \mathbf{g}_i^T(\mathbf{x}_k) \cdot \mathbf{s} + \sigma \leqq 0, i \in B_k$ or $i \in B_k(\epsilon)$, that corresponds to a *linear* inequality $g_i(\mathbf{x}) \leqq b_i$, without affecting the manner in which the feasible-directions method operates or invalidating any of its properties.

Section 9.7

9-21. Show that the nonlinear program

$$\text{Max } f(\mathbf{x}) = f(x_1, x_2) = x_2$$
$$\text{subject to } x_1^2 + x_2 \leqq 0$$
$$\text{and} \qquad x_1 \geqq 0$$

can be solved via the barrier approach, using the barrier function $B(\mathbf{x}) = \sum_i 1/g_i(\mathbf{x})$ and any decreasing sequence of positive values $\{r_k\}$ that converges to zero. This can be done by determining $\hat{\mathbf{x}}(r)$, the point at which the unconstrained function $C(\mathbf{x}, r) \equiv f(\mathbf{x}) + rB(\mathbf{x})$ is maximized, and then showing that as r approaches 0 the unconstrained optimum $\hat{\mathbf{x}}(r)$ approaches the optimal solution of the original problem.

9-22. Using the barrier function $B(\mathbf{x}) = \sum_i 1/g_i(\mathbf{x})$, write the unconstrained maximization function $C(\mathbf{x}, r)$ for the mathematical program

$$\text{Max } f(x_1, x_2) = x_1$$
$$\text{subject to } x_1 - x_2 \leqq 0$$
$$\text{and } x_2 - 2 \leqq 0$$

and solve algebraically to determine the optimal values of x_1 and x_2 as functions of the positive-valued parameter r. Verify that as r is decreased toward zero, the objective value $f(\mathbf{x})$ at the unconstrained optimum approaches its maximum feasible value, while the value of $rB(\mathbf{x})$ converges to zero.

9-23. Can a nonlinear program (9-36) satisfying the assumptions stated in Section 9.7 have an unbounded optimal solution, that is, a finite feasible solution $\mathbf{x}^*$ such that $f(\mathbf{x}^*) = \infty$? If so, what would happen if the barrier algorithm were applied to such a problem?

9-24. Consider a nonlinear program of the form (9-36) that satisfies the assumptions of Section 9.7 and has a unique optimal solution $\mathbf{x}^*$ with a finite objective value $f(\mathbf{x}^*) = f^*$. Suppose this problem is to be solved via a barrier algorithm, using the barrier function $B(\mathbf{x}) = \sum_i 1/g_i(\mathbf{x})$ and a suitably chosen sequence of positive numbers $\{r_k\}$ that decreases to zero in the limit. Let $\mathbf{x}_k$ denote the solution to the kth unconstrained optimization problem

$$\text{Max } C(\mathbf{x}, r_k) \equiv f(\mathbf{x}) + r_k B(\mathbf{x}).$$

It is then possible to prove that the sequence of unconstrained maxima $\{\mathbf{x}_k\}$ converges to the desired optimal solution $\mathbf{x}^*$ by proceeding as follows. Given any $\epsilon > 0$, let $\mathbf{y}$ be a point in the interior of the feasible region such that

$$f(\mathbf{y}) + \frac{\epsilon}{2} > f^*$$

(by assumption, such a point exists within a sufficiently small neighborhood of $\mathbf{x}^*$). Inasmuch as each $g_i(\mathbf{y})$, $i = 1, \ldots, m$, has a negative value, we can then choose a term of the sequence $\{r_k\}$, call it r_K, that is small enough to satisfy

$$0 < r_K < -\frac{\epsilon}{2m}\left[\max_i\{g_i(\mathbf{y})\}\right].$$

(a) By developing a chain of inequalities, prove that

$$f^* \geqq C(\mathbf{x}_k, r_k) \geqq C(\mathbf{x}_K, r_K) > f^* - \epsilon$$

for all $k \geqq K$. It then follows immediately that $\lim_{k\to\infty} C(\mathbf{x}_k, r_k) = f^*$.
(b) As a corollary to part (a), prove that

$$\lim_{k\to\infty} f(\mathbf{x}_k) = f^* \quad \text{and} \quad \lim_{k\to\infty} r_k B(\mathbf{x}_k) = 0.$$

Section 9.9

9-25. (a) In general, would it be reasonable to use a barrier algorithm to solve a nonlinear program having a large number of linear inequality constraints? Why or why not?
(b) Consider the nonlinear program

$$\begin{aligned} \text{Max } z &= f(\mathbf{x}) \\ \text{subject to } g_i(\mathbf{x}) &\leqq 0, \qquad i = 1, \ldots, m, \\ \boldsymbol{\alpha}_i\mathbf{x} &\leqq b_i, \qquad i = m+1, \ldots, s, \\ \text{and } \boldsymbol{\alpha}_i\mathbf{x} &= b_i, \qquad i = s+1, \ldots, t, \end{aligned}$$

where each $\boldsymbol{\alpha}_i$ is a row vector of constant coefficients. Describe how the barrier approach could be modified for application to this problem in such a way that the subproblem solved at each iteration is linearly constrained rather than unconstrained. Sketch out whatever proofs are needed to justify your "modified barrier algorithm." What would be the pros and cons of using the gradient projection method, as opposed to the reduced gradient or some other simplex-based algorithm, for solving the linearly constrained subproblems?

9-26. (a) Prove that if $h(\mathbf{x})$ is a concave function of $\mathbf{x} = (x_1, \ldots, x_n)$ and $p(y)$ is a concave and nondecreasing univariate function, then $\phi(\mathbf{x}) \equiv p[h(\mathbf{x})]$ is a concave function of $\mathbf{x}$ [for a nondecreasing function $p(y)$, $y_1 \leqq y_2$ implies $p(y_1) \leqq p(y_2)$].

(b) Use the result of part (a) to prove that for any $\epsilon > 0$ the general barrier function

$$B(\mathbf{x}) = \sum_{i=1}^{m} \frac{-1}{[-g_i(\mathbf{x})]^{\epsilon}}$$

is concave over the interior of the feasible region of any convex program (9-36).

9-27. With respect to the convex program (9-36) and its dual problem (9-49), prove or disprove via counterexample each of the following statements:
(a) If the primal problem has an unbounded optimal solution, then the dual has no feasible solution.
(b) If the primal problem has no feasible solution, then the dual has an unbounded optimum.

Section 9.10

9-28. Using the standard "quadratic" penalty function (9-54), write the unconstrained optimization function $D(x, r)$ associated with the following nonlinear program:

$$\text{Max } f(x) = 4x - x^2$$
$$\text{subject to } x \leqq 1.$$

Show, via elementary calculus, that for any given positive value of r the function $D(x, r)$ takes on its (global) maximum value at $x = (2r + 1)/(r + 1)$. Then show that as r approaches 0 this global maximum approaches the optimal solution of the original problem.

9-29. Given any nonlinear program (9-36) whose feasible region includes a nonempty interior, let $B(\mathbf{x})$ and $P(\mathbf{x})$ be a suitably defined barrier function and penalty function, respectively, and prove rigorously that the limiting relationship

$$\lim_{r \to 0} rB(\mathbf{x}) \equiv \lim_{r \to 0} \frac{1}{r} P(\mathbf{x})$$

holds for all $\mathbf{x}$ in the feasible region.

9-30. Describe the behavior of the penalty method when it is applied to a problem (9-36) having no feasible solutions. Does Lemma 9.4 hold? What can be said about the limiting values of $f(\mathbf{x}_k)$, $P(\mathbf{x}_k)$, and $D(\mathbf{x}_k, r_k)$ as r_k approaches zero? Where does Theorem 9.3 break down? Illustrate your answers by applying the penalty method to the problem

$$\text{Max } f(x) = x$$
$$\text{subject to } x \leqq 0 \text{ and } 1 - x \leqq 0,$$

using the penalty function $P(x) = -\sum_i [\max\{g_i(x), 0\}]^2$ and performing three iterations with r-values .5, .1, and .01.

9-31. Show how the penalty function (9-54) could be modified so as to make the penalty algorithm directly applicable to nonlinear programs of the following form:

$$\begin{aligned} \text{Max } z &= f(\mathbf{x}), && \mathbf{x} = (x_1, \ldots, x_n), \\ \text{subject to } g_i(\mathbf{x}) &\leqq 0, && i \in I, \\ \text{and } g_i(\mathbf{x}) &= 0, && i \in E. \end{aligned}$$

Given some specific problem satisfying this format, would the modified algorithm generate the same sequence of solution points $\{\mathbf{x}_k\}$ as the "standard" penalty method would if the latter were applied (using an analogous penalty function and the same sequence of r-values) to the equivalent problem

$$\begin{aligned} &\text{Max } z = f(\mathbf{x}) \\ \text{subject to } \quad & g_i(\mathbf{x}) \leqq 0, \qquad i \in I, i \in E, \\ \text{and } \; & -g_i(\mathbf{x}) \leqq 0, \qquad i \in E. \end{aligned}$$

References

1. J. ABADIE, ed., *Nonlinear Programming*, North-Holland, Amsterdam, 1967.

2. J. ABADIE, ed., *Integer and Nonlinear Programming*, North-Holland, Amsterdam, 1970.

3. K. ARROW, L. HURWICZ, and H. UZAWA, *Studies in Linear and Nonlinear Programming*, Stanford University Press, Stanford, Calif., 1958.

4. E. M. L. BEALE, "On Quadratic Programming," *Naval Res. Log. Quart.*, *6*, no. 3, Sept. 1959, pp. 227–243.

5. E. M. L. BEALE, "Numerical Methods," in Abadie [1], 1967, pp. 135–205.

6. C. W. CARROLL, "The Created Response Surface Technique for Optimizing Nonlinear Restrained Systems," *Operations Res.*, *9*, no. 2, March–April 1961, pp. 169–184.

7. A. CHARNES and C. LEMKE, "Minimization of Nonlinear Separable Convex Functionals," *Naval Res. Log. Quart.*, *1*, no. 4, Dec. 1954, pp. 301–312.

8. H. B. CURRY, "The Method of Steepest Descent for Nonlinear Minimization Problems," *Quart. Appl. Math.*, *2*, no. 3, Oct. 1944, pp. 258–261.

9. G. DANTZIG, *Linear Programming and Extensions*, Princeton University Press, Princeton, N.J., 1963.

10. S. DREYFUS and M. FREIMER, "A New Approach to the Duality Theory of Mathematical Programming," in *Applied Dynamic Programming*, R. Bellman and S. Dreyfus, Princeton University Press, Princeton, N.J., 1962, Appendix II.

11. R. DUFFIN, E. PETERSON, and C. ZENER, *Geometric Programming*, Wiley, New York, 1967.

12. B. CURTIS EAVES, "On Quadratic Programming," *Management Sci.*, *17*, no. 11, July 1971, pp. 698–711.

13. A. V. FIACCO and G. P. MCCORMICK, "The Sequential Unconstrained Minimization Technique for Nonlinear Programming, a Primal-Dual Method," *Management Sci.*, *10*, no. 2, Jan. 1964, pp. 360–366.

14. A. V. FIACCO and G. P. MCCORMICK, "Computational Algorithm for the Sequential Unconstrained Minimization Technique for Nonlinear Programming," *Management Sci.*, *10*, no. 4, July 1964, pp. 601–617.

15. A. V. FIACCO and G. P. MCCORMICK, *Nonlinear Programming: Sequential Unconstrained Minimization Techniques*, Wiley, New York, 1968.

16. M. FRANK and P. WOLFE, "An Algorithm for Quadratic Programming," *Naval Res. Log. Quart.*, *3*, no. 1, March 1956, pp. 95–110.

17. A. M. GEOFFRION, "Strictly Concave Parametric Programming, Part I: Basic Theory," *Management Sci.*, *13*, no. 3, Nov. 1966, pp. 244–253.

18. A. M. GEOFFRION, "Strictly Concave Parametric Programming, Part II: Additional Theory and Computational Considerations," *Management Sci.*, *13*, no. 5, Jan. 1967, pp. 359–370.

19. R. GRAVES and P. WOLFE, eds., *Recent Advances in Mathematical Programming*, McGraw-Hill, New York, 1963.

20. G. HADLEY, *Linear Algebra*, Addison-Wesley, Reading, Mass., 1961.

21. G. HADLEY, *Linear Programming*, Addison-Wesley, Reading, Mass., 1962.

22. G. HADLEY, *Nonlinear and Dynamic Programming*, Addison-Wesley, Reading, Mass., 1964.

23. F. B. HILDEBRAND, *Introduction to Numerical Analysis*, McGraw-Hill, New York, 1956.

24. R. HOOKE and T. A. JEEVES, " 'Direct Search' Solution of Numerical and Statistical Problems," *J. Assoc. Computing Machinery*, *8*, no. 2, April 1961, pp. 212–229.

25. J. E. KELLEY, "The Cutting-Plane Method for Solving Convex Programs," *J. Soc. Ind. Appl. Math.*, *8*, no. 4, Dec. 1960, pp. 703–712.

26. J. KIEFER, "Sequential Minimax Search for a Maximum," *Proc. Amer. Math. Soc.*, *4*, no. 3, June 1953, pp. 502–506.

27. T. C. KOOPMANS, ed., *Activity Analysis of Production and Allocation*, Wiley, New York, 1951.

28. H. KUHN and A. W. TUCKER, "Nonlinear Programming," in *Proceedings of the Second Berkeley Symposium on Mathematical Statistics and Probability*, J. Neyman, ed., University of California Press, Berkeley, Calif., 1951, pp. 481–492.

29. L. LASDON, *Optimization Theory for Large Scale Systems*, Macmillan, New York, 1970.

30. C. E. LEMKE, "A Method of Solution for Quadratic Programs," *Management Sci.*, *8*, no. 4, July 1962, pp. 442–453.

31. D. G. LUENBERGER, *Introduction to Linear and Nonlinear Programming*, Addison-Wesley, Reading, Mass., 1973.

32. D. L. MARQUARDT, "An Algorithm for Least-Squares Estimation of Nonlinear Parameters," *J. Soc. Ind. Appl. Math.*, *11*, no. 2, June 1963, pp. 431–441.

33. G. P. MCCORMICK, "Anti-Zig-Zagging by Bending," *Management Sci.*, *15*, no. 5, Jan. 1969, pp. 315–320.

34. G. P. MCCORMICK, "The Variable Reduction Method for Nonlinear Programming," *Management Sci.*, *17*, no. 3, Nov. 1970, pp. 146–160.

35. C. E. MILLER, "The Simplex Method for Local Separable Programming," in Graves and Wolfe [19], 1963, pp. 89–100.

36. A. RALSTON, *A First Course in Numerical Analysis*, McGraw-Hill, New York, 1965.

37. J. B. ROSEN, "The Gradient Projection Method for Nonlinear Programming, Part I: Linear Constraints," *J. Soc. Ind. Appl. Math.*, *8*, no. 1, March 1960, pp. 181–217.

38. J. B. ROSEN, "The Gradient Projection Method for Nonlinear Programming, Part II: Nonlinear Constraints," *J. Soc. Ind. Appl. Math.*, *9*, no. 4, Dec. 1961, pp. 514–532.

39. W. RUDIN, *Principles of Mathematical Analysis*, McGraw-Hill, New York, 1964.

40. T. L. SAATY and J. BRAM, *Nonlinear Mathematics*, McGraw-Hill, New York, 1964.

41. B. V. SHAH, R. J. BUEHLER, and O. KEMPTHORNE, "Some Algorithms for Minimizing

a Function of Several Variables," *J. Soc. Ind. Appl. Math.*, *12*, no. 1, March 1964, pp. 74–92.

42. D. SIMMONS, *Linear Programming for Operations Research*, Holden-Day, San Francisco, 1972.

43. H. A. SPANG, "A Review of Minimization Techniques for Nonlinear Functions," *S.I.A.M. Rev.*, *4*, no. 4, Oct. 1962, pp. 343–365.

44. H. THEIL and C. VAN DE PANNE, "Quadratic Programming as an Extension of Classical Quadratic Maximization," *Management Sci.*, *7*, no. 1, Oct. 1960, pp. 1–20.

45. G. B. THOMAS, JR., *Calculus and Analytic Geometry*, Addison-Wesley, Reading, Mass., 1951.

46. D. M. TOPKIS and A. F. VEINOTT, JR., "On the Convergence of Some Feasible Direction Algorithms for Nonlinear Programming," *S.I.A.M. J. Control*, *5*, no. 2, May 1967, pp. 268–279.

47. A. W. TUCKER, "Linear and Nonlinear Programming," *Operations Res.*, *5*, no. 2, April 1957, pp. 244–257.

48. P. WEGNER, "A Nonlinear Extension of the Simplex Method," *Management Sci. 7*, no. 1, Oct. 1960, pp. 43–55.

49. D. J. WILDE, "Differential Calculus in Nonlinear Programming," *Operations Res.*, *10*, no. 6, Nov.–Dec. 1962, pp. 764–773.

50. D. J. WILDE, *Optimum Seeking Methods*, Prentice-Hall, Englewood Cliffs, N.J., 1964.

51. D. J. WILDE and C. S. BEIGHTLER, *Foundations of Optimization*, Prentice-Hall, Englewood Cliffs, N.J., 1967.

52. P. WOLFE, "The Simplex Method for Quadratic Programming," *Econometrica*, *27*, no. 3, July 1959, pp. 382–398.

53. P. WOLFE, "A Duality Theorem for Nonlinear Programming," *Quart. Appl. Math.*, *19*, no. 3, Oct. 1961, pp. 239–244.

54. P. WOLFE, "Methods of Nonlinear Programming," in Abadie [1], 1967, pp. 97–131.

55. P. WOLFE, "On the Convergence of Gradient Methods under Constraints," *IBM Research Paper RC-1752*, Jan. 1967.

56. W. I. ZANGWILL, "A Decomposable Nonlinear Programming Approach," *Operations Res.*, *15*, no. 6, Nov.–Dec. 1967, pp. 1068–1087.

57. W. I. ZANGWILL, "Nonlinear Programming via Penalty Functions," *Management Sci.*, *13*, no. 5, Jan. 1967, pp. 344–358.

58. W. I. ZANGWILL, "The Convex Simplex Method," *Management Sci.*, *14*, no. 3, Nov. 1967, pp. 221–238.

59. W. I. ZANGWILL, "Convergence Conditions for Nonlinear Programming Algorithms," *Management Sci.*, *16*, no. 1, Sept. 1969, pp. 1–13.

60. W. I. ZANGWILL, *Nonlinear Programming: A Unified Approach*, Prentice-Hall, Englewood Cliffs, N.J., 1969.

61. G. ZOUTENDIJK, *Methods of Feasible Directions*, American Elsevier, New York, 1960.

62. G. ZOUTENDIJK, "Nonlinear Programming: A Numerical Survey," *S.I.A.M. J. Control*, *4*, no. 1, Feb. 1966, pp. 194–210.

63. G. ZOUTENDIJK, "Nonlinear Programming, Computational Methods," in Abadie [2], 1970, pp. 37–86.

Index